Dynamics in One Non-Archimedean Variable

GRADUATE STUDIES
IN MATHEMATICS 198

Dynamics in One Non-Archimedean Variable

Robert L. Benedetto

AMERICAN
MATHEMATICAL
SOCIETY
Providence, Rhode Island

2010 *Mathematics Subject Classification.* Primary 37P40;
Secondary 37P30, 37P50.

For additional information and updates on this book, visit
www.ams.org/bookpages/gsm-198

Library of Congress Cataloging-in-Publication Data

Names: Benedetto, Robert L., 1972– author.
Title: Dynamics in one non-archimedean variable / Robert L. Benedetto.
Description: Providence, Rhode Island : American Mathematical Society, [2019] | Series: Graduate studies in mathematics ; volume 198 | Includes bibliographical references and index.
Identifiers: LCCN 2018044194 | ISBN 9781470446888 (alk. paper)
Subjects: LCSH: Geometry, Analytic–Textbooks. | p-adic analysis–Textbooks. | Analytic spaces–Textbooks. | AMS: Dynamical systems and ergodic theory – Arithmetic and non-Archimedean dynamical systems – Non-Archimedean Fatou and Julia sets. msc | Dynamical systems and ergodic theory – Arithmetic and non-Archimedean dynamical systems – Height functions; Green functions; invariant measures. msc | Dynamical systems and ergodic theory – Arithmetic and non-Archimedean dynamical systems – Dynamical systems on Berkovich spaces. msc
Classification: LCC QA551 .B4255 2019 | DDC 515/.39—dc23
LC record available at https://lccn.loc.gov/2018044194

Printed in the United States of America.

♾ The paper used in this book is acid-free and falls within the guidelines
established to ensure permanence and durability.
Visit the AMS home page at `https://www.ams.org/`

10 9 8 7 6 5 4 3 2 1 24 23 22 21 20 19

For Danielle

Contents

List of Notation

$\phi(\zeta) = \|\cdot\|_{\phi(\zeta)}$ .... image of $\zeta \in \mathbb{P}^1_{\mathrm{an}}$ under $\phi \in \mathbb{C}_v(z)$; p. 153
W_{I} .... Type I points of the Berkovich affinoid W; p. 155
$\mathrm{int}\, X$ .... interior of the set X; p. 155
$\phi_{\#}(\boldsymbol{v})$ .... image of Berkovich direction $\boldsymbol{v}$ under ϕ; p. 159
$\deg_{\zeta,\boldsymbol{v}}(\phi)$ .... local degree of ϕ in Berkovich direction $\boldsymbol{v}$; p. 159
$\deg_{\zeta}(\phi)$ .... local degree of ϕ at $\zeta \in \mathbb{P}^1_{\mathrm{an}}$; p. 162
$\mathrm{Ram}(\phi)$ .... ramification locus of ϕ; p. 170
$\mathrm{Inj}(\phi)$ .... injectivity locus of ϕ; p. 170
$\mathcal{F}_{\phi,\mathrm{an}}$ .... (Berkovich) Fatou set of ϕ; p. 183
$\mathcal{J}_{\phi,\mathrm{an}}$ .... (Berkovich) Julia set of ϕ; p. 183
$\{s < |\zeta - a| < r\}$ .... open Berkovich annulus $D_{\mathrm{an}}(a,r) \setminus \overline{D}_{\mathrm{an}}(a,s)$; p. 188
$\mathcal{K}_{\phi,\mathrm{an}}$ .... (Berkovich) filled Julia set of ϕ; p. 194
$\mathfrak{A}_{\phi}(a)$ .... attracting basin of $a \in \mathbb{P}^1(\mathbb{C}_v)$ under ϕ; p. 206
$N(\phi, \boldsymbol{v})$ .... multiplicity with which ϕ fixes $\boldsymbol{v}$; p. 231
$\mathrm{wdeg}_W(\psi)$ .... common inner and outer Weierstrass degree of ψ on thin annulus W; p. 232
$\mathrm{ord}_{\boldsymbol{v}}(\psi)$ .... order of vanishing of ψ in the direction $\boldsymbol{v}$; p. 232
$L_{\phi}(z)$ .... iterative logarithm of ϕ; p. 235
$\mathcal{Q}(\phi)$ .... quasi-periodicity domain of ϕ; p. 235
$\mathrm{dist}_{\mathrm{I}}(x, S)$ .... distance from $x \in \mathbb{C}_v$ to $S \cap \mathbb{C}_v$; p. 258
C_{t} .... set of tame critical points in $\mathcal{J}_{\phi,\mathrm{an}} \cap \mathbb{P}^1(\mathbb{C}_v)$; p. 260
C_{w} .... set of wild critical points in $\mathcal{J}_{\phi,\mathrm{an}} \cap \mathbb{P}^1(\mathbb{C}_v)$; p. 260
$\{G_n(\xi)\}_{n\geq 1}$ .... Trucco's geometric sequence of ξ; p. 279
$\mathrm{lcg}(\zeta, \zeta')$ .... largest common geometric level of ζ and ζ'; p. 280
μ_+, μ_- .... Jordan decomposition of the measure $\mu = \mu_+ - \mu_-$; p. 324
$|\mu|$ .... total variation measure of the measure μ; p. 324
$\mathrm{supp}(\mu)$ .... support of the measure μ; p. 324
$\int_E f\, d\mu$ .... integral of f on E with respect ot μ; p. 325
$[a]$ .... delta-measure at a; p. 325
$M_{\mathrm{Rad}}(X)$ .... space of signed Radon measures on X; p. 325
M_{Rad} .... $M_{\mathrm{Rad}}(\mathbb{P}^1_{\mathrm{an}})$; p. 325
$\|\mu\|$ .... norm of Radon measure μ; p. 325
$C_c(X)$ .... space of continuous, compactly supported functions $f : X \to \mathbb{R}$; p. 325

Preface

This book grew from my lecture notes for a short course I gave at the Arizona Winter School in March 2010. My aim has been to present an exposition accessible to students in their second or third year of graduate school, while also providing a one-stop reference for active researchers in arithmetic and non-archimedean dynamics. In particular, I do not assume the reader has any prior familiarity with the basics of dynamics, with non-archimedean analysis, or with Berkovich spaces. I *do* assume the reader has already seen some non-archimedean fields, usually the p-adic numbers and hopefully the complete, algebraically closed p-adic field $\mathbb{C}_p$, but even those topics are reviewed briefly in Chapter 2.

The theory of p-adic dynamics, and more generally, non-archimedean dynamics, is modeled on the theory of complex dynamics. Naturally, a student of complex dynamics is expected to have already completed a full course on complex analysis. For the sake of argument, though, one could probably learn quite a bit of complex dynamics with minimal prior complex analysis coursework if one were willing to accept certain analytic facts on faith. It would of course be unwise to try to approach complex dynamics that way, but my point is that it *could* be done.

The same is true of non-archimedean dynamics: mastery of the subject requires fluency in non-archimedean analysis, but getting started requires far less background. In practice, after all, most students of non-archimedean dynamics learn non-archimedean analysis along the way, not as a prerequisite. Or at least, that was how *I* learned it.

Thus, this book interleaves the basics of non-archimedean analysis and of elementary dynamics with the main topics of non-archimedean dynamics. For example, the more advanced theory relies heavily on the Berkovich projective line, discussed in Chapter 6. However, a lot can be done with less background, and therefore the fundamentals of non-archimedean dynamics appear earlier, in Chapter 4.

In addition, even in the exposition of non-archimedean analysis (Chapter 3) and Berkovich's theory (Chapters 6 and 7), I defer the more involved proofs to later chapters (specifically, Chapters 14–16). My aim is to acquaint the reader with the requisite analysis as quickly as possible before returning to our primary topic of dynamics. At the same time, a reader who wishes to see those proofs can easily find them by flipping to the back of the book.

As is the case for any project, there are many people who helped make this book possible. I would therefore like to thank the University of Arizona and the organizers of the 2010 Arizona Winter School: Matt Papanikolas, Rachel Pries, Fernando Rodriguez-Villegas, David Savitt, and Dinesh Thakur, as well as the staff of the University of Arizona. I am similarly grateful to Ina Mette, Marcia Almeida, and the rest of the publishing staff at the American Mathematical Society, especially Jennifer Wright Sharp for her careful reading of the manuscript. Many thanks also to Matt Baker, Xander Faber, Charles Favre, and Juan Rivera-Letelier, whose expert input was enormously valuable for organizing the material, improving certain proofs, and presenting important results and concepts clearly. Further thanks are due to Ryan Alvarado, John Benedetto, Amalia Culiuc, Jeff Diller, Fan Shilei, and the anonymous referees who provided mathematical expertise, found errors in early drafts, or otherwise made helpful suggestions. I am particularly grateful to Joe Silverman, who read a draft of the manuscript very closely and taught a course out of it; his countless excellent suggestions improved the presentation substantially. In addition, I gratefully acknowledge the support of NSF grants DMS-1201341 and DMS-1501766. Above all, I thank my wonderful wife Danielle for her constant love and support.

Introduction

A brief history

The study of non-archimedean dynamics began in fits and starts, with a handful of papers dating back to the growth of complex dynamics in the 1970s and 1980s, such as [**10, 94, 110**]. The most far-reaching investigation of this time was the 1983 paper [**74**] by the complex dynamicists Herman and Yoccoz, on local linearizations of analytic functions at fixed points over arbitrary non-archimedean fields. We briefly discuss this problem, also known as the problem of "small denominators", in Section 4.2.

Starting around 1990, due to interest in part from theoretical physicists and in part from number theorists working in the then-nascent field of arithmetic dynamics, there were several preliminary investigations of the dynamics of p-adic power series on disks [**3, 99**] and of p-adic polynomials, especially in degree 2 [**111, 136, 139**]. Results on such topics appeared with increasing frequency from the mid-1990s to the early 2000s, such as [**1, 35, 47, 48, 71, 91–93, 95, 133**]. Also in the mid-1990s, broader studies of the dynamics of p-adic rational functions, including some of the first discussions of p-adic Julia sets and dynamics on p-adic Fatou sets in greater generality, appeared in [**76, 142**]. Further study of the finer properties of Julia sets in the p-adic projective line $\mathbb{P}^1(\mathbb{C}_p)$ followed in the work of Hsia [**77**] and Bézivin [**27, 28**], which we discuss in Sections 12.4 and 12.5.

The development of a more complete theory of one-variable p-adic dynamics began in earnest in the late 1990s, with the author's thesis [**12**] (which appeared in [**13–15**]), which was followed soon after by the pivotal 2000 thesis of Rivera-Letelier (which appeared in [**114**]). Together, these two theses introduced the fundamental results on the dynamics of non-archimedean Fatou components, which we discuss in Chapters 9–11. The work of Rivera-Letelier, in particular, marked the beginning of the use of Berkovich spaces in non-archimedean dynamics. (More precisely, Rivera-Letelier built a version of the Berkovich projective line $\mathbb{P}^1_{\mathrm{an}}$ on his own in [**114, 115, 117**], and it was not recognized as $\mathbb{P}^1_{\mathrm{an}}$ in disguise until about 2004. We briefly discuss his construction and terminology in Section B.4.) One of the simplest examples of the non-archimedean analytic spaces introduced by Vladimir Berkovich in 1990, the Berkovich projective line has now become a central object in the study of dynamics in one non-archimedean variable.

The next few years saw further developments in the study of both Fatou sets and Julia sets in the non-archimedean setting, such as [**16–19, 23, 115–117**]. Most of these results appear in various forms in Chapters 8–12. During this time, the theory expanded from p-adic fields to more general non-archimedean fields, in part because of the natural need to generalize, but also in part due to interest from some complex dynamicists, such as in [**83**]. In particular, the (non-archimedean) field of formal Puiseux series (see Example 2.2) has proven to be a useful setting for understanding the degeneration of complex dynamical systems at the boundary of moduli space; see, for example, [**43, 44, 84, 85**].

A huge burst of activity came in 2004. Three different pairs of mathematicians, working independently, constructed a canonical invariant measure on $\mathbb{P}^1_{\mathrm{an}}$ for the dynamics of a one-variable non-archimedean rational function. This equilibrium measure, supported on the Julia set, had been constructed years before in the complex setting [**34, 65, 100**] via potential theory, as we note in Section 1.6.4. The work of Chambert-Loir and Thuillier [**40, 138**], of Baker and Rumely [**7**], and of Favre and Rivera-Letelier [**60–62**] built the analogous objects in the non-archimedean setting, especially the requisite Laplacian operator and potential theory on $\mathbb{P}^1_{\mathrm{an}}$. (A few years later, in [**63**], Favre and Rivera-Letelier presented numerous ergodic- and entropy-theoretic properties of the equilibrium measure.) The use of Berkovich spaces is essential in the non-archimedean case, as the equilibrium measures need to be defined not just on the set $\mathbb{P}^1(\mathbb{C}_v)$ of "classical" points, but also at other points of $\mathbb{P}^1_{\mathrm{an}}$. In addition, all three pairs proved an arithmetic Galois equidistribution theorem with respect to their equilibrium measure. We discuss all of these topics in Chapter 13.

Not incidentally, the equidistribution theorem concerns not just the equilibrium measure, but also another essential object in arithmetic dynamics, the canonical height. Although we only discuss the canonical height rather briefly (in Section 13.7), it has been a significant motivation for number-theoretic interest in non-archimedean dynamics, from arithmetic dynamics' early days in the 1990s through the present. The canonical height in the context of arithmetic dynamics was introduced by Call and Silverman in their 1993 foundational paper on the subject [**37**] as a broad generalization of the theory of canonical heights on elliptic curves, abelian varieties, and K3 surfaces. The same paper also describes the canonical height as a sum of canonical local heights, which we discuss in Section 13.6 and which are more directly related to non-archimedean dynamics. As a consequence, number-theoretic questions of a dynamical flavor often require advanced results from non-archimedean dynamics, such as in [**6, 24, 25, 55, 68, 78**], to give a small selection of examples.

What's in this book

Anyone seeking to get started in non-archimedean dynamics has faced at least two major obstacles. First, the essential results of the theory are scattered across dozens of papers and unpublished notes, proven in varying degrees of generality (e.g., just for p-adic fields or only for polynomials of low degree), and sometimes using inconsistent or outdated terminology. The central aim of this book, therefore, is to provide a single resource collecting all of the fundamentals of the field in one place, with uniform, current terminology, and working over an arbitrary non-archimedean field whenever possible. Second, the need for Berkovich's theory of analytic spaces, even just for the projective line, presents another obstacle for many prospective non-archimedean analysts, whether of a dynamical bent or not. Thus, a second aim of this book is to provide an accessible but also thorough exposition of the Berkovich projective line $\mathbb{P}^1_{\text{an}}$.

Part 1, Background, contains a variety of essential background that some readers may wish to skip:

- Chapter 1 mainly concerns the most elementary basics of discrete dynamics as well as a few scattered facts about morphisms of the projective line $\mathbb{P}^1(K) = K \cup \{\infty\}$ over an arbitrary field K.
- Chapter 2 is a high-speed primer on non-archimedean fields, including some basic properties of disks in a complete and algebraically closed non-archimedean field $\mathbb{C}_v$.
- Chapter 3 is substantially more advanced, presenting numerous results on non-archimedean power series and Laurent series. So as not to bog the exposition down with technical proofs relying on strategies that are never used again in the book, most of the proofs are deferred to Chapter 14.

Part 2, Elementary Non-Archimedean Dyanmics, is about dynamics on the classical non-archimedean projective line $\mathbb{P}^1(\mathbb{C}_v)$, *without* using Berkovich spaces:

- Chapter 4 introduces several important concepts and objects that are of fundamental importance in non-archimedean dynamics with or without Berkovich spaces: the classification of periodic points, the notions of good and bad reduction, Lattès maps, and dynamics on non-archimedean disks.
- Chapter 5 concerns Fatou and Julia sets in $\mathbb{P}^1(\mathbb{C}_v)$. Although their Berkovich analogues are even more significant, these classical Fatou and Julia sets are much easier to define, and they are still important objects in the theory. Moreover, they can be recovered from their Berkovich siblings simply by intersecting with $\mathbb{P}^1(\mathbb{C}_v)$.

Part 3, The Berkovich Line, is a detailed introduction to the Berkovich projective line $\mathbb{P}^1_{\text{an}}$. For the same reasons as in Chapter 3, the proofs of many of the results in this part are deferred to later in the book.

- Chapter 6 defines $\mathbb{P}^1_{\text{an}}$ and derives many of its fundamental properties and structures.
- Chapter 7 describes how a rational function $\phi \in \mathbb{C}_v(z)$ acts on $\mathbb{P}^1_{\text{an}}$, including a detailed discussion of local degrees and ramification.

Part 4, Dynamics of the Berkovich Line, which is the heart of the book, is a thorough analysis of dynamics on $\mathbb{P}^1_{\text{an}}$, including deep theorems on Fatou and Julia sets, as well as the construction of the equilibrium measure. Many of the results in these chapters were originally stated and proven only for the p-adic field $\mathbb{C}_p$, but whenever possible, the presentation here is for arbitrary complete, algebraically closed non-archimedean fields $\mathbb{C}_v$.

- Chapter 8 defines the Berkovich space versions of the Fatou and Julia sets and describes their basic properties.
- Chapter 9 gives Rivera-Letelier's classification of dynamics on periodic Fatou components.
- Chapter 10 concerns a number of finer properties of dynamics on periodic Fatou components.
- Chapter 11 is about wandering domains, i.e., components of the Fatou set whose iterates never repeat. Some of the results here forbid wandering domains under many hypotheses, while others show that they can exist in certain other circumstances.
- Chapter 12 concerns repelling periodic points, which always lie in the Julia set. Most of the biggest results here are proofs that the

repelling periodic points are *dense* in the Julia set, both in the Berkovich setting and in the classical setting.

- Chapter 13 gives a lightning tour of some relevant non-archimedean potential theory in Sections 13.1–13.4, culminating in a careful construction of the equilibrium measure in Section 13.5. We also present the equidistribution theorem, but only a sketch of its proof, in Section 13.7. (For details on non-archimedean potential theory and a full proof of the equidistribution theorem, we refer the reader to the excellent and thorough textbook [**8**].)

Finally, Part 5, Proofs from Non-Archimedean Analysis, contains those proofs deferred earlier.

- Chapter 14 supplies the missing proofs from Chapter 3.
- Chapter 15 supplies the missing proofs from Chapter 6.
- Chapter 16 supplies the missing proofs from Chapter 7.

There are also two appendices.

- Appendix A describes theories of Fatou components that do not use Berkovich spaces. One of these theories yields the exact same components, but simply intersected with the classical points $\mathbb{P}^1(\mathbb{C}_v)$.
- Appendix B gives three other ways of constructing the Berkovich projective line, including a sketch of the completely different construction and terminology Rivera-Letelier used in his early papers [**114, 115, 117**].

There are exercises included at the end of each chapter. Some of these ask for proofs left to the reader in the main text, others lead the reader through the analysis of relevant or illuminating examples, and others concern curious phenomena and counterexamples. The difficulty of the exercises varies wildly; many are straightforward, but some are quite challenging.

Some things not in this book

The theory of non-archimedean dynamics has grown enough that no single textbook can cover everything at the level of detail found here. In addition, the field is an active research area and continues to evolve. Thus, a number of topics had to be omitted or presented only minimally, as we now describe.

As mentioned above regarding Chapter 13, although we give a full construction of the equilibrium measure after a reasonably detailed overview of the relevant potential theory, we only sketch the proof of the equidistribution theorem. For further details, the reader may consult either the original papers [**7, 40, 61, 138**] or the comprehensive textbook [**8**]. Various other non-archimedean dynamics results, most of which appear in this book too, can also be found in [**8**, Sections 10.5–10.10]. Incidentally, other expositions of portions of the field appear in [**29**] (for the older theory, before the introduction of Berkovich spaces) and [**38**] (mostly on Rivera-Letelier's thesis work in [**114**]), although those topics are also fully treated in this book.

Also as mentioned above, non-archimedean local linearization (or the problem of small denominators) is another topic that we only touch on in Section 4.2. Interested readers may wish to consult [**74, 95–98**].

Yet another topic that we only mention briefly is the study of families of non-archimedean dynamical systems, especially one-parameter families. We touch on their use in [**17, 19**] for constructing certain wandering domains in our sketched proof of Theorem 11.17, as well as on their use in [**118**] for constructing wild recurrent critical points at the end of Section 11.3. On the other hand, we do not even attempt to discuss more recent work [**90, 132**] on non-archimedean analogues of the 1983 Mañé–Sad–Sullivan λ-lemma [**101**] and associated results on J-stability.

The theory of the equilibrium measure has led to an explosion of research that relies on its use, and our hope is that this book will equip the reader well to study such results. Examples include Rumely's minimal resultant locus and crucial set [**46, 125**], bifurcation measures on parameter spaces for familes [**6, 55, 56, 68**], Lyapunov exponents of non-archimedean dynamical systems [**67, 79, 108, 109**], results on ergodic theory and entropy such as [**4, 63**], and much more.

Some possible paths through this book

This book was written with multiple audiences in mind, and readers with particular interests may wish to skip certain portions to focus on others. Here are some options.

- If you seek a thorough introduction to the Berkovich projective line but are not really interested in the dynamics, we suggest Chapters 2, 3, 14, 6, 15, 7, and 16. You may also be interested in Appendix B.
- If you already know the basics of non-archimedean fields and want a fast path to learning about the equilibrium measure, we suggest Chapters 1, 6, 7, and 13, possibly supplemented by Chapter 3 if you could use a refresher on non-archimedean analysis.
- If you are already fairly familiar with the Berkovich projective line and want to learn results about components of non-archimedean Fatou sets, we suggest Chapters 8–11. If you also want to know about

fixed-point theorems and their applications to non-archimedean Julia sets, see Chapter 12.

- If you want to learn about non-archimedean dynamics that can be done without Berkovich spaces, try Chapters 1–5 and Appendix A.
- If you want to learn the important fundamentals of the theory of non-archimedean dynamics, read the whole book.

Part 1

Background

Chapter 1

Basic Dynamics on $\mathbb{P}^1(K)$

Given an arbitrary field K and a rational function $\phi \in K(z)$, we can of course view ϕ as a morphism $\phi : \mathbb{P}^1(K) \to \mathbb{P}^1(K)$. In particular, for any field L containing K (such as K itself), ϕ acts as a function from $\mathbb{P}^1(L) := L \cup \{\infty\}$ to itself. We wish to study the dynamics of this action.

1.1. Elementary discrete dynamics

A (*discrete*) *dynamical system* on a set X is a function $\phi : X \to X$. For any integer $n \geq 0$, $\phi^n : X \to X$ denotes the *nth iterate* of ϕ,

$$\phi^n := \underbrace{\phi \circ \phi \circ \cdots \circ \phi}_{n \text{ times}},$$

that is, $\phi^0 = \mathrm{id}_X$ and $\phi^{n+1} = \phi \circ \phi^n$.

The (*forward*) *orbit* of a point $x \in X$ is $\mathrm{Orb}^+_\phi(x) := \{\phi^n(x) : n \geq 0\}$, and the *backward orbit* of x is $\mathrm{Orb}^-_\phi(x) := \bigcup_{n \geq 0} \phi^{-n}(x)$. The *grand orbit* of x is the union $\mathrm{GO}_\phi(x)$ of the backward orbits of all points in the forward orbit of x.

A point $x \in X$ is said to be *fixed* (under ϕ) if $\phi(x) = x$. More generally, x is *periodic* (under ϕ, of period n) if $\phi^n(x) = x$ for some positive integer $n \geq 1$; in that case, we also say x is an *n-periodic point* of ϕ. Even more generally, x is *preperiodic* (under ϕ) if $\phi^n(x) = \phi^m(x)$ for distinct integers $n > m \geq 0$.

If x is a periodic point of period n, the forward orbit of x is called a *periodic cycle* of period n or an n-periodic cycle. The smallest positive integer $n \geq 1$ such that $\phi^n(x) = x$ is called the *minimal period* or *exact period* of x under ϕ. Several elementary observations are in order:

1. The relation "y is in the grand orbit of x" is an equivalence relation. More precisely, x and y share the same grand orbit if and only if there are integers $m, n \geq 0$ such that $\phi^m(x) = \phi^n(y)$.
2. Every point in the forward orbit of a periodic point is periodic of the same minimal period. Thus, the period is really a property of the periodic cycle.
3. A point is preperiodic if and only if its forward orbit is finite.
4. If ϕ is onto (as is the case for the dynamical systems of this book), and if x has finite backward orbit, then x also has finite forward orbit. In that case, in fact, the grand orbit of x consists solely of a periodic cycle; see also Definition 1.13 and Theorem 1.19.

We often consider the following case. Let K be a field, and let X be the projective line $X := \mathbb{P}^1(K) := K \cup \{\infty\}$. Let $\phi(z) \in K(z)$ be a rational function, and write $\phi(z) = f(z)/g(z)$, where the polynomials $f, g \in K[z]$ have no common factors. Then ϕ maps X to itself by evaluation, i.e., by $\phi(x) = f(x)/g(x)$. If $x \in K$ is a root of g, then $\phi(x) = \infty$. Finally, $\phi(\infty)$ is ∞ if $\deg(f) > \deg(g)$, or 0 if $\deg(f) < \deg(g)$, or a/b if $\deg(f) = \deg(g)$, where $a, b \in K$ are the lead coefficients of f, g, respectively.

Example 1.1. Let $X := \mathbb{P}^1(\mathbb{C}) = \mathbb{C} \cup \{\infty\}$, and let $\phi(z) := z^2$. The forward orbit of the point -2 is $\{-2, 4, 16, 256, \ldots\}$. More generally, the forward orbit of any $x \in \mathbb{P}^1(\mathbb{C})$ is $\{x^{2^n} : n \geq 0\}$. The points $0, 1, \infty$ are fixed by ϕ. The grand orbit of 0 is simply $\{0\}$, and that of ∞ is simply $\{\infty\}$. Every other point $x \in \mathbb{C} \smallsetminus \{0\}$ has backward orbit

$$\left\{\zeta \cdot \sqrt[2^n]{x} : n \geq 0, \, \zeta^{2^n} = 1\right\}.$$

In particular, the grand orbit of 1 is

$$\mu_{2^\infty} := \left\{\zeta : \zeta^{2^n} = 1 \text{ for some } n \geq 0\right\}.$$

The preperiodic points of ϕ are 0, ∞, and all roots of unity. A primitive mth root of unity is periodic if and only if m is odd; its minimal period is the smallest integer n such that $2^n \equiv 1 \pmod{m}$.

More generally, for a fixed integer $d \in \mathbb{Z} \smallsetminus \{0, 1, -1\}$, the preperiodic points of $\phi(z) = z^d$ are still 0, ∞, and the roots of unity. A primitive mth root of unity is now periodic if and only if $d \nmid m$, and the minimal period is the multiplicative order of d modulo m.

The previous example is misleadingly simple; everything works out nicely because z^2 is an endomorphism of the multiplicative group $\mathbb{G}_m$; see [**131**,

Chapter 6], especially Section 6.1. The following seemingly slight variation is far more complicated and far more illustrative of the general situation.

Example 1.2. Let $X := \mathbb{P}^1(\mathbb{C})$, and let $\phi(z) := z^2 - 1$. The forward orbit of the point -2 is $\{-2, 3, 8, 63, 3968, \ldots\}$; there is no general closed formula for $\phi^n(x)$. The point ∞ is fixed, and its grand orbit is just $\{\infty\}$, but no other point has finite backward orbit. The two roots of $z^2 - z - 1 = 0$ (that is, of $\phi(z) - z = 0$) are the only other fixed points of ϕ. The points 0 and -1 form a periodic cycle of period 2, and 1 is preperiodic, as $\phi(1) = 0$ is periodic. (In fact, there are no other rational preperiodic points besides $0, \pm 1, \infty$, but there are infinitely many preperiodic points in $\mathbb{C}$.) In general, it is difficult to tell whether a given point $x \in X$ is preperiodic.

1.2. Morphisms and coordinate changes

Definition 1.3. Let $\phi : X \to X$ and $\psi : Y \to Y$ be dynamical systems. A *morphism* (of dynamical systems) from ϕ to ψ is a function $h : X \to Y$ such that $h \circ \phi = \psi \circ h$; that is, so that the diagram

$$\begin{array}{ccc} X & \xrightarrow{\phi} & X \\ {\scriptstyle h}\downarrow & & {\scriptstyle h}\downarrow \\ Y & \xrightarrow{\psi} & Y \end{array}$$

commutes.

If h is a morphism from ϕ to ψ, then

$$h \circ \phi^n = \psi^n \circ h \quad \text{for all } n \geq 0.$$

Thus, h preserves the structure of the dynamical system: it maps fixed points of ϕ to fixed points of ψ, periodic points to periodic points, preperiodic points to preperiodic points, forward orbits onto forward orbits, backward orbits into backward orbits, and grand orbits into grand orbits. If h is onto, then backward and grand orbits are mapped onto their counterparts. If h is one-to-one, then the minimal period of all periodic points is preserved by ϕ; see Exercise 1.1.

If h is one-to-one and onto, then its inverse h^{-1} is a morphism from ψ to ϕ. Thus, h is an isomorphism of dynamical systems, and we have

$$\psi = h \circ \phi \circ h^{-1}.$$

If $X = Y$ in this case, then h can be considered to be a change of coordinates on X. That is, a coordinate change on the underlying space corresponds to a conjugation of the map ϕ.

Example 1.4. Let $X := \mathbb{P}^1(\mathbb{C})$ and $\phi(z) := z^2$, as in Example 1.1. Let $Y = X$ and $\psi(z) = z^2 - 2$. Then the map $h : X \to Y$ given by $h(z) = z + z^{-1}$ is a morphism from ϕ to ψ, because

$$h \circ \phi(z) = z^2 + z^{-2} = (z + z^{-1})^2 - 2 = \psi \circ h(z).$$

The preperiodic points of ψ are precisely points of the form $h(x)$, where x is a preperiodic point of ϕ. That is, the preperiodic points are

$$\infty = h(0) = h(\infty)$$

and any point of the form $\zeta + \zeta^{-1}$, where ζ is a root of unity. (The latter kind of points can be written $2\cos(2\pi q)$, for $q \in \mathbb{Q}$ rational.) Because ϕ is onto, the periodic points of ψ are also precisely the images under h of the periodic points of ϕ; however, the minimal periods may change. For example, a primitive cube root of unity ζ_3 is periodic of minimal period 2 under ϕ, but $h(\zeta_3) = -1$ is fixed by ψ.

More generally, for any integer $d \geq 2$, the dth *Chebyshev polynomial* is the unique polynomial T_d is such that

$$T_d(z + z^{-1}) = z^d + z^{-d}.$$

Note that T_d has integer coefficients and also satisfies the functional equation $T_d(2\cos\theta) = 2\cos(d\theta)$. In addition,

$$T_n \circ T_m = T_{mn} = T_m \circ T_n,$$

meaning that the Chebyshev polynomials form a rare family of maps of arbitrary degree (rather than all being powers of some ϕ) that commute under composition, just as the power maps z^n commute. Another such family appears in Section 4.4. See [**131**, Section 6.2] for more on the dynamics of Chebyshev polynomials.

Morphisms as in Example 1.4 are rare. Usually, the only morphisms of dynamical systems of particular interest to us are coordinate changes.

Example 1.5. Let $X := \mathbb{P}^1(\mathbb{C})$, and let $\psi(z) := z^2/(2z+2)$. One can check that the fixed points of ψ are exactly 0, -2, and ∞. Further investigation reveals that the only preimage of -2 is -2 itself, and similarly, the only preimage of 0 is 0. Inspired by this observation, we write down the unique linear fractional transformation $h(z)$ mapping $0 \mapsto 0$, $\infty \mapsto -2$, and $1 \mapsto \infty$, which is $h(z) = 2z/(1-z)$. We can check that $\psi = h \circ \phi \circ h^{-1}$, where $\phi(z) = z^2$. Thus, ψ is just the squaring map under a change of coordinates. In particular, this is one of the rare dynamical systems for which we can write down a simple closed form for ψ^n, namely

$$\psi^n(z) = h\Big(\big(h^{-1}(z)\big)^{2^n}\Big) = \frac{2z^{2^n}}{(z+2)^{2^n} - z^{2^n}}.$$

Example 1.6. Let K be a field of characteristic different from 2 with algebraic closure $\overline{K}$, let $X := \mathbb{P}^1(\overline{K}) = \overline{K} \cup \{\infty\}$, and let

$$\phi(z) = Az^2 + Bz + C \in K[z]$$

be a quadratic polynomial. Conjugate by $h_1(z) := z + B/(2A)$ to complete the square and obtain

$$\psi_1(z) = Az^2 + D, \quad \text{where } D := C - \frac{B(B-2)}{4A^2}.$$

To make the polynomial monic, conjugate by $h_2(z) := Az$, giving

$$\psi_2(z) = z^2 + c, \quad \text{where } c := AD = AC - \frac{B(B-2)}{4A}.$$

The reader can check that if $h \in \mathrm{PGL}(2, K)$ is a linear fractional transformation conjugating $z^2 + c_1$ to $z^2 + c_2$, then $c_1 = c_2$; see Exercise 1.5. Thus, over any field not of characteristic 2, the set of quadratic polynomial dynamical systems is parametrized by K itself, with the parameter $c \in K$ corresponding to the conjugacy class of $z^2 + c$.

The study of the family maps of the form $z^2 + c$ is therefore common in complex dynamics, as it yields all quadratic polynomials. (However, it is *not* a suitable family over fields of characteristic 2; see Exercise 1.6.) The corresponding moduli spaces of higher degree polynomials and rational functions are far more complicated; we refer the interested reader to [**131**, Chapter 4] as a starting point for learning about such spaces.

1.3. Degrees, multiplicities, and multipliers

Let K be a field, and let $\phi \in K(z)$ be a rational function. Writing $\phi = f/g$, where $f, g \in K[z]$ are relatively prime polynomials, we define the *degree* of ϕ to be

$$\deg(\phi) := \max\{\deg(f), \deg(g)\}.$$

This notion of degree coincides with the geometric degree of ϕ. In other words, for any point $x \in \mathbb{P}^1(\overline{K})$, x has exactly $\deg(\phi)$ preimages under ϕ, counting multiplicity. It is easy to check that

$$\deg(\phi \circ \psi) = \big(\deg(\phi)\big) \cdot \big(\deg(\psi)\big) \quad \text{for any } \phi, \psi \in K(z);$$

see Exercise 1.11. As a result, we also have $\deg(\phi^n) = (\deg(\phi))^n$.

The fixed points of ϕ are precisely the roots of the equation $\phi(z) - z = 0$, and hence of the polynomial $f(z) - zg(z) = 0$, along with, perhaps, the point ∞. Thus, ϕ has exactly $1 + \deg(\phi)$ fixed points in $\mathbb{P}^1(\overline{K})$, counted with appropriate multiplicity. To be clear, the *multiplicity* of a fixed point $x \neq \infty$ is its multiplicity as a root of the polynomial $f(z) - zg(z)$. If ∞ is fixed, its multiplicity is that of 0 as a root of $1/\phi(1/z)-z=0$, or equivalently, as a root of the polynomial

$$z^d g(1/z) - z^{d+1} f(1/z).$$

After all, $1/\phi(1/z)$ is the conjugate of ϕ by $1/z$, which moves ∞ to 0. As we will see in Exercise 1.12, the multiplicity of a fixed point is invariant under coordinate change, and hence it makes sense to define the multiplicity of a fixed point at infinity in this way.

Alternatively, to consider $\mathbb{P}^1$ holistically, we should use homogeneous coordinates $[x, y]$ and write ϕ as

$$\phi(x, y) = [f(x, y), g(x, y)],$$

where $f, g \in K[x, y]$ are homogeneous of degree d. (Indeed, the new f can be written in terms of the old f as $y^d f(x/y)$, and similarly the new g is $y^d g(x/y)$.) The old equation $\phi(z) - z = 0$ now becomes

$$yf(x, y) - xg(x, y) = 0.$$

Since $yf(x, y) - xg(x, y)$ is a homogeneous polynomial of degree $d+1$, it has $d+1$ roots in $\mathbb{P}^1$, counting multiplicity. For any integer $n \geq 1$, by considering ϕ^n in place of ϕ, we see that ϕ has exactly $1 + (\deg(\phi))^n$ periodic points in $\mathbb{P}^1(\overline{K})$ of period n (that is, of minimal period dividing n), as long as ϕ^n is neither constant nor the identity function. The latter condition is obviously true if $\deg(\phi) \geq 2$, which we usually assume.

If we denote by M_n the number of periodic points of minimal period n, then barring the coalescing of higher-period periodic points on smaller-period points, we would have

$$\sum_{m|n} M_m = d^{n+1}.$$

Then by the Möbius inversion formula, we have

$$M_n = \sum_{m|n} \mu(n/m) \cdot (d^m + 1).$$

Thus, for example, a rational function ϕ of degree 2 has exactly

$$(2^6 + 1) - (2^3 + 1) - (2^2 + 1) + (2^1 + 1) = 54$$

points of minimal period 6, assuming none of the periodic points involved coalesce on points of smaller period. (Note that 54 is divisible by 6, giving nine cycles of period 6.)

However, periodic points *can* coalesce. The simplest example arises for $\phi(z) = z^2 - z$. We have the expected $2^1 + 1 = 3$ fixed points (at 0, 2, and

∞), and the previous paragraph predicts $(2^2+1)-3=2$ points of period 2. However, we compute

$$\phi^2(z) - z = z^4 - 2z^3,$$

giving 2-periodic points only at the same three points $0, 2, \infty$ that were already fixed. The problem is that 0 is a triple root of $\phi^2(z)-z$; the two extra roots that were supposed to be points of period 2 are instead copies of a fixed point. That is, the two points of the expected 2-cycle have coalesced at a fixed point. More generally, this can happen any time the multiplier of a periodic point, which we define in the next paragraph, is a root of unity.

Given a periodic point $x \in \mathbb{P}^1 \smallsetminus \{\infty\}$ of minimal period $n \geq 1$, the *multiplier* of x is $\lambda := (\phi^n)'(x)$. If $x = \infty$, then we define the multiplier to be

$$\lambda := (\psi^n)'(0),$$

where $\psi(z) = 1/(\phi(1/z))$; that is, change coordinates to move $x=\infty$ to 0, and compute the multiplier of 0 under ψ as before.

It is a simple consequence of the chain rule that the multiplier is unaffected by coordinate change. That is, for any $h \in \mathrm{PGL}(2,K)$, the multiplier of a periodic point x of ϕ is the same as that of the periodic point $h(x)$ of $h \circ \phi \circ h^{-1}$; see Exercise 1.12. The chain rule also gives

$$(\phi^n)'(x) = \prod_{i=0}^{n-1} \phi'\big(\phi^i(x)\big) \tag{1.1}$$

(at least, assuming none of $x, \phi(x), \ldots, \phi^n(x)$ is ∞), from which it follows that every point in the periodic cycle of x has the same multiplier. Thus, like the period, the multiplier is really a property of the periodic cycle, rather than of the individual periodic point.

Note that a periodic point $x \in \mathbb{P}^1$ of minimal period n has multiplicity at least 2 if and only if its multiplier is exactly 1. This claim is clear if $x \neq \infty$, because such a periodic point is a multiple root of $\phi^n(z) - z$ if and only if it is also a root of the derivative $(\phi^n)'(z) - 1$ of that same polynomial. It can also be checked at $x = \infty$, either by changing coordinates or by working in homogeneous coordinates.

More generally, a periodic point $x \in \mathbb{P}^1$ of minimal period n has multiplicity at least 2 *as an mn-periodic point* if and only if its multiplier is an mth root of unity. It is in precisely this situation that the phenomenon of coalescing of periodic points occurs: if x has minimal period n, with multiplier $\lambda \neq 1$ a primitive mth root of unity, then some mn-periodic points coalesce at x. See [**131**, Section 4.1] and [**9**, Section 6.2] for more on this phenomenon.

The following result is well known in complex dynamics (even in his 1919 *Mémoire*, Julia himself called it "bien connue" [**82**, Section 9]), but it holds over any field.

Theorem 1.7 (The Holomorphic Fixed-Point Formula). *Let K be a field, and let $\phi \in K(z)$ be a rational function of degree $d \geq 0$ with no fixed points of multiplier exactly 1. Let $x_1, \ldots, x_{d+1} \in \mathbb{P}^1(\overline{K})$ be the (distinct) fixed points of ϕ, and let $\lambda_1, \ldots, \lambda_{d+1} \in \overline{K} \setminus \{1\}$ be their multipliers. Then*

$$\sum_{i=1}^{d+1} \frac{1}{1-\lambda_i} = 1.$$

Proof. After a $\overline{K}$-rational change of coordinates, assume that ∞ is not a fixed point of ϕ. (Note that if K is a finite field, all points of $\mathbb{P}^1(K)$ might be fixed; that is why we may have to pass to $\overline{K}$ for the coordinate change.) Let ω be the meromorphic differential form

$$\omega := \frac{dz}{\phi(z) - z} \in \Omega^1(\mathbb{P}^1(\overline{K})), \tag{1.2}$$

which is holomorphic except for simple poles at $x_1, \ldots, x_{d+1}$ and at ∞; see Exercise 1.17. Clearly, the residue at x_i is $\phi'(x_i) - 1 = \lambda_i - 1$. In the coordinate $w = 1/z$, we have

$$\omega = -\frac{dw}{w^2\phi(1/w) - w} = \frac{dw}{w + O(w^2)},$$

because $\phi(1/w)$ is holomorphic at $w = 0$. Thus, the residue of ω at $w = 0$ is 1, and hence the residue of ω at $z = \infty$ is also 1.

Recall that the Residue Theorem states that for a meromorphic differential form on a smooth projective curve, the sum of the residues at all the poles is 0. See [**127**, Section II.7] or Exercise 1.18 for a proof over an arbitrary field, using the partial fraction expansion of $f(t)$ for the differential form $f(t)\,dt$. In particular, we may apply the Residue Theorem to our form ω, giving

$$1 + \sum_{i=1}^{d+1} \frac{1}{\lambda_i - 1} = 0,$$

from which the desired formula is immediate. □

Remark 1.8. Theorem 1.7 can be generalized to the case that some fixed point x_i has multiplier $\lambda_i = 1$ if one replaces the summand $1/(1-\lambda_i)$ with the residue of the differential form ω at the point x_i.

1.4. Critical points and exceptional sets

Let $\phi(z) \in K(z)$ be a nonconstant rational function. Then for any point $x \in \mathbb{P}^1(\overline{K})$, one can choose $\eta, \theta \in \mathrm{PGL}(2, \overline{K})$ such that

$$\eta\bigl(\phi(x)\bigr) = \theta(x) = 0.$$

The function $\psi := \eta \circ \phi \circ \theta^{-1}$ clearly satisfies $\psi(0) = 0$, and hence ψ may be written as

$$\psi(z) = z^e h(z),$$

for some integer $e \geq 1$ and some rational function $h(z) \in \overline{K}(z)$ for which $h(0) \neq 0$. The integer e is the *multiplicity* or *ramification index* of ϕ at x, and it is independent of the choices of η and θ; see Exercise 1.19. If $e \geq 2$, we say that x is a *critical point*, or *ramification point* of ϕ, and that ϕ is *ramified* at x. If $e = \deg(\phi) \geq 2$, then we say ϕ is *totally ramified* at x.

Of course, if neither x nor $\phi(x)$ is the point at ∞, then the multiplicity e is simply the order of vanishing of the function $\phi(z) - \phi(x)$ at $z = x$. Thus, such a point x is critical for ϕ if and only if $\phi'(x) = 0$.

If x is a critical point of ϕ and we set $\psi = \eta \circ \phi \circ \theta^{-1}$ as above, then the order of vanishing $\rho \geq 1$ of the derivative $\psi'(z)$ at $z = 0$ is not usually given a name, but since we occasionally need to discuss it, we call it the *order* of the critical point x of ϕ. (Any noncritical point has order zero.) Like the multiplicity, the order of a critical point is independent of the choices of η and θ. If $\operatorname{char} K = 0$, then the multiplicity $e \geq 2$ and the order $\rho \geq 1$ of a critical point x are related by $\rho = e - 1$. The same is true if $\operatorname{char} K = p > 0$ and $p \nmid e$. However, if $\operatorname{char} K = p > 0$ and $p | e$, then $\rho \geq e$; see Exercise 1.20.

However, if $\operatorname{char} K = p > 0$ and $\phi(z) \in K(z^p)$, then $\phi'(z) = \psi'(z) = 0$. In that case, by the above definitions, every point $x \in \mathbb{P}^1$ is a critical point with multiplicity divisible by p. Maps of this sort are said to be *inseparable*; any other map is *separable*. A nonconstant rational function ϕ over a characteristic p field K can be written uniquely as $\psi(z^{p^j})$, where ψ is separable and $j \geq 0$; we say that ϕ is *totally inseparable* if $\deg(\psi) = 1$ or, equivalently in this notation, if $\deg(\phi) = p^j$.

The critical points turn out to be very important for understanding the dynamics of the map ϕ. Their number is dictated by the following result.

Theorem 1.9 (The Riemann–Hurwitz formula for morphisms of $\mathbb{P}^1$). *Let K be a field, let $\phi(z) \in K(z)$ be a separable rational function of degree $d \geq 1$, and let $R \subseteq \mathbb{P}^1(\overline{K})$ be the set of all critical points of ϕ. For each $x \in R$, define $\rho_x \geq 1$ to be the order of x as a critical point of ϕ. Then*

$$\sum_{x \in R} \rho_x = 2d - 2.$$

Proof. This is [**72**, Corollary IV.2.4] for the special case of a map from $\mathbb{P}^1$ to itself. See also [**9**, Theorem 2.7.1] for an elementary proof in the case $K = \mathbb{C}$; we sketch a generalization of this elementary proof in Exercise 1.23. □

Corollary 1.10. *Let K be a field, and let $\phi(z) \in K(z)$ be a separable rational function of degree $d \geq 1$. Then ϕ has at most $2d-2$ distinct critical points.*

Proof. This is immediate from Theorem 1.9 because $\rho_x \geq 1$ for all $x \in R$. □

Corollary 1.11. *Let K be a field, and let $\phi(z) \in K(z)$ be a rational function of degree $d \geq 2$ that is not totally inseparable. Then ϕ has at most two totally ramified critical points. Furthermore, if ϕ is separable and has two totally ramified critical points, then it has no other critical points.*

Proof. We first consider the case that ϕ is separable. In the notation of Theorem 1.9, at a totally ramified critical point $x \in R$, the multiplicity of ϕ at x is d and, hence, the order ρ_x of the critical point satisfies $\rho_x \geq d-1$. The conclusion of the corollary is now immediate from Theorem 1.9.

Now consider $\phi(z) = \psi(z^{p^j})$, where $p = \operatorname{char} K > 0$, $j \geq 0$, and ψ is separable and of degree $d' \geq 2$. Then ϕ is totally ramified at a point $x \in \mathbb{P}^1(\overline{K})$ if and only if ψ is totally ramified at x^{p^j}. However, by the previous paragraph, ψ is totally ramified at at most two points. Finally, the Frobenius map $x \mapsto x^p$ is bijective from the set $\mathbb{P}^1(\overline{K})$ to itself, since $\operatorname{char} K = p$. (Of course, the Frobenius map is not an isomorphism as an algebraic morphism, but only as a map of sets.) Thus, there are at most two points $x \in \mathbb{P}^1(\overline{K})$ for which ψ is totally ramified at x^{p^j}, and we are done. □

Remark 1.12. Let $\phi(z) \in K(z)$. If $\operatorname{char} K = 0$, or more generally, if no critical point of ϕ has multiplicity divisible by $p := \operatorname{char} K$, then the Riemann–Hurwitz formula for ϕ can be rephrased as

$$\sum_{x \in R} (e_x - 1) = 2\deg(\phi) - 2, \tag{1.3}$$

where $R \subseteq \mathbb{P}^1(\overline{K})$ is the set of ramification points and e_x is the multiplicity of ϕ at x. However, equation (1.3) fails if $p \mid e_x$ for any $x \in R$, since in that case $\rho_x \geq e_x$. We still have $\sum_{x \in R}(e_x - 1) \leq 2\deg(\phi) - 2$, but the right-hand side can be substantially larger than the left. We can even have $\rho_x > \deg(\phi)$; see, for example, Exercise 1.21.

Definition 1.13. Let K be a field, and let $\phi \in K(z)$ be a rational function. A point $x \in \mathbb{P}^1(\overline{K})$ is said to be an *exceptional* point of ϕ if the grand orbit of x in $\mathbb{P}^1(\overline{K})$ is finite. The set E_ϕ of all exceptional points of ϕ is called the *exceptional set* of ϕ.

Exceptional points are so named in part because they are generally quite rare. However, there are a number of simple ways that a nonconstant map $\phi \in K(z)$ could have exceptional points, as follows.

Example 1.14. If $\deg(\phi) = 1$, then the periodic points of ϕ are exceptional. If some iterate ϕ^n has at least three distinct fixed points — as would be the case if ϕ has even one periodic point that is not fixed—then $\phi^n(z) = z$, and hence all of $\mathbb{P}^1(\overline{K})$ is exceptional.

Example 1.15. If $\phi \in K[z]$ is a polynomial, then $\phi^{-1}(\infty) = \{\infty\}$. Thus, ∞ is an exceptional point of ϕ.

If, moreover, $\phi(z) = z^d$ for some $d \geq 1$, then $\phi^{-1}(0) = \{0\}$ as well, and hence both 0 and ∞ are exceptional points of ϕ.

Example 1.16. If $\phi(z) = z^{-d}$ for some $d \geq 1$, then $\phi^{-1}(0) = \{\infty\}$ and $\phi^{-1}(\infty) = \{0\}$. Thus, both 0 and ∞ are exceptional points of ϕ.

Example 1.17. If $\operatorname{char} K = p > 0$, then K contains a finite subfield $\mathbb{F}_p$, and hence $\overline{K}$ contains a copy of $\overline{\mathbb{F}}_p$. If $\phi(z) = z^{p^m}$ for some $m \geq 1$, then the exceptional set of ϕ is $\mathbb{P}^1(\overline{\mathbb{F}}_p)$; see Exercise 1.25.

Example 1.18. If $\psi = h \circ \phi \circ h^{-1}$ is conjugate to $\phi \in K(z)$ and if ϕ has exceptional set E_ϕ, then ψ has exceptional set $h(E_\phi)$.

The following result shows that up to conjugacy, Examples 1.14–1.17 describe all maps that have nonempty exceptional sets.

Theorem 1.19. *Let K be a field, and let $\phi \in K(z)$ be a rational function of degree $d \geq 2$ with exceptional set $E_\phi \subseteq \mathbb{P}^1(\overline{K})$.*

(a) *If $\#E_\phi > 2$, then K contains a finite field $\mathbb{F}_p$, and ϕ is conjugate over $\overline{K}$ to z^{p^m} for some $m \geq 1$. In particular, $\#E_\phi = \infty$.*

(b) *If $\#E_\phi = 2$, then ϕ is conjugate over $\overline{K}$ to either z^d or z^{-d}.*

(c) *If $\#E_\phi = 1$, then ϕ is conjugate over $\overline{K}$ to a polynomial.*

Proof. We begin by claiming that ϕ is totally ramified at every exceptional point. To see this, given $x \in E_\phi$, its grand orbit $\operatorname{GO}(x)$ is finite by definition of exceptional. Moreover, by definition of grand orbit, ϕ maps $\operatorname{GO}(x)$ onto itself; since $\#\operatorname{GO}(x) < \infty$, the mapping is bijective. Thus, for some $n \geq 1$, ϕ^n fixes every point of $\operatorname{GO}(x)$. In particular, $\phi^{-n}(x) = \{x\}$. Thus, ϕ^n is totally ramified at x, and hence ϕ itself must also be totally ramified at x, as claimed.

(a) By Corollary 1.11, if $\#E_\phi > 2$, then ϕ is totally inseparable of degree $d = p^m$ for some $m \geq 1$. Because $\phi'(z) = 0$, all the fixed points of ϕ have multiplier 0. Since $0 \neq 1$, then there are $d + 1$ distinct fixed points; see Section 1.3. In particular, ϕ has at least three distinct fixed points, and so we may move them to 0, 1, and ∞ by some $\overline{K}$-rational change of coordinates. Thus, we have $\phi(z) = \eta(z^{p^m})$, where $m \geq 1$ and $\eta \in \mathrm{PGL}(2, \overline{K})$. Since ϕ fixes each of 0, 1, ∞, it follows that η also fixes each of 0, 1, ∞. Therefore, $\eta(z) = z$, and hence $\phi(z) = z^{p^m}$. The rest of part (a) follows immediately.

(b) If $\#E_\phi = 2$, then after a change of coordinates over $\overline{K}$, we may assume that $E_\phi = \{0, \infty\}$. On the one hand, if $\phi(0) = 0$, then $\phi^{-1}(0) = \{0\}$ and $\phi^{-1}(\infty) = \{\infty\}$; therefore $\phi(z) = az^d$ for some $a \in \overline{K}^\times$. Conjugating again, we get $a^{1/(d-1)}\phi(a^{-1/(d-1)}z) = z^d$. On the other hand, if $\phi(0) \neq 0$, then $\phi^{-1}(0) = \{\infty\}$ and $\phi^{-1}(\infty) = \{0\}$, and hence $\phi(z) = az^{-d}$. This time, the conjugation $a^{-1/(d+1)}\phi(a^{1/(d+1)}z)$ gives z^{-d}, as desired.

(c) If $\#E_\phi = 1$, change coordinates over $\overline{K}$ to move the unique point of E_ϕ to ∞. Then $\phi^{-1}(\infty) = \{\infty\}$, and hence ϕ is a polynomial. □

1.5. Dynamics in degree less than 2

Most of the interesting dynamics of rational functions occurs when the degree of the function is at least 2. For completeness, however, we now describe what happens in degrees 0 and 1.

If $\phi(z) \in K(z)$ has degree 0, then ϕ is simply a constant function $\phi(z) = c$, and we have $\phi^n(z) = c$ for all $n \geq 1$. The point c is fixed, and there are no other dynamical features to speak of.

For the remainder of this section, then, suppose that $\phi(z) \in K(z)$ has degree 1; that is, $\phi \in \mathrm{PGL}(2, K)$. In other words, ϕ can be represented by a 2×2 matrix, and iterating ϕ corresponds to taking powers of the matrix. Thus, ϕ^n can be described fully using elementary linear algebra. We can understand the iterates of such a function dynamically, as well. After all, ϕ has at least one fixed point, and, unless $\phi(z) = z$, at most $1 + \deg(\phi) = 2$ fixed points. We consider two cases: that ϕ has two distinct fixed points, or that it has only one.

If ϕ has (at least) two distinct fixed points, then after a $\overline{K}$-rational change of coordinates, we can move one of those points to 0 and the other to ∞. This change corresponds to conjugating the matrix representation of ϕ by a matrix of eigenvectors; one eigenvector is moved to $(1, 0)$, corresponding to ∞, and the other to $(0, 1)$, corresponding to 0. Thus, $\phi(z)$ is now of the form az, for some $a \in \overline{K}^\times$. Then, for any $n \geq 1$, $\phi^n(z) = a^n z$.

Hence, if a is a root of unity, then some iterate of ϕ is the identity function. Otherwise, ϕ has no preperiodic points besides the two original fixed points. Note also that $\phi(z) = az$ is conjugate to $\psi(z) = bz$ if and only if either $a = b$ or $a = 1/b$. After all, any conjugation must map fixed points to fixed points; hence, either fix both 0 and ∞ or else exchange them. Conjugating by $z \mapsto cz$ leaves $\phi(z) = az$ unaffected, whereas conjugating by $z \mapsto c/z$ gives $\psi(z) = (1/a)z$.

Finally, if ϕ has only one fixed point, then that point must have multiplier 1; see Section 1.3. By a $\overline{K}$-rational change of coordinates, we can move the fixed point to ∞, so that $\phi(z) = az + b$. However, the multiplier of ∞ for $az + b$ is $1/a$, and thus we must have $\phi(z) = z + b$. Moreover, $b \neq 0$, since ϕ has only one fixed point. Conjugating by $z \mapsto z/b$, ϕ then becomes $\phi(z) = z + 1$. This coordinate change corresponds to the case that the associated 2×2 matrix has a repeated eigenvalue with eigenspace of dimension 1. The matrix can then only be reduced to Jordan canonical form but not diagonalized. Clearly, then, $\phi^n(z) = z + n$ for all $n \geq 0$. Thus, if $\operatorname{char} K = p > 0$, then ϕ^p is the identity function; otherwise, if $\operatorname{char} K = 0$, then ϕ has no preperiodic points besides the fixed point at ∞.

1.6. An overview of complex dynamics

Much of the study of non-archimedean dynamics is modeled on the theory of complex dynamics. In this section, we present some highlights of the complex theory, without proofs. For more extensive background on the subject, the reader may consult any of a number of excellent expositions of the subject, including [**9**, **39**, **104**].

1.6.1. Classifying periodic points. Let $\phi(z) \in \mathbb{C}(z)$, and let $x \in \mathbb{P}^1(\mathbb{C})$ be a periodic point of ϕ with multiplier $\lambda \in \mathbb{C}$. We say that x is *attracting* if $|\lambda| < 1$, *repelling* if $|\lambda| > 1$, or *indifferent* (or *neutral*) if $|\lambda| = 1$. If $\lambda = 0$, we say that x is *superattracting*. If $|\lambda| = 1$, we say that x is *rationally indifferent* if λ is a root of unity, or it is *irrationally indifferent* otherwise.

The terms "attracting" and "repelling" are appropriate, as follows. Suppose $x \in \mathbb{P}^1(\mathbb{C})$ is periodic of period $m \geq 1$. If x is attracting, then it has an open neighborhood $U \subseteq \mathbb{P}^1(\mathbb{C})$ such that for all $y \in U$, $\lim_{n\to\infty} \phi^{mn}(y) = x$. If x is repelling, then it has an open neighborhood $U \subseteq \mathbb{P}^1(\mathbb{C})$ such that for all $y \in U \smallsetminus \{x\}$, there is some $n \geq 1$ such that $\phi^{mn}(y) \notin U$; see [**9**, Section 6.3–6.4], [**39**, Sections II.2–II.3], or [**104**, Section 8].

If $\deg(\phi) \geq 2$, then it is a simple consequence of the Holomorphic Fixed-Point Formula that ϕ has a fixed point that is either repelling or of multiplier 1; see Exercise 1.27.

Let $x \in \mathbb{P}^1(\mathbb{C})$ be a periodic point of ϕ of period $m \geq 1$ and of multiplier $\lambda \in \mathbb{C}$. If $|\lambda| \neq 0, 1$, and even in certain circumstances for $|\lambda| = 1$, the map

ϕ^m is locally conjugate to the map $z \mapsto \lambda z$. In this case, we say ϕ^m is *locally linearizable* at x; see Section 4.2, especially Remark 4.4.

1.6.2. Fatou and Julia sets. The *chordal metric* on $\mathbb{P}^1(\mathbb{C})$ is the Euclidean distance under the usual embedding of the Riemann sphere in $\mathbb{R}^3$. More precisely, it is given by the formula

$$d_{\mathbb{C},\text{chord}}(z,w) := \frac{|z-w|}{\sqrt{(1+|z|^2)(1+|w|^2)}}.$$

Alternatively, the *spherical metric*, defined by

$$d_{\mathbb{C},\text{sph}}(z,w) := \arcsin\big(d_{\mathbb{C},\text{chord}}(z,w)\big),$$

is the distance over the surface of the sphere itself. The two metrics are equivalent, and hence they may be used interchangeably for our purposes.

We recall the following definitions from analysis.

Definition 1.20. Let (X, d_X) and (Y, d_Y) be metric spaces, and let $\mathcal{G}$ be a set of continuous functions from X to Y.

(a) We say that $\mathcal{G}$ is *equicontinuous* on X if, for every $\varepsilon > 0$, there is some $\delta > 0$ such that for every $f \in \mathcal{G}$ and every $x, y \in X$ satisfying $d_X(x,y) < \delta$, we have $d_Y(f(x), f(y)) < \varepsilon$.

(b) We say that $\mathcal{G}$ is *normal* on X if every sequence$\{f_n\}_{n\geq 1}$ in $\mathcal{G}$ has a subsequence that converges uniformly on compact subsets of X.

The key to Definition 1.20(a) is that the choice of δ does *not* depend on the function $f \in \mathcal{G}$. By the Arzelà–Ascoli theorem (see, for example, [**42**, Theorem VII.1.23]), normality is equivalent to equicontinuity when X is an open subset of $\mathbb{P}^1(\mathbb{C})$ and Y is compact.

Definition 1.21. Let $\phi \in \mathbb{C}(z)$. The *Fatou set* of ϕ, denoted $\mathcal{F}_\phi$, is the set of all points $x \in \mathbb{C}_v$ having a neighborhood on which the family of iterates $\{\phi^n : n \geq 0\}$ is equicontinuous, or equivalently normal, with respect to the spherical metric $d_{\mathbb{C},\text{sph}}$. The *Julia set* of ϕ, denoted $\mathcal{J}_\phi$, is the complement $\mathbb{P}^1(\mathbb{C}) \smallsetminus \mathcal{F}_\phi$.

The Fatou set $\mathcal{F}_\phi$ is, by definition, an open subset of $\mathbb{P}^1(\mathbb{C})$. It is also invariant under ϕ, meaning that $\phi^{-1}(\mathcal{F}_\phi) = \mathcal{F}_\phi$. Similarly, the Julia set $\mathcal{J}_\phi$ is an invariant closed subset of $\mathbb{P}^1(\mathbb{C})$. Moreover, the Fatou and Julia sets of an iterate ϕ^n of ϕ coincide with those of ϕ, and if $\psi = \eta \circ \phi \circ \eta^{-1}$ for some $\eta \in \text{PGL}(2, \mathbb{C})$, then $\mathcal{F}_\psi = \eta(\mathcal{F}_\phi)$.

It is straightforward to show that all attracting periodic points of ϕ belong to $\mathcal{F}_\phi$ and that all repelling periodic points of ϕ belong to $\mathcal{J}_\phi$. It is also true, but less obvious, that any rationally indifferent periodic point of ϕ also belongs to $\mathcal{J}_\phi$, provided $\deg(\phi) \geq 2$. It follows that every $\phi \in \mathbb{C}(z)$

of degree at least 2 has nonempty Julia set, since it must have a fixed point that is either repelling or has multiplier 1; see Section 1.6.1.

Theorem 1.22. *Let $\phi \in \mathbb{C}(z)$ be a rational function of degree $d \geq 2$. Then*

(a) *$\mathcal{J}_\phi$ is the smallest closed subset of $\mathbb{P}^1(\mathbb{C})$ that is both nonempty and invariant under ϕ.*

(b) *Let $a \in \mathcal{J}_\phi$. Then $\mathcal{J}_\phi$ is the closure of the backward orbit of a.*

(c) *$\mathcal{J}_\phi$ is the closure of the set of repelling periodic points of ϕ.*

(d) *If $\mathcal{J}_\phi$ has nonempty interior, then $\mathcal{J}_\phi = \mathbb{P}^1(\mathbb{C})$.*

(e) *$\mathcal{J}_\phi$ is a perfect set; that is, it is closed and has no isolated points. In particular, it is uncountable.*

Proof. See [**9**, Section 4.2], [**39**, Section III.1], [**104**, Sections 4,14]. □

1.6.3. Components of the Fatou set. A rational function $\phi \in \mathbb{C}(z)$ maps any connected component U of its Fatou set $\mathcal{F}_\phi$ to another component (or to itself). Thus, ϕ induces a dynamical system on its set of Fatou components. By the work of Fatou, Julia, Siegel, Herman, and others, every periodic component U of minimal period m falls into one of four types:

- *Attracting.* U contains a unique attracting periodic point a, also of period m, and every point in U approaches a under iteration of ϕ^m.
- *Parabolic.* The boundary of U contains a unique rationally indifferent periodic point a of period j dividing m (in fact, the multiplier is a primitive ℓth root of unity, with $j\ell = m$), and every point in U is attracted to a under iteration of ϕ^m.
- *Siegel disk.* U is simply connected and contains a unique periodic point a, which is irrationally indifferent and of period m. Moreover, ϕ^m is analytically conjugate on U to the rotation $z \mapsto \lambda z$ of the open unit disk, where λ is the multiplier of a.
- *Herman ring.* U is topologically an annulus, and ϕ^m is analytically conjugate on U to some irrational rotation of an annulus.

Moreover, all attracting and parabolic cycles of components contain critical points, implying that there are at most $2d-2$ such component cycles, where $d := \deg(\phi)$, although the *length* of those cycles can be arbitrarily large. By a perturbation argument, Fatou showed in 1919 [**54**] that there are at most $6d-6$ nonrepelling cycles, and hence at most $6d-6$ cycles of attracting components, parabolic components, and Siegel disks; see [**39**, Section III.2] for a proof. In the mid-1980s, as described in [**45**], Sullivan proved that there are at most $4d+2$ cycles of Herman rings. However, he never published this result because in 1987, using a method known as quasi-conformal surgery, Shishikura sharpened the bounds to at most $2d-2$ periodic cycles of Fatou components, including at most $d-2$ cycles of Herman rings [**129**].

Still, the credit for pioneering the use of quasi-conformal maps in complex dynamics goes to Sullivan, who used them to prove in 1985 that complex rational functions have "no wandering domains", meaning that all Fatou components of a complex rational function are preperiodic [**134**]. Together, the classification, bounds, and no wandering domains theorems give a very clear description of the possible dynamics on the Fatou set over $\mathbb{C}$. For details, see [**9**, Chapters 6–9], [**39**, Chapters II,IV–VI], or [**104**, Sections 10,11,15,16].

1.6.4. The equilibrium measure. Another feature of complex dynamics is the existence of a Borel probability measure μ_ϕ on the Riemann sphere that is *invariant* with respect to the rational function $\phi \in \mathbb{C}(z)$. That is,

$$\mu_\phi\big(\phi^{-1}(E)\big) = \mu_\phi(E)$$

for every measurable set $E \subseteq \mathbb{P}^1(\mathbb{C})$. The support of μ_ϕ is precisely the Julia set of ϕ. The measure μ_ϕ was first defined for polynomials by Brolin in [**34**], and later for rational functions independently by Lyubich [**100**] and by Freire, Lopes, and Mañé [**65**]. Potential theory on the Riemann sphere is a key tool in the construction.

Exercises for Chapter 1

1.1. Let $\phi : X \to X$ and $\psi : Y \to Y$ be dynamical systems, let $h : X \to Y$ be a morphism from ϕ to ψ, and let $x \in X$. Prove the following claims made after Definition 1.3.

(a) If x is a fixed (resp., or periodic, or preperiodic) point of ϕ, prove that $h(x)$ is a fixed (resp., or periodic, or preperiodic) point of ψ.

(b) If x is periodic under ϕ of minimal period n, and if h is one-to-one, prove that $h(x)$ also has minimal period n under ψ.

(c) If x is periodic under ϕ of minimal period n, give an example for which h is not one-to-one, and the minimal period of $h(x)$ under ϕ is *not* n.

(d) Prove that h maps forward orbits onto forward orbits; that is,

$$h\big(\operatorname{Orb}_\phi^+(x)\big) = \operatorname{Orb}_\psi^+\big(h(x)\big).$$

(e) Prove that h maps backward orbits into backward orbits and grand orbits into grand orbits; that is,

$$h(\operatorname{Orb}_\phi^-(x)) \subseteq \operatorname{Orb}_\psi^-(h(x)) \quad \text{and} \quad h(\operatorname{GO}_\phi(x)) \subseteq \operatorname{GO}_\psi(h(x)).$$

(f) If h is onto, prove that

$$h(\mathrm{Orb}_\phi^-(x)) = \mathrm{Orb}_\psi^-(h(x)) \quad \text{and} \quad h(\mathrm{GO}_\phi(x)) = \mathrm{GO}_\psi(h(x)).$$

Give an example for which h is not onto and for which both of these equalities fail.

1.2. Let $\phi, \psi : X \to X$ be nonconstant maps.

(a) Prove that there is a nonconstant morphism from the dynamical system $\phi \circ \psi : X \to X$ to the dynamical system $\psi \circ \phi : X \to X$.

(b) Let $X = \mathbb{P}^1(\mathbb{C})$, $\phi(z) = z^2$, and $\psi(z) = z^2 - 1$. Prove that $\phi \circ \psi$ is *not* conjugate to $\psi \circ \phi$.

1.3. For each $d \geq 1$, let T_d be the dth Chebyshev polynomial; see Example 1.4.

(a) Prove that T_d exists and is unique. That is, for each $d \geq 1$, there is a unique polynomial T_d such that

$$T_d(z + z^{-1}) = z^d + z^{-d}.$$

(b) For every $d \geq 1$, prove that T_d has integer coefficients, and that

$$T_d(2\cos\theta) = 2\cos(d\theta).$$

(c) Compute T_d for $d = 1, 2, 3, 4, 5, 6$.

(d) For every $m, n \geq 1$, prove that $T_m \circ T_n = T_{mn}$.

(e) Fix $d, n \geq 1$, let ζ be a primitive dth root of unity, and let

$$\xi := \zeta + \zeta^{-1}.$$

Find the minimal integers M, N such that $N > M \geq 0$ and

$$T_d^M(\xi) = T_d^N(\xi).$$

1.4. Let $\phi, \psi \in K(z)$ be rational functions, both of degree at least 2 (but not necessarily the same degree) such that

$$\phi \circ \psi = \psi \circ \phi.$$

Prove that ϕ and ψ have the same set of preperiodic points in $\mathbb{P}^1(\overline{K})$.

1.5. Let K be a field of characteristic not equal to 2, and let $h \in \mathrm{PGL}(2, K)$. Let $c_1, c_2 \in K$, and define $f_i(z) = z^2 + c_i \in K[z]$ for $i = 1, 2$. Suppose that

$$f_1 = h \circ f_2 \circ h^{-1}.$$

Prove that $c_1 = c_2$.

1.6. Let K be an algebraically closed field, and let $f \in K[z]$ be a quadratic polynomial.

(a) Prove that there is some $a \in K$ such that f is conjugate (via an element of $\mathrm{PGL}(2, K)$) to the polynomial $z^2 + az$.

(b) If $z^2 + az$ and $z^2 + bz$ are conjugate via an element of $\mathrm{PGL}(2, K)$, prove that $a = \pm b$.

(c) If $\operatorname{char} K = 2$, prove that $z^2 + z$ is not conjugate to $z^2 + c$ for any $c \in K$.

(d) If $\operatorname{char} K = 2$, prove that $z^2 + c$ is conjugate to z^2 for every $c \in K$.

1.7. Let K be a field of characteristic not equal to 3, and let $f \in K[z]$ be a cubic polynomial.

(a) Prove that there exist $a, b \in K$ such that f is conjugate (via an element of $\mathrm{PGL}(2, K)$) to either $az^3 + bz + 1$ or to $az^3 + bz$.

(b) Prove that no polynomial in $K[z]$ of the form $az^3 + bz + 1$ is conjugate to any other polynomial of the form $cz^3 + dz + 1$ or $cz^3 + dz$.

(c) If two cubic polynomials of the form $az^3 + bz, cz^3 + dz \in K[z]$ are conjugate, prove that $b = d$ and that a/c is a square in $K^\times$.

1.8. Let K be an algebraically closed field, and let $\phi \in K(z)$ be a rational function with $\deg(\phi) - 2$. Prove that ϕ is conjugate either to

$$\psi(z) = z + \frac{1}{z}$$

or to a map of the form

$$\psi(z) = \frac{z^2 + \lambda z}{\mu z + 1}$$

with $\lambda, \mu \in K$.

(*Suggestion*: If there is only one fixed point, move it to ∞. If there are at least two, move them to 0 and ∞. Note that this observation appears in the proof of [**105**, Lemma 3.1].)

1.9. Let $\phi \in K(z)$ be a rational function of degree $\deg(\phi) = 2$, and let $\lambda_1, \lambda_2, \lambda_3 \in \overline{K}$ be the multipliers of the three fixed points of ϕ. Prove that

$$\lambda_1\lambda_2\lambda_3 = \lambda_1 + \lambda_2 + \lambda_3 - 2.$$

(*Suggestion*: Use the Holomorphic Fixed-Point Formula.)

1.10. Let K be a field of characteristic not equal to 2 or 3, and let $\phi \in K(z)$ be a rational function with $\deg(\phi) = 2$. If the multipliers $\lambda_1, \lambda_2, \lambda_3$ of the three fixed points of ϕ are all distinct, prove that there exist $a, b \in K$ and $h \in \mathrm{PGL}(2, K)$ such that

$$h \circ \phi \circ h^{-1}(z) = \frac{2z^2 + az + a}{-z^2 + (4 - a)z + b}.$$

(*Suggestion*: Use $a = -\lambda_1\lambda_2\lambda_3$ and $b = a - \lambda_1\lambda_2 - \lambda_1\lambda_3 - \lambda_2\lambda_3$. Note that this is the generic case of [**102**, Theorem 1].)

1.11. Let $\overline{K}$ be an algebraically closed field. Let $\phi = f/g \in \overline{K}(z)$ for relatively prime polynomials $f, g \in \overline{K}[z]$. Let $d := \max\{\deg(f), \deg(g)\}$. Prove that every point $x \in \mathbb{P}^1(\overline{K})$ has exactly d preimages in $\mathbb{P}^1(\overline{K})$, counting multiplicity.

1.12. Let $\phi \in K(z)$, let $h \in \mathrm{PGL}(2, K)$, and define $\psi := h \circ \phi \circ h^{-1}$. Let $x \in K$ be a fixed point of ϕ of multiplicity $m \geq 1$ and multiplier $\lambda \in K$. Prove that $h(x)$ is a fixed point of ψ of multiplicity m and multiplier λ.

1.13. Let $n \geq 1$. Prove equation (1.1): if none of $x, \phi(x), \ldots, \phi^n(x)$ is ∞, then

$$(\phi^n)'(x) = \prod_{i=0}^{n-1} \phi'(\phi^i(x)).$$

1.14. Let K be a field of characteristic $p \geq 0$ with algebraic closure $\overline{K}$, let $\phi \in K(z)$, and let $x \in \mathbb{P}^1(\overline{K})$ be a fixed point of ϕ of multiplicity $m \geq 2$. For each $n \geq 1$ with $p \nmid n$, prove that the multiplicity of x as a fixed point of ϕ^n is also m.

(*Suggestion*: Move x to 0 and write $f = z + z^m g$.)

1.15. Let K be a field of characteristic $p \geq 2$, and let $\phi(z) = z + z^p \in K(z)$. For each $n \geq 1$, prove that the multiplicity of $x = 0$ as a fixed point of ϕ^{p^n} is p^{p^n}.

1.16. Let K be a field with algebraic closure $\overline{K}$, and let $\psi \in K(z)$. Suppose that ψ has only finitely many *distinct* periodic points in $\mathbb{P}^1(\overline{K})$. Prove that $\deg(\psi) \leq 1$.

(*Suggestion*: Use Exercise 1.14. Note that when $\operatorname{char} K = 0$, a much stronger result is true: ψ must have periodic points of all minimal periods, with a small finite list of possible exceptions; see [**9**, Theorems 6.2.1, 6.2.2].)

1.17. Let K be a field with algebraic closure $\overline{K}$, let $\phi \in \overline{K}(z)$ be a rational function, and let

$$\omega := \frac{dz}{\phi(z) - z} \in \Omega^1(\mathbb{P}^1(\overline{K}))$$

be the meromorphic differential form of equation (1.2). Prove that ω is indeed holomorphic except for, at worst, simple poles at $x_1, \ldots, x_{d+1}$ and at ∞, as claimed in the proof of Theorem 1.7.

1.18. Let K be an arbitrary field, and let $\omega \in \Omega^1(\mathbb{P}^1(\overline{K}))$ be a meromorphic differential form on $\mathbb{P}^1(\overline{K})$. Prove the Residue Theorem for ω: let $x_1, \ldots, x_n \in \mathbb{P}^1(\overline{K})$ be the poles of ω, and let $a_1, \ldots, a_n$ be the corresponding residues; then $a_1 + \cdots + a_n = 0$. (Recall that the residue of ω at $x \in \mathbb{P}^1(\overline{K})$ is defined as follows. Let z be a uniformizer at x, i.e., a rational function with a simple zero at x, and expand ω as a formal Laurent series

$$\omega = \sum_{n \geq m} a_n z^n \, dz$$

at x. Then the residue of ω at x is $a_{-1} \in \overline{K}$, which is independent of the choice of uniformizer.)

(*Suggestion*: Write $\omega = f(t)\,dt$ and consider the partial fraction expansion of $f(t)$. Use $t - x$ as the uniformizer at $x \in \overline{K}$ and $1/t$ as the uniformizer at $x = \infty$.)

1.19. Let $\phi(z) \in K(z)$, and let $x \in \mathbb{P}^1(\overline{K})$. In this exercise, you will prove that the definitions of order and multiplicity of ϕ at x are well defined.

Let $\eta_1, \eta_2, \theta_1, \theta_2 \in \mathrm{PGL}(2, \overline{K})$ such that

$$\eta_1(\phi(x)) = \theta_1(x) = \eta_2(\phi(x)) = \theta_2(x) = 0,$$

and define the functions

$$\psi_1 := \eta_1 \circ \phi \circ \theta_1^{-1} \quad \text{and} \quad \psi_2 := \eta_2 \circ \phi \circ \theta_2^{-1}.$$

For any rational function $f(z) \in \overline{K}(z)$, let $\mathrm{ord}_0(f)$ denote the order of vanishing of f at 0. That is, writing $f(z) = z^n h(z)$ with $n \in \mathbb{Z}$, $h \in \overline{K}(z)$, and $h(0) \neq 0$, we set $\mathrm{ord}_0(f) := n$.

(a) Prove that $\mathrm{ord}_0(\psi_1) = \mathrm{ord}_0(\psi_2)$.

(b) Prove that $\mathrm{ord}_0(\psi_1') = \mathrm{ord}_0(\psi_2')$.

1.20. Let $\phi(z) \in K(z)$ be a separable rational function, let $x \in \mathbb{P}^1(\overline{K})$, let $e \geq 1$ be the multiplicity of ϕ at x, and let $\rho \geq 0$ be the order of the critical point (if any) at x. Let $p := \operatorname{char} K \geq 0$.

(a) If $p \nmid e$, prove that $\rho = e - 1$.

(b) If $p | e$, prove that $e \leq \rho \leq 2\deg(\phi) - 2$, and that $p \nmid (\rho + 1)$.

1.21. Fix $n \geq 2$, and suppose $\operatorname{char} K = p > 0$.

(a) Let $\phi \in K[z]$ be a separable polynomial of degree np. Prove that equation (1.3) fails for ϕ.

(b) Let $\phi(z) = z^{np} + z$. Prove that ϕ has a critical point only at ∞, with multiplicity $np = \deg(\phi)$, but with order of the critical point $\rho_\infty = 2np - 2 > \deg(\phi)$.

1.22. Fix a prime p. For any choice of integers e, ρ, d with

- $p \leq e \leq d$,
- $e \leq \rho \leq 2d - 2$
- $p | e$, and
- $p \nmid (\rho + 1)$,

is there an example of a rational function $\phi(z) \in \mathbb{F}_p(z)$ such that $\deg(\phi) = d$, ϕ has multiplicity e at 0, and the critical point at 0 has order ρ? That is, are there any restrictions on $\deg(\phi)$ and the multiplicity and order of ϕ at 0 other than those noted in Remark 1.12 and Exercise 1.20?

1.23. Let $\phi(z) \in K(z)$ be a separable rational function of degree $d \geq 1$. Write $\phi = f/g$ with $f, g \in K[z]$ relatively prime. Prove the Riemann–Hurwitz formula for ϕ (Theorem 1.9) by analyzing the quotient rule formula $(f/g)' = (f'g - g'f)/g^2$ and by separately considering the order of the critical point (if any) at ∞.

1.24. Let K be a field, and let $\phi \in K(z)$ or $\phi \in K[[z]]$ be a rational function or formal power series with $\phi(0) = 0$.

(a) If $\operatorname{ord}_0(\phi(z) - z - az^n) \geq n + 1$ for some $n \geq 2$ and $a \in K$, prove that

$$\operatorname{ord}_0(\phi^j(z) - z - jaz^n) \geq n + 1 \quad \text{for all } j \geq 1.$$

(b) If $\operatorname{char} K = p > 0$ and $\operatorname{ord}_0(\phi(z) - z) \geq n$ for some $n \geq 2$, prove that

$$\operatorname{ord}_0(\phi^p(z) - z) \geq n + 1.$$

(c) If $\operatorname{char} K = p > 0$ and $\operatorname{ord}_0(\phi(z) - z) \geq 2$, prove that

$$\lim_{n \to \infty} \operatorname{ord}_0(\phi^{p^n}(z) - z) = \infty.$$

(As in Exercise 1.19, $\operatorname{ord}_0(f)$ denotes the order of vanishing of f at 0.)

1.25. Prove the claim of Example 1.17: if K contains a finite subfield $\mathbb{F}_p$ and if $\phi(z) = z^{p^m}$ for some $m \geq 1$, then the exceptional set of ϕ is $\mathbb{P}^1(\overline{\mathbb{F}}_p)$.

1.26 (Symbolic dynamics)**.** Let X be a compact Hausdorff topological space, let $f : X \to X$ be a continuous function, and let $\{Y_\alpha : \alpha \in I\}$ be a nonempty set of nonempty compact subsets of X. Suppose that

$$\phi(Y_\beta) \supseteq \bigcup_{\alpha \in I} Y_\alpha \quad \text{for every } \beta \in I.$$

(a) Let $\Sigma := \prod_{n \geq 0} I$, called the set of *symbol sequences*. For any symbol sequence $A = (\alpha_0, \alpha_1, \alpha_2, \ldots) \in \Sigma$, define

$$Z_A := \bigl\{x \in X : f^n(x) \in Y_{\alpha_n} \text{ for all } n \geq 0\bigr\},$$

and for any $n \geq 0$, define

$$C_{A,n} := \bigl\{x \in X : f^i(x) \in Y_{\alpha_i} \text{ for all } i = 0, \ldots, n\bigr\},$$

which is called a *cylinder set*. Prove that $C_{A,n}$ and Z_A are nonempty compact subsets of X.

(b) Suppose further that the sets Y_α are pairwise disjoint. For any $A, B \in \Sigma$, prove that the cylinder sets $C_{A,n}$ and $C_{B,n}$ intersect if and only if A and B agree from the $i = 0$ through $i = n$ entries. In addition, prove that Z_A and Z_B intersect if and only if $A = B$.

1.27. Use the Holomorphic Fixed-Point Formula (Theorem 1.7) to prove the following claim from Section 1.6.1. Let $\phi(z) \in \mathbb{C}(z)$ be a rational function with $\deg(\phi) \geq 2$. Then ϕ has a fixed point that is either repelling or of multipler 1.

Chapter 2

Some Background on Non-Archimedean Fields

In this chapter we introduce non-archimedean fields and their basic properties. Our presentation here is rather terse, as we assume the reader already has some familiarity with such fields. The reader desiring a gentler introduction to the subject may wish to consult [**70, 86**] for the special case of p-adic fields.

2.1. Absolute values

Definition 2.1. An *absolute value* on a field K is a function $|\cdot| : K \to [0, \infty)$ satisfying the following properties.

- $|\cdot|$ is *non-degenerate*: for any $x \in K$, we have $|x| = 0$ if and only if $x = 0$.
- $|\cdot|$ is *multiplicative*: for any $x, y \in K$, we have $|xy| = |x||y|$.
- $|\cdot|$ satisfies the *triangle inequality*: for any $x, y \in K$, we have $|x + y| \leq |x| + |y|$.

The *trivial* absolute value on K is given by $|0| = 0$ and $|x| = 1$ for all $x \in K^\times$; any other absolute value is said to be *nontrivial.* We are not interested in the trivial absolute value, and so from now on, we set the convention that **all of our absolute values are nontrivial**.

A nontrivial absolute value $|\cdot|$ is said to be *non-archimedean* if it satisfies the stronger *ultrametric* triangle inequality

$$|x+y| \leq \max\{|x|, |y|\}. \tag{2.1}$$

In that case, K is said to be a *non-archimedean field.* If $|x| \neq |y|$ and $|\cdot|$ is non-archimedean, one has equality in (2.1); see Exercise 2.1. Colloquially, then, all triangles are isoceles in a non-archimedean field.

Any non-archimedean field K has a *ring of integers*

$$\mathcal{O}_K := \{x \in K : |x| \leq 1\},$$

which is indeed a ring under the multiplication and addition inherited from K, and which has a unique maximal ideal

$$\mathcal{M}_K := \{x \in K : |x| < 1\};$$

see Exercise 2.2. The quotient field

$$k := \mathcal{O}_K/\mathcal{M}_K$$

is called the *residue field* of K, and the characteristic $\operatorname{char} k$ of the residue field is called the *residue characterstic* of K. If K has positive characteristic $p > 0$, then $\operatorname{char} k = p$ as well; however, if $\operatorname{char} K = 0$, then $\operatorname{char} k$ can be 0 or any prime. We say that K is a field of *equal characteristic* if $\operatorname{char} k = \operatorname{char} K$; otherwise, we say K is of *mixed characteristic.* The set $|K^\times| \subseteq (0,\infty)$ of absolute values attained by nonzero elements of K is a subgroup of $(0,\infty)$ called the *value group* of K; see Exercise 2.3. If $|K^\times|$ is a discrete subgroup of $(0,\infty)$—or equivalently, if it is cyclic—we say that K is *discretely valued.* In that case, recalling our convention that absolute values are nontrivial, there is a largest real number ε in $(0,1)\cap|K|$. An element $\pi \in K$ such that $|\pi| = \varepsilon$ is called a *uniformizer* for K; note that $|K^\times| = |\pi|^{\mathbb{Z}}$. A discretely valued field cannot be algebraically closed; for example, if $\pi \in K$ is a uniformizer, there is no element $x \in K$ such that $x^2 = \pi$.

We almost always consider non-archimedean fields K that are *complete,* that is, for which all $|\cdot|$-Cauchy sequences in K converge in K. Any non-archimedean field K has a completion K_v formed by taking the set of $|\cdot|$-Cauchy sequences modulo an appropriate equivalence. There is a natural embedding $K \hookrightarrow K_v$, and the absolute value $|\cdot|$ on K extends canonically to K_v. The value group of K_v coincides with that of K, and any uniformizer of K is still a uniformizer of K_v. In addition, the residue fields of K and K_v are canonically isomorphic; see Exercise 2.4.

If $K = K_v$ is a *complete* non-archimedean field and L/K is an *algebraic* extension, then the absolute value on K extends canonically to L. In particular, for any $\alpha \in L$ with minimal polynomial

$$x^n + a_{n-1}x^{n-1} + \cdots + a_0 \in K[x],$$

the absolute value of α is $|\alpha| := |a_0|^{1/n}$. Clearly, the value group $|L^\times|$ contains $|K^\times|$ as a subgroup. The index

$$e := \big[|L^\times| : |K^\times|\big],$$

called the *ramification degree of* L/K, is of course either a positive integer or ∞; if $e = 1$, we say that L/K is *unramified*. In addition, there is a natural inclusion of the residue field k of K in the residue field ℓ of L; the associated *residue field extension degree* $[\ell : k]$ is usually denoted f. We have the identity

$$[L : K] = ef;$$

see Exercise 2.5.

A non-archimedean field K is locally compact (in the topology induced by $|\cdot|$) if and only if its ring of integers $\mathcal{O}_K$ is compact. Equivalently, K is locally compact if and only if it is complete, discretely valued, and has finite residue field. If K is locally compact and if L/K is a *finite* extension, then L is also locally compact. However, if K is a complete non-archimedean field and L is an *infinite* extension, then L cannot be locally compact; see Exercise 2.7.

Thus, for any complete non-archimedean field K, its algebraic closure $\overline{K}$ may not be complete, but it still has the absolute value $|\cdot|$. We can then form the completion $\mathbb{C}_v$ of $\overline{K}$ with respect to $|\cdot|$. This field $\mathbb{C}_v$ is, fortunately, both complete *and* algebraically closed; in fact, it is the smallest field extension of K with both properties. The value group $|\mathbb{C}_v^\times| = |\overline{K}^\times|$ is dense in $(0, \infty)$, and therefore neither $\mathbb{C}_v$ nor $\overline{K}$ is discretely valued. In addition, the residue field of $\mathbb{C}_v$, which is canonically isomorphic to that of $\overline{K}$, is an algebraic closure of the residue field of K; see Exercise 2.8.

Example 2.2. Let $\mathbb{F}$ be a field. The associated field $\mathbb{F}((t))$ of formal Laurent series over $\mathbb{F}$ can be equipped with a non-archimedean absolute value $|\cdot|$ by fixing a real number $\varepsilon \in (0, 1)$ and defining $|\sum a_n t^n| = \varepsilon^{n_0}$, where n_0 is the smallest integer $n \in \mathbb{Z}$ such that $a_n \neq 0$. The value group is $|\mathbb{F}((t))^\times| = \varepsilon^{\mathbb{Z}}$; thus, $\mathbb{F}((t))$ is discretely valued, and $t \in \mathbb{F}((t))$ is a uniformizer, with $|t| = \varepsilon$. The associated ring of integers is simply the ring of power series

$$\mathbb{F}[[t]] := \left\{ \sum_{n\geq 0} a_n t^n \ : \ a_n \in \mathbb{F} \right\} = \{ f \in \mathbb{F}((t)) \ : \ |f| \leq 1 \}$$

under the usual operations of addition and multiplication. The maximal ideal $t\mathbb{F}[[t]]$ consists of all power series with zero constant term, and the residue field $\mathbb{F}[[t]]/t\mathbb{F}[[t]]$ is canonically isomorphic to $\mathbb{F}$. In particular, $\mathbb{F}((T))$ is a complete non-archimedean field of equal characteristic, because $\operatorname{char} \mathbb{F} = \operatorname{char} \mathbb{F}((t))$.

Let $\overline{\mathbb{F}}$ denote an algebraic closure of $\mathbb{F}$. Then the field $\overline{\mathbb{F}}((t))$ is an unramified extension of $\mathbb{F}((t))$, and in fact it is a maximal unramified extension. In addition, the algebraic closure $\overline{\mathbb{F}((t))}$ of $\mathbb{F}((t))$ properly contains $\overline{\mathbb{F}}((t))$, and the extension $\overline{\mathbb{F}((t))}/\overline{\mathbb{F}}((t))$ is infinitely ramified.

The completion $\mathbb{C}_v$ of $\overline{\mathbb{F}((t))}$ has an associated value group

$$|\mathbb{C}_v^\times| = |\overline{\mathbb{F}((t))}^\times| = \varepsilon^{\mathbb{Q}}.$$

Although this group is dense in $(0, \infty)$, it is *not* the whole interval. On the other hand, the residue field of both $\overline{\mathbb{F}((t))}$ and $\mathbb{C}_v$ is an algebraic closure $\overline{\mathbb{F}}$ of $\mathbb{F}$. Like $\mathbb{F}((t))$, the fields $\overline{\mathbb{F}((t))}$ and $\mathbb{C}_v$ are of equal characteristic, since $\operatorname{char} \overline{\mathbb{F}((t))} = \operatorname{char} \mathbb{C}_v = \operatorname{char} \overline{\mathbb{F}} = \operatorname{char} \mathbb{F}$.

The original field $\mathbb{F}((t))$ and its finite extensions are locally compact if and only if $\mathbb{F}$ is finite. However, none of the infinite extensions of $\mathbb{F}((t))$ are locally compact.

If $\operatorname{char} \mathbb{F} = 0$, then the field $\mathbb{C}_v$ is called the field of *formal Puiseux series* over $\mathbb{F}$, and we can describe it more explicitly as follows. First, one can show, using the characteristic 0 assumption, that $\overline{\mathbb{F}((t))}$ can be written as the direct limit

$$\overline{\mathbb{F}((t))} = \varinjlim_{m} \overline{\mathbb{F}}((t^{1/m}))$$

under the obvious inclusions of the various fields $\overline{\mathbb{F}}((t^{1/m}))$. That is, any particular element of $\overline{\mathbb{F}((t))}$ belongs to $\overline{\mathbb{F}}((t^{1/m}))$ for some $m \geq 1$. An arbitrary element $x \in \mathbb{C}_v$, therefore, can be written as a *Puiseux series*, which is a series of the form

$$\sum_{n \geq 0} \alpha_n t^{q_n},$$

where $\{q_n\}_{n \geq 0} \subseteq \mathbb{Q}$ is a strictly increasing sequence of rational numbers such that $q_n \nearrow \infty$, and where $\alpha_n \in \overline{\mathbb{F}}$; see Exercise 2.9.

One the other hand, if $\operatorname{char} \mathbb{F} = p > 0$, then $\overline{\mathbb{F}((t))}$ contains elements other than the Puiseux series described in the previous paragraph. For example, the polynomial $x^p - x + t^{-1} \in \mathbb{F}((t))[x]$ has no roots in any field of the form $\overline{\mathbb{F}}((t^{1/m}))$. Thus, the completion $\mathbb{C}_v$ is strictly larger than the field $\varinjlim \overline{\mathbb{F}}((t^{1/m}))$ of Puiseux series over such fields; see Exercise 2.10.

Example 2.3. Let $p \geq 2$ be a prime number. The p-adic absolute value $|\cdot|_p$ on $\mathbb{Q}$ is defined by

$$\left|\frac{m}{n} p^r\right| := p^{-r}, \quad \text{for } m, n, r \in \mathbb{Z} \text{ with } p \nmid mn,$$

and $|0|_p := 0$. In particular, all integers $n \in \mathbb{Z}$ satisfy $|n| \leq 1$, positive powers of p are small, and negative powers of p are big. The p-adic field $\mathbb{Q}_p$ is the completion of $\mathbb{Q}$ with respect to $|\cdot|_p$, and it can be thought of as the set of all Laurent series in the "variable" p with "coefficients" in $\{0, 1, \ldots, p-1\}$. That is,

$$\mathbb{Q}_p := \Big\{ \sum_{n \geq n_0} a_n p^n \ : \ n_0 \in \mathbb{Z} \text{ and } a_n \in \{0, 1, \ldots, p-1\} \Big\},$$

except that the addition and multiplication operations involve carrying of digits to higher powers of p.

The absolute value $|\cdot|_p$ extends to $\mathbb{Q}_p$ by

$$\Big| \sum a_n p^n \Big|_p := p^{-n_0},$$

where n_0 is the smallest integer $n \in \mathbb{Z}$ such that $a_n \neq 0$. The value group is $|\mathbb{Q}_p^\times|_p = p^{\mathbb{Z}}$. In particular, $\mathbb{Q}_p$ is discretely valued, and the element $p \in \mathbb{Q}_p$ is a uniformizer, with $|p|_p = 1/p$. The associated ring of integers

$$\mathbb{Z}_p := \{x \in \mathbb{Q}_p : |x|_p \leq 1\}$$

is the subset consisting of power series (as opposed to Laurent series) in p; it is also the completion of $\mathbb{Z}$ with respect to $|\cdot|_p$. The maximal ideal $p\mathbb{Z}_p$ consists of all power series with zero constant term, and the residue field $\mathbb{Z}_p/p\mathbb{Z}_p$ is isomorphic to $\mathbb{F}_p$, the field of p elements. In particular, $\mathbb{Q}_p$ is a complete non-archimedean field of mixed characteristic, because $\operatorname{char} \mathbb{Q}_p = 0$, but $\operatorname{char} \mathbb{F}_p = p > 0$.

The completion of an algebraic closure of $\mathbb{Q}_p$ is denoted $\mathbb{C}_p$. As was the case for the function field of Example 2.2, the value group $|\mathbb{C}_p^\times|_p = p^{\mathbb{Q}}$ is dense in $(0, \infty)$ but is not the whole interval. Although $\mathbb{Q}_p$ and its finite extensions are locally compact, $\mathbb{C}_p$ is complete but not locally compact. Examples of the many elements of $\mathbb{C}_p \smallsetminus \overline{\mathbb{Q}}_p$ include

$$\sum_{n=1}^{\infty} p^{1/n} p^n \quad \text{and} \quad \sum_{n=1}^{\infty} \alpha_n p^n,$$

where $\alpha_n \in \overline{\mathbb{Q}}_p$ is a primitive $(p^n - 1)$-st root of unity.

The function fields and p-adic fields of Examples 2.2 and 2.3 are the principal examples of non-archimedean fields to have in mind throughout this book. Sometimes we need to work over a particular such field, but most of our theory and results apply in a general setting. Thus, we set the following notation throughout this book:

$\mathbb{C}_v$ denotes a complete, algebraically closed field equipped with a nontrivial non-archimedean absolute value $\vert\cdot\vert$.

2.2. Disks

Let $\mathbb{C}_v$ be a complete, algebraically closed non-archimedean field with absolute value $|\cdot|$. We denote the *open disk* and *closed disk* centered at $a \in \mathbb{C}_v$ and of radius $r > 0$ by

$$(2.2) \qquad \begin{aligned} D(a,r) &:= \{x \in \mathbb{C}_v : |x-a| < r\}, \quad \text{and} \\ \overline{D}(a,r) &:= \{x \in \mathbb{C}_v : |x-a| \le r\}, \end{aligned}$$

respectively. We use the convention that **all disks have positive radius**.

Disks in non-archimedean metric spaces behave somewhat counterintuitively, as the following proposition demonstrates.

Proposition 2.4.

(a) *Let $a, b \in \mathbb{C}_v$ and $s \ge r > 0$ such that $a \in D(b,s)$. Then*

$$D(a,r) \subseteq D(b,s) \quad \textit{and} \quad D(a,s) = D(b,s).$$

(b) *Let $a, b \in \mathbb{C}_v$ and $s \ge r > 0$ such that $a \in \overline{D}(b,s)$. Then*

$$\overline{D}(a,r) \subseteq \overline{D}(b,s) \quad \textit{and} \quad \overline{D}(a,s) = \overline{D}(b,s).$$

(c) *Let $D_1, D_2 \subseteq \mathbb{C}_v$ be disks such that $D_1 \cap D_2 \neq \varnothing$. Then*

$$\textit{either} \quad D_1 \subseteq D_2 \quad \textit{or} \quad D_2 \subseteq D_1.$$

(d) *All disks in $\mathbb{C}_v$ are both open and closed* topologically.

(e) *$\mathbb{C}_v$ is totally disconnected as a topological space; that is, its only nonempty connected subsets are singletons.*

Proof.

(a) Given $x \in D(a,r)$, we have $|x-a| < r \le s$ and $|a-b| < s$. Thus,

$$|x-b| \le \max\{|x-a|, |a-b|\} < s,$$

and hence $x \in D(b,s)$. The reverse inclusion in the case $r = s$ is similar.

(b) This is similar to (a).

(c) Pick $c \in D_1 \cap D_2$. By parts (a) and (b), the point c is a center of each disk; that is, each D_i can be written as either $D(c, r_i)$ or $\overline{D}(c, r_i)$. After possibly exchanging D_1 and D_2, either $r_1 > r_2$, or else $r_1 = r_2$ with either D_1 closed or D_2 open (or both). Then $D_1 \supseteq D_2$.

(d) Given $x \in \mathbb{C}_v \smallsetminus D(a,r)$. If the two disks $D(x,r)$ and $D(a,r)$ intersect, then one contains the other by part (c), and hence they coincide by part (a). That contradicts our assumption that $x \notin D(a,r)$, and therefore

$$D(x,r) \subseteq \mathbb{C}_v \smallsetminus D(a,r).$$

Thus, the open disk $D(a,r)$ is (topologically) closed.

On the other hand, given $x \in \overline{D}(a,r)$, the disks $\overline{D}(x,r)$ and $\overline{D}(a,r)$ intersect at x. One therefore contains the other by part (c), and hence they coincide by part (b). Thus,

$$D(x,a) \subseteq \overline{D}(x,r) = \overline{D}(a,r),$$

proving that the closed disk $\overline{D}(a,r)$ is open.

(e) Let $X \subseteq \mathbb{C}_v$ be a set containing two distinct points a, b; we must show that X is disconnected. Let $r = |a-b| > 0$. Then by part (d), the two sets

$$U := X \cap D(a,r) \quad \text{and} \quad V := X \smallsetminus D(a,r)$$

are open subsets of X. They are both nonempty because $a \in U$ and $b \in V$. Moreover, we clearly have $U \cap V = \varnothing$. Therefore, X is disconnected. □

Proposition 2.4(a)–(b) shows that disks in $\mathbb{C}_v$ do not have well defined centers; indeed, every point of a disk can be called its center. The radius of a disk in $\mathbb{C}_v$, however, *is* well defined, because $\mathbb{C}_v$ is *not* discretely valued.

Proposition 2.5. *Let $D \subseteq \mathbb{C}_v$ be a disk, and define*

$$r = \operatorname{diam}(D) := \sup\{|x-y| : x, y \in D\}.$$

(a) *If $r \in |\mathbb{C}_v^\times|$ and there exist $x, y \in D$ such that $|x-y| = r$, then for any $a \in D$, we have $D = \overline{D}(a,r)$. Moreover, this is the only way to write D as a disk; it cannot be written as an open disk, nor as a disk of radius other than r.*

(b) *If $r \in |\mathbb{C}_v^\times|$ and $|x-y| < r$ for all $x, y \in D$, then for any $a \in D$, we have $D = D(a,r)$. Moreover, this is the only way to write D as a disk; it cannot be written as a closed disk, nor as a disk of radius other than r.*

(c) *If $r \notin |\mathbb{C}_v^\times|$, then for any $a \in D$, we have*

$$D = \overline{D}(a,r) = D(a,r).$$

Moreover, this is the only way to write D as a disk; it cannot be written as a disk of radius other than r.

Proof. For any $a \in D$, Proposition 2.4(a)–(b) says that D may be written as either $D(a,s)$ or $\overline{D}(a,s)$ for some $s > 0$. Before considering the cases (a)–(c), we claim that s can only be r.

Given any $x, y \in D \subseteq \overline{D}(a,s)$, then

$$|x-y| \leq \max\{|x-a|, |y-a|\} \leq s.$$

Thus, $r \leq s$, by the definition of r as the diameter of D. However, if $r < s$, then $D \supseteq D(a,s)$. Because $\mathbb{C}_v$ is not discretely valued, there is some $c \in \mathbb{C}_v$

such that $r < |c| < s$. Therefore, $a, a + c \in D(a, s) \subseteq D$, and hence

$$r = \operatorname{diam}(D) \geq |(a + c) - a| = |c| > r,$$

a contradiction. It follows that $r = s$, as claimed.

We now consider the individual cases.

(a) We only need to show that $D \neq D(a, r)$. Given $x, y \in D$ such that $|x - y| = r$, we have

$$\max\{|x - a|, |y - a|\} \geq |x - y| = r,$$

and thus $|x - a| = r$ or $|y - a| = r$. Either way, $D \neq D(a, r)$, as desired.

(b) We only need to show that $D \neq \overline{D}(a, r)$. Pick $c \in \mathbb{C}_v$ such that $|c| = r$. If $D = \overline{D}(a, r)$, then choosing $x = a$ and $y = a + c$ gives $x, y \in D$ with $|x - y| = r$, contradicting the hypotheses.

(c) We only need to show that $D(a, r) = \overline{D}(a, r)$. The forward inclusion is obvious. For the reverse inclusion, given $x \in \overline{D}(a, r)$, we have $|x - a| \leq r$ but also $|x - a| \in |\mathbb{C}_v|$, because $x - a \in \mathbb{C}_v$. Thus, $|x - a| \neq r$, giving $|x - a| < r$ and hence $x \in D(a, r)$, as desired. □

Closed disks with radius in $|\mathbb{C}_v^\times|$ (as in Proposition 2.5(a)) are called *rational closed disks*. Similarly, open disks with radius in $|\mathbb{C}_v^\times|$ (as in Proposition 2.5(b)) are called *rational open disks*. Disks with radius not in $|\mathbb{C}_v^\times|$ (as in Proposition 2.5(c)) are called *irrational disks*. Any rational closed disk $\overline{D}(a, r)$ can be written as a disjoint union of open disks $D(b_\alpha, r)$ of the same radius, with one such disk for each element α of the residue field k. Indeed, after translating, we may assume $a = 0$. Then, after scaling by $z \mapsto cz$, where $r = |c|$ for some $c \in \mathbb{C}_v^\times$, we may assume that $r = 1$. The resulting rational closed disk $\overline{D}(0, 1)$ is the ring of integers $\mathcal{O}$ of $\mathbb{C}_v$, and the open disk $D(0, 1)$ is the maximal ideal $\mathcal{M}$. By the definition of the residue field as $\overline{k} := \mathcal{O}/\mathcal{M}$, we have

$$\overline{D}(0, 1) = \bigcup_{\alpha \in \overline{k}} b_\alpha + \mathcal{M},$$

where $b_\alpha \in \mathcal{O}$ is a coset representative for $\alpha \in \overline{k}$. However, each set $b_\alpha + \mathcal{M}$ is simply $D(b, 1)$, as desired.

We close this section with a phenomenon that does not arise in the archimedean context.

Definition 2.6. A field K with absolute value $|\cdot|$ is said to be *spherically complete* if for every decreasing sequence of closed disks

$$D_1 \supseteq D_2 \supseteq D_3 \supseteq \cdots$$

contained in K, the intersection $\bigcap_{n \geq 1} D_n$ is nonempty.

Lemma 2.7. *Let K be a field equipped with an absolute value $|\cdot|$, and let*

$$\overline{D}(a_1, r_1) \supseteq \overline{D}(a_2, r_2) \supseteq \overline{D}(a_3, r_3) \supseteq \cdots$$

be a decreasing sequence of disks in K.

(a) *If $\lim_{n\to\infty} r_n = r_N$ for some $N \geq 1$, then*

$$\bigcap_{n\geq 1} \overline{D}(a_n, r_n) = \overline{D}(a_N, r_N).$$

(b) *If K is complete and $\lim_{n\to\infty} r_n = 0$, then $\bigcap_{n\geq 1} \overline{D}(a_n, r_n)$ is a singleton.*

(c) *If K is locally compact, then K is spherically complete.*

(d) *If K is spherically complete, then K is complete.*

Proof. This is left to the reader; see Exercise 2.12. □

By Lemma 2.7(c), the archimedean fields $\mathbb{R}$ and $\mathbb{C}$ are spherically complete. However, it is possible for a complete non-archimedean field K to allow the existence of decreasing sequences of disks with radii decreasing to a *positive* limit and with empty intersection. For example, the complete field $\mathbb{C}_p$ is not spherically complete, as illustrated by the following decreasing sequence of disks. Define $D_n := \overline{D}(a_n, r_n)$, where

$$a_n := \sum_{j=1}^{n} p^{1-1/j} \quad \text{and} \quad r_n := |p|_p^{1-1/n}.$$

The intersection $\bigcap_{n\geq 1} D_n$ is empty, because $r_n \searrow |p|_p$, and thus any point in the intersection would have to have all of the infinitely many terms $p^{1-1/j}$ in its p-adic expansion. However, that is impossible, since $\lim_{j\to\infty} p^{1-1/j}$ diverges in $\mathbb{C}_p$. Similarly, the function fields $\mathbb{C}_v$ of Example 2.2 are complete but not spherically complete.

Remark 2.8. Proposition 2.4 holds in any non-archimedean metric space, and *all* the results of this section hold if we work with disks in $\overline{K}$ instead of in $\mathbb{C}_v$. However, if K is discretely valued, the analogue of Proposition 2.5 fails if we work with disks in K. That is,

$$\overline{D}_K(a, r) := \overline{D}(a, r) \cap K \quad \text{and} \quad D_K(a, r) := D(a, r) \cap K$$

do not have well-defined radii. After all, if π is a uniformizer of K, then

$$\overline{D}_K(a, |\pi|^n) = \overline{D}_K(a, r) = D_K(a, r) = D_K(a, |\pi|^{n-1})$$

for any $a \in K$, any integer $n \in \mathbb{Z}$, and any radius $r \in (|\pi|^n, |\pi|^{n-1})$. On the other hand, if K is *not* discretely valued, then even if it is not algebraically closed, Proposition 2.5 still applies to $\overline{D}_K(a, r)$ and $D_K(a, r)$. Indeed, our proof relied only on the fact that $|\mathbb{C}_v^\times|$ is dense in $(0, \infty)$.

Exercises for Chapter 2

2.1. Let K be a field with non-archimedean absolute value $|\cdot|$.

(a) For any $x, y \in K$ with $|x| \neq |y|$, prove that $|x + y| = \max\{|x|, |y|\}$.

(b) For any $x_1, \ldots, x_n \in K$, prove that

$$|x_1 + \cdots + x_n| \leq \max\{|x_1|, \ldots, |x_n|\}.$$

(c) For any $x_1, \ldots, x_n \in K$ with $|x_n| > |x_i|$ for all $1 \leq i \leq n-1$, prove that

$$|x_1 + \cdots + x_n| = |x_n|.$$

2.2. Let K be a non-archimedean field, and let $\mathcal{O}_K$ and $\mathcal{M}_K$ be the subsets of K defined in Section 2.1. Prove the following claims made in that section:

(a) $\mathcal{O}_K$ is a subring of K.

(b) $\mathcal{M}_K$ is a maximal ideal of $\mathcal{O}_K$.

(c) Every proper ideal of $\mathcal{O}_K$ is contained in $\mathcal{M}_K$.

(d) If $\operatorname{char} K = p > 0$, then $\operatorname{char} k = p$ also, where $k = \mathcal{O}_K/\mathcal{M}_K$.

2.3. Let K be a field with an absolute value $|\cdot|$. Prove that the value group $|K^\times|$ is indeed a subgroup of $(0, \infty)$ under multiplication. Moreover, prove that $|K^\times|$ is cyclic if and only if it is a discrete subset of $(0, \infty)$.

2.4. Let K be a field with an absolute value $|\cdot|$. In this exercise, we show how to construct the completion K_v of K with respect to $|\cdot|$. Recall that a sequence $\{a_n\}_{n \geq 1}$ is $|\cdot|$-*Cauchy* if for every $\varepsilon > 0$, there exists $N \geq 1$ such that for all $m, n \geq N$, we have $|a_m - a_n| < \varepsilon$.

Let C be the set of all $|\cdot|$-Cauchy sequences in K, and let I be the set of all sequences in K that converge to 0 with respect to $|\cdot|$.

(a) Prove that C is a commutative ring with unity under termwise addition and multiplication of sequences, and that I is a maximal ideal of C. Define $K_v := C/I$.

(b) Prove that the map $K \to C$ given by $a \mapsto \{a, a, a, \ldots\}$ induces an injective ring homomorphism $i : K \hookrightarrow K_v$.

(c) Prove that the function $C \to \mathbb{R}$ by $\{a_n\}_{n \geq 1} \mapsto \lim_{n \to \infty} |a_n|$ induces a well-defined function $|\cdot|_v : K_v \to \mathbb{R}$.

(d) Prove that $|\cdot|_v$ is an absolute value on K_v, that K_v is complete with respect to $|\cdot|_v$, and that $|i(a)|_v = |a|$ for all $a \in K$.

(e) Prove that $|\cdot|_v$ is non-archimedean if and only if $|\cdot|$ is. In that case, prove that the residue field k of K is canonically isomorphic to the residue field k_v of K_v, and that if $\pi \in K$ is a uniformizer of K, then $i(\pi)$ is a uniformizer of K_v.

2.5. Let K be a complete non-archimedean field with absolute value $|\cdot|$, and let L/K be an algebraic extension.

(a) Prove that there is a unique extension of $|\cdot|$ to an absolute value on L, and that this absolute value is also non-archimedean. (*Suggestion*: If $\alpha \in L$ has minimal polynomial $x^n + \cdots + a_0 \in K[z]$, define $|\alpha| := |a_0|^{1/n}$.)

(b) Prove that there is a canonical inclusion of the residue field k of K in the residue field ℓ of L.

(c) Let

$$e := \big[|L^\times| : |K^\times|\big] \quad \text{and} \quad f := [\ell : k]$$

be the ramification degree and the residue field extension degree of L/K, respectively. Prove that $[L : K] = ef$.

2.6. Give an example of a finite extension $L/\mathbb{Q}$ and a non-archimedean absolute value $|\cdot|$ on $\mathbb{Q}$ that has more than one distinct absolute value on L. Thus, the result of Exercise 2.5(a) fails if we remove the hypothesis that K is complete.

(*Suggestion*: Any choice of a prime that splits in L will work as a counterexample here.)

2.7. Let K be a non-archimedean field with absolute value $|\cdot|$, and consider the topology on K induced by $|\cdot|$.

(a) Prove that the following are equivalent:
- K is locally compact.
- $\mathcal{O}_K$ is compact.
- K is complete, discretely valued, and has finite residue field.

(b) Suppose that K is locally compact and that L/K is a finite extension. Prove that L is also locally compact.

(c) Suppose that K is locally compact and that L/K is an infinite extension. Prove that L is *not* locally compact.

2.8. Let K be a complete non-archimedean field with absolute value $|\cdot|$, let $\overline{K}$ be an algebraic closure of K, and extend $|\cdot|$ to $\overline{K}$ as in Exercise 2.5. Let $\mathbb{C}_v$ be the completion of $\overline{K}$ with respect to $|\cdot|$, and extend $|\cdot|$ to $\mathbb{C}_v$, as in Exercise 2.4. Let k, $\hat{k}$, and $\hat{k}'$ be the residue fields of K, $\overline{K}$, and $\mathbb{C}_v$, respectively.

(a) Prove that $\mathbb{C}_v$ is algebraically closed.

(b) Prove that $\hat{k}$ is an algebraic closure of k via the map $k \to \hat{k}$ induced by the inclusion $K \hookrightarrow \overline{K}$.

(c) Prove that the map $\hat{k} \to \hat{k}'$ induced by the inclusion $\overline{K} \hookrightarrow \mathbb{C}_v$ is an isomorphism.

2.9. In this exercise, you will prove some claims made in Example 2.2. Let $\mathbb{F}$ be a field with $\operatorname{char}\mathbb{F} = 0$.

(a) Prove that the algebraic closure of $\mathbb{F}((t))$ is
$$\overline{\mathbb{F}((t))} = \varinjlim_{m} \overline{\mathbb{F}}((t^{1/m})).$$

(b) Prove that the completion of $\overline{\mathbb{F}((t))}$ is the set of Puiseux series
$$\Big\{ \sum_{n\geq 0} \alpha_n t^{q_n} \ : \ \{q_n\}_{n\geq 0} \subseteq \mathbb{Q} \text{ strictly increases to } \infty, \text{ and } \alpha_n \in \overline{\mathbb{F}} \Big\}.$$

2.10. Prove the following claim made near the end of Example 2.2. Let $\mathbb{F}$ be a field with $\operatorname{char}\mathbb{F} = p > 0$, and let $f := x^p - x + t^{-1} \in \mathbb{F}((t))[x]$. Prove that f has no roots in the field $\varinjlim_{m} \overline{\mathbb{F}}((t^{1/m}))$.

2.11. Fix a prime $p \geq 2$, and let $|\cdot|_p : \mathbb{Q} \to [0, \infty)$ be the function given in Example 2.3.

(a) Prove that $|\cdot|_p$ is a non-archimedean absolute value on $\mathbb{Q}$.

(b) Prove that
$$\mathbb{Q}_p := \Big\{ \sum_{n\geq n_0} a_n p^n \ : \ n_0 \in \mathbb{Z} \text{ and } a_n \in \{0, 1, \ldots, p-1\} \Big\}$$
is the completion of $\mathbb{Q}$ with respect to $|\cdot|_p$.

(c) Prove that $\mathbb{Q}_p$ has value group $|\mathbb{Q}_p^\times|_p = p^{\mathbb{Z}}$ and residue field isomorphic to $\mathbb{F}_p$.

(d) Prove that the expressions
$$\sum_{n=1}^{\infty} p^{1/n} p^n \quad \text{and} \quad \sum_{n=1}^{\infty} \alpha_n p^n,$$
where $\alpha_n \in \mathbb{Q}_p$ is a primitive $(p^n - 1)$-st root of unity, belong to $\mathbb{C}_p \smallsetminus \overline{\mathbb{Q}}_p$.

2.12. Prove Lemma 2.7.

2.13. Let K be a non-archimedean field, not necessarily complete or algebraically closed. Define open and closed disks $D_K(a, r)$ and $\overline{D}_K(a, r)$ in K as in Remark 2.8.

(a) Prove the analogue of Proposition 2.4 for disks in K.

(b) State and prove an analogue of Proposition 2.5. That is, give necessary and sufficient conditions for two disks in K to coincide, based on the radius and center of each, and whether each is written as an open or closed disk. You should expect a different set of conditions depending on whether or not the value group $|K^\times|$ is dense in $(0, \infty)$.

2.14. Let $a \in \mathbb{C}_v$, and let S be a set of disks in $\mathbb{C}_v$ containing a. Define

$$U := \bigcup_{D \in S} D.$$

(a) Prove that U is either an open disk, a closed disk, or all of $\mathbb{C}_v$.

(b) If all disks in S are open, prove that U is not a rational closed disk.

2.15. Let K be a non-archimedean field with absolute value $|\cdot|$, residue field k, and positive residue characteristic $p := \operatorname{char} k > 0$.

(a) For any integer $n \geq 0$, prove that

$$\big|n!\big| = |p|^M, \quad \text{where } M := \sum_{i \geq 1} \left\lfloor \frac{n}{p^i} \right\rfloor.$$

(b) Let $a, b, j \geq 0$ be nonnegative integers with $a \geq \max\{1, b\}$. Prove that

$$\left|\binom{p^j a}{p^j b}\right| = \left|\binom{a}{b}\right|.$$

(c) Let $a, b, j \geq 0$ with $a \geq \max\{1, b\}$ and $\max\{|a|, |b|\} = 1$. Prove that

$$\left|\binom{p^j a}{p^j b}\right| \leq |a|.$$

(d) Let $0 < r \leq 1$ be a real number, and let $n \geq 1$ be a positive integer. Write $n = p^j m$ with $j \geq 0$ and $p \nmid m$. Prove that

$$\max\left\{\left|\binom{n}{i}\right| r^i : 1 \leq i \leq n\right\} = \max\left\{|p|^{j-i} r^{p^i} : 0 \leq i \leq j\right\}.$$

2.16. Suppose the residue characteristic of $\mathbb{C}_v$ is $p \geq 2$.

(a) Let $x \in \mathbb{C}_v$, and let $n \geq 1$ be an integer. If $n = p$ or $p \nmid n$, prove that

$$\left|1 + x + x^2 + \cdots + x^{n-1}\right| \leq \max\left\{|n|, |x-1|^{n-1}\right\}.$$

(*Suggestion*: Write $x = 1 + y$ and simplify.)

(b) Let $n \geq 2p$ be an integer divisible by p. Prove that there is some $x \in \mathbb{C}_v$ for which the inequality of part (a) fails.

2.17. Let K be a non-archimedean field with absolute value $|\cdot|$, residue field k, and positive residue characteristic $p := \operatorname{char} k > 0$.

(a) Let $n \geq 0$ and $1 \leq j \leq p^n$ be integers. Prove that

$$\left|\frac{1}{p^n}\binom{p^n}{j} - \frac{(-1)^{j-1}}{j}\right| \leq |p|^n \max\left\{|ij|^{-1} : 1 \leq i \leq j-1\right\}.$$

(b) Fix a real number $0 < \gamma < 1$. Prove that

$$\lim_{j \to \infty} \gamma^j \max\left\{|ij|^{-1} : 1 \leq i \leq j-1\right\} = 0.$$

(The bounds of this exercise will be needed in Section 10.2.)

2.18. Let K be a non-archimedean field with absolute value $|\cdot|$, residue field k, and positive residue characteristic $p := \operatorname{char} k > 0$.

Define $\rho : (0,\infty) \to (0,\infty)$ by

$$\rho(s) = \begin{cases} |p|^{-1/(p-1)} s & \text{if } s > 1, \\ \left(|p|^{-n-1/(p-1)} s\right)^{1/p^n} & \text{if } |p|^n < s \le |p|^{n-1}, \quad \text{for } n \ge 1. \end{cases}$$

For any $w \in K$ and $j \ge 0$, define

$$\binom{w}{j} = \frac{w(w-1)\cdots(w-j+1)}{j!}.$$

(a) Prove that ρ is continuous and satisfies

$$\rho(|p|^n) = |p|^{-1/(p^n(p-1))}$$

for integers $n \ge 0$.

(b) Prove that for any $j \ge 1$ and any $w \in K$ with $|w| \le 1$, we have

$$\prod_{i=1}^{j-1} \left|1 - \frac{w}{i}\right| \le \rho(|w|)^{j-1}.$$

(c) Prove that for any $j \ge 1$ and any $w \in K$, we have

$$\left|\binom{w}{j}\right| \le \frac{|w|}{|j|}\rho(|w|)^{j-1}.$$

(d) Prove that for any $j \ge 1$ and any $w \in K$, we have

$$\left|\frac{1}{w}\binom{w}{j} - \frac{(-1)^{j-1}}{j}\right| \le \frac{|w|}{|j|} \max\left\{|p|^{-i} : 1 \le p^i \le j-1\right\}\rho(|w|)^{j-1}.$$

Chapter 3

Power Series and Laurent Series

Non-archimedean analysis is vast area of study, and we only scratch the surface in this chapter's brief survey. Besides the friendlier introductions [**70**, **86**] already mentioned in Chapter 2, the reader interested in topics such as arbitrary continuous or differentiable functions may wish to consult [**122**, **126**]. Our particular focus is on rational functions, and more generally on power series and Laurent series, as they act on disks and annuli. Our presentation here therefore falls in the realm of *rigid analysis*. The treatise [**30**] gives a fully general treatment of rigid analysis over arbitrary non-archimedean fields, while [**66**, **69**] present more accessible treatments of much of the same theory.

For our purposes, many of the tools provided by rigid analysis are subsumed by Berkovich's theory of analytic spaces, which we discuss for the case of $\mathbb{P}^1$ beginning in Chapter 6. However, since Berkovich's theory for $\mathbb{P}^1$ relies on certain facts from rigid analysis, the results of this chapter will be essential later. On the other hand, since our aim is to present the necessary background as efficiently as possible, we have deferred many of the proofs to Chapter 14.

Throughout this chapter, K denotes a complete non-archimedean field, and $\mathbb{C}_v$ denotes the completion of an algebraic closure of K.

3.1. Convergence of power series

The convergence of series in non-archimedean fields is much simpler than in $\mathbb{R}$ or $\mathbb{C}$, because of the following elementary result.

Proposition 3.1. *Let $\{a_n\}_{n\geq 0} \subseteq K$ be a sequence. Then*

$$\sum_{n\geq 0} a_n \text{ converges in } K \quad \text{if and only if } \lim_{n\to\infty} a_n = 0.$$

Proof. Let $s_n := a_0 + \cdots + a_n$. If the series converges in K, then the sequence $\{s_n\}_{n\geq 0}$ of partial sums is Cauchy. Given $\varepsilon > 0$, pick $N \geq 0$ so that $|s_n - s_m| < \varepsilon$ for all $m, n \geq N$. Then choosing $m = n-1$, we have $|a_n| < \varepsilon$ for all $n \geq N+1$, as desired.

Conversely, and more interestingly, suppose $\lim_{n\to\infty} a_n = 0$. It suffices to show that $\{s_n\}_{n\geq 0}$ is Cauchy. Given $\varepsilon > 0$, pick $N \geq 0$ such that $|a_n| < \varepsilon$ for all $n \geq N$. Then for any $n \geq m \geq N$, we have

$$|s_n - s_m| = |a_{m+1} + a_{m+2} + \cdots + a_n| \leq \max\left\{|a_{m+1}|, \ldots, |a_n|\right\} < \varepsilon. \quad \square$$

Theorem 3.2. *Let $a \in K$, and let $\{c_n\}_{n\geq 0} \subseteq K$ be a sequence. Define $f(z)$ to be the power series*

$$f(z) := \sum_{n\geq 0} c_n (z-a)^n \in K[[z-a]].$$

(a) *For any $r > 0$, if $\lim_{n\to\infty} |c_n| r^n = 0$, then f converges on $\overline{D}(a,r)$.*

(b) *For any $r \in |\mathbb{C}_v^\times|$, $\lim_{n\to\infty} |c_n| r^n = 0$ if and only if f converges on $\overline{D}(a,r)$.*

(c) *The region of convergence*

$$\left\{x \in \mathbb{C}_v : f(x) \text{ converges}\right\}$$

is either all of $\mathbb{C}_v$, a disk containing a, or the singleton $\{a\}$.

Proof. See Section 14.1. $\square$

If $r \notin |\mathbb{C}_v^\times|$ or if we are considering the open disk $D(a,r)$, the analogue of part (b) of Theorem 3.2 is not quite so simple. Fortunately, the convergence condition for rational closed disks usually suffices for our purposes.

Just as in complex analysis, given a power series f centered at $a \in \mathbb{C}_v$, we can find a new power series for f centered at another point b in the domain of f. In fact, because any point of a disk can be considered its center, we have the following more precise statement.

Proposition 3.3. *Let $D \subseteq \mathbb{C}_v$ be a disk, let $a, b \in D$, and let $f(z) \in K[[z-a]]$ be a power series converging on D. Then there is a power series $g(z) \in \mathbb{C}_v[[z-b]]$ converging on D such that $g(z) = f(z)$ for all $z \in D$. Moreover, the coefficients of g lie in the completion of the field $K(b-a)$.*

Proof. See Section 14.1. $\square$

3.2. Power series rings

In complex analysis, a function f that is analytic on an open disk $D \subseteq \mathbb{C}$ can be described by a power series converging on D. Unfortuately, no local condition such as analyticity is enough to produce useful conclusions in non-archimedean analysis. For example, the function $f : \mathbb{C}_v \to \mathbb{C}_v$ given by

$$f(z) = \begin{cases} 0 & \text{if } |z| < 1, \\ 1 & \text{if } |z| \geq 1 \end{cases}$$

is locally analytic, and indeed locally constant, at every point of $\mathbb{C}_v$. However, it follows from the upcoming Proposition 3.10 that no power series centered at 0 can equal f on any disk containing $\overline{D}(0,1)$. Thus, rather than studying locally analytic functions on a given disk $D \subseteq \mathbb{C}_v$, we consider the set of power series that converge on D.

Definition 3.4. Let $D \subseteq \mathbb{C}_v$ be a disk containing a point a. The *ring of convergent power series* on D is

$$\mathcal{A}_D := \big\{ f \in \mathbb{C}_v[[z-a]] : f \text{ converges on } D \big\}.$$

If $D = D(a,r)$ is open, we also write $\mathcal{A}(a,r) := \mathcal{A}_D$. If $D = \overline{D}(a,r)$ is closed, we write $\overline{\mathcal{A}}(a,r) := \mathcal{A}_D$.

We consider two power series $f, g \in \mathcal{A}_D$ centered at $a \in D$ to be equal if and only if all their coefficients match. As we will see in Proposition 3.10, this condition holds if and only if f and g agree as functions on D.

We make a few observations. First, by Proposition 3.3, a power series $f \in \mathcal{A}_D$ may be written as a convergent power series centered at any point $b \in D$. Our notation $\mathcal{A}_D$, with no reference to the choice of center, is therefore justified. Second, if D and E are disks such that $E \subseteq D \subseteq \mathbb{C}_v$, then clearly $\mathcal{A}_D \subseteq \mathcal{A}_E$. Third, by Theorem 3.2(b), we have

$$\overline{\mathcal{A}}(a,r) = \left\{ \sum_{n \geq 0} c_n (z-a)^n \ : \ \lim_{n \to \infty} |c_n| r^n = 0 \right\}$$

for any $a \in \mathbb{C}_v$ and $r \in |\mathbb{C}_v^\times|$. (However, the corresponding condition for rational open or irrational disks is more complicated; see Exercise 3.6.) Fourth, as our next result states, $\mathcal{A}_D$ is indeed a ring.

Theorem 3.5. *Let $D \subseteq \mathbb{C}_v$ be a disk. Then $\mathcal{A}_D$ is a commutative ring with unity, under the operations of addition and multiplication of power series.*

Proof. See Section 14.1. □

The ring $\mathcal{A}_D$ admits a norm associated with any subdisk $E \subseteq D$, as follows.

Definition 3.6. Let $D \subseteq \mathbb{C}_v$ be a disk, and let $E \subseteq D$ be a subdisk. Write $E = \overline{D}(a,r)$ or $E = D(a,r)$, and define a function $\|\cdot\|_E : \mathcal{A}_D \to \mathbb{R} \cup \{\infty\}$, also denoted $\|\cdot\|_{\zeta(a,r)} : \mathcal{A}_D \to \mathbb{R} \cup \{\infty\}$, by

$$\|f\|_E = \|f\|_{\zeta(a,r)} := \sup\left\{|c_n|r^n \ : \ n \geq 0\right\}$$

for any $f = \sum_{n\geq 0} c_n(z-a)^n \in \mathcal{A}_D$.

If $E \subsetneq D$ is a *proper* subdisk or if $E = D$ is a *rational closed* disk, then $\lim_{n\to\infty} |c_n|r^n = 0$, and hence the supremum in Definition 3.6 is attained and finite. However, if $E = D$ is a rational open or irrational disk, then the supremum may be attained and finite, finite but unattained, or infinite, depending on the particular coefficients of f. For this reason, we usually consider $\|\cdot\|_E$ on the ring $\mathcal{A}_D$ only in the case that $E \neq D$ or that $E = D$ is rational closed.

We leave it to the reader to check that if we recenter the power series f at a different point $b \in E$, then the value of $\|f\|_E$ is the same as before, even though the coefficients of f have changed; see Exercise 3.2(a). Thus, $\|\cdot\|_E$ is well defined. In fact, as we will see in Proposition 3.17, $\|\cdot\|_E$ is simply the sup-norm relative to the set E.

It is immediate from Definition 3.6 that $\|\cdot\|_E$ is the same if $E = D(a,r)$ is open as it is if $E = \overline{D}(a,r)$ is the corresponding closed disk. Moreover, if $E := D(a,r)$ and $E' := D(b,r)$ are subdisks of D such that $\overline{D}(a,r) = \overline{D}(b,r)$, then we must have

$$\|f\|_E = \|f\|_{\zeta(a,r)} = \|f\|_{E'}$$

for all $f \in \mathcal{A}_D$. That is, $\|\cdot\|_E$ and $\|\cdot\|_{E'}$ coincide as functions on $\mathcal{A}_D$, even though the open disks E and E' may be disjoint. Thus, although the set of these disk-induced norms is in one-to-one correspondence with the set of *closed* disks in $\mathbb{C}_v$, it is *not* in one-to-one correspondence with the set of *open* disks. This observation will be important when we discuss Berkovich spaces. In fact, the use of the letter ζ is meant to evoke a complex variable, precisely because it turns out to be an actual point in the Berkovich projective line, which is a space containing $\mathbb{P}^1(\mathbb{C}_v)$ but with nicer topological properties, reminiscent of the Riemann sphere $\mathbb{P}^1(\mathbb{C})$.

For the moment, however, we treat $\|\cdot\|_{\zeta(a,r)}$ solely as a norm. First, however, we must confirm that it *is* a norm (in most circumstances), and that it is bounded by $\|\cdot\|_D$.

Proposition 3.7. *Let $E \subseteq D \subseteq \mathbb{C}_v$ be disks. Then*

(a) $\|c\|_E = |c|$ *for any constant* $c \in \mathbb{C}_v$*, and*

(b) $\|f\|_E \leq \|f\|_D$ *for any* $f \in \mathcal{A}_D$.

Moreover, if either $E = D$ is rational closed or $E \neq D$ is a proper subdisk, then $\|\cdot\|_E$ is a non-archimedean multiplicative norm on $\mathcal{A}_D$, meaning that for all $f, g \in \mathcal{A}_D$,

(c) $\|f\|_E \in [0, \infty)$, *with* $\|f\|_E = 0$ *if and only if* $f = 0$.

(d) $\|f + g\|_E \leq \max\{\|f\|_E, \|g\|_E\}$.

(e) $\|fg\|_E = \|f\|_E \cdot \|g\|_E$.

Proof. See Section 14.1. □

The zeros of non-archimedean power series are generally easier to study than their complex counterparts, thanks to the Weierstrass preparation theorem (Theorem 14.2). Although we have shunted its statement and proof to Section 14.2, the result lends its name to the following useful terminology.

Definition 3.8. Let $a \in \mathbb{C}_v$, let $r > 0$, let $D \subseteq \mathbb{C}_v$ be a disk containing a of radius r, and let

$$f = \sum_{n \geq 0} c_n (z - a)^n \in \mathcal{A}_D$$

be a nonzero power series converging on D. The *Weierstrass degree* of f on D is the quantity $\operatorname{wdeg}_D(f)$, defined as follows.

(a) If $D = \overline{D}(a, r)$ is a rational closed disk, then $\operatorname{wdeg}_D(f)$ is the *largest* integer $d \geq 0$ such that

$$|c_d| r^d = \max_{n \geq 0} |c_n| r^n.$$

(b) If $D = D(a, r)$ is a rational open or irrational disk, then $\operatorname{wdeg}_D(f)$ is the *smallest* integer $d \geq 0$ such that

$$|c_d| r^d = \max_{n \geq 0} |c_n| r^n,$$

or ∞ if there is no such integer.

We mainly focus on the case of rational closed disks, although we occasionally need to refer to the Weierstrass degree of a power series on an open or irrational disk. In the case that D is a rational closed disk and $f \neq 0$, the maximum in Definition 3.8(a) is indeed attained at some maximal $d \geq 0$, because the convergence of f on $\overline{D}(a, r)$ guarantees that $\lim_{n \to \infty} |c_n| r^n = 0$.

Regardless of whether D is rational closed, rational open, or irrational, if we recenter the power series f at another point $b \in \overline{D}(a, r)$, then the Weierstrass degree of f is unchanged; see Exercise 3.2(b). Thus, it is legitimate for us to discuss $\operatorname{wdeg}_D(f)$ without specific reference to a choice of center $a \in D$.

We can of course differentiate a convergent power series $f \in \mathcal{A}_D$ formally. We have the following simple result.

Proposition 3.9. *Let $D \subseteq \mathbb{C}_v$ be a disk, and let $f \in \mathcal{A}_D$ be a power series converging on D.*

(a) *The formal power series f' also belongs to $\mathcal{A}_D$.*

(b) *If $\operatorname{wdeg}_D(f)$ is not divisible by the residue characteristic of $\mathbb{C}_v$, then $\operatorname{wdeg}_D(f') = \operatorname{wdeg}_D(f) - 1$.*

Proof. This is left to the reader; see Exercise 3.3. □

The following two results require the Weierstrass preparation theorem, and so again we defer their proofs to Section 14.2. The first says that two convergent power series coincide if and only if they are equal as functions. In fact, it is stronger: if the two power series have the same values on infinitely many points of a rational closed disk, then they are the same.

Proposition 3.10. *Let $a \in \mathbb{C}_v$, let $r \in |\mathbb{C}_v^\times|$, and let $f, g \in \overline{\mathcal{A}}(a, r)$ be convergent power series on the rational closed disk $\overline{D}(a, r)$. Write*

$$f(z) = \sum_{n \geq 0} b_n (z-a)^n \quad \text{and} \quad g(z) = \sum_{n \geq 0} c_n (z-a)^n,$$

and suppose that $f(x) = g(x)$ for infinitely many points $x \in \overline{D}(a, r)$. Then $f = g$ as power series; that is, $b_n = c_n$ for all $n \geq 0$.

Proof. See Section 14.2. □

Proposition 3.10 would be false if we removed the hypothesis that D is a rational closed disk. Indeed, for any open or irrational disk $D \subseteq \mathbb{C}_v$, there exist nonzero power series $f \in \mathcal{A}_D \smallsetminus \{0\}$ with infinitely many zeros in D. (Of course, for any rational closed disk $E \subseteq D$, only finitely many of those zeros can lie in E.) See Exercise 3.5.

We close this section with a result characterizing those power series in $\mathcal{A}_D$ having multiplicative inverses.

Proposition 3.11. *Let $D \subseteq \mathbb{C}_v$ be a disk. Then the unit group of $\mathcal{A}_D$ is*

$$\begin{aligned} \mathcal{A}_D^\times &= \{f \in \mathcal{A}_D : f \text{ has Weierstrass degree } 0 \text{ on } D\} \\ &= \{f \in \mathcal{A}_D : f \text{ has no zeros in } D\}. \end{aligned}$$

Proof. See Section 14.2. □

3.3. Newton polygons

In complex analysis, one can define the order of vanishing of an analytic function at a point b either as the smallest derivative that does not vanish at b or as the smallest order term appearing in the power series expansion of the function at b. For power series over fields of characteristic $p > 0$, however, the pth derivative is already identically zero, and therefore the first definition should not be used. Fortunately, the second definition still works perfectly well.

Definition 3.12. Let $a \in K$, and let $f(z) \in K[[z-a]]$ be a power series converging on a disk $D \subseteq \mathbb{C}_v$ containing a. We say that a point $b \in D$ is a *zero* of f if $f(b) = 0$. Recentering the power series at b as

$$f(z) = \sum_{n \geq 0} c'_n (z-b)^n,$$

we say that the *multiplicity* $\operatorname{ord}_b(f)$ of the zero of f at b, also known as the *order of vanishing* of f at b, is the smallest integer $m \geq 0$ such that $c'_m \neq 0$, or $m = \infty$ if f is identically zero.

We leave it to the reader to check that ord_b is a valuation; that is,

$$\operatorname{ord}_b(fg) = \operatorname{ord}_b(f) + \operatorname{ord}_b(g)$$

and

$$\operatorname{ord}_b(f+g) \geq \min\{\operatorname{ord}_b(f), \operatorname{ord}_b(g)\};$$

see Exercise 3.8.

Definition 3.12 can be used to rephrase the Weierstrass preparation theorem: the number of zeros of a power series f on a disk D, counted with multiplicity, is the Weierstrass degree of f on D. In fact, we can say much more, as follows.

Theorem 3.13 (Newton polygons). *Let $a \in K$, let $D \subseteq \mathbb{C}_v$ be a disk containing a, and let $f \in K[[z-a]]$ be a nonzero power series defined over K and converging on D.*

(a) *Let $d \geq 0$ be the Weierstrass degree of f on D. Then f has exactly d zeros in D counting multiplicity.*

(b) *If $d < \infty$, then there is a polynomial $g \in K[x]$ of degree d whose zeros are exactly the zeros of f in D, and with the same multiplicities.*

(c) *Let $r > 0$ such that $\overline{D}(a,r) \subseteq D$, with either $\overline{D}(a,r) \subsetneq D$ or $r \in |\mathbb{C}_v^\times|$. Let i be the Weierstrass degree of f on $D(a,r)$, and let j be the Weierstrass degree of f on $\overline{D}(a,r)$. Then*

 (i) $0 \leq i \leq j < \infty$.

 (ii) *f has exactly $j-i$ zeros $x \in \mathbb{C}_v$ satisfying $|x-a| = r$, counting multiplicity.*

 (iii) *There is a polynomial $g_r \in K[x]$ of degree $j-i$ whose zeros are exactly the zeros x of f with $|x-a| = r$, and with the same multiplicities.*

Proof. See Section 14.3. □

The term "Newton polygon" in the name of Theorem 3.13 actually refers to an object we now describe. As preparation, fix an element $\pi \in K$ with $0 < |\pi| < 1$. If K is discretely valued, we usually choose π to be a uniformizer. Define $v : \mathbb{C}_v \to \mathbb{R} \cup \{\infty\}$ by

$$v(x) := \log|x| / \log|\pi|;$$

note that $\log|\pi| < 0$. Like ord_b on the ring $K[[z-a]]$, the function v is an example of a *valuation* on the field $\mathbb{C}_v$. For our purposes, its key properties are that $v(\pi^n) = n$, and if π is a uniformizer for K, then $v(K) = \mathbb{Z} \cup \{\infty\}$.

Given a nonconstant power series $f(z) = \sum_{n\geq 0} c_n(z-a)^n$, we construct a region in the Cartesian plane $\mathbb{R}^2$ as follows. For each $n \geq 0$, let R_n be the vertical ray extending upward from the point $(n, v(c_n))$. Define P_f, the Newton polygon of f at a, to be the *convex hull* of $\bigcup_{n\geq 0} R_n$. That is, P_f is the intersection of all convex sets containing $\bigcup_{n\geq 0} R_n$.

Clearly, P_f is a convex region in the plane, extending upward infinitely but with piecewise linear boundary along the bottom and sides. In spite of the moniker "polygon", it is of course technically an unbounded planar region with piecewise linear boundary.

The polygon P_f gives a convenient way to visualize the content of Theorem 3.13. For the purposes of this discussion, call a radius $r > 0$ *pivotal* for f if there is at least one zero $x \in \mathbb{C}_v$ of f with $|x-a| = r$. (In most of the literature, such radii are called "critical" radii, but to avoid confusion with zeros of f', we use the term "pivotal" instead.) The reader can check that the following two conditions are equivalent:

(1) r is a pivotal radius, with f having i zeros in $D(a,r)$ and $j > i$ zeros in $\overline{D}(a,r)$.

(2) $(i, v(c_i))$ and $(j, v(c_j))$ are vertices of the Newton polygon P_f at opposite ends of a boundary segment of slope $-\log r / \log|\pi|$;

see Exercise 3.9, and see also [**70**, Section 6.5] or [**122**, Section VI.2.2].

It is customary to call the quantity $j - i$ the *length* of the boundary segment, even though, strictly speaking, it is really the length of the segment's projection onto the x-axis. Note that Theorem 3.13 says that all the zeros of f with that particular distance r from a are the zeros of a degree $j-i$ polynomial in $K[z]$. In particular, for any such zero $\alpha \in \mathbb{C}_v$, we have $[K(\alpha) : K] \leq j - i$.

Less commonly noted, for *any* radius $r > 0$, whether pivotal or not, the Newton polygon can also tell us the value of $t := \max_{n\geq 0} |c_n| r^n$, as least as long as f converges on some rational closed disk containing $\overline{D}(a,r)$.

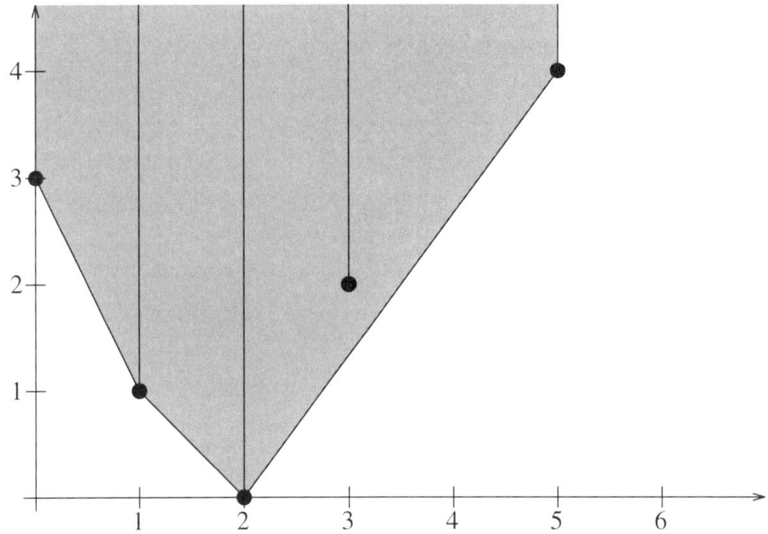

Figure 3.1. The Newton polygon for $\pi^3 + \pi z + z^2 + \pi^2 z^3 + \pi^4 z^5$

Indeed, because f converges on $\overline{D}(a, r)$, there is a unique line in the plane of slope $-\log r / \log|\pi|$ that intersects the boundary, but not the interior, of P_f. This intersection is either one of the boundary segments or else one of the boundary vertices. Either way, the y-intercept of this line is exactly $\log t / \log|\pi|$; see Exercise 3.12.

Example 3.14. Fix $\pi \in K$ with $0 < |\pi| < 1$, and let $f(z) \in K[z]$ be the polynomial

$$f(z) = \pi^3 + \pi z + z^2 + \pi^2 z^3 + \pi^4 z^5.$$

The associated Newton polygon P_f appears as the shaded region in Figure 3.1. The black dots in Figure 3.1 are the various points $(n, v(c_n))$; note that not all of them lie on the boundary of the polygon.

The first segment, lying above the interval $[0, 1]$ on the x-axis, has length 1 and slope -2. Thus, f has exactly one root in $\mathbb{C}_v$ with absolute value $|\pi|^2$.

Similarly, the second segment has length 1 and slope -1, and hence f has exactly one root of absolute value $|\pi|$.

Finally, the third segment, of length 3, tells us that f has exactly three roots of absolute value $|\pi|^{-4/3}$.

Thus, we know the absolute values of all five roots of f. Further, we see that the two roots of absolute value $|\pi|^2$ and $|\pi|$ lie in K, while the three of absolute value $|\pi|^{-4/3}$ lie in a cubic ramified extension of K.

Moreover, any line just touching the boundary of P_f and of slope $m \le -2$ must pass through the point $(0,3)$ and hence have y-intercept 3. Therefore, the value of $t = \max_{n\ge 0} |c_n| r^n$ is $|\pi|^3$ for all $0 < r \le |\pi|^2$, because the constant term of f is dominant for such r.

Next, we have $t = |\pi| r$ for $|\pi|^2 \le r \le |\pi|$, because the z-term of f is dominant for such r, or because a line of slope $m \in [-2,-1]$ just touching the boundary of P_f has y-intercept $1-m$.

For $m \in [-1, 4/3]$, the y-intercept is $-2m$, and indeed, we have $t = r^2$ for $|\pi| \le r \le |\pi|^{-4/3}$, because the z^2-term of f is dominant for such r.

Finally, for $m > 4/3$, the y-intercept is $5m-4$, and at the same time, the z^5-term of f is dominant for $r \ge |\pi|^{-4/3}$, giving $t = |\pi|^4 r^5$.

The z^3-term is never dominant at any radius, which is why the point $(3,2)$ lies in the interior of the Newton polygon, rather than on its boundary.

3.4. Images of disks under power series

The following result is a corollary of Theorem 3.13.

Theorem 3.15. *Let $D \subseteq \mathbb{C}_v$ be a disk of radius $r > 0$ containing a point $a \in \mathbb{C}_v$, and let $f(z) = \sum_{n\ge 0} c_n (z-a)^n$ be a nonconstant power series converging on D. Let $d := \operatorname{wdeg}_D(f - c_0)$, and assume that $d < \infty$. Let $t := |c_d| r^d > 0$. Then*

(a) *$f(D)$ is a disk of the same type (rational open, rational closed, or irrational) as D, centered at c_0 and of radius t;*

(b) *$f : D \to f(D)$ is everywhere d-to-1, counting multiplicity.*

Proof. Clearly $d \ge 1$ and $t > 0$, since $f - c_0$ is nontrivial but has trivial constant term. By definition of d, we have $|c_n| r^n \le t$ for all $n \ge 1$. Moreover, $t \in |\mathbb{C}_v^\times|$ if and only if $r \in |\mathbb{C}_v^\times|$.

If $D = \overline{D}(a,r)$ is rational closed, then $|c_n(x-a)^n| \le |c_n| r^n \le t$ for any $x \in D$ and any $n \ge 1$; hence $|f(x) - c_0| \le t$. Conversely, for any point $y \in \overline{D}(c_0, t)$, Theorem 3.13 tells us that $f - y_0$ has exactly d zeros in D, counting multiplicity.

Similarly, if $D = D(a,r)$ is open, then $|c_n(x-a)^n| < |c_n| r^n \le t$ for any $x \in D$ and any $n \ge 1$; hence $|f(x) - c_0| < t$. Conversely, for any point $y \in D(c_0, t)$, Theorem 3.13 again tells us that $f - y_0$ has exactly d zeros in D, counting multiplicity. □

Remark 3.16. If the disk D in Theorem 3.15 is rational closed, then of course $d := \operatorname{wdeg}_D(f)$ is necessarily finite. However, if D is open, it is possible that $d = \infty$, in which case the situation is a little more complicated. Specifically, if $\{|a_n| r^n\}_{n\ge 1}$ is unbounded, then $f(D) = \mathbb{C}_v$, and $f : D \to \mathbb{C}_v$

is infinite-to-one. On the other hand, if $t := \sup_{n\geq 1} |a_n|r^n$ is finite but not attained, then $f(D)$ is the open disk $D(a,t)$, and again, $f : D \to D(a,t)$ is infinite-to-one. In this second case, even though D and $f(D)$ are both open disks, it is possible for one to be rational and the other to be irrational; see Exercise 3.13 for further details.

Among its many uses, Theorem 3.15 allows us to recognize the norm $\|\cdot\|_D$ of Definition 3.6 as the sup-norm on D.

Proposition 3.17. *Let $D \subseteq \mathbb{C}_v$ be a disk, and let $f \in \mathcal{A}_D$ be a convergent power series of finite Weierstrass degree on D. Then*

$$\|f\|_D = \sup\big\{|f(x)| : x \in D\big\}.$$

Proof. See Section 14.4. □

The norm $\|\cdot\|_D$ also allows us to rephrase a portion of Theorem 3.15 in the following more compact fashion.

Corollary 3.18. *Let $D \subseteq \mathbb{C}_v$ be a disk, let $a \in D$, and let $f \in \mathcal{A}_D$ be a nonconstant power series converging on D of finite Weierstrass degree. Then*

$$f(D) = \begin{cases} \overline{D}\big(f(a), \|f - f(a)\|_D\big) & \text{if } D \text{ is closed, or} \\ D\big(f(a), \|f - f(a)\|_D\big) & \text{if } D \text{ is open.} \end{cases}$$

Proof. This is immediate from Theorem 3.15. □

Even without the use of Theorem 3.15, and hence without the machinery of Newton polygons and the Weierstrass Preparation Theorem, we have the following characterization of $\|f\|_D$ in the case that D is an open disk and $\operatorname{wdeg}_D(f) < \infty$.

Proposition 3.19. *Let $D = D(a,r) \subseteq \mathbb{C}_v$ be an open disk, and let $f \in \mathcal{A}_D$ be a convergent power series of finite Weierstrass degree on D. Let $\{x_n\}_{n\geq 1} \subseteq D$ be any sequence in D satisfying*

$$\lim_{n\to\infty} |x_n - a| = r.$$

Then

$$\|f\|_D = \lim_{n\to\infty} |f(x_n)|.$$

Proof. Write $f(z) = \sum_{i\geq 0} c_i(z-a)^i$, and let $d := \operatorname{wdeg}_D(f)$, so that

$$|c_d|r^d \geq |c_i|r^i \quad \text{for all } i \geq 0,$$

with strict inequality for $i < d$. We may therefore pick another positive radius $s < r$ such that $|c_d|s^d \geq |c_i|s^i$ for all $i \geq 0$, with strict inequality for

$i \neq d$. By hypothesis, there is some $N \geq 1$ so that for all $n \geq N$, we have $|x_n - a| > s$. Therefore,

$$\lim_{n\to\infty} |f(x_n)| = \lim_{n\to\infty} |c_d| \cdot |x_n - a|^d = |c_d| r^d = \|f\|_D. \qquad \square$$

The following two further consequences of Theorem 3.15 are very useful in the study of non-archimedean dynamics.

Proposition 3.20. *Let $D \subseteq \mathbb{C}_v$ be a disk of radius $r > 0$, and let $f(z)$ be a nonconstant power series converging on D. Suppose that the image $f(D)$ is a disk of radius $s > 0$. Then for all $x, y \in D$,*

$$\big|f(x) - f(y)\big| \leq \frac{s}{r} \cdot |x - y|. \tag{3.1}$$

Moreover, if the Weierstrass degree of f on D is 1, then we have equality in (3.1).

Proof. Given $x, y \in D$, write $f(z) = \sum_{n\geq 0} c_n (z - y)^n$ as a power series centered at y. Then by Theorem 3.15,

$$s := \text{radius of } f(D) = \max_{n\geq 1} \left\{ |c_n| r^n \right\},$$

giving

$$\begin{aligned} \big|f(x) - f(y)\big| &= \left| \sum_{n\geq 1} c_n (x^n - y^n) \right| = \left| \sum_{n\geq 1} c_n \sum_{i=0}^{n-1} x^i y^{n-1-i} \right| \cdot |x - y| \\ &\leq \max_{n\geq i+1\geq 1} \left\{ |c_n| |x|^i |y|^{n-1-i} \right\} \cdot |x - y| \qquad (3.2) \\ &\leq \max_{n\geq 1} \left\{ |c_n| r^{n-1} \right\} \cdot |x - y| = \frac{s}{r} |x - y|. \end{aligned}$$

Finally, if $\mathrm{wdeg}_D(f) = 1$, then the first inequality in (3.2) is an equality, because the $n = 1$ term has strictly larger absolute value than any of the $n \geq 2$ terms. The second inequality is also equality in that case, because the second maximum is attained by the $n = 1$ term (although this time, if D is open, some other terms may have the same size, too). $\square$

Corollary 3.21. *Let $D \subseteq \mathbb{C}_v$ be a disk of radius $r > 0$, and let $f(z)$ be a power series converging on D. Then for all $x, y \in D$,*

$$\big|f(x) - f(y)\big| \leq \frac{1}{r} \|f\|_D \cdot |x - y|.$$

Proof. If f is constant or if the image of f is all of $\mathbb{C}_v$, the result is vacuous. Otherwise, $f(D)$ is a disk of some radius $s > 0$, and we have $s \leq \|f\|_D$. The result then follows from Proposition 3.20. $\square$

Bijective power series have inverses also given by power series and defined over the same complete field, as the following result shows.

Proposition 3.22. *Let $a \in K$, let $D \subseteq \mathbb{C}_v$ be a disk containing a, and let $f \in K[[z-a]]$ be a power series converging on D, defined over K, and of Weierstrass degree 1. Then f has an inverse function $g : f(D) \to D$ given by a power series $g \in K[[z - f(a)]]$ converging on $f(D)$ and defined over K.*

Proof. See Section 14.4. □

The following result concerns *inverse* images of disks under *polynomials.*

Theorem 3.23. *Let $D \subseteq \mathbb{C}_v$ be a disk, and let $\phi \in \mathbb{C}_v[z]$ be polynomial of degree $d \geq 1$. Then $\phi^{-1}(D)$ is a disjoint union*

$$\phi^{-1}(D) = D_1 \cup \cdots \cup D_m$$

of disks of the same type as D, with $1 \leq m \leq d$. Moreover, for each $i = 1, \ldots, m$, there is an integer $1 \leq d_i \leq d$ such that every point in D has exactly d_i preimages in D_i, and such that $d_1 + \cdots + d_m = d$.

Proof. See Section 14.4. □

3.5. $\mathbb{P}^1(\mathbb{C}_v)$-disks and affinoids

Since we are interested in the action of rational functions on the projective line, we need the following more general definition of a disk in $\mathbb{P}^1(\mathbb{C}_v)$ to allow for disks containing the point ∞.

Definition 3.24. A $\mathbb{P}^1(\mathbb{C}_v)$-*disk* is either D or $\mathbb{P}^1(\mathbb{C}_v) \smallsetminus D$, where $D \subseteq \mathbb{C}_v$ is a disk. A $\mathbb{P}^1(\mathbb{C}_v)$-disk is

- *rational closed* if it is either $D \subseteq \mathbb{C}_v$ where D is rational closed, or $\mathbb{P}^1(\mathbb{C}_v) \smallsetminus D$ where $D \subseteq \mathbb{C}_v$ is rational open;
- *rational open* if it is either $D \subseteq \mathbb{C}_v$ where D is rational open, or $\mathbb{P}^1(\mathbb{C}_v) \smallsetminus D$ where $D \subseteq \mathbb{C}_v$ is rational closed;
- *irrational* if it is either D or $\mathbb{P}^1(\mathbb{C}_v) \smallsetminus D$, where $D \subseteq \mathbb{C}_v$ is irrational.

We sometimes refer to a $\mathbb{P}^1(\mathbb{C}_v)$-disk simply as a "disk".

A linear fractional transformation $\eta \in \mathrm{PGL}(2, \mathbb{C}_v)$ maps $\mathbb{P}^1(\mathbb{C}_v)$-disks to $\mathbb{P}^1(\mathbb{C}_v)$-disks, preserving the type (rational closed, rational open, or irrational); see Exercise 3.16. Indeed, a set $X \subseteq \mathbb{P}^1(\mathbb{C}_v)$ is a $\mathbb{P}^1(\mathbb{C}_v)$-disk if and only if it is the image of a disk $D \subseteq \mathbb{C}_v$ of the same type under some linear fractional transformation. Moreover, we have the following stronger result.

Proposition 3.25. *Let D be a $\mathbb{P}^1(\mathbb{C}_v)$-disk, and let $\phi(z) \in \mathbb{C}_v(z)$ be a nonconstant rational function. Then the image $\phi(D)$ is either $\mathbb{P}^1(\mathbb{C}_v)$ or a $\mathbb{P}^1(\mathbb{C}_v)$-disk of the same type as D. In the latter case, there is some integer $d \geq 1$ such that every point of $\phi(D)$ has exactly d preimages in D, counting multiplicity.*

Proof. See Section 14.4. □

The following result is a simple corollary of Proposition 2.4.

Proposition 3.26. *Let $D, E \subseteq \mathbb{P}^1(\mathbb{C}_v)$ be $\mathbb{P}^1(\mathbb{C}_v)$-disks, and suppose that $D \cap E \neq \varnothing$. Then either*

$$D \cup E = \mathbb{P}^1(\mathbb{C}_v), \quad D \subseteq E, \quad \textit{or} \quad E \subseteq D.$$

Proof. If $D \cup E \neq \mathbb{P}^1(\mathbb{C}_v)$, pick a point $a \in \mathbb{P}^1(\mathbb{C}_v) \smallsetminus (D \cup E)$. Choose $\eta \in \mathrm{PGL}(2, \mathbb{C}_v)$ such that $\eta(a) = \infty$. Then $\eta(D)$ and $\eta(E)$ are both disks in $\mathbb{C}_v$. By Proposition 2.4(c), we have either

$$\eta(D) \subseteq \eta(E) \quad \text{or} \quad \eta(E) \subseteq \eta(D).$$

The conclusion is now immediate by applying η^{-1}. □

To handle intersections of $\mathbb{P}^1(\mathbb{C}_v)$-disks and their inverse images under rational functions, we need the following more general kind of domain.

Definition 3.27. A *connected affinoid* is a nonempty intersection of finitely many $\mathbb{P}^1(\mathbb{C}_v)$-disks $D_1, \ldots, D_n$. If all of the disks $D_1, \ldots, D_n$ are closed (resp., open, rational closed, rational open, irrational), we say that the connected affinoid

$$D_1 \cap \cdots \cap D_n$$

is also *closed* (resp., *open*, *rational closed*, *rational open*, *irrational*).

Equivalently, a connected affinoid is either $\mathbb{P}^1(\mathbb{C}_v)$ or else a $\mathbb{P}^1(\mathbb{C}_v)$-disk with finitely many subdisks removed. In particular, we consider $\mathbb{P}^1(\mathbb{C}_v)$ itself to be a connected affinoid, as it is an intersection of zero $\mathbb{P}^1(\mathbb{C}_v)$-disks. This convention, as well as our convention that the disks in the intersection can be arbitrary, rather than rational closed, conflicts with the usual definition in rigid analysis; see Section 14.7. However, the above definition is more convenient for our purposes.

Note also that in spite of their name, connected affinoids are subsets of the totally disconnected space $\mathbb{P}^1(\mathbb{C}_v)$, and hence they are *not* connected as topological spaces. However, they behave in many ways like connected sets; see Proposition 3.28, Proposition 3.29, and Section 14.7.

Proposition 3.28. *Let $U, V \subseteq \mathbb{P}^1(\mathbb{C}_v)$ be connected affinoids, and suppose that $U \cap V \neq \varnothing$. Then $U \cap V$ and $U \cup V$ are also connected affinoids.*

If U and V are both of the same type (rational closed, etc.), then $U \cap V$ and $U \cup V$ are also of that same type.

Proof. Write $U = D_1 \cap \cdots \cap D_m$ and $V = E_1 \cap \cdots \cap E_n$ as nonempty intersections of $\mathbb{P}^1(\mathbb{C}_v)$-disks.

Clearly, $U \cap V$ is a finite intersection of $\mathbb{P}^1(\mathbb{C}_v)$-disks. Thus, because it is nonempty by hypothesis, it is a connected affinoid. Moreover, if the disks D_i and E_j are all of the same type, then $U \cap V$ is of that type by definition.

Similarly, $U \cup V = \bigcap_i \bigcap_j (D_i \cup E_j)$. For each i, j, we have

$$D_i \cap E_j \supseteq U \cap V \neq \varnothing,$$

and therefore $D_i \cup E_j$ is either $\mathbb{P}^1(\mathbb{C}_v)$ or a $\mathbb{P}^1(\mathbb{C}_v)$-disk, by Proposition 3.26. Thus, $U \cup V$ can be written as a finite intersection of $\mathbb{P}^1(\mathbb{C}_v)$-disks. Moreover, if U and V are of the same type, then all the disks in this intersection are also of that type. Finally, $U \cup V \supseteq U \neq \varnothing$; hence, $U \cup V$ is a connected affinoid. □

We also have the following analogue of Theorems 3.15 and 3.23.

Proposition 3.29. *Let $U \subseteq \mathbb{P}^1(\mathbb{C}_v)$ be a connected affinoid, and let $\phi(z) \in \mathbb{C}_v(z)$ be a rational function of degree $d \geq 1$. Then*

(a) *$\phi(U)$ is a connected affinoid of the same type, if any, as U.*

(b) *$\phi^{-1}(U)$ is a disjoint union $V_1 \cup \cdots \cup V_m$ of connected affinoids, each of the same type, if any, as U, and with $1 \leq m \leq d$.*

Moreover, for each $i = 1, \ldots, m$, there is an integer $1 \leq d_i \leq d$ such that every point in U has exactly d_i preimages in V_i, counting multiplicity, and such that $d_1 + \cdots + d_m = d$.

Proof. See Section 14.4 for a sketch. □

3.6. Laurent series on open annuli

Besides power series on disks, we also need to study Laurent series on annuli. For the sake of clarity, our exposition here concerns only open annuli, but the results of this section may be generalized to arbitrary annuli, punctured disks, the complement of a disk, and the complement of a point. We will not use the material in this section until Section 7.1, and hence the reader may wish to skip this section for now.

An *open annulus* in $\mathbb{C}_v$ is a connected affinoid of the form

$$U = D(a, R) \smallsetminus \overline{D}(a, r),$$

with $a \in \mathbb{C}_v$ and $R > r > 0$. We frequently abuse notation and write

$$D(a, R) \smallsetminus \overline{D}(a, r) = \{r < |z - a| < R\}.$$

Definition 3.30. Let $a \in \mathbb{C}_v$, let $0 < r < R$ be real numbers, let U be the annulus $U = \{r < |z - a| < R\}$, and let

$$f(z) = \sum_{n \in \mathbb{Z}} c_n (z - a)^n \in \mathbb{C}_v[[z - a, (z - a)^{-1}]].$$

We say f is a *Laurent series converging on* U if $f(x)$ converges for all $x \in U$. Moreover, if $f \neq 0$, we define

(a) the *inner Weierstrass degree* of f on U to be the largest integer $M \in \mathbb{Z}$ such that $|c_M|r^M = \sup_{n\in\mathbb{Z}} |c_n|r^n$, or $-\infty$ if there is no such integer; and

(b) the *outer Weierstrass degree* of f on U to be the smallest integer $N \in \mathbb{Z}$ such that $|c_N|R^N = \sup_{n\in\mathbb{Z}} |c_n|R^n$, or ∞ if there is no such integer.

We denote by $\mathcal{A}_U$ the set of all convergent Laurent series on U.

As in complex analysis, but unlike the *formal* Laurent series encountered in the study of rings and fields, a convergent Laurent series in this context may include infinitely many nontrivial terms of negative exponent.

We leave it to the reader to verify that the series f converges on U if and only if

$$\lim_{n\to\pm\infty} |c_n|s^n = 0 \quad \text{for all } s \in (r, R),$$

and that the inner Weierstrass degree is less than or equal to the outer Weierstrass degree; see Exercise 3.17.

Proposition 3.31. *Let U be an annulus as in Definition* 3.30. *Then $\mathcal{A}_U$ is a commutative ring with unity, and the group of units of $\mathcal{A}_U$ is*

$$(\mathcal{A}_U)^\times = \big\{f \in \mathcal{A}_U : f(x) \neq 0 \text{ for all } x \in U\big\}. \tag{3.3}$$

Proof. See Section 14.5. □

Proposition 3.32. *Let $U \subseteq \mathbb{C}_v$ be an open annulus, and let $\phi(z) \in \mathbb{C}_v(z)$ be a rational function.*

(a) *If ϕ has no poles in U, then $\phi \in \mathcal{A}_U$ has a Laurent series expansion converging on U.*

(b) *If ϕ has no poles or zeros in U, then the inner and outer Weierstrass degrees of ϕ on U coincide.*

Proof. See Section 14.5. □

3.7. Images of annuli

Just as for images of disks, we need to understand the images of annuli under analytic and rational functions.

Theorem 3.33. *Let $a \in \mathbb{C}_v$, let $0 < r < R$, let U be the open annulus $U = \{r < |z-a| < R\}$, and let $f(z) \in \mathcal{A}_U$ be a nonconstant Laurent series convergent on U.*

Write $f(z) = \sum_{n\in\mathbb{Z}} c_n(z-a)^n$, *and suppose that* $f - c_0$ *has finite inner and outer Weierstrass degrees* $M \le N \in \mathbb{Z}$, *respectively. Let*

$$s := |c_M| r^M \quad \text{and} \quad t := |c_N| R^N.$$

(a) *If* $M < N$, *then* $f(U) = D(c_0, \max\{s,t\})$.

(b) *If* $M = N \ge 1$, *then* $f(U) = \{s < |z - c_0| < t\}$, *and the mapping* $f : U \to f(U)$ *is everywhere* M*-to-*1.

(c) *If* $M = N \le -1$, *then* $f(U) = \{t < |z - c_0| < s\}$, *and the mapping* $f : U \to f(U)$ *is everywhere* $(-M)$*-to-*1.

Moreover, in cases (b) *and* (c), *we have*

$$\left|f(x) - c_0\right| = \left|c_M(x-a)^M\right| \quad \text{for all } x \in U.$$

Proof. See Section 14.6. □

When the inner and outer Weierstrass degrees coincide, cases (b) and (c) of Theorem 3.33 imply that the image annulus $f(U)$ is rational open if and only if the original annulus U is rational as well. In case (a), if U is rational open, then the disk $f(U)$ is also rational open, although the converse may fail at one of the two ends of the annulus. Also in case (a), the degree of the mapping $f : U \to f(U)$ is not always uniform, in contrast with cases (b) and (c). That is, different points of $f(U)$ may have different numbers of preimages in U, as the following example shows.

Example 3.34. Fix $R > 1 > r$, let

$$U := D(0,R) \smallsetminus \overline{D}(0,r) = \{r < |x| < R\},$$

and let $f(z) = z^{-1} + z$. According to Theorem 3.33(c), U has image

$$f(U) = D(0,S), \quad \text{where} \quad S := \max\{R, 1/r\} > 1.$$

How many preimages does an arbitrary point $y \in D(0,S)$ have in U? Solving $f(z) = y$ gives the equation

$$z^2 - yz + 1 = 0.$$

The associated Newton polygon shows that this equation has two roots of absolute value 1 if $|y| \le 1$, but it has one root of absolute value $|y|$ and one of absolute value $|y|^{-1}$ if $|y| > 1$. Setting $s := \min\{R, 1/r\}$, each point in $D(0,s)$ has two preimages in U, but each point in $D(0,S) \smallsetminus \overline{D}(0,s)$ has only one. Thus, if $R \ne 1/r$, then $f : U \to f(U)$ does not have uniform degree.

Remark 3.35. Let $U = \{r < |z-a| < R\}$, and let $f(z) \in \mathcal{A}_U$. Suppose f has the same inner and outer Weierstrass degree d, and write $d = \pm e$, with $e \ge 1$. Then Theorem 3.33 says that $f(U)$ is an annulus of the form

$\{s < |z - b| < S\}$ with the properties that

- $f : U \to f(U)$ is everywhere e-to-1, and
- $S/s = (R/r)^e$.

In particular, if f has degree ± 1, then $f(U)$ is an annulus with the same ratio R/r of its radii as U. This observation is reminiscent of the corresponding fact from complex analysis: two annuli in $\mathbb{C}$ are analytically homeomorphic if and only if the ratios of their inner and outer radii coincide.

Exercises for Chapter 3

3.1. Let $D \subseteq \mathbb{C}_v$ be a disk. Prove that the ring $\mathcal{A}_D$ is complete with respect to $\|\cdot\|_D$. That is, if $\{f_n\}_{n\geq 1} \subseteq \mathcal{A}_D$ is a $\|\cdot\|_D$-Cauchy sequence, prove that $f_n \to f$ with respect to $\|\cdot\|_D$.

3.2. Let $D \subseteq \mathbb{C}_v$ be a disk of radius $r > 0$, let $a, b \in D$, and let $f \in \mathcal{A}_D$. Let

$$\sum_{n\geq 0} c_n(z-a)^n \quad \text{and} \quad \sum_{n\geq 0} d_n(z-b)^n$$

be the power series expansions of f about a and b, respectively.

(a) Prove that $\|f\|_D$ is well defined, by showing that

$$\sup\left\{|c_n|r^n : n \geq 0\right\} = \sup\left\{|d_n|r^n : n \geq 0\right\}.$$

Work directly from the definitions; in particular, do not use Proposition 3.17 or any related results.

(b) Prove that the Weierstrass degree of f on D is the same when we center f at a as when we center f at b.

3.3. Prove Proposition 3.9.

3.4. Let $D \subseteq \mathbb{C}_v$ be a disk. Prove the product rule for convergent power series on D: that $(fg)' = f'g + fg'$ for all $f, g \in \mathcal{A}_D$.

3.5. Let $a \in \mathbb{C}_v$ and $r > 0$, and define $D := D(a, r)$. Let $\{b_n\}_{n\geq 1}$ be a sequence of points in D with $|b_n - a| \nearrow r$. Prove that there is a nonzero convergent power series $f \in \mathcal{A}_D \smallsetminus \{0\}$ such that $f(b_n) = 0$ for all $n \geq 1$. Note that f may have other zeros besides the points b_n.

(*Suggestion*: Write f as an infinite product of the form $\prod_n (1 - (x/b_n)^{e_n})$ for appropriately chosen exponents $e_n \geq 1$.)

3.6. Let $a \in \mathbb{C}_v$ and $r > 0$. Let D be the open disk $D := D(a, r) \subseteq \mathbb{C}_v$, with associated ring $\mathcal{A}_D = \mathcal{A}(a, r)$ of convergent power series.

(a) Prove that

$$\mathcal{A}(a,r) = \left\{ \sum_{n\geq 0} c_n(z-a)^n \;:\; \lim_{n\to\infty} |c_n|s^n = 0 \text{ for all } s \in (0,r) \right\}.$$

(b) Let $\mathcal{A}'(a,r) := \left\{ \sum_{n\geq 0} c_n(z-a)^n \;:\; \{|c_n|r^n\}_{n\geq 0} \text{ is bounded} \right\}.$

Prove that

$$\mathcal{A}(a,r) \supsetneq \mathcal{A}'(a,r) \supsetneq \left\{ \sum_{n\geq 0} c_n(z-a)^n \;:\; \lim_{n\to\infty} |c_n|r^n = 0 \right\}.$$

3.7. Let $a \in \mathbb{C}_v$ and $r > 0$. Let D be the open disk $D := D(a,r) \subseteq \mathbb{C}_v$, and let

$$\mathcal{A}'(a,r) := \left\{ \sum_{n\geq 0} c_n(z-a)^n \;:\; \{|c_n|r^n\}_{n\geq 0} \text{ is bounded} \right\}$$

as in Exercise 3.6(b).

(a) Prove that $\mathcal{A}'(a,r)$ is a ring under addition and multiplication of $\mathbb{C}_v$-valued functions.

(b) Define $\|\cdot\|'_D : \mathcal{A}'_D \to \mathbb{R} \cup \{\infty\}$ by

$$\|f\|'_D := \|f\|_D \qquad \text{for all } f \in \mathcal{A}'(a,r).$$

Prove that $\|\cdot\|'_D$ is a multiplicative norm on $\mathcal{A}'(a,r)$, i.e., that for all $f, g \in \mathcal{A}'(a,r)$,

(i) $\|f\|'_D \in [0,\infty)$, with $\|f\|'_D = 0$ if and only if $f = 0$,

(ii) $\|f+g\|'_D \leq \max\{\|f\|'_D, \|g\|'_D\}$,

(iii) $\|fg\|'_D = \|f\|'_D \|g\|'_D$.

(c) Prove that $\mathcal{A}'(a,r)$ is complete with respect to $\|\cdot\|'_D$.

3.8. Let $D \subseteq \mathbb{C}_v$ be a disk, let $b \in D$, and let $f, g \in \mathcal{A}_D$. Prove the following claims made just after Definition 3.12.

(a) $\operatorname{ord}_b(fg) = \operatorname{ord}_b(f) + \operatorname{ord}_b(g)$.

(b) $\operatorname{ord}_b(f+g) \geq \min\{\operatorname{ord}_b(f), \operatorname{ord}_b(g)\}$.

3.9. Prove that conditions (1) and (2) on page 54, regarding Newton polygons and numbers of zeros, are equivalent.

3.10. Let $D, E \subseteq \mathbb{C}_v$ be disks, let $f \in \mathcal{A}_D$, let $h \in \mathcal{A}_E$, and suppose that $f(D) \subseteq E$. Prove that $h \circ f \in \mathcal{A}_D$.

(*Suggestion*: First consider the case that D and E are both rational closed.)

3.11. Let $D \subseteq \mathbb{C}_v$ be a disk, and let $b \in D$. For any $f \in \mathcal{A}_D$ and $n \geq 0$, let $f^{(n)}$ denote the formal nth derivative of f. Define $\operatorname{ord}'_b(f)$ to be the smallest integer $n \geq 0$ such that $f^{(n)}(b) \neq 0$, or ∞ if there is no such n.

(a) Prove that

$$\operatorname{ord}'_b(f+g) \geq \min\{\operatorname{ord}'_b(f), \operatorname{ord}'_b(g)\}$$

and

$$\operatorname{ord}'_b(fg) \geq \operatorname{ord}'_b(f) + \operatorname{ord}'_b(g)$$

for all $f, g \in \mathcal{A}_D$.

(b) If $\operatorname{char} \mathbb{C}_v = 0$, prove that $\operatorname{ord}'_b(f) = \operatorname{ord}_b(f)$ for all $f \in \mathcal{A}_D$.

(c) If $\operatorname{char} \mathbb{C}_v > 0$, find $f, g \in \mathcal{A}_D$ such that

$$\operatorname{ord}'_b(fg) > \operatorname{ord}'_b(f) + \operatorname{ord}'_b(g).$$

3.12. Let $f(z) = \sum_{n \geq 0} c_n (z-a)^n$, let $r > 0$, and assume that f converges on some rational closed disk containing $\overline{D}(a,r)$. Define

$$t := \max_{n \geq 0} |c_n| r^n,$$

and let P_f be the Newton polygon for f with respect to some $\pi \in K$ with $0 < |\pi| < 1$. Prove the claim made on page 55: that there is a unique line in the plane of slope $-\log r / \log |\pi|$ intersecting the boundary but not the interior of P_f, and that the y-intercept of this line is $\log t / \log |\pi|$.

3.13. Fix $\pi \in \mathbb{C}_v$ with $0 < |\pi| < 1$.

(a) Let $f(z) = \sum_{n \geq 1} a_n z^n$ with $|a_n| = |\pi|^{1/n}$. Show that $f \in \mathcal{A}(0,1)$ with infinite Weierstrass degree, and with $f(D(0,1)) = D(0,1)$.

(b) Given a radius $r > 0$, let $\{b_n\}_{n \geq 0} \subseteq \mathbb{C}_v$ be a sequence with $|b_n| \nearrow r$, and define $g(z) = \sum_{n \geq 1} b_n z^n$. Show that $g \in \mathcal{A}(0,1)$ with infinite Weierstrass degree, and with $g(D(0,1)) = D(0,r)$.

(c) Given a radius $r > 0$, show that there is a convergent power series $h \in \mathcal{A}(0,r)$ with infinite Weierstrass degree, and with $h(D(0,r)) = D(0,1)$.

(d) Show that there is a convergent power series $F \in \mathcal{A}(0,1)$ with infinite Weierstrass degree, and with $F(D(0,1)) = \mathbb{C}_v$.

3.14. Prove Corollary 3.18 without the hypothesis that $\operatorname{wdeg}_D(f)$ is finite. For the purposes of this exercise, define a disk of infinite radius to be $\mathbb{C}_v$.

3.15. Use Proposition 3.19 to provide an alternate proof of Proposition 3.17, first for open disks and then for closed disks, *without* using the Weierstrass preparation theorem or Newton polygons.

3.16. Let D be a $\mathbb{P}^1(\mathbb{C}_v)$-disk.

(a) Let $\eta \in \operatorname{PGL}(2, \mathbb{C}_v)$. Prove that $\eta(D)$ is also a $\mathbb{P}^1(\mathbb{C}_v)$-disk, and of the same type as D (rational closed, rational open, or irrational).

(b) Prove that there are a disk $E \subseteq \mathbb{C}_v$ and a linear fractional transformation $\eta \in \operatorname{PGL}(2, \mathbb{C}_v)$ such that $D = \eta(E)$.

3.17. Let $a \in \mathbb{C}_v$, let $0 < r < R$, let $U := \{r < |z-a| < R\}$, and let $f = \sum_{n\in\mathbb{Z}} c_n(z-a)^n$ be a Laurent series with $c_n \in \mathbb{C}_v$.

(a) Prove that $f \in \mathcal{A}_U$ if and only if

$$\lim_{n\to\pm\infty} |c_n|s^n = 0 \quad \text{for all } s \in (r, R).$$

(b) Prove that the inner Weierstrass degree of f on U is less than or equal to the outer Weierstrass degree.

3.18. Let $U = D(a,R) \smallsetminus \{a\} \subseteq \mathbb{C}_v$ be a punctured open disk, and let $\mathcal{A}_U$ be the set of all Laurent series centered at a that converge on U. Prove that $\mathcal{A}_U$ is a commutative ring with unity, with group of units

$$(\mathcal{A}_U)^\times = \{f \in \mathcal{A}_U : f(x) \neq 0 \text{ for all } x \in U\}.$$

(*Suggestion*: Write U as a union of open annuli, and apply Proposition 3.31.)

3.19. Repeat Exercise 3.18 for the cases that $U = \mathbb{C}_v \smallsetminus \{a\}$ or that $U = \mathbb{C}_v \smallsetminus \overline{D}(r,a)$.

3.20. Let $a \in \mathbb{C}_v$, let $0 < r < R$ be real numbers, let U be the open annulus $U := \{r < |z-a| < R\}$, and let $f, g \in \mathbb{C}_v[z]$ be nonzero polynomials. Suppose that f has Weierstrass degree $m \geq 0$ on both $\overline{D}(a,r)$ and $D(a,R)$, and that g has Weierstrass degree $n \geq 0$ on both $\overline{D}(a,r)$ and $D(a,R)$.

(a) Prove that the inner and outer Weierstrass degrees of fg on U are both equal to $m+n$.

(b) Prove that the inner and outer Weierstrass degrees of f/g on U are both equal to $m-n$.

3.21. Let $a \in \mathbb{C}_v$, let $0 < r < R$ be real numbers, let U be the open annulus $U := \{r < |z-a| < R\}$, and let ϕ be a rational function with no zeros or poles in U. Prove that the inner and outer Weierstrass degrees of ϕ on U are both equal to

$$(\text{number of zeros of } \phi \text{ in } \overline{D}(a,r)) - (\text{number of poles of } \phi \text{ in } \overline{D}(a,r)),$$

counted with multiplicity.

3.22. Let $U \subseteq \mathbb{C}_v$ be an open annulus, let ϕ be a rational function with no zeros or poles in U, and let m be the common inner and outer Weierstrass degree of ϕ on U. If m is not divisible by the residue characteristic of $\mathbb{C}_v$, prove that the derivative ϕ' has inner and outer Weierstrass degree $m-1$ on U.

3.23. Fix $0 < r < 1$, and let $U = \{r < |z| < 1\}$. For each of the following functions $f \in \mathcal{A}_U$, compute the inner and outer Weierstrass degrees of f on

U, find the image $f(U)$, and confirm that the conclusions of Theorem 3.33 hold for f on U.

(a) $f(z) = 1 + z^d$, for $d \geq 1$.

(b) $f(z) = 1 + z^{-d}$, for $d \geq 1$.

(c) $f(z) = cz^{-a} + z^b$, for $a, b \geq 1$ and $|c| \leq r^{a+b}$.

(d) $f(z) = cz^{-a} + z^b$, for $a, b \geq 1$ and $r^{a+b} < |c| < 1$.

(e) $f(z) = cz^{-a} + z^b$, for $a, b \geq 1$ and $|c| \geq 1$.

3.24. Let $U = \{r < |z - a| < R\}$ be an open annulus, and let $f \in \mathcal{A}_U$ be a nonconstant Laurent series converging on U. However, do *not* assume that f has finite inner and outer Weierstrass degrees on U. Prove that the image $f(U)$ is one of the following:

- an open disk $D(b, s)$,
- an open annulus $\{t < |z - b| < s\}$, or
- the whole affine line $\mathbb{C}_v$.

(*Suggestion*: Consider images $f(V)$ for slightly smaller annuli $V \subsetneq U$.)

3.25. Fix $0 < r < R$, and let $U = \{r < |z| < R\}$. For each of the three kinds of regions listed in Exercise 3.24, show that there is a convergent Laurent series $f \in \mathcal{A}_U$ such that the image $f(U)$ is that kind of region.

Part 2

Elementary Non-Archimedean Dyanmics

Chapter 4

Fundamentals of Non-Archimedean Dynamics

We are now ready to begin our study of the dynamics of a rational function defined over a non-archimedean field. As we will see, there are both similarities and contrasts with the theory of complex dynamics from Section 1.6.

Throughout this chapter, $\mathbb{C}_v$ denotes a complete and algebraically closed non-archimedean field.

4.1. Classifying periodic points

Definition 4.1. Let $\phi(z) \in \mathbb{C}_v(z)$ be a rational function, and let $x \in \mathbb{P}^1(\mathbb{C}_v)$ be a periodic point with multiplier $\lambda \in \mathbb{C}_v$. We say that x is

- *attracting* if $|\lambda| < 1$,
- *repelling* if $|\lambda| > 1$,
- *indifferent* (or *neutral*) if $|\lambda| = 1$.

In addition, we say that x is

- *superattracting* if $\lambda = 0$,
- *rationally indifferent* if λ is a root of unity,
- *irrationally indifferent* if $|\lambda| = 1$ and λ is not a root of unity.

The following two propositions are analogues of the complex dynamics properties of Section 1.6.1, with some slight changes that arise in the non-archimedean setting.

Proposition 4.2. *Let $\phi(z) \in \mathbb{C}_v(z)$ be a rational function. Then ϕ has a nonrepelling fixed point.*

Proof. We are immediately done if ϕ has a fixed point of multiplier 1. In addition, if ϕ is constant, then the constant value is a superattracting fixed point of ϕ. Excluding these cases, then, we may assume that ϕ has no fixed points of multiplier 1, and that $\deg(\phi) \geq 1$.

Observe that $|1/(1-\lambda)| = |\lambda|^{-1} < 1$ for any $\lambda \in \mathbb{C}_v$ with $|\lambda| > 1$. Therefore, if all fixed points of ϕ were repelling, taking absolute values of both sides of the Holomorphic Fixed-Point Formula (Theorem 1.7) would give $1 > 1$. Thus, there must be at least one nonrepelling fixed point. □

Proposition 4.3. *Let $\phi(z) \in \mathbb{C}_v(z)$ be a rational function, and let $x \in \mathbb{P}^1(\mathbb{C}_v)$ be a periodic point of period $m \geq 1$.*

(a) *If x is attracting, then there is an open set $U \subseteq \mathbb{P}^1(\mathbb{C}_v)$ containing x such that for all $y \in U$,*

$$\lim_{n\to\infty} \phi^{mn}(y) = x.$$

(b) *If x is repelling, then there is an open set $U \subseteq \mathbb{P}^1(\mathbb{C}_v)$ containing x such that for all $y \in U \smallsetminus \{x\}$, there is some $n \geq 1$ such that $\phi^{mn}(y) \notin U$.*

(c) *If x is nonrepelling, then there is an open $\mathbb{P}^1(\mathbb{C}_v)$-disk $U \subseteq \mathbb{P}^1(\mathbb{C}_v)$ containing x such that $\phi^m(U) \subseteq U$. If x is indifferent, then the mapping $\phi^m : U \to U$ is bijective.*

Proof. Considering ϕ^m in place of ϕ, we may assume that $m = 1$, and after a change of coordinates, we may assume that $x = 0$. Expanding ϕ as a power series in $\mathbb{C}_v[[z]]$ (which is possible by Theorem 3.5 and Proposition 3.11), we write

$$\phi(z) = \lambda z + a_2 z^2 + a_3 z^3 + \cdots,$$

which converges on some rational closed disk $\overline{D}(0,r)$.

If $\lambda = 0$, i.e., if 0 is superattracting, let $j \geq 2$ be the smallest integer such that $a_j \neq 0$, and choose $s \in (0,r)$ small enough so that $|a_n|s^n < |a_j|s^j$ for all $n > j$. (The choice of such s is possible because the power series converges on $\overline{D}(0,r)$.) Because $j \leq 2$, we may decrease s as necessary so that $|a_j|s^{j-1} < 1$. Let $\varepsilon := |a_j|s^{j-1} < 1$ and $U := D(0,s)$. For any $z \in \overline{D}(0,s)$, we have

$$|\phi(z)| = |a_j z^j| \leq |a_j| s^{j-1}|z| = \varepsilon|z|.$$

In particular, $\phi(z) \in U$, and therefore, by induction, $|\phi^n(z)| \leq \varepsilon^n |z|$ for all $n \geq 0$. Hence, $\lim_{n\to\infty} \phi^n(z) = 0$, proving the superattracting portion of part (a).

If $\lambda \neq 0$, we can choose $s \in (0, r)$ small enough so that $|a_n|s^n < |\lambda| s$ for all $n > 1$. Set $U := D(0, s)$. Then for all $z \in U$, we have

$$|\phi(z)| = |\lambda| \cdot |z|.$$

If $|\lambda| \leq 1$, then $\phi(z) \in U$, and hence by induction, $|\phi^n(z)| = |\lambda|^n \cdot |z|$ for all $n \geq 0$. Thus, $\phi(U) \subseteq U$. If $|\lambda| = 1$, then $\phi : U \to U$ is bijective by Theorem 3.15, proving part (c). On the other hand, if $|\lambda| < 1$, we obtain $\lim_{n\to\infty} \phi^n(z) = 0$, proving part (a).

If $|\lambda| > 1$ and $y \in U \smallsetminus \{0\}$, set n to be the smallest integer for which $|\lambda|^n |y| \geq |s|$. Then $\phi^i(y) \in U$ for all $i = 0, \ldots, n-1$, but $|\phi^n(y)| = |\lambda|^n |y|$, and hence $\phi^n(y) \notin U$, proving part (b). □

4.2. Local conjugacies at fixed points (optional)

Two maps ϕ and ψ are *locally conjugate* at points a and b, respectively, if there are neighborhoods U and V containing a and b, respectively, and a power series h converging on U such that

$$h : U \to V \text{ is bijective}, \quad h(a) = b, \quad \text{and } h \circ \phi \circ h^{-1}(z) = \psi(z) \text{ on } V.$$

If $b = 0$ and $\psi(z) = \lambda z$, we say h is a *local linearization* of ϕ at the (necessarily fixed) point a.

Proposition 4.3(a)–(b) can be strengthened to produce a local linearization at any fixed point of multiplier λ with $|\lambda| \neq 0, 1$. This result was first proven over $\mathbb{C}$ by Kœnigs in 1884 [**87**]; see [**9**, Sections 6.2 and 6.3], [**39**, Sections II.2 and II.3], or [**104**, Section 8] for modern presentations. The same proof works essentially verbatim over a complete non-archimedean field K. In fact, the resulting linearization $h(z)$ is defined over K itself, with no need to extend to $\mathbb{C}_v$; see Exercise 4.1.

Moreover, at a repelling fixed point $x \in \mathbb{P}^1(\mathbb{C})$, the inverse of the linearization h can actually be extended—by a clever trick attributed to Poincaré—to a meromorphic function $\tilde{h} : \mathbb{C} \to \mathbb{P}^1(\mathbb{C})$. More precisely, $U := \tilde{h}(\mathbb{C})$ is an open neighborhood x, and

$$\tilde{h}(\lambda z) = \phi(\tilde{h}(z)) \quad \text{for all } z \in \mathbb{C};$$

see [**104**, Corollary 8.10]. (However, $\tilde{h}$ is not injective if $\deg(\phi) \geq 2$.) The analogous result also holds when working over a non-archimedean field $\mathbb{C}_v$ in place of $\mathbb{C}$. Bézivin proved the p-adic version of this result in [**27**, Lemme 1]; essentially the same proof also works for arbitrary $\mathbb{C}_v$; see Exercise 4.2.

As for superattracting points, clearly there cannot be a local linearization, for then the nonconstant map ϕ would be locally conjugate to a constant. Instead, the best that could be true is that if the power series for $\phi(z)$ at 0 is of the form $c_j z^j + O(z^{j+1})$ for some $j \geq 2$, then ϕ is locally conjugate to z^j. Böttcher [**32**] proved the existence of this conjugation over $\mathbb{C}$ in 1904; see [**9**, Section 6.10], [**39**, Section II.4], or [**104**, Section 9]. Once again, the same proof works over $\mathbb{C}_v$, with some modifications required when j is divisible by $\operatorname{char} \mathbb{C}_v$; see Exercise 4.3.

Next we turn to irrationally indifferent points over $\mathbb{C}_v$. Herman and Yoccoz [**74**] proved a very general result, which applies to arbitrary non-archimedean fields and maps on spaces of arbitrary dimension, on the existence of local linearizations at such points. In the case of $\mathbb{P}^1$, if $\operatorname{char} \mathbb{C}_v = 0$, their result settles the question completely: for *any* irrationally indifferent point, there is a neighborhood on which a local linearization is defined. However, if $\operatorname{char} \mathbb{C}_v = p > 0$, the question remains open. Lindahl [**95**–**98**] completely settled the question in characteristic 2 for quadratic polynomials and gave some sufficient conditions in slightly greater generality, but beyond his results, very little is known about local linearization in positive characteristic.

Finally, unless $\deg(\phi) = 1$, there cannot be a local linearization at a rationally indifferent point over any field. Indeed, since $\lambda^e = 1$ for some $e \geq 1$, if $\phi(z)$ were locally conjugate to λz, then $\phi^e(z)$ would be locally conjugate to z, implying that $\phi^e(z) = z$, and hence $\deg(\phi) = 1$.

Remark 4.4 (Local linearization over $\mathbb{C}$)**.** Proposition 4.3(c) and the discussion above beg the question of what happens at indifferent periodic points over $\mathbb{C}$. This question leads to a huge series of other questions involving local linearization, Diophantine approximation, and more; we only scratch the surface here. If 0 is a rationally indifferent fixed point of $\phi \in \mathbb{C}(z)$, then there are finitely many repelling directions extending radially from 0, and in between those directions, points are attracted to 0, albeit rather slowly. (The attracting dynamics is conjugate to the dynamics of $z \mapsto z+1$ near ∞.) This arrangement of attracting "petals" separated by repelling rays is often called the "Leau–Fatou flower"; see [**9**, Section 6.5], [**39**, Section II.5], or [**104**, Section 10] for more information.

If 0 is an irrationally indifferent fixed point of $\phi \in \mathbb{C}(z)$, the situation essentially boils down to the question of whether ϕ is locally linearizable at 0. If it is locally linearizable, then the dynamics on U are easily described in terms of the multiplier λ and the conjugating function h; the maximal such open set U is always simply connected, and it is a *Siegel disk*, as mentioned in Section 1.6.3. Otherwise, the dynamics are far more complicated, as the the fixed point 0 belongs to the Julia set in this case.

The question of which of the two cases applies is related to Diophantine approximation. Specifically, write $\lambda = e^{2\pi i\theta}$, where $\theta \in \mathbb{R}$ is irrational. If θ is badly approximable by rationals (for a certain very precise description of "badly"), then by a result of Bryuno [**36**], ϕ is locally linearizable at 0, and hence 0 lies in a Siegel disk. Conversely, at least in the case that ϕ is a quadratic polynomial, if θ does not satisfy Bryuno's bad approximability condition, Yoccoz proved [**141**] that ϕ is not locally linearizable at 0, and hence 0 lies in the Julia set.

The question of existence of the linearization in this setting is often called the "problem of small denominators". Although the results of Bryuno, Yoccoz, and others tell us quite a bit, a full understanding of complex dynamics near an irrationally indifferent fixed point remains open in general. For more on the subject, see [**9**, Sections 6.6–7], [**39**, Sections II.6,V.1], or [**104**, Section 11].

Remark 4.5. In complex dynamics, according to Remark 4.4, the existence of a local linearization at an indifferent periodic point x is exactly tied to whether x lies in the Fatou or Julia set. By contrast, Proposition 4.3(c) says that in non-archimedean dynamics, *all* indifferent periodic points lie in the Fatou set; yet there is no local linearization at a rationally indifferent periodic point. Even for irrationally indifferent points, at which there *is* a linearization, the maximum possible domain of the linearization is usually far smaller than the region U of Proposition 4.3(c). Indeed, as we shall see in Section 10.1, U often contains infinitely many periodic points. Thus, for an indifferent periodic point in non-archimedean dynamics, the associated component of the Fatou set properly contains the domain of linearization, whereas in the complex setting, the two regions coincide.

4.3. Good reduction

We set the following notation for this section.

K	a non-archimedean field
$\lvert\cdot\rvert$	the absolute value on K
$\mathcal{O}_K$	the ring of integers of K
$\mathcal{M}_K$	the maximal ideal of $\mathcal{O}_K$
k	the residue field $\mathcal{O}_K/\mathcal{M}_K$
$\overline{K}$	an algebraic closure of K
$\overline{k}$	an algebraic closure of k
$a \mapsto \overline{a}$	the natural reduction map $\mathcal{O}_K \twoheadrightarrow k$.

Reduction induces a map $\mathbb{P}^1(K) \to \mathbb{P}^1(k)$, also called the reduction map, defined by $[x, y] \mapsto [\overline{x}, \overline{y}]$, for $x, y \in \mathcal{O}_K$ chosen so that at least one of x, y has absolute value 1. If we view K as a subset of $\mathbb{P}^1(K)$ by identifying $x \in K$ with $[x,1] \in \mathbb{P}^1(K)$, and similarly for k and $\mathbb{P}^1(k)$, then the reduction map on $\mathbb{P}^1(K)$ sends every point $a \in \mathcal{O}_K \subseteq \mathbb{P}^1(K)$ to the corresponding residue class $\overline{a}$ in $k \subseteq \mathbb{P}^1(k)$, and it sends every point in $\mathbb{P}^1(K) \setminus \mathcal{O}_K$ to the point $\overline{\infty} = [\overline{1}, \overline{0}] \in \mathbb{P}^1(K)$.

It is common in number theory to ask whether an object defined over K makes sense modulo $\mathcal{M}_K$. The following definition for dynamical systems was first proposed in the introduction of [**106**]. To state it, we define the reduction $\overline{f}(x,y) \in k[x,y]$ of a polynomial $f(x,y) \in \mathcal{O}_K[x,y]$ by replacing each coefficient in $\mathcal{O}_K$ by its reduction in k.

Definition 4.6. Let $\phi(z) \in K(z)$ be a rational function with homogenous presentation

$$\phi([x,y]) = [f(x,y), g(x,y)],$$

where $f, g \in \mathcal{O}_K[x,y]$ are homogeneous polynomials of degree $d = \deg(\phi)$ with no common irreducible factors in $K[x,y]$ and such that at least one coefficient of f or g has absolute value 1. We say that ϕ has *explict good reduction* if any of the following three equivalent conditions holds:

(a) the only common zero of $\overline{f}$ and $\overline{g}$ in $\overline{k} \times \overline{k}$ is $(0,0)$;

(b) $\overline{f}$ and $\overline{g}$ have no common zeros in $\mathbb{P}^1(\overline{k})$;

(c) $\overline{f}$ and $\overline{g}$ are relatively prime in $k[x,y]$.

If there is some linear fractional transformation $\eta \in \mathrm{PGL}(2,K)$ such that $\eta \circ \phi \circ \eta^{-1}$ has explicit good reduction, we say that ϕ has *good reduction* (over K).

If ϕ does not have good reduction, we say it has *bad reduction.* If there is some linear fractional transformation $\eta \in \mathrm{PGL}(2,\overline{K})$ such that $\eta \circ \phi \circ \eta^{-1}$ has explicit good reduction, we say that ϕ has *potential good reduction.*

If ϕ has explicit good reduction, as exhibited by the homogeneous presentation $\phi = [f,g]$ of Definition 4.6, we define the reduced map

$$\overline{\phi} := [\overline{f}, \overline{g}] : \mathbb{P}^1(\overline{k}) \longrightarrow \mathbb{P}^1(\overline{k}).$$

In that case, the conditions of Definition 4.6 say that $\deg(\overline{\phi}) = \deg(\phi)$; in addition, if $z, w \in \mathbb{P}^1(\mathbb{C}_v)$ have the same reduction $\overline{z} = \overline{w} \in \mathbb{P}^1(\overline{k})$, then $\overline{\phi(z)} = \overline{\phi(w)}$, and this common point is $\overline{\phi}(\overline{z})$.

There are three ways that $\phi \in K(z)$ can fail to have explicit good reduction. First, a residue class of $\mathbb{P}^1(\mathbb{C}_v)$ could contain both a zero and a pole of ϕ. Second, writing $\phi = f/g$ with $f, g \in \mathcal{O}_K[z]$, we could have $\overline{f} \neq 0$ but $\overline{g} = 0$. Third, with the same notation, we could have $\overline{g} \neq 0$ but $\overline{f} = 0$. (The second and third possibilities are mututally exclusive, but neither excludes the first.)

Example 4.7. Let $\pi \in K$ with $0 < |\pi| < 1$, and fix an integer $d \geq 2$. Then $\phi_1(z) := \pi z^d \in \mathbb{C}_v[z]$ does not have explicit good reduction because in homogeneous coordinates it is $[\pi x^d, y^d]$, the reduction of which is $[0, y^d]$, and both 0 and y^d are zero at the point $[1, 0]$.

However, choosing a $(d-1)$-st root $\alpha \in \overline{K}$ of π and conjugating by $\eta(z) = \alpha z$ gives

$$\psi(z) := \eta \circ \phi_1 \circ \eta^{-1}(z) = z^d \in \mathbb{C}_v[z].$$

In homogeneous coordinates, $\psi[x,y] = [x^d, y^d]$, whose reduction is again $\overline{\psi}[x,y] = [x^d, y^d]$. Since $x^d, y^d \in k[x,y]$ have no common zeros in $\mathbb{P}^1(\overline{k})$, ψ has explicit good reduction, and therefore ϕ_1 has potential good reduction. In fact, if $d = 2$, or more generally if π has a $(d-1)$-st root in K, then we may choose $\alpha \in K$, and hence ϕ_1 has good reduction.

Similarly, $\phi_2(z) := z^d/\pi$ does not have explicit good reduction, because in homogeneous coordinates it is $[x^d, \pi y^d]$, whose reduction $[x^d, 0]$ is undefined at $[0, 1]$. However, conjugating by $z \mapsto z/\alpha$, where $\alpha \in \overline{K}$ is again a $(d-1)$-st root of π, gives the explicit good reduction map $\psi(z) = z^d$. Thus, ϕ_2 also has potential good reduction, with good reduction if $d = 2$, or more generally if π has a $(d-1)$-st root in K.

On the other hand, $\phi_3(z) := (z^2 - 1)/\pi$ has homogeneous presentation $[x^2 - y^2, \pi y^2]$. The reduction is $[x^2 - y^2, 0]$, which is undefined at both $[-1, 1]$ and $[1, 1]$, and hence ϕ_3 does not have explicit good reduction. Unlike ϕ_1 and ϕ_2, however, if the residue characteristic of K is not 2, then ϕ_3 does not even have potential good reduction. Indeed, ϕ_3 has a fixed point at each of the roots of $z^2 - \pi z - 1 = 0$, and the condition on the residue characteristic guarantees that both of these fixed points are repelling. As we will soon see, maps of explicit good reduction cannot have repelling periodic points, and hence maps of potential good reduction cannot have such points either.

Proposition 4.8. *Let $\phi_1, \phi_2 \in K(z)$ be rational functions with explicit good reduction. Then $\phi_1 \circ \phi_2$ also has explicit good reduction.*

Proof. Write $\phi_1 = [f_1, g_1]$ and $\phi_2 = [f_2, g_2]$. Then the reductions of the two coordinates of $\phi_1 \circ \phi_2$ are

$$\overline{F}(x,y) := \overline{f}_1(\overline{f}_2(x,y), \overline{g}_2(x,y))$$

and

$$\overline{G}(x,y) := \overline{g}_1(\overline{f}_2(x,y), \overline{g}_2(x,y)).$$

If $\overline{F}(x,y) = \overline{G}(x,y) = 0$ for some point $(x,y) \in \overline{k} \times \overline{k}$, then because ϕ_1 has explicit good reduction, we must have $\overline{f}_2(x,y) = \overline{g}_2(x,y) = 0$. Thus, because ϕ_2 has explicit good reduction, we have $(x,y) = (0,0)$, as desired. □

Warning 4.9. It is possible for $\phi_1 \circ \phi_2$ to have explicit good reduction even if both ϕ_1 and ϕ_2 have bad reduction; see Exercise 4.9.

The criterion of Definition 4.6 is easy enough to check in practice, but for polynomials and for linear fractional transformations, it is especially easy, as the following result shows.

Proposition 4.10. *Let $\phi \in K(z)$ be a nonconstant rational function.*

(a) *If ϕ is a polynomial, then ϕ has explicit good reduction if and only if all of its coefficients lie in $\mathcal{O}_K$, and its lead coefficient has absolute value* 1.

(b) *If* $\deg(\phi) = 1$, *then ϕ has explicit good reduction if and only if* $\phi \in \mathrm{PGL}(2, \mathcal{O}_K)$.

Proof. This is left to the reader; see Exercises 4.10 and 4.11. □

Remark 4.11 (The homogeneous resultant)**.** For an arbitary rational function $\phi \in K(z)$, we can determine whether it has explicit good reduction by computing the resultant of the numerator and denominator of ϕ.

Recall that if $f, g \in K[z]$ are polynomials of degrees d and e, respectively, then the *resultant* of f and g is the quantity

$$\mathrm{Res}(f, g) := a_d^e b_e^d \prod_{\alpha} \prod_{\beta} (\alpha - \beta), \tag{4.1}$$

where a_d is the lead coefficient of f, b_e is the lead coefficient of g, and the products are taken over all roots $\alpha \in \overline{K}$ of f and $\beta \in \overline{K}$ of g, each repeated according to multiplicity. The resultant can be computed directly from the coefficients of f and g as the determinant of the *Sylvester matrix*, which is the $(d + e) \times (d + e)$ matrix

$$\begin{bmatrix}
a_0 & a_1 & a_2 & \cdots & a_{d-1} & a_d & 0 & 0 & \cdots & 0 \\
0 & a_0 & a_1 & \cdots & a_{d-2} & a_{d-1} & a_d & 0 & \cdots & 0 \\
\vdots & \ddots & \ddots & \cdots & \ddots & \ddots & \ddots & \ddots & \ddots & \vdots \\
0 & \cdots & 0 & a_0 & a_1 & a_2 & \cdots & a_{d-1} & a_d & 0 \\
0 & \cdots & 0 & 0 & a_0 & a_1 & \cdots & a_{d-2} & a_{d-1} & a_d \\
b_0 & b_1 & b_2 & \cdots & b_{e-1} & b_e & 0 & 0 & \cdots & 0 \\
0 & b_0 & b_1 & \cdots & b_{e-2} & b_{e-1} & b_e & 0 & \cdots & 0 \\
\vdots & \ddots & \ddots & \cdots & \ddots & \ddots & \ddots & \ddots & \ddots & \vdots \\
0 & \cdots & 0 & b_0 & b_1 & b_2 & \cdots & b_{e-1} & b_e & 0 \\
0 & \cdots & 0 & 0 & b_0 & b_1 & \cdots & b_{e-2} & b_{e-1} & b_e
\end{bmatrix}, \tag{4.2}$$

where $f(z) = a_0 + \cdots + a_d z^d$ and $g(z) = b_0 + \cdots + b_e z^e$.

However, given homogeneous polynomials $f, g \in K[x, y]$ *of the same degree* $d \geq 1$, it makes more sense to consider their *homogeneous resultant.* Such f and g can be factored as

$$f(x, y) = \prod_{i=1}^{d} (x, y) \wedge (\alpha_i, \beta_i) \quad \text{and} \quad g(x, y) = \prod_{i=1}^{d} (x, y) \wedge (\gamma_i, \delta_i),$$

where the points $[\alpha_i, \beta_i], [\gamma_i, \delta_i] \in \mathbb{P}^1(\overline{K})$ are the zeros of f and g, respectively, and where in general we define

$$(s, t) \wedge (u, v) := vs - ut.$$

Then the homogeneous resultant is

$$\operatorname{Res}(f, g) := \prod_{i=1}^{d} \prod_{j=1}^{d} (\alpha_i, \beta_i) \wedge (\gamma_j, \delta_j).$$

Again, the homogeneous resultant can be computed directly from the coefficients of

$$f(x, y) = a_0 y^d + a_1 x y^{d-1} + \cdots + a_d x^d$$

and

$$g(x, y) = b_0 y^d + b_1 x y^{d-1} + \cdots + b_d x^d$$

as the determinant of the $2d \times 2d$ matrix (4.2), with $e = d$.

The resultant $\operatorname{Res}(f, g)$ is zero if and only if f and g have a common zero in $\mathbb{P}^1(\overline{K})$. Similarly, the homogeneous resultant is zero if and only if

$$(\alpha_i, \beta_i) \wedge (\gamma_i, \delta_j) = 0 \quad \text{for some } i, j,$$

which happens exactly when $[\alpha_i, \beta_i]$ and $[\gamma_j, \delta_j]$ are the same point in $\mathbb{P}^1(\overline{K})$. Thus, if we choose $f, g \in \mathcal{O}_K[x, y]$ as in Definition 4.6, i.e., with at least one coefficient having absolute value 1, then their homogeneous resultant is zero modulo the maximal ideal $\mathcal{M}_K$ if and only if some $(\alpha_i, \beta_i) \wedge (\gamma_j, \delta_j)$ is zero modulo the maximal ideal, or equivalently, if and only if f and g have a common zero in $\mathbb{P}^1(\overline{k})$. In other words, ϕ fails to have explicit good reduction if and only if the homogeneous resultant belongs to $\mathcal{M}$.

For example, the map $\phi(z) = (10z^2 + 4z - 11)/(3z - 2)$ can be written in homogeneous coordinates as $[10x^2 + 4xy - 11y^2, 3xy - 2y^2]$, and this presentation fits the conditions of Definition 4.6 for $\mathbb{Q}_p$ for each prime p of $\mathbb{Z}$. The associated Sylvester matrix is

$$\begin{bmatrix} -11 & 4 & 10 & 0 \\ 0 & -11 & 4 & 10 \\ -2 & 3 & 0 & 0 \\ 0 & -2 & 3 & 0 \end{bmatrix},$$

which has determinant $-350 = -2 \cdot 5^2 \cdot 7$. Thus, ϕ does not have explicit good reduction over $\mathbb{Q}_p$ for $p = 2, 5, 7$. While the problem at 2 and 5 was easy to predict due to the coefficient of 10 on the only z^2-term, the problem at $p = 7$ is less obvious. In retrospect, one can check that $\overline{\phi}$ wants to have both a zero and a pole at $\overline{z} = 3$ modulo 7. In addition, we can be sure that at all *other* primes p, the map ϕ has explicit good reduction.

Remark 4.12. The original definition of good reduction in [**106**], as well as in much of the literature that followed, actually coincides with our definition of *explicit* good reduction. However, in the context of varieties defined over K, the term "good reduction" allows for K-rational coordinate changes and, hence, is analogous to our Definition 4.6.

4.4. Lattès maps

In Example 1.4, we saw an important family of commuting maps, the Chebyshev polynomials. In this section, we present an even more important commuting family, arising from elliptic curves. We assume in this section that the reader has a least a basic familiarity with elliptic curves. For relevant background, see [**130**, Chapter III] or [**131**, Section 6.3].

Definition 4.13. Let K be a field, and let $\phi \in K(z)$ be a rational function with $\deg(\phi) \geq 2$. Suppose that there are an elliptic curve E and finite morphisms $\psi : E \to E$ and $\alpha : E \to \mathbb{P}^1$, all defined over K, such that α is separable, and the diagram

$$\begin{array}{ccc} E & \xrightarrow{\psi} & E \\ \alpha\downarrow & & \alpha\downarrow \\ \mathbb{P}^1 & \xrightarrow{\phi} & \mathbb{P}^1 \end{array}$$

commutes; that is, α is a separable morphism of dynamical systems, from ψ to ϕ. Then we say ϕ is a *Lattès* map.

At first glance, the conditions of Definition 4.13 may appear so restrictive that Lattès maps would be very rare. However, the group structure on any elliptic curve E guarantees that any finite endomorphism of E has an associated Lattès map, just as any power map $z \mapsto z^n$, being an endomorphism of the multiplicative group $\mathbb{G}_m$, has an associated Chebyshev map T_n; see Example 1.4. Specifically, let $\alpha : E \to \mathbb{P}^1$ be the quotient map identifying a point P on E with its negative $-P$. That is, if we write E in Weierstrass form as

$$E : y^2 + a_1xy + a_3y = x^3 + a_2x^2 + a_4x + a_6, \tag{4.3}$$

then the map $\alpha : E \to \mathbb{P}^1$ is simply the x-coordinate map $\alpha(x, y) := x$. Note that α is finite and separable, even if $\operatorname{char} K = 2$, because $P \neq -P$ for all but finitely many points on E.

If $\psi : E \to E$ is an endomorphism, meaning that ψ is a morphism in the sense of algebraic geometry and, in addition, that $\psi(O) = O$, then because ψ necessarily respects the group law on E, we have $\psi(-P) = -\psi(P)$ for all points $P \in E$. Since α identifies P with $-P$, it follows that ψ induces a morphism

$$\phi : \mathbb{P}^1 \to \mathbb{P}^1 \quad \text{by } \phi\big(\alpha(P)\big) := \alpha\big(\psi(P)\big)$$

for all $P \in E$. Thus, $\phi \in K(z)$ is a rational function of degree $d := \deg(\psi)$, and hence it is a Lattès map if $d \geq 2$. In this construction, if $\psi = [m]$ is the multiplication-by-m map on E, for some integer $m \geq 2$, then the associated Lattès map ϕ is sometimes called a *flexible* Lattès map; see [**131**, Section 6.5]. In fact, the class of flexible Lattès maps is a bit bigger, consisting of Lattès maps corresponding to endomorphisms of the form

$$\psi(P) = [m]P + T$$

and the x-coordinate map $\alpha(x, y) = x$, where $T \in E$ satisfies $T + T = O$.

Even more generally, if E has complex multiplication (i.e., if there are endomorphisms of E *besides* the multiplication-by-m maps), then any extra endomorphism $\psi : E \to E$ defined over K also induces a so-called *rigid* Lattès map $\phi \in K(z)$ with $\phi(\alpha(P)) = \alpha(\psi(P))$ for all $P \in E$. Moreover, some curves with complex multiplication—specifically those with j-invariant 0 or 1728—admit other morphisms $\alpha : E \to \mathbb{P}^1$, allowing for yet more rigid Lattès maps; see [**131**, Sections 6.4–6.6] for more on general Lattès maps. For simplicity, however, we restrict our discussion to flexible Lattès maps with $T = O$; that is, we consider only $\phi \in K(z)$ with $\phi(\alpha(P)) = \alpha([m](P))$, where $\alpha(P) = \alpha(-P)$.

In principle, one can write out flexible Lattès map explicitly. For example, for $m = 2$, given a Weierstrass form for E as in equation (4.3), then if $\psi = [2]$ is the multiplication-by-2 map on E, the associated Lattès map is

$$\phi(z) = \frac{z^4 - b_4 z^2 - 2b_6 z - b_8}{4z^3 + b_2 z^2 + 2b_4 z + b_6}, \tag{4.4}$$

where

$$b_2 := a_1^2 + 4a_2, \quad b_4 := 2a_4 + a_1 a_3, \quad b_6 := a_3^2 + 4a_6, \quad \text{and}$$
$$b_8 := a_1^2 a_6 + 4a_2 a_6 - a_1 a_3 a_4 + a_2 a_3^2 - a_4^2,$$

as is standard notation in the study of elliptic curves. (See [**130**, Sections III.1–2].)

If $\psi : E \to E$ is multiplication-by-m for some $m \geq 3$, then again one can write out ϕ explicitly, although of course the formula becomes rather complicated for $m \geq 3$, as $\deg(\phi) = \deg([m]) = m^2$. In addition, since $[m] \circ [n] = [mn] = [n] \circ [m]$, the associated Lattès maps on $\mathbb{P}^1$ also commute, and hence, by Exercise 1.4, they have the same set of preperiodic points.

Thus, when studying flexible Lattès maps, it is common to use the degree 4 map given in equation (4.4).

Example 4.14. Choose $a \in \mathbb{C}_v$ so that $|a| = |1 + 432a| = 1$. (For example, if K has residue characteristic anything other than $p = 433$, we can choose $a = 1$; and if the residue characteristic *is* 433, we can choose $a = -1$.) Let E be given by $y^2 + xy = x^3 + a$, which has good reduction by our choice of a. Then the Lattès map associated with $[2] : E \to E$ (via the x-coordinate morphism) is

$$\phi(z) = \frac{z^4 - 8az - a}{4z^3 + z^2 + 4a},$$

which has explicit good reduction.

Example 4.15. Choose $a \in \mathbb{C}_v$ with $0 < |a| < 1$, and let E be given by $y^2 + xy = x^3 + a$. Then the Lattès map associated with $[2] : E \to E$ (via the x-coordinate morphism) is

$$\phi(z) = \frac{z^4 - 8az - a}{4z^3 + z^2 + 4a},$$

but this time ϕ does not have explicit good reduction: $\overline{z} = 0$ is a root of both the reduced numerator $\overline{z}^4$ and the reduced denominator $4\overline{z}^3 + \overline{z}^2$. In fact, ϕ has bad reduction; see Exercise 4.14.

In spite of the possibility of bad reduction, the following result shows that all periodic points of a Lattès map have integer multipliers, and hence—in non-archimedean dynamics—are nonrepelling. Recall that for any integer $N \in \mathbb{Z}$, the set of *N-torsion points* is

$$E[N] = \bigl\{P \in E : [N]P = O\bigr\}.$$

Theorem 4.16. *Let E be an elliptic curve defined over $\mathbb{C}_v$, let $\alpha : E \to \mathbb{P}^1$ be the quotient map identifying P with $-P$, let $m \geq 2$ be an integer, and let $\phi \in \mathbb{C}_v(z)$ be the flexible Lattès map satisfying $\phi(\alpha(P)) = \alpha([m](P))$.*

(a) *The set of n-periodic points of ϕ is*

$$\alpha\bigl(E[m^n - 1] \cup E[m^n + 1]\bigr).$$

(b) *Let $c \in \mathbb{P}^1(\mathbb{C}_v)$ be periodic for ϕ of minimal period n. Then the multiplier of c is*

$$\begin{gathered} m^n, \text{ if } c \in \alpha\bigl(E[m^n - 1] \smallsetminus E[2]\bigr), \\ -m^n, \text{ if } c \in \alpha\bigl(E[m^n + 1] \smallsetminus E[2]\bigr), \\ m^{2n}, \text{ if } c \in \alpha\bigl(E[m^n + 1] \cap E[2]\bigr). \end{gathered}$$

Proof.

(a) This is left to the reader; see Exercise 4.15.

(b) Choose a Weierstrass form for E as in equation (4.3). Change coordinates if necessary so that $a_1, a_2, a_3, a_4, a_6 \in \mathcal{O}$.

Step 1. In this step we give a coarse description of the group law of E on a neighborhood of the identity point O. (The reader may recognize that we are deriving a rough approximation the formal group of E; see [**130**, Chapter IV] for more on the formal group of an elliptic curve.)

Substituting $z = -x/y$ and $w = -1/y$ in (4.3) gives the equation $w = f(z,w)$ for E, where

$$f(z,w) = z^3 + (a_1 z + a_2 z^2)w + (a_3 + a_4 z)w^2 + a_6 w^3. \tag{4.5}$$

Note that the identity point O of E is $(z,w) = (0,0)$. Repeatedly substituting $w = f(z,w)$ into itself gives

$$w = f(z,w) = f\big(z, f(z,w)\big) = f\Big(z, f\big(z, f(z,w)\big)\Big) = \cdots. \tag{4.6}$$

We leave it to the reader to check that for $z, w \in D(0,1)$, taking the appropriate limit in (4.6) gives a power series $g(z)$ such that

$$w = g(z) \in z^3 + z^4\mathcal{O}[[z]]; \tag{4.7}$$

see Exercise 4.16(a). In particular, for any $z \in D(0,1)$, there is one and only one value $w \in D(0,1)$ such that $w = f(z,w)$, and this unique value is $w = g(z)$.

Consider the formal power series ring $R := \mathcal{O}[[z_1, z_2]]$ and the ideal $I := \langle z_1^2, z_1 z_2, z_2^2 \rangle$ of R. The line L through the points $P_1 := (z_1, g(z_1))$ and $P_2 := (z_2, g(z_2))$ is given by $w = Az + B$, where

$$A = \frac{g(z_2) - g(z_1)}{z_2 - z_1} \in R \quad \text{and} \quad B = g(z_1) - Az_1 \in R$$

if $P_1 \neq P_2$, or by an appropriate tangency condition if $P_1 = P_2$. In either case, we have $A, B \in I$; see Exercise 4.16(b). Substituting $w = Az + B$ into $w = f(z,w)$ gives a cubic equation

$$D_3 z^3 + D_2 z^2 + D_1 z + D_0 = 0. \tag{4.8}$$

The reader can check that $D_2 \in I$ and $D_3 \in 1 + I$; see Exercise 4.16(c). The three roots of (4.8) are z_1, z_2, z_3, where $P_3 := (z_3, g(z_3))$ is the third point of intersection of E and L. Thus, we have

$$z_1 + z_2 + z_3 = -\frac{D_2}{D_3} \in I.$$

That is, the z-coordinate of $P_3 = -(P_1 + P_2)$ is given by a power series G, with

$$z_3 = G(z_1, z_2) \in -z_1 - z_2 + I.$$

Since $P_1+P_2 = -(-(P_1+P_2)+O)$, the z-coordinate of P_1+P_2 is also given by a power series F, where

$$F(z_1,z_2) = G\big(G(z_1,z_2),0\big) \in z_1+z_2+I. \tag{4.9}$$

We leave it to the reader to deduce that for any $N \in \mathbb{Z}$, there is a power series

$$H_N(z) \in Nz + z^2\mathcal{O}[[z]] \tag{4.10}$$

such that for any $z \in D(0,1)$, the z-coordinate of the point $[N]P$ is $H_N(z)$, where $P = (z, g(z))$ and $[N] : E \to E$ is multiplication-by-N; see Exercise 4.16(d).

Step 2. Let $N \in \mathbb{Z}$ with either $N \geq 3$ or $N \leq -2$, and let $Q \in E[N-1]$ be an $(N-1)$-torsion point. Assume that $-Q \neq Q$, and hence that Q is not a ramification point of $\alpha : E \to \mathbb{P}^1$. Let ϕ_N be the flexible Lattès map for the multiplication-by-N map $[N] : E \to E$. Note that $\alpha(Q)$ is a fixed point of ϕ_N; we claim it has multiplier N.

Define $\eta : D(0,1) \to E$ by

$$\eta(z) := \big(z, g(z)\big) + Q$$

in our (z,w)-coordinates, where g is the power series of (4.7). By the defining property of the power series H_N of (4.10), we have

$$\eta \circ H_N = [N] \circ \eta. \tag{4.11}$$

Choose $\theta \in \mathrm{PGL}(2, \mathbb{C}_v)$ such that $\theta(\alpha(Q)) = 0$. We have

$$D(0,1) \xrightarrow{\eta} E \xrightarrow{\alpha} \mathbb{P}^1(\mathbb{C}_v) \xrightarrow{\theta} \mathbb{P}^1(\mathbb{C}_v).$$

Define $h : D(0,1) \to \mathbb{P}^1(\mathbb{C}_v)$ by $h := \theta \circ \alpha \circ \eta$. Choose $r \in (0,1)$ small enough that $\infty \notin h(D(0,r))$. Then $h : D(0,r) \to \mathbb{C}_v$ is given by a convergent power series.

Define $\psi(z) := \theta \circ \phi_N \circ \theta^{-1} \in \mathbb{C}_v(z)$. By (4.11) and the defining property of the Lattès map ϕ_N, we have $h \circ H_N = \psi \circ h$. Then, differentiating at 0,

$$h'\big(H_N(0)\big) \cdot H_N'(0) = \psi'\big(h(0)\big) \cdot h'(0). \tag{4.12}$$

By our choice of θ, we have $h(0) = 0$. In addition,

$$\psi(0) = \theta \circ \phi_N\big(\alpha(Q)\big) = \theta \circ \alpha([N]Q) = \theta(\alpha(Q)) = 0.$$

Therefore, (4.12) becomes $N \cdot h'(0) = \psi'(0) \cdot h'(0)$. Since θ and η are one-to-one and since Q is not a ramification point of α, we have $h'(0) \neq 0$. Thus, $\psi'(0) = N$, proving our claim.

Step 3. Given $c \in \mathbb{P}^1(\mathbb{C}_v)$ of minimal period n for the Lattès map ϕ for multiplication-by-m, we may choose $Q \in E[m^n - 1] \cup E[m^n + 1]$ with $\alpha(Q) = c$, by part (a).

If $Q \in E[m^n - 1] \smallsetminus E[2]$, then let $N := m^n$, so that the Lattès map ϕ_N of Step 2 is ϕ^n, and Q is an $(N-1)$-torsion point. By the claim of Step 2, the multiplier of c as an n-periodic point of ϕ is $N = m^n$.

If $Q \in E[m^n + 1] \smallsetminus E[2]$, then let $N := -m^n$. Then Q is an $(N-1)$-torsion point, and by Exercise 4.17 the Lattès map ϕ_N of Step 2 is ϕ^n. Again by Step 2, the multiplier of c as an n-periodic point of ϕ is $N = -m^n$.

We leave the final case, that $Q \in E[m^n + 1] \cap E[2]$, to the reader; see Exercise 4.19. □

The above proof of Theorem 4.16 relies on non-archimedean analysis. For a proof over an arbitrary field, using the invariant differential of the elliptic curve E, see [**131**, Proposition 6.52].

4.5. Dynamics on disks

In order to study dynamics on $\mathbb{P}^1(\mathbb{C}_v)$, we often need to understand the dynamics on certain subsets, especially disks.

Proposition 4.17. *Let $r > 0$, let $a \in \mathbb{C}_v$, and let U be either $D(a,r)$ or $\overline{D}(a,r)$. Let $\phi \in \mathbb{C}_v[[z-a]]$ be a power series converging on U such that $\phi - a$ has finite Weierstrass degree $d := \operatorname{wdeg}_U(\phi - a)$. Suppose that $\phi(U) \cap U \neq \varnothing$, and that either*

(a) *$\phi(U) \neq U$, or*
(b) *$d \neq 1$, or*
(c) *$d = 1$ and $|\phi'(a) - 1| = 1$.*

Then the disk U contains a fixed point of ϕ.

Proof. The statement is trivial if ϕ is constant, and hence we assume $d \geq 1$. We assume U is a closed disk; the case that U is an open disk is similar. After a change of coordinates, we may also assume that $a = 0$.

Write $\phi(z) = \sum_{i \geq 0} c_i z^i$, and let $s = |c_d| r^d = \max\{|c_i| r^i : i \geq 1\}$, so that $\phi(U) = \overline{D}(c_0, s)$. By considering the Newton polygon of

$$\phi(z) - z = c_0 + (c_1 - 1)z + \sum_{i \geq 2} c_i z^i,$$

it suffices to show that $|c_1 - 1| r \geq |c_0|$ or, if $d \geq 2$, that $|c_d| r^d \geq |c_0|$.

Since $\phi(U) \cap U \neq \varnothing$, we have either $\phi(U) \subseteq U$ or $\phi(U) \supseteq U$. Thus, we consider four cases: that $\phi(U) \subsetneq U$, that $\phi(U) \supsetneq U$, that $d \geq 2$, or that $d = 1$ and $|\phi'(0) - 1| = 1$.

If $\phi(U) \subsetneq U$, then $|c_1| r \leq s < r$, implying that $|c_1| < 1$, and hence $|c_1 - 1| = 1$. Thus, $|c_1 - 1| r = r > s \geq |c_0|$, as desired.

If $\phi(U) \supsetneq U$, then $|c_d|r^d = s \geq |c_0|$, and we are done if $d \geq 2$. If $d = 1$, then $|c_1| = s/r > 1$, and hence $|c_1 - 1| = s/r$. Again, then, we have $|c_1 - 1|r = s \geq |c_0|$.

If $d \geq 2$, then by the preceding cases, we are done unless $\phi(U) = U$. In that case, we have $|c_d|r^d = r \geq |c_0|$, as desired.

Finally, if $d = 1$ and $|\phi'(0) - 1| = 1$, then we have $|c_1 - 1| = 1$ and, in addition, $|c_1| \leq 1$, so that $s = |c_1|r \leq r$. Thus, $|c_1 - 1|r = r \geq s \geq |c_0|$, and we are done. □

Proposition 4.17 can be rephrased as follows: if an analytic function ϕ on a disk U fails to have a fixed point, then either $U \cap \phi(U) = \varnothing$, or ϕ must map U bijectively onto itself with $|\phi'(a) - 1| < 1$ for all $a \in U$. Even in the latter case, however, ϕ may *still* have a fixed point in U. For example, consider the maps

$$p^2 + (1 - p^2)z, \quad 2p^2 + (1 - p^2)z - p^2 z^2, \quad p + (1 - p^2)z \in \mathbb{C}_p[[z]]$$

on the disk $\overline{D}(0, 1)$. All three are bijective mappings from $\overline{D}(0, 1)$ to itself with $|\phi'(z) - 1| < 1$ everywhere on the disk. However, the first has exactly one fixed point in the disk, at 1; the second has two, at 1 and -2; and the third has none.

Note that when ϕ *does* have a fixed point in U, the fixed point need not be unique. In some cases, however, we can both guarantee uniqueness and classify the fixed point, as the following result shows.

Theorem 4.18. *Let $r > 0$, let $a \in \mathbb{C}_v$, let $U = D(a, r)$, and let $\phi \in \mathbb{C}_v[[z - a]]$ be a power series converging on U such that $\phi - a$ has finite Weierstrass degree $d = \operatorname{wdeg}_U(\phi - a)$. Suppose either that*

(a) *$\phi(U) \subsetneq U$, or*

(b) *$\phi : U \to U$ is onto, with $d \geq 2$.*

Then there is a unique attracting fixed point $b \in U$, and for every $x \in U$, we have $\lim_{n\to\infty} \phi^n(x) = b$. Moreover, in case (b)*, we must have $r \in |\mathbb{C}_v^\times|$, i.e., U is a* rational *open disk.*

On the other hand, if

(c) *$\phi(U) \supsetneq U$ and $d = 1$,*

then there is a unique repelling fixed point $b \in U$, and for every $x \in U \smallsetminus \{b\}$, there is some $n \geq 1$ such that $\phi^n(x) \notin U$.

Finally, if

(d) *$\phi : U \to U$ is one-to-one and onto,*

then any fixed points in U are indifferent.

Proof. In each of the first three cases, Proposition 4.17 gives us the existence of a fixed point $b \in U$. After recentering and changing coordinates, we may assume that $a = b = 0$. Write $\phi(U) = D(0, s)$.

(a) By Proposition 3.20, we have $|\phi(x)| \le (s/r)|x|$ for all $x \in U$. Because $\phi(U) \subsetneq U$, we have $s < r$, and therefore $\lim_{n\to\infty} \phi^n(x) = 0$ for all $x \in U$. The fact that 0 is both attracting and the unique fixed point in U is then immediate.

(b) Write $\phi(z) = \sum_{i\ge 0} c_i z^i$. Our assumptions so far imply that $c_0 = 0$ and that $|c_i|r^i \le r$ for all $i \ge 1$. (In fact, $|c_i|r^i < r$ for $1 \le i \le d-1$, and $|c_d|r^d = r$, but we do not need those facts here.) Thus, for any positive $t < r$, we have $|c_i|t^i < t$ for all $i \ge 1$. In other words, setting $V = D(0, t)$, we have $\phi(V) \subsetneq V$. By part (a), all points of V are attracted to a unique attracting fixed point, which must be 0, since we already know 0 is fixed. Since U is the union of all such disks V for $t < r$, we are done.

(c) By Proposition 3.20, we have $|\phi(x)| = (s/r)|x|$ for all $x \in U$. Because $\phi(U) \supsetneq U$, we have $s > r$, and therefore for any

$$x \in D(0, r) \smallsetminus \{0\},$$

there is some $n \ge 1$ such that $|\phi^n(x)| > r$, as desired. The fact that 0 is both repelling and the unique fixed point in U is then immediate.

(d) We have $\phi : U \to U$ onto, with $\operatorname{wdeg}_U(\phi) = 1$. Then, by Proposition 3.20,

$$\big|\phi(x) - \phi(y)\big| = |x - y| \quad \text{for all} \quad x, y \in U,$$

precluding the possibility of an attracting or repelling fixed point lying in U, by Proposition 4.3(a)–(b). □

One consequence of Theorem 4.18 is that there are only two qualitative types of dynamics posssible for a convergent power series $\phi : U \to U$ from an open disk U into itself. The first is that everything in U is ultimately drawn to a unique attracting fixed point, and the second is that $\phi : U \to U$ is bijective. When we discuss Rivera-Letelier's classification in Section 9.4, after having developed a notion of components of the Fatou set, we will generalize this dichotomy to all periodic Fatou components, and we will put further restrictions on the resulting dynamical behavior.

The reader may have noticed that Theorem 4.18 does not address the case that $\phi(U) \supsetneq U$ with $\operatorname{wdeg}_U(\phi) \ge 2$. In addition, it does not specify the number or behavior of the indifferent fixed points in part (d). These cases are explored in Exercises 4.22, 4.24, and 4.25.

The study of dynamics on open disks is particularly helpful for understanding maps of good reduction, as our next result illustrates.

Proposition 4.19. *Let $\phi \in \mathbb{C}_v(z)$ be a rational function of explicit good reduction and of degree $d \geq 2$. Let $a \in \mathbb{P}^1(\mathbb{C}_v)$, let $U \subseteq \mathbb{P}^1(\mathbb{C}_v)$ be the residue class containing a, and let $V \subseteq \mathbb{P}^1(\mathbb{C}_v)$ be the residue class containing $\phi(a)$. Then $\phi(U) = V$. Moreover, if $U = V$, then U contains a fixed point of ϕ.*

Proof.

Step 1. We claim that if $\phi \in \mathbb{C}_v(z)$ has explicit good reduction, and if $\phi(0) \in D(0,1)$, then $\phi(D(0,1)) = D(0,1)$, and $D(0,1)$ contains a fixed point of ϕ.

To see this, write

$$\phi(z) = \frac{f(z)}{g(z)} = \frac{b_0 + b_1 z + \cdots + b_d z^d}{c_0 + c_1 z + \cdots + c_d z^d},$$

with $b_i, c_i \in \mathcal{O}$ for all i. (Note that b_d or c_d may be zero.) Because ϕ has explicit good reduction, we may assume after multiplying f and g by an appropriate constant that $|b_j| = |c_k| = 1$ for some $j, k \geq 0$. In addition, we must have $|c_0| = 1$, or else $\operatorname{wdeg}_U g \geq 1$, and g would have a root in $U = D(0,1)$, contradicting explicit good reduction. Moreover, because $b_0/c_0 = \phi(0) \in D(0,1)$, we have $|b_0| < 1$.

Thus, $|g(x)| = 1$ for all $x \in D(0,1)$, and hence $|\phi(x)| = |f(x)| < 1$, proving that $\phi(D(0,1)) \subseteq D(0,1)$. Conversely, for all $y \in D(0,1)$, we have $\phi(z) - y = h_y(z)/g(z)$, where

$$h_y(z) = f(z) - yg(z) = \sum_{i=0}^{d} (b_i - c_i y) z^i.$$

Since $|b_0 - c_0 y| < 1 = |b_j - c_j y|$, the Newton polygon shows that h_y has a root in $D(0,1)$. Hence, $\phi(D(0,1)) = D(0,1)$.

In addition, $\phi(z) - z = h(z)/g(z)$, where

$$h(z) = f(z) - zg(z) = b_0 + \left[\sum_{i=1}^{d} (b_i - c_{i-1}) z^i\right] - c_d z^{d+1}.$$

If $\|h\|_{D(0,1)} < 1$, then the reductions $\overline{f}, \overline{g}$ would satisfy $\overline{f} = z\overline{g}$, again contradicting explicit good reduction. Therefore, $\|h\|_{D(0,1)} = 1$; but then because $|b_0| < 1$, the Newton polygon shows that h has a root $a \in D(0,1)$. That is, a is the desired fixed point, proving the claim.

Step 2. Choose $h_1, h_2 \in \mathrm{PGL}(2, \mathcal{O})$ such that $h_1(a) = h_2(\phi(a)) = 0$. By Propositions 4.8 and 4.10(a), $h_2 \circ \phi \circ h_1^{-1}$ has explicit good reduction. By the claim of Step 1, the same function maps $D(0,1)$ onto itself. Applying h_1 and h_2^{-1}, it follows that $\phi(U) = V$.

Step 3. If $U = V$, choose $h \in \mathrm{PGL}(2, \mathcal{O})$ such that $h(a) = 0$. As in Step 2, $h \circ \phi \circ h^{-1}$ has explicit good reduction and maps $D(0,1)$ to itself.

By the claim, it fixes a point b of $D(0,1)$ Applying h, it follows that ϕ fixes $h^{-1}(b) \in U$. □

We close this section with an algorithm for determining whether a given non-archimedean rational function has potential good reduction, and in that case, providing a coordinate exhibiting explicit good reduction. See also [**21**, **22**].

Theorem 4.20. *Let $\phi \in \mathbb{C}_v(z)$ be a rational function of degree $d \geq 2$.*

(a) *If ϕ has any repelling periodic points, then ϕ does not have potential good reduction.*

(b) *If ϕ has an indifferent fixed point $x_0 \in \mathbb{P}^1(\mathbb{C}_v)$, then choose*

$$x_1 \in \phi^{-1}(x_0) \smallsetminus \{x_0\} \quad \text{and} \quad x_2 \in \phi^{-1}(x_1).$$

(c) *If all fixed points of ϕ are attracting, then they are distinct; choose three distinct attracting fixed points $x_0, x_1, x_2 \in \mathbb{P}^1(\mathbb{C}_v)$.*

In cases (b) *and* (c)*, let $h \in \mathrm{PGL}(2, \mathbb{C}_v)$ be the unique linear fractional transformation such that $h(x_0) = 0$, $h(x_1) = 1$, and $h(x_2) = \infty$, and let $\psi = h \circ \phi \circ h^{-1}$. Then ϕ has potential good reduction if and only if ψ has explicit good reduction.*

Proof.

(a) If ϕ has potential good reduction, then for the purposes of statement (a), we may assume, after a $\mathrm{PGL}(2, \mathbb{C}_v)$-change of coordinates, that ϕ has explicit good reduction. If ϕ has a periodic point at $a \in \mathbb{P}^1(\mathbb{C}_v)$, then since any iterate ϕ^n also has explicit good reduction by Proposition 4.8, we may assume that a is fixed. By Proposition 4.19, the residue class U containing a satisfies $\phi(U) = U$. If $\phi : U \to U$ is bijective, then a must be indifferent by Theorem 4.18(d); otherwise, a is attracting, by Theorem 4.18(a).

(b) If ψ has explicit good reduction, then ϕ has potential good reduction, and we are done. Conversely, suppose ϕ has potential good reduction. After a change of coordinates, we may assume that ϕ already has explicit good reduction. We need to show that ψ has explicit good reduction. It suffices to show that x_0, x_1, and x_2 lie in distinct residue classes of $\mathbb{P}^1(\mathbb{C}_v)$.

Let U_i be the residue class containing x_i, for $i = 0, 1, 2$. By Proposition 4.19, $\phi(U_0) = \phi(U_1)$ is the residue class containing

$$\phi(x_0) = \phi(x_1) = x_0,$$

and hence $\phi(U_0) = \phi(U_1) = U_0$. By Theorem 4.18(b), $\phi : U_0 \to U_0$ must be injective, or else x_0 could not be an *indifferent* fixed point. Thus, $x_1 \notin U_0$, and hence $U_1 \neq U_0$.

If $x_2 \in U_0$, then Proposition 4.19 implies $\phi(U_0) = U_1 \neq U_0$, a contradiction. Similarly, if $x_2 \in U_1$, then $\phi(U_1) = U_1 \neq U_0$, which is again a contradiction. Thus, the residue class U_2 containing x_2 is also distinct from U_0 and U_1, as desired.

(c) As in the proof of (b), we may assume that ϕ has explicit good reduction, and we must show that the residue classes U_0, U_1, U_2 containing x_0, x_1, x_2 are distinct. By Proposition 4.19, we have

$$\phi(U_i) = U_i \quad \text{for} \quad i = 0, 1, 2.$$

Hence by Theorem 4.18(d), the Weierstrass degree of ϕ on each U_i must be at least 2. By Theorem 4.18(b), each U_i contains a *unique* attracting fixed point. In particular, U_0, U_1, and U_2 must be distinct, as desired. □

The criteria in cases (b) and (c) of Theorem 4.20 can be generalized to certain other combinations; see Exercise 4.27. However, not all seemingly comparable conditions work, for two main reasons. First, given a map ϕ of explicit good reduction with an attracting periodic point a, there are infinitely many elements of the backward orbit of a in U, even though there can be no other periodic points in the residue class U of a. Second, given a map ϕ of explicit good reduction with an indifferent fixed point a, there can be many other periodic points—even fixed points—in U, even though there can be no preimages of a in the residue class U of a. See Exercises 4.28–4.30 for further details.

Exercises for Chapter 4

4.1. Let K be a complete non-archimedean field, and let

$$\phi(z) = \sum_{n=1}^{\infty} c_1 z^n \in K[[z]]$$

be a power series converging on some disk $D(0, r)$, with $|c_1| \neq 0, 1$. Prove that ϕ is locally conjugate over K to $c_1 z$.

(*Suggestion*: If $|c_1| < 1$, let $h_m(z) = c_1^{-m}\phi^m(z)$ for each $m \geq 1$, and show that $h := \lim_{m\to\infty} h_m$ converges in $K[[z]]$ and works as the conjugacy on a small enough disk. If $|c_1| > 1$, consider ϕ^{-1}.)

4.2. Let K be a complete non-archimedean field, let

$$\phi(z) = \sum_{n=1}^{\infty} c_1 z^n \in K[[z]]$$

be a power series converging on some disk $D(0, r)$, with $|c_1| > 1$. Prove that there is a function F defined over K and meromorphic on $\mathbb{C}_v$ (i.e., $F = f/g$, where $f, g \in K[[z]]$ converge on $\mathbb{C}_v$ and g is not identically zero) such that

$$F(c_1 z) = \phi\big(F(z)\big) \quad \text{for all } z \in \mathbb{C}_v.$$

(*Suggestion*: Using the conjugacy h of Exercise 4.1, define F on successively larger domains by $F(z) = \phi^n(h^{-1}(c_1^{-n} z))$.)

4.3. Let K be a complete non-archimedean field, let $p := \operatorname{char} K \geq 0$, and let $j \geq 2$ be an integer. Let

$$\phi(z) = \sum_{n=j}^{\infty} c_n z^n \in K[[z]]$$

be a power series converging on some disk $D(0, r)$, with $c_j \neq 0$.

(a) If $p \nmid j$, prove that ϕ is locally conjugate over K to $c_j z^j$. (*Suggestion*: Define $h_m(z) = (\phi^m(z))^{1/j^m}$ for each $m \geq 1$. Show that $h_m \in K[[z]]$, and that the sequence $\{h_m\}$ approaches a limit h converging on a small enough disk.)

(b) If $p^e \| j$ for some integer $e \geq 1$, but $c_n \neq 0$ for some $n \geq j$ with $p^e \nmid j$, prove that ϕ is *not* locally conjugate, even over $\mathbb{C}_v$, to either $c_j z^j$ or z^j.

(c) If $\phi = \psi(z^{p^e})$ for some $\psi \in K[[z]]$, and either $j > p^e$ or $|c_j| \neq 1$, prove that ϕ is locally conjugate over K to $c_j z^j$.

(d) If $\phi = \psi(z^{p^e})$ for some $\psi \in K[[z]]$, and $j = p^e$ with c_j a root of unity, prove that ϕ is *not* locally conjugate, via a conjugacy of the form $h(z) = z + O(z^2)$, to either $c_j z^j$ or z^j, even over $\mathbb{C}_v$. However, prove that if we allow

$$h(z) = az + O(z^2) \quad \text{with} \quad 0 < |a| < 1,$$

then ϕ *is* locally conjugate to $c_j z^j$ over K.

4.4. With notation as in Exercise 4.3, if ϕ is locally conjugate over K to $c_j z^j$, prove that ϕ is locally conjugate over $K(c_j^{1/(j-1)})$ to z^j.

4.5. Verify the claims made at the start of Section 4.3 about the reduction map $\mathbb{P}^1(K) \to \mathbb{P}^1(k)$, as follows.

(a) Prove that the map is defined: every point in $\mathbb{P}^1(K)$ can be written as $[x, y]$ with $x, y \in \mathcal{O}_K$ such that $\max\{|x|, |y|\} = 1$.

(b) Prove that the map is well defined: if $x_1, y_1, x_2, y_2 \in \mathcal{O}_K$ with

$$[x_1, y_1] = [x_2, y_2] \quad \text{and} \quad \max\big\{|x_1|, |y_1|\big\} = \max\big\{|x_2|, |y_2|\big\} = 1,$$

then $\overline{[x_1, y_1]} = \overline{[x_2, y_2]}$.

(c) For $a \in \mathcal{O}_K \subseteq \mathbb{P}^1(K)$, prove that the reduction $\overline{a} \in \mathbb{P}^1(k)$ lies in $k \subseteq \mathbb{P}^1(k)$ and agrees with the original reduction

$$\overline{a} \in k = \mathcal{O}_K / \mathcal{M}_K.$$

(d) For $a \in \mathbb{P}^1(K) \setminus \mathcal{O}_K$, prove that $\overline{a} = \overline{\infty} \in \mathbb{P}^1(k)$.

4.6. Let $U \subseteq \mathbb{P}^1(\mathbb{C}_v)$ be a $\mathbb{P}^1(\mathbb{C}_v)$-disk containing at most one of $0, 1, \infty$. Prove that U is completely contained in a single residue class.

4.7. Verify that explicit good reduction (Definition 4.6) is a well-defined notion, as follows.

(a) Given $\phi \in K(z)$, prove that there are homogeneous polynomials $f, g \in \mathcal{O}_K[x, y]$ of degree $d = \deg(\phi)$ with no common irreducible factors in $K[x, y]$, with at least one coefficient of f or g having absolute value 1, and with $\phi(z) = f(z, 1)/g(z, 1)$.

(b) Given ϕ, f, g as in part (a), prove that the three conditions of Definition 4.6 are equivalent.
   (i) The only common zero of $\overline{f}$ and $\overline{g}$ in $\overline{k} \times \overline{k}$ is $(0, 0)$.
   (ii) $\overline{f}$ and $\overline{g}$ have no common zeros in $\mathbb{P}^1(\overline{k})$.
   (iii) $\overline{f}$ and $\overline{g}$ are releatively prime in $k[x, y]$.

(c) Given $\phi \in K(z)$, and given two pairs f_1, g_1 and f_2, g_2 of homogeneous polynomials satisfying the conditions of part (a), prove that there is a constant $c \in \mathcal{O}_K^\times$ such that $f_2 = cf_1$ and $g_2 = cf_2$.

(d) For ϕ, f_1, g_1, f_2, g_2 as in part (c) above, prove that the conditions of part (b) hold for the pair f_1, g_1 if and only if they hold for f_2, g_2, and that in that case the reduced map

$$\overline{\phi} = [\overline{f}_1, \overline{g}_1] : \mathbb{P}^1(\overline{k}) \to \mathbb{P}^1(\overline{k})$$

coincides with $[\overline{f}_2, \overline{g}_2]$.

(e) If ϕ has explicit good reduction, and if $z, w \in \mathbb{P}^1(\mathbb{C}_v)$ have the same reduction $\overline{z} = \overline{w} \in \mathbb{P}^1(\overline{k})$, prove that $\overline{\phi(z)} = \overline{\phi(w)} = \overline{\phi}(\overline{z})$.

4.8. For readers with a backgound in algebraic geometry, this exercise provides a scheme-theoretic definition of good reduction.

(a) Given $\phi \in K(z)$, prove that ϕ has explicit good reduction if and only if it extends from a $\operatorname{Spec} K$-morphism of schemes $\mathbb{P}^1_K \to \mathbb{P}^1_K$ to a $\operatorname{Spec} \mathcal{O}_K$-morphism of schemes $\mathbb{P}^1_{\mathcal{O}_K} \to \mathbb{P}^1_{\mathcal{O}_K}$.

(b) Prove that ϕ has good reduction if and only if there is a $\operatorname{Spec} \mathcal{O}_K$-scheme X which is $\operatorname{Spec} K$-isomorphic to $\mathbb{P}^1_{\mathcal{O}_K}$ such that ϕ extends to a $\operatorname{Spec} \mathcal{O}_K$-morphism from X to itself.

4.9. For any two positive integers $d_1, d_2 \geq 1$, find rational functions $\phi_1, \phi_2 \in K(z)$ with $\deg(\phi_i) = d_i$ such that ϕ_1 and ϕ_2 both have bad reduction, but $\phi_1 \circ \phi_2$ has explicit good reduction.

4.10. Prove Proposition 4.10(a): A nonconstant polynomial $\phi \in K[z]$ has explicit good reduction if and only if $\phi \in \mathcal{O}_K[z]$ and the lead coefficient has absolute value 1.

4.11. Prove Proposition 4.10(b): $\phi \in \mathrm{PGL}(2, K)$ has explicit good reduction if and only if $\phi \in \mathrm{PGL}(2, \mathcal{O}_K)$.

4.12. Let $f_1, f_2 \in K[x, y]$ be homogeneous polynomials of degree $d \geq 1$, and let $g_1, g_2 \in K[x, y]$ be homogeneous polynomials of degree $e \geq 1$. Define $\Phi, \Psi : K^2 \to K^2$ by $\Phi := (f_1, f_2)$ and $\Psi := (g_1, g_2)$. Prove that

$$\mathrm{Res}\left(\Phi \circ \Psi\right) = \mathrm{Res}(\Phi)^e \, \mathrm{Res}(\Psi)^{d^2},$$

where Res denotes the homogeneous resultant.

(*Note*: This is a special case of [**89**, Theorem IX.3.13].)

4.13. Let E be an elliptic curve in Weierstrass form as in (4.3), and let ϕ be the flexible Lattès map of equation (4.4) for [2]. Prove that the resultant of ϕ is equal to Δ^2, where

$$\Delta = -b_2^2 b_8 - 8b_4^3 - 27b_6^2 + 9b_2 b_4 b_6$$

is the discriminant of E.

4.14. Let $a \in \mathbb{C}_v$ with $0 < |a| < 1$, and let

$$\phi(z) = \frac{z^4 - 8a^4 z - a^4}{4z^3 + z^2 + 4a^4}$$

be the Lattès map of Example 4.15. Prove that ϕ has bad reduction.

(*Suggestion*: Theorem 4.20 may be useful.)

4.15. Prove Theorem 4.16(a).

4.16. In this exercise, you will prove some claims made in Step 1 of the proof of Theorem 4.16(b). All notation here comes from that proof.

(a) By taking an appropriate limit of the expressions in equation (4.6), prove that there is a power series

$$g(z) \in z^3 + z^4 \mathcal{O}[[z]]$$

such that $g(z) = f(z, g(z))$, where $f(z, w)$ is given by equation (4.5).

(b) Prove that there are power series $A, B \in I \subseteq R := \mathcal{O}[[z_1, z_2]]$ such that for any $z_1, z_2 \in D(0, 1)$, the line L through $(z_1, g(z_1))$ and $(z_2, g(z_2))$—or tangent to E at this point if $z_1 = z_2$—is given by $w = Az + B$.

(c) Prove that $D_2 \in I$ and $D_3 \in 1 + I$, where D_2 and D_3 are as in equation (4.8).

(d) Let $N \in \mathbb{Z}$. Use the existence of the power series $F \in z_1 + z_2 + I$ of equation (4.9) to prove that there is a power series

$$H_N(z) \in Nz + z^2 \mathcal{O}[[z]]$$

such that

$$[N]\big(z, g(z)\big) = \Big(H_N(z), g\big(H_N(z)\big)\Big) \quad \text{for any } z \in D(0,1),$$

where $[N] : E \to E$ is multiplication-by-N.

4.17. Let E be an elliptic curve defined over a field K, let $\alpha : E \to \mathbb{P}^1$ be the two-to-one quotient map identifying P with $-P$, and let $m \geq 2$ be an integer. Let $\phi_1 \in K(z)$ be the flexible Lattès map corresponding to $[m]$, and let $\phi_2 \in K(z)$ be the flexible Lattès map corresponding to $[-m]$. That is,

$$\phi_1(\alpha(P)) = \alpha([m]P) \quad \text{and} \quad \phi_2(\alpha(P)) = \alpha([-m]P).$$

Prove that $\phi_1 = \phi_2$.

4.18. Let $F \in R := \mathcal{O}[[z_1, z_2]]$ be the power series of equation (4.9), in the proof of Theorem 4.16(b). Prove that

$$F \in z_1 + z_2 + J, \quad \text{where } J := \langle z_1 z_2 \rangle \subsetneq I.$$

(*Note*: To prove Theorem 4.16(b), we needed only the weaker statement that $F \in z_1 + z_2 + I$, where $I := \langle z_1^2, z_1 z_2, z_2^2 \rangle$. F is called the *formal group law* of the elliptic curve E.)

4.19. Prove Theorem 4.16(b) in the case that $c \in \alpha(E[m^n + 1] \cap E[2])$.

4.20. Suppose $\operatorname{char} \mathbb{C}_v \neq 2$. Let E be an elliptic curve defined over $\mathbb{C}_v$, let $\alpha : E \to \mathbb{P}^1$ be the quotient map identifying P with $-P$, let $N \geq 2$, and let ϕ be the flexible Lattès map associated with $[N] : E \to E$.

(a) Prove that the set of critical points of ϕ is

$$\alpha\big(E[2N] \smallsetminus E[2]\big).$$

(*Suggestion*: Use the fact that $[N] : E \to E$ is unramified.)

(b) Prove that every critical point of ϕ is preperiodic but not periodic.

4.21. Do Exercise 4.20 over in the case that $\operatorname{char} \mathbb{C}_v = 2$.

4.22. In this exercise you will investigate the possibilities left open by part (d) of Theorem 4.18. Let $U := D(0,1)$.

(a) Let $n \geq 0$ be an integer. Find a power series $\phi \in \mathbb{C}_v[[z]]$ converging on U with exactly n indifferent fixed points in U.

(b) Find another $\phi \in \mathbb{C}_v[[z]]$ converging on U whose number of indifferent fixed points in U is countably infinite.

(c) Find another $\phi \in \mathbb{C}_v[[z]]$ converging on U whose number of indifferent fixed points in U is uncountable.

4.23. Prove Proposition 4.17 in the case that $U = D(a, r)$ is an open disk.

4.24 (The missing case from Theorem 4.18)**.** Let $r > 0$, let $a \in \mathbb{C}_v$, let $U = D(a, r)$, and let $\phi \in \mathbb{C}_v[[z - a]]$ be a power series converging on U with finite Weierstrass degree $d = \mathrm{wdeg}_U(\phi)$. Suppose that

$$\phi(U) \supsetneq U \quad \text{and} \quad d \geq 2.$$

Prove that ϕ has exactly d fixed points in U, counting multiplicity.

4.25. Let $U := D(0, 1)$. In each of the following scenarios, find $\phi \in \mathbb{C}_v[[z]]$ convergent on U with the given properties.

(a) ϕ has some repelling, some attracting, and some indifferent fixed points in U.

(b) ϕ has some some attracting and some indifferent, but no repelling, fixed points in U.

(c) ϕ has some some attracting and some repelling, but no indifferent, fixed points in U.

(d) ϕ has some some indifferent and some repelling, but no attracting, fixed points in U.

(e) ϕ has some some attracting, but no indifferent or repelling, fixed points in U.

(f) ϕ has some some indifferent, but no attracting or repelling, fixed points in U.

(g) ϕ has some some repelling, but no attracting or indifferent, fixed points in U.

(h) ϕ has no fixed points in U.

4.26. Let $f \in \mathbb{C}_v[z]$ be a polynomial of degree $d \geq 2$ with lead coefficient c, and let $b \in \mathbb{C}_v$ be a fixed point of f. Choose $a \in \mathbb{C}_v$ with $a^{d-1} = c$, and define

$$h(z) := a(z - b) \quad \text{and} \quad g := h \circ f \circ h^{-1}.$$

Note that h is chosen so that g is monic and fixes 0. Prove that f has potential good reduction if and only if g has explicit good reduction.

(*Suggestion*: Use Proposition 4.10(a). This exercise is [**15**, Corollary 4.6].)

4.27. Let $\phi \in \mathbb{C}_v(z)$ be a rational function of degree at least 2. Let x_0, x_1, x_2 be three distinct points in $\mathbb{P}^1(\mathbb{C}_v)$ such that one of the following three conditions holds:

(a) x_0, x_1, and x_2 are all attracting periodic points.

(b) x_0 and x_1 are attracting periodic points, and x_2 is an indifferent periodic point.

(c) x_0 is an attracting periodic point, x_1 is an indifferent fixed point, and $\phi^n(x_2) = x_1$ for some $n \geq 1$.

Let $h \in \mathrm{PGL}(2, \mathbb{C}_v)$ be the unique linear fractional transformation such that

$$h(x_0) = 0, \quad h(x_1) = 1, \quad \text{and} \quad h(x_2) = \infty,$$

and let $\psi := h \circ \phi \circ h^{-1}$. Prove that ϕ has potential good reduction if and only if ψ has explicit good reduction.

4.28. Fix $a \in \mathbb{C}_v$ with $0 < |a| < 1$, and let $\phi(z) := z^2 + (1-a)z$. Let $x_0 = 0$, $x_1 = a$, and $x_2 = \infty$.

(a) Prove that x_0 and x_1 are indifferent fixed points of ϕ, while x_2 is an attracting fixed point.

(b) Prove that ϕ has explicit good reduction.

(c) Let $h(z) := a^{-1}z$, and let $\psi = h \circ \phi \circ h^{-1}$. (Note that h is the linear fractional transformation with $h(x_0) = 0$, $h(x_1) = 1$, and $h(x_2) = \infty$.) Prove that ψ does *not* have explicit good reduction.

4.29. Fix $a \in \mathbb{C}_v$ with $0 < |a| < 1$, and let $\phi(z) := \frac{z^2 - az}{z+1}$. Let $x_0 = 0$, $x_1 = a$, and $x_2 = \infty$.

(a) Prove that x_0 is an attracting fixed point of ϕ, x_2 is an indifferent fixed point, and $x_1 \in \phi^{-1}(x_0) \setminus \{x_0\}$.

(b) Prove that ϕ has explicit good reduction.

(c) Let $h(z) := a^{-1}z$, and let $\psi = h \circ \phi \circ h^{-1}$. (Note that h is the linear fractional transformation with $h(x_0) = 0$, $h(x_1) = 1$, and $h(x_2) = \infty$.) Prove that ψ does *not* have explicit good reduction.

4.30. For each of the following scenarios, give an example of a rational function $\phi \in \mathbb{C}_v(z)$ of degree at least 2 and three distinct points $x_0, x_1, x_2 \in \mathbb{P}^1(\mathbb{C}_v)$ such that ϕ has explicit good reduction, but

$$\psi := h \circ \phi \circ h^{-1}$$

does *not* have explicit good reduction, where $h \in \mathrm{PGL}(2, \mathbb{C}_v)$ is the linear fractional transformation such that $h(x_0) = 0$, $h(x_1) = 1$, and $h(x_2) = \infty$.

(a) x_0, x_1, and x_2 are all indifferent periodic points.
(Try to do it when they are all fixed or again when they all belong to the same cycle.)

(b) x_0 and x_1 are indifferent periodic points, and $\phi(x_2) = x_0$.
(Try to do it when x_0 and x_1 are both fixed or again when they both belong to the same cycle.)

(c) x_0 is an attracting periodic point, x_1 is an indifferent periodic point that is *not* fixed, and $\phi(x_2) = x_1$.

(d) x_0 and x_1 are attracting periodic points, and $\phi(x_2) = x_0$

Chapter 5

Fatou and Julia Sets

5.1. The spherical metric

Definition 5.1. The *spherical metric* on $\mathbb{P}^1(\mathbb{C}_v)$ is the function

$$d_{\text{sph}} : \mathbb{P}^1(\mathbb{C}_v) \times \mathbb{P}^1(\mathbb{C}_v) \to [0, \infty)$$

given by

$$d_{\text{sph}}\big([x,y],[u,v]\big) := \frac{|xv - yu|}{\max\{|x|,|y|\}\max\{|u|,|v|\}} \tag{5.1}$$

for $[x,y],[u,v] \in \mathbb{P}^1(\mathbb{C}_v)$. We sometimes call d_{sph} the *chordal metric* on $\mathbb{P}^1(\mathbb{C}_v)$.

In affine coordinates, for $z, w \in \mathbb{C}_v \subseteq \mathbb{P}^1(\mathbb{C}_v)$, we have

$$d_{\text{sph}}(z,w) := \frac{|z-w|}{\max\{1,|z|\}\max\{1,|w|\}}. \tag{5.2}$$

We leave it to the reader to check that the two definitions agree on $\mathbb{C}_v \times \mathbb{C}_v$; see Exercise 5.1. We will see in Proposition 5.3(a) that d_{sph} is indeed a metric on $\mathbb{P}^1(\mathbb{C}_v)$.

Warning 5.2. $\mathbb{P}^1(\mathbb{C}_v)$-disks are generally *not* disks in the metric space $(\mathbb{P}^1(\mathbb{C}_v), d_{\text{sph}})$. As we will see in Proposition 5.3, for example, any d_{sph}-disk containing the points 0 and 1 must have radius at least 1 and therefore is the whole of $\mathbb{P}^1(\mathbb{C}_v)$.

On the other hand, any d_{sph}-disk is either $\mathbb{P}^1(\mathbb{C}_v)$ or a $\mathbb{P}^1(\mathbb{C}_v)$-disk, and any $\mathbb{P}^1(\mathbb{C}_v)$-disk contained in a single residue class is a d_{sph}-disk. In addition, the *topology* induced by the metric d_{sph} coincides with the topology generated by the set of $\mathbb{P}^1(\mathbb{C}_v)$-disks.

Proposition 5.3. *Let $\mathcal{O}$ and $\mathcal{M}$ be the ring of integers and the maximal ideal, respectively, of $\mathbb{C}_v$.*

(a) *d_{sph} is a non-archimedean metric on $\mathbb{P}^1(\mathbb{C}_v)$. That is, for all $a,b,c \in \mathbb{P}^1(\mathbb{C}_v)$,*
   (i) *$d_{\mathrm{sph}}(a,b) \in [0,\infty)$, with $d_{\mathrm{sph}}(a,b) = 0$ if and only if $a = b$.*
   (ii) *$d_{\mathrm{sph}}(b,a) = d_{\mathrm{sph}}(a,b)$.*
   (iii) *$d_{\mathrm{sph}}(a,c) \le \max\{d_{\mathrm{sph}}(a,b), d_{\mathrm{sph}}(b,c)\}$.*

(b) *$d_{\mathrm{sph}}(z,w) = |z-w|$ for all $z,w \in \mathcal{O}$.*

(c) *$d_{\mathrm{sph}}(z,w) = 1$ for all $z,w \in \mathbb{P}^1(\mathbb{C}_v)$ that have different reductions modulo $\mathcal{M}$, i.e., for which $\overline{z} \ne \overline{w}$.*

(d) *$d_{\mathrm{sph}}(z,w) < 1$ for all $z,w \in \mathbb{P}^1(\mathbb{C}_v)$ with the same reduction $\overline{z} = \overline{w}$.*

(e) *$d_{\mathrm{sph}}(\eta(z),\eta(w)) = d_{\mathrm{sph}}(z,w)$ for all $z,w \in \mathbb{P}^1(\mathbb{C}_v)$ and all $\eta \in \mathrm{PGL}(2,\mathcal{O})$.*

Proof.

(b) This is obvious from Definition 5.1.

(e) The group $\mathrm{PGL}(2,\mathcal{O})$ consists of linear fractional transformations $(az+b)/(cz+d)$ with $a,b,c,d \in \mathcal{O}$ and with $|ad-bc| = 1$. By Exercise 5.3, $\mathrm{PGL}(2,\mathcal{O})$ is generated by maps of the form $z \mapsto z+\alpha$ for $\alpha \in \mathcal{O}$, maps of the form $z \mapsto \beta z$ for $\beta \in \mathcal{O}^\times$, and $z \mapsto 1/z$. The reader may check in Exercise 5.5 that the spherical metric is unaffected by each such generator.

(a) For properties (i) and (ii), see Exercise 5.2. For (iii), given $a,b,c \in \mathbb{P}^1(\mathbb{C}_v)$, note that the residue field of $\mathbb{C}_v$ contains more than three (in fact, infinitely many) elements, because $\mathbb{C}_v$ is algebraically closed. Thus, there is some $w \in \mathbb{P}^1(\mathbb{C}_v)$ such that $\overline{w} \ne \overline{a}, \overline{b}, \overline{c}$. Choose $\eta \in \mathrm{PGL}(2,\mathcal{O})$ such that $\eta(w) = \infty$; see Exercise 5.4. By Proposition 4.10(b), $\eta^{-1} \in \mathrm{PGL}(2,\mathcal{O})$ has explicit good reduction, and hence $\overline{\eta(a)}, \overline{\eta(b)}, \overline{\eta(c)} \ne \overline{\infty}$. Therefore, $\eta(a), \eta(b), \eta(c) \in \mathcal{O}$. Inequality (iii) is now immediate from parts (b) and (e).

(d) After a $\mathrm{PGL}(2,\mathcal{O})$-change of coordinates η, we may assume that $w = 0$. As above, η has explicit good reduction by Proposition 4.10(b), and hence $z \in \mathcal{M}$. Thus, by part (b), $d_{\mathrm{sph}}(z,w) = |z-0| < 1$.

(c) After a $\mathrm{PGL}(2,\mathcal{O})$-change of coordinates, we may assume that $w = \infty$. By Proposition 4.10(b) again, we have $z \in \mathcal{O}$. Writing z as $[z,1]$ and ∞ as $[1,0]$, Definition 5.1 gives $d_{\mathrm{sph}}(z,\infty) = 1/1 = 1$. □

Morton and Silverman proved the following result about Lipschitz constants of rational functions in [**107**, Proposition 5.2].

Proposition 5.4. *Let $\phi \in \mathbb{C}_v(z)$. Then there is a constant $C \ge 1$ such that*

$$d_{\mathrm{sph}}(\phi(z),\phi(w)) \le C d_{\mathrm{sph}}(z,w) \quad \text{for all } z,w \in \mathbb{P}^1(\mathbb{C}_v).$$

Moreover, if ϕ has explicit good reduction, then we can choose $C = 1$.

Proof. Define $r_1, r_2, r_3, r \in (0,1]$ by

$$\begin{aligned} r_1 &:= \min\{d_{\mathrm{sph}}(\alpha,\beta) : \alpha \in \phi^{-1}(0) \quad \text{and} \quad \beta \in \phi^{-1}(\infty)\}, \\ r_2 &:= \min\{d_{\mathrm{sph}}(\alpha,\beta) : \alpha \in \phi^{-1}(1) \quad \text{and} \quad \beta \in \phi^{-1}(\infty)\}, \\ r_3 &:= \min\{d_{\mathrm{sph}}(\alpha,\beta) : \alpha \in \phi^{-1}(0) \quad \text{and} \quad \beta \in \phi^{-1}(1)\}, \\ r &:= \min\{r_1, r_2, r_3\}. \end{aligned}$$

Set $C := 1/r$, and note that $C \geq 1$. Given $z, w \in \mathbb{P}^1(\mathbb{C}_v)$, suppose $d_{\mathrm{sph}}(z,w) \geq r$. Since $d_{\mathrm{sph}}(\phi(z),\phi(w)) \leq 1$ by Proposition 5.3(c)–(d), we have

$$d_{\mathrm{sph}}\big(\phi(z),\phi(w)\big) \leq C d_{\mathrm{sph}}(z,w).$$

On the other hand, if $d_{\mathrm{sph}}(z,w) < r$, then Proposition 5.3(c) yields $\overline{z} = \overline{w}$, because $r \leq 1$. Replacing ϕ by $\phi \circ \eta$ for some appropriate $\eta \in \mathrm{PGL}(2,\mathcal{O})$, which does not affect the value of r or C, by Proposition 5.3(e), we may assume that $w = 0$, and hence that $z \in D(0,r)$. By Proposition 5.3(b) and our choice of r, the image $\phi(D(0,r))$ cannot contain more than one of $0, 1, \infty$. By Proposition 3.25, $\phi(D(0,r))$ is a $\mathbb{P}^1(\mathbb{C}_v)$-disk containing at most one of $0, 1, \infty$. It follows that $\phi(D(0,r))$ is completely contained in the residue class $\overline{\phi(0)}$; see Exercise 4.6.

Choosing $\theta \in \mathrm{PGL}(2,\mathcal{O})$ such that $\theta(\phi(0)) = 0$ and considering $\theta \circ \phi$ in place of ϕ, we may assume that $\phi(0) = 0$, and hence that

$$\phi\big(D(0,r)\big) \subseteq D(0,1).$$

Thus, by Proposition 3.20, we have $|\phi(z)| \leq (1/r)|z|$. Since $C = 1/r$ and $w = 0$ and because Proposition 5.3(b) tells us that $|\phi(z)| = d_{\mathrm{sph}}(\phi(z),\phi(w))$ and $|z| = d_{\mathrm{sph}}(z,w)$, we are done.

To prove the final statement, assuming that ϕ has explicit good reduction, it suffices to show that $r = 1$. First, for any $\alpha \in \phi^{-1}(0)$ and $\beta \in \phi^{-1}(\infty)$, we have $\overline{\phi(\alpha)} \neq \overline{\phi(\beta)}$, and therefore explicit good reduction implies $\overline{\alpha} \neq \overline{\beta}$. Thus, $d_{\mathrm{sph}}(\alpha,\beta) = 1$, by Proposition 5.3(c), and hence $r_1 = 1$.

Second, $\theta_2(z) := 1-z$ and $\theta_3(z) := z/(z-1)$ have explicit good reduction, and therefore $\theta_2 \circ \phi$ and $\theta_3 \circ \phi$ do too, by Proposition 4.8. Applying the argument of the previous paragraph to $\theta_2 \circ \phi$ and $\theta_3 \circ \phi$, we obtain $r_2 = 1$ and $r_3 = 1$, respectively. Thus, $r = 1$, as desired. □

5.2. Fatou and Julia sets

In this section, we define the Fatou and Julia sets of a rational function $\phi \in \mathbb{C}_v(z)$, using equicontinuity; recall Definition 1.20(a). As in complex dynamics, the Fatou set of ϕ is the region of orderly behavior, and the Julia set is the region of chaotic behavior.

Definition 5.5. Let $\phi \in \mathbb{C}_v(z)$ be a rational function. The (*classical*) *Fatou set* of ϕ, denoted $\mathcal{F}_{\phi,\mathrm{I}}$, is the set of all points $x \in \mathbb{P}^1(\mathbb{C}_v)$ having a neighborhood on which the family of iterates $\{\phi^n : n \geq 0\}$ is equicontinuous with respect to d_{sph}. The (*classical*) *Julia set* of ϕ, is the complement $\mathcal{J}_{\phi,\mathrm{I}} := \mathbb{P}^1(\mathbb{C}_v) \smallsetminus \mathcal{F}_{\phi,\mathrm{I}}$ of the Fatou set.

A few comments are in order. First, the equicontinuity of $\{\phi^n : n \geq 0\}$ is as a family of maps from $U \subseteq \mathbb{P}^1(\mathbb{C}_v)$ to $\mathbb{P}^1(\mathbb{C}_v)$, where *both* U and $\mathbb{P}^1(\mathbb{C})_v$ are equipped with the spherical metric. Second, Definition 5.5 conveys the idea that nearby points stay close under iteration if they lie in the Fatou set, but not if they lie in the Julia set. Third, we use the word "classical" and the subscript I to distinguish $\mathcal{F}_{\phi,\mathrm{I}}$ and $\mathcal{J}_{\phi,\mathrm{I}}$ from their analogues in the Berkovich projective line, which we define in Section 8.1. Finally, Lemma 5.7 below shows that we can use a stronger *uniformly equi-Lipschitz* property to study equicontinuity of families of non-archimedean rational functions.

Lemma 5.6. *Let $U \subseteq \mathbb{P}^1(\mathbb{C}_v)$, let $\mathcal{G}$ be a set of functions from U to $\mathbb{P}^1(\mathbb{C}_v)$, and let $\eta \in \mathrm{PGL}(2, \mathbb{C}_v)$. For each $f \in \mathcal{G}$, let $\theta_f \in \mathrm{PGL}(2, \mathbb{C}_v)$. Define*

$$\mathcal{G}' := \{\theta_f \circ f \circ \eta : f \in \mathcal{G}\}.$$

Suppose there is a constant $C \geq 1$ such that for all $f \in \mathcal{G}$, we have

$$C^{-1} d_{\mathrm{sph}}(x, y) \leq d_{\mathrm{sph}}(\theta_f(x), \theta_f(y)) \leq C d_{\mathrm{sph}}(x, y) \tag{5.3}$$

for all $x, y \in \mathbb{P}^1(\mathbb{C}_v)$. Then $\mathcal{G}$ is equicontinuous on U with respect to d_{sph} if and only if $\mathcal{G}'$ is equicontinuous on $\eta^{-1}(U)$ with respect to d_{sph}.

Proof. By applying Proposition 5.4 to η and η^{-1}, we may assume that C is large enough so that inequality (5.3) applies not only to each θ_f, but also to η. It suffices to show that if $\mathcal{G}$ is equicontinuous, then so is $\mathcal{G}'$; the converse follows by the symmetry of the hypotheses.

Given $\varepsilon > 0$, pick $\delta > 0$ for the family $\mathcal{G}$ corresponding to ε/C. For any $g \in \mathcal{G}'$, we have $g = \theta_f \circ f \circ \eta$ for some $f \in \mathcal{G}$. For any $x, y \in \eta^{-1}(U)$ satisfying $d_{\mathrm{sph}}(x, y) < \delta/C$, then because $\eta(x), \eta(y) \in U$ and $d_{\mathrm{sph}}(\eta(x), \eta(y)) < \delta$, we have

$$d_{\mathrm{sph}}\big(g(x), g(y)\big) \leq C d_{\mathrm{sph}}\Big(f\big(\eta(x)\big), f\big(\eta(y)\big)\Big) < C(\varepsilon/C) = \varepsilon. \qquad \square$$

Lemma 5.7. *Let $x_0 \in \mathbb{P}^1(\mathbb{C}_v)$, and let $\mathcal{G} \subseteq \mathbb{C}_v(z)$ be a family of rational functions. Then $\mathcal{G}$ is equicontinuous with respect to the spherical metric on some neighborhood of x_0 if and only if there is a $\mathbb{P}^1(\mathbb{C}_v)$-disk D containing x_0 and a constant $C > 0$ such that*

$$d_{\mathrm{sph}}\big(f(x), f(y)\big) \leq C d_{\mathrm{sph}}(x, y)$$

for all $x, y \in D$ and all $f \in \mathcal{G}$.

Proof. Pick $\eta \in \mathrm{PGL}(2, \mathcal{O})$ such that $\eta(0) = x_0$. For each $f \in \mathcal{G}$, pick $\theta_f \in \mathrm{PGL}(2, \mathcal{O})$ such that $\theta_f(f(x_0)) = 0$. By Proposition 5.3(e), we can use a constant of $C = 1$ in Lemma 5.6 for each θ_f, and hence we may assume without loss that $x_0 = 0$, and that $f(0) = 0$ for all $f \in \mathcal{G}$.

If C and D exist as in Lemma 5.7, then $\mathcal{G}$ is clearly equicontinuous on D, and we are done.

Conversely, suppose $\mathcal{G}$ is equicontinuous on some neighborhood U of 0. Pick $\delta > 0$ such that for all $x, y \in U$ with $d_{\mathrm{sph}}(x, y) < \delta$ and all $f \in \mathcal{G}$, we have $d_{\mathrm{sph}}(f(x), f(y)) < 1$. Without loss, δ is small enough that $\delta < 1$ and $D(0, \delta) \subseteq U$. Set $C := 1/\delta$ and $D := D(0, \delta)$.

Thus, for any $f \in \mathcal{G}$ and any $x \in D$, we have

$$d_{\mathrm{sph}}(f(x), 0) = d_{\mathrm{sph}}(f(x), f(0)) < 1,$$

and hence $\overline{f(x)} = \overline{0}$ by Proposition 5.3(c). That is, $f(D) \subseteq D(0, 1)$. Then, by Proposition 3.20, for all $x, y \in D$,

$$d_{\mathrm{sph}}\big(f(x), f(y)\big) = |f(x) - f(y)| \leq \frac{1}{\delta} \cdot |x - y| = C d_{\mathrm{sph}}(x, y). \qquad \square$$

The following lemma gives yet another characterization of equicontinuity for non-archimedean rational functions, our third so far. It involves the reduction map from $\mathbb{P}^1(\mathbb{C}_v)$ to $\mathbb{P}^1(\overline{k})$.

Lemma 5.8. *Let $x_0 \in \mathbb{P}^1(\mathbb{C}_v)$, and let $\mathcal{G} \subseteq \mathbb{C}_v(z)$ be a family of rational functions. Then $\mathcal{G}$ is equicontinuous with respect to the spherical metric on some neighborhood of x_0 if and only if there is a $\mathbb{P}^1(\mathbb{C}_v)$-disk D containing x_0 such that $\overline{f(x)} = \overline{f(x_0)}$ for every $f \in \mathcal{G}$ and every $x \in D$.*

Proof. As in the proof of Lemma 5.7, we may assume without loss that $x_0 = 0$ and that $f(0) = 0$ for all $f \in \mathcal{G}$.

If $\mathcal{G}$ is equicontinuous on a neighborhood of 0, choose a disk D and a constant C for $\mathcal{G}$ and 0 as in Lemma 5.7. Assume without loss that $C \geq 1$, and that $D \subseteq D(0, C^{-1})$. Then for all $x \in D$ and all $f \in \mathcal{G}$, we have

$$d_{\mathrm{sph}}(f(x), 0) = d_{\mathrm{sph}}(f(x), f(0)) \leq C d_{\mathrm{sph}}(x, 0) < 1,$$

and hence, by Proposition 5.3(e), $\overline{f(x)} = \overline{f(x_0)}$.

Conversely, given D satisfying the condition of Lemma 5.8, shrink D if necessary so that $D \subseteq D(0, 1)$. Write $D = \overline{D}(0, r)$ with $0 < r < 1$, or $D = D(0, r)$ with $0 < r \leq 1$. Set $C := r^{-1}$. Then for any $f \in \mathcal{G}$ we have $f(D(0, r)) \subseteq D(0, 1)$. Hence, by Proposition 3.20, for any $x, y \in D(0, r)$,

$$d_{\mathrm{sph}}\big(f(x), f(y)\big) = \big|f(x) - f(y)\big| \leq r^{-1}|x - y| = C d_{\mathrm{sph}}(x, y). \qquad \square$$

Remark 5.9 (Normality vs. equicontinuity)**.** The Arzelà–Ascoli theorem fails over fields like $\mathbb{C}_v$ that are not locally compact. For example, let

$\{a_n\}_{n\geq 0}$ be any sequence of points in $\mathbb{P}^1(\mathbb{C}_v)$ with no accumulation points. Then the family of constant functions $\{a_n : n \geq 0\}$ is equicontinuous but not normal on any open subset $U \subseteq \mathbb{P}^1(\mathbb{C}_v)$.

It turns out that if ϕ is defined over a locally compact subfield K of the non-archimedean field $\mathbb{C}_v$, then one can prove that the family $\{\phi^n : n \geq 0\}$ is indeed normal on a neighborhood of every point of the classical Fatou set. However, the proof requires some of the more advanced dynamical theorems yet to come; see Exercise 10.3. In addition, normality can fail even on the classical Fatou set if ϕ is not defined over any such locally compact subfield. In light of these complications, the classical Fatou set of a rational function over a non-archimedean field is usually defined in the literature via equicontinuity instead of normality.

We close this remark by pointing out that our definitions of the Berkovich Fatou and Julia sets in Section 8.1 involve neither equicontinuity nor normality. Instead, those definitions are inspired by the alternate description of the classical Fatou and Julia sets in Theorem 5.19.

Proposition 5.10. *Let $\phi \in \mathbb{C}_v(z)$ be a rational function with Fatou set $\mathcal{F}_{\phi,\mathrm{I}}$ and Julia set $\mathcal{J}_{\phi,\mathrm{I}}$.*

(a) *$\mathcal{F}_{\phi,\mathrm{I}}$ is an open subset of $\mathbb{P}^1(\mathbb{C}_v)$, and $\mathcal{J}_{\phi,\mathrm{I}}$ is a closed subset of $\mathbb{P}^1(\mathbb{C}_v)$.*

(b) *$\phi^{-1}(\mathcal{F}_{\phi,\mathrm{I}}) = \phi(\mathcal{F}_{\phi,\mathrm{I}}) = \mathcal{F}_{\phi,\mathrm{I}}$, and $\phi^{-1}(\mathcal{J}_{\phi,\mathrm{I}}) = \phi(\mathcal{J}_{\phi,\mathrm{I}}) = \mathcal{J}_{\phi,\mathrm{I}}$.*

(c) *For every positive integer $m \geq 1$,*

$$\mathcal{F}_{\phi^m,\mathrm{I}} = \mathcal{F}_{\phi,\mathrm{I}}, \quad \text{and} \quad \mathcal{J}_{\phi^m,\mathrm{I}} = \mathcal{J}_{\phi,\mathrm{I}}.$$

(d) *For any $\eta \in \mathrm{PGL}(2,\mathbb{C}_v)$, if we set $\psi = \eta \circ \phi \circ \eta^{-1}$, then*

$$\mathcal{F}_{\psi,\mathrm{I}} = \eta\big(\mathcal{F}_{\phi,\mathrm{I}}\big), \quad \text{and} \quad \mathcal{J}_{\psi,\mathrm{I}} = \eta\big(\mathcal{J}_{\phi,\mathrm{I}}\big).$$

Proof. (a) This is immediate from Definition 5.5, because $\mathcal{F}_{\phi,\mathrm{I}}$ is defined by a property of neighborhoods.

(b) To show $\phi^{-1}(\mathcal{F}_{\phi,\mathrm{I}}) \subseteq \mathcal{F}_{\phi,\mathrm{I}}$, given any $x_0 \in \phi^{-1}(\mathcal{F}_{\phi,\mathrm{I}})$, choose a corresponding $\mathbb{P}^1(\mathbb{C}_v)$-disk D containing $\phi(x_0)$ for the family

$$\mathcal{G} := \{\phi^n : n \geq 0\},$$

as in Lemma 5.8. Pick a $\mathbb{P}^1(\mathbb{C}_v)$-disk D' containing x_0 and which is contained in the residue class of x_0 so that $\phi(D') \subseteq D$. Then $\phi^n(D')$ is contained in the residue class of $\phi^n(x_0)$ for all $n \geq 0$. Therefore, by Lemma 5.8, the family $\mathcal{G}$ is equicontinuous on D'.

For the reverse inclusion, given any $x_0 \in \mathcal{F}_{\phi,\mathrm{I}}$, we wish to show that $\phi(x_0) \in \mathcal{F}_{\phi,\mathrm{I}}$. Pick a $\mathbb{P}^1(\mathbb{C}_v)$-disk D as in Lemma 5.8 for x_0. Then $\phi(D)$ is a disk containing $\phi(x_0)$ such that $\phi^n(\phi(D))$ is contained in the residue class of $\phi^{n+1}(x_0)$ for all $n \geq 0$. Thus, by the same lemma, the family $\{\phi^n\}$ is equicontinuous on $\phi(D)$.

The remaining equalities

$$\phi(\mathcal{F}_{\phi,\mathrm{I}}) = \mathcal{F}_{\phi,\mathrm{I}} \quad \text{and} \quad \phi^{-1}(\mathcal{J}_{\phi,\mathrm{I}}) = \phi(\mathcal{J}_{\phi,\mathrm{I}}) = \mathcal{J}_{\phi,\mathrm{I}}$$

follow immediately from $\phi^{-1}(\mathcal{F}_{\phi,\mathrm{I}}) = \mathcal{F}_{\phi,\mathrm{I}}$.

(c) The inclusion $\mathcal{F}_{\phi^m,\mathrm{I}} \supseteq \mathcal{F}_{\phi,\mathrm{I}}$ is immediate, because the family $\mathcal{G}_m := \{(\phi^m)^n : n \geq 0\}$ is contained in $\mathcal{G} = \{\phi^n : n \geq 0\}$. As for the reverse inclusion, let $C' \geq 1$ be the maximum of the Lipschitz constants from Proposition 5.4 for ϕ^i, for $i = 0, \ldots, m-1$. Given $x_0 \in \mathcal{F}_{\phi^m,\mathrm{I}}$ and $\varepsilon > 0$, pick a $\mathbb{P}^1(\mathbb{C}_v)$-disk D containing x_0 and a constant $C > 0$ as in Lemma 5.7 for the family $\mathcal{G}_m$ at x_0. Then for any $x, y \in D$ and any integer $j \geq 0$, writing $j = mn + i$ for some $i = 0, \ldots, n-1$, we have

$$d_{\mathrm{sph}}\big(\phi^j(x), \phi^j(y)\big) \leq C' d_{\mathrm{sph}}\big(\phi^{mn}(x), \phi^{mn}(y)\big) \leq C' C d_{\mathrm{sph}}(x, y),$$

yielding equicontinuity by Lemma 5.7 and proving the reverse inclusion $\mathcal{F}_{\phi^m,\mathrm{I}} \subseteq \mathcal{F}_{\phi,\mathrm{I}}$. Taking complements, we also obtain $\mathcal{J}_{\phi^n,\mathrm{I}} = \mathcal{J}_{\phi,\mathrm{I}}$.

(d) The Fatou portion of (d) is immediate by applying Proposition 5.4 to η and η^{-1} and then applying Lemma 5.6. The Julia portion then follows by taking complements. □

Proposition 5.11. *Let $\phi \in \mathbb{C}_v(z)$ be a rational function of potential good reduction. Then*

$$\mathcal{F}_{\phi,\mathrm{I}} = \mathbb{P}^1(\mathbb{C}_v) \quad \textit{and} \quad \mathcal{J}_{\phi,\mathrm{I}} = \varnothing.$$

Proof. After a $\mathrm{PGL}(2, \mathbb{C}_v)$-change of coordinates, we may assume that ϕ has explicit good reduction; by Proposition 5.10(d), this change does not affect our desired conclusion.

Given $x \in \mathbb{P}^1(\mathbb{C}_v)$, let D be any $\mathbb{P}^1(\mathbb{C}_v)$-disk containing x. By Proposition 4.8, ϕ^n has explicit good reduction for all $n \geq 0$. Therefore, by Proposition 5.4,

$$d_{\mathrm{sph}}\big(\phi^n(a), \phi^n(b)\big) \leq d_{\mathrm{sph}}(a, b)$$

for all $a, b \in D$ and all $n \geq 0$. Thus, by Lemma 5.7, $x \in \mathcal{F}_{\phi,\mathrm{I}}$. □

There are also maps without potential good reduction that still have empty classical Julia set. The following result provides an easy way to produce such maps only over fields of positive characteristic.

Proposition 5.12. *Suppose that* $\operatorname{char} \mathbb{C}_v = p > 0$. *Let $\phi \in \mathbb{C}_v(z^p)$ be an inseparable rational function. Then all periodic points of ϕ are superattracting, and $\mathcal{J}_{\phi,\mathrm{I}} = \varnothing$.*

Proof. Because $\phi' = 0$, the fact that all periodic points are superattracting is immediate.

Write $\phi(z) = \psi(z^p)$ for some $\psi \in \mathbb{C}_v(z)$. Let $C \geq 1$ be the constant from Proposition 5.4 for ψ. We claim that for all $x, y \in \mathbb{P}^1(\mathbb{C}_v)$,

$$\text{if } d_{\text{sph}}(x,y) \leq C^{-1/(p-1)}, \quad \text{then } d_{\text{sph}}\big(x^p, y^p\big) \leq C^{-1} d_{\text{sph}}(x,y).$$

From the claim, it is immediate that $d_{\text{sph}}(\phi(x), \phi(y)) \leq d_{\text{sph}}(x,y)$, from which equicontinuity follows by Lemma 5.7.

To prove the claim, given $x, y \in \mathbb{P}^1(\mathbb{C}_v)$ with $d_{\text{sph}}(x,y) < C^{-1/(p-1)}$, simply observe that

$$\begin{aligned} d_{\text{sph}}(x^p, y^p) &= \frac{|x^p - y^p|}{\max\{1, |x|^p\} \max\{1, |y|^p\}} = \frac{|x-y|^p}{\max\{1, |x|\}^p \max\{1, |y|\}^p} \\ &= d_{\text{sph}}(x,y)^p \leq C^{-1} d_{\text{sph}}(x,y). \end{aligned}$$

□

Remark 5.13. Propositions 5.11 and 5.12 illustrate just two of many ways in which the classical Julia set of a non-archimedean rational function can be empty, in contrast with the archimedean case. On the other hand, an *archimedean* rational function can have empty *Fatou* set, as noted implicitly in Theorem 1.22(d). The most famous examples of such archimedean maps are the Lattès maps presented in Section 4.4. However, there are many other such maps; see, for example, [**39**, Theorem V.1.2] as well as [**73, 112, 113**].

The following result on non-archimedean periodic points appeared, along with Corollary 5.15 below, as [**14**, Proposition 1.1 and Corollary 1.3].

Theorem 5.14. *Let $\phi \in \mathbb{C}_v(z)$ be a rational function, and let $x \in \mathbb{P}^1(\mathbb{C}_v)$ be a periodic point. If x is repelling, then $x \in \mathcal{J}_{\phi,\text{I}}$. If x is nonrepelling, then $x \in \mathcal{F}_{\phi,\text{I}}$.*

Proof. Let $m \geq 1$ be the period of x. If x is repelling, let D be any $\mathbb{P}^1(\mathbb{C}_v)$-disk containing x, and let $C > 0$ be any constant. We wish to prove that the family $\{\phi^n\}$ fails the conditions of Lemma 5.7 for C and D. Shrinking D if necessary, we may assume, by Proposition 4.3(b), that for every $y \in D \smallsetminus \{x\}$, there is some $n \geq 0$ such that $\phi^{mn}(y) \notin D$. Further shrinking D, we may also assume that D is a d_{sph}-disk (that is, a disk with respect to the spherical metric) of d_{sph}-radius $0 < r < 1$.

Pick $y \in D$ such that $0 < d_{\text{sph}}(x,y) < C^{-1}r$. By the previous paragraph, there is some $n \geq 0$ such that $\phi^{mn}(y) \notin D$, and hence

$$d_{\text{sph}}\big(\phi^{mn}(x), \phi^{mn}(y)\big) = d_{\text{sph}}\big(x, \phi^{mn}(y)\big) \geq r > C d_{\text{sph}}(x,y),$$

disproving equicontinuity. Thus, $x \in \mathcal{J}_{\phi,\text{I}}$.

On the other hand, if x is nonrepelling, then by Proposition 4.3(c), there is an open $\mathbb{P}^1(\mathbb{C}_v)$-disk D containing x such that $\phi^m(D) \subseteq D$. After a change of coordinates—which respects the Fatou set by Proposition 5.10(d)—we may assume that $D \subseteq D(0,1)$. Moreover, by Proposition 5.10(c), we may also assume that $m = 1$. Thus,

$$\phi^n(D) \subseteq D \subseteq D(0,1) \quad \text{for all } n \geq 0.$$

By Propositions 3.20 and 5.3(b), we have

$$d_{\mathrm{sph}}\big(\phi^n(a), \phi^n(b)\big) \leq d_{\mathrm{sph}}(a,b) \quad \text{for all } a, b \in D \text{ and } n \geq 0.$$

Hence, by Lemma 5.7, $x \in \mathcal{F}_{\phi,\mathrm{I}}$. □

As noted in Remark 4.4, the analogue of Theorem 5.14 is true in complex dynamics (by essentially the same proof) only for attracting and repelling points. However, rationally indifferent points in complex dynamics are always in the Julia set (provided $\deg(\phi) \geq 2$), while irrationally indifferent points may be either in the Fatou set or Julia set, depending on more subtle factors. Moreover, we have the following result that draws a further contrast between complex and non-archimdean dynamics; cf. Remark 5.13.

Corollary 5.15. *Let $\phi(z) \in \mathbb{C}_v(z)$ be a rational function. Then $\mathcal{F}_{\phi,\mathrm{I}} \neq \varnothing$.*

Proof. All nonrepelling points lie in the Fatou set by Theorem 5.14, and therefore the result follows from Proposition 4.2. □

Corollary 5.16. *Let $\phi(z) \in \mathbb{C}_v(z)$ be a rational function of degree $d \geq 2$. Then the exceptional set of ϕ is contained in the Fatou set of ϕ.*

Proof. If ϕ has more than two exceptional points, then by Theorem 1.19(a), ϕ is conjugate to z^{p^m}, where $m \geq 1$ and $p = \operatorname{char} \mathbb{C}_v > 0$. Thus, by Proposition 5.12, we have $\mathcal{F}_{\phi,\mathrm{I}} = \mathbb{P}^1(\mathbb{C}_v)$, which clearly contains the exceptional set.

If ϕ has exactly two exceptional points, then by Theorem 1.19(b), ϕ is conjugate to either z^d or z^{-d}. The exceptional points are 0 and ∞, both of which are superattracting periodic points, of period either 1 or 2. Thus, by Theorem 5.14, both belong to the Fatou set.

Similarly, if ϕ has exactly one exceptional point, then by Theorem 1.19(a) ϕ is conjugate to a polynomial, and the exceptional point ∞ is a superattracting fixed point, which again is Fatou by Theorem 5.14. □

5.3. Further properties of Fatou and Julia sets

Montel's theorem, also known as the Montel–Carathéodory theorem, is the following statement in complex analysis. Suppose a family of (complex) meromorphic functions $\mathcal{G}$ on a domain $U \subseteq \mathbb{C}$ omits three points of $\mathbb{P}^1(\mathbb{C})$; that is, there are three distinct points $a, b, c \in \mathbb{P}^1(\mathbb{C})$ such that for all $f \in \mathcal{G}$, we have $f(U) \cap \{a, b, c\} = \varnothing$. Then in fact $\mathcal{G}$ must be normal on U. (See, for example, [**42**, Theorem XII.4.1].) By the Arzelà–Ascoli theorem, this three-point criterion can be viewed equivalently as a sufficient condition for equicontinuity. In [**77**, Theorem 2.2], Hsia proved the following two-point non-archimedean analogue of Montel's theorem.

Theorem 5.17 (Hsia). *Let $D \subseteq \overline{D}(0,1) \subseteq \mathbb{C}_v$ be a disk, and let $\mathcal{G} \subseteq \mathcal{A}_D$ be a set of power series converging on D. Let $a \in \mathbb{C}_v$, and suppose that for all $f \in \mathcal{G}$, the image $f(D)$ does not contain a. Then $\mathcal{G}$ is equicontinuous on D with respect to the spherical metric.*

Proof. First, we claim that we may assume $a = 0$. To see this, define $\mathcal{G}' := \{f - a : f \in \mathcal{G}\}$, and apply Proposition 5.4 to produce a Lipschitz constant $C \geq 1$ for both $z \mapsto z - a$ and $z \mapsto z + a$. The claim then follows from Lemma 5.6.

By Lemma 5.7, it suffices to find a constant C such that

$$d_{\text{sph}}\big(f(x), f(y)\big) \leq C d_{\text{sph}}(x,y) \quad \text{for all } x, y \in D \text{ and } f \in \mathcal{G}.$$

Write $D = D(b,r)$ or $D = \overline{D}(b,r)$, where $b \in \overline{D}(0,1)$ and $0 < r \leq 1$. Set $C := r^{-1}$. For any $f \in \mathcal{G}$, the image $f(D)$ is a disk not containing 0, and therefore $f(D) \subseteq D(c_f, |c_f|)$ for some $c_f \in \mathbb{C}_v^\times$. Hence, for any $x, y \in D$, we have $|f(x)| = |f(y)| = |c_f|$. Therefore, by Proposition 3.20, we have

$$\big|f(x) - f(y)\big| \leq r^{-1}|c_f| \cdot |x - y|.$$

It follows that

$$\begin{aligned} d_{\text{sph}}\big(f(x), f(y)\big) &= \frac{|f(x) - f(y)|}{\max\{|f(x)|, 1\} \max\{|f(y)|, 1\}} \\ &\leq \frac{r^{-1}|c_f| \cdot |x-y|}{\max\{|c_f|^2, 1\}} \leq r^{-1}|x - y| = C d_{\text{sph}}(x,y), \end{aligned}$$

as desired. □

Corollary 5.18. *Let $D \subseteq \mathbb{P}^1(\mathbb{C}_v)$ be a $\mathbb{P}^1(\mathbb{C}_v)$-disk, and let $\mathcal{G} \subseteq \mathbb{C}_v(z)$ be a set of rational functions. Let $a, b \in \mathbb{P}^1(\mathbb{C}_v)$ be distinct, and suppose that for all $f \in \mathcal{G}$, the image $f(D)$ does not contain a or b. Then $\mathcal{G}$ is equicontinuous on D with respect to the spherical metric.*

Proof. Pick $\theta, \eta \in \mathrm{PGL}(2, \mathbb{C}_v)$ such that $\theta(D) \subseteq \overline{D}(0,1)$ and $\eta(b) = \infty$. Let $\mathcal{G}' = \{\eta \circ f \circ \theta^{-1} : f \in \mathcal{G}\}$, which is a set of rational functions on $\theta(D)$ omitting $\eta(a) \in \mathbb{C}_v$ and ∞, and hence is a set of power series converging on $\theta(D)$ and omitting $\eta(a)$. By Theorem 5.17, $\mathcal{G}'$ is equicontinuous on $\theta(D)$. Choosing a constant $C \geq 1$ for Proposition 5.4 that applies to θ, η, and their inverses, the result follows from Lemma 5.6. □

Thus, we have the following alternate characterization of the Fatou set of a non-archimedean rational function.

Theorem 5.19. *Let $\phi \in \mathbb{C}_v(z)$ be a rational function, and let $x \in \mathbb{P}^1(\mathbb{C}_v)$. Then $x \in \mathcal{F}_{\phi,\mathrm{I}}$ if and only if there is a $\mathbb{P}^1(\mathbb{C}_v)$-disk D containing x such that*

$$\mathbb{P}^1(\mathbb{C}_v) \smallsetminus \left[\bigcup_{n\geq 0} \phi^n(D) \right] \quad \text{has more than one element.}$$

In that case, it is possible to choose the $\mathbb{P}^1(\mathbb{C}_v)$-disk D so that the set $\mathbb{P}^1(\mathbb{C}_v) \smallsetminus \left[\bigcup_{n\geq 0} \phi^n(D) \right]$ is uncountable.

To prove Theorem 5.19, we need the following lemma.

Lemma 5.20. *Let $\phi \in \mathbb{C}_v(z)$ be a rational function, and let D be a $\mathbb{P}^1(\mathbb{C}_v)$-disk. Suppose that $\phi(D) \supsetneq D$. Then there is some $n \geq 0$ such that $\phi^n(D)$ is not contained in any residue class of $\mathbb{P}^1(\mathbb{C}_v)$.*

Proof. After a change of coordinates, we may assume that $0 \in D$. We also assume that $D = D(0,r)$ is an open disk; the case that D is closed is similar.

If the conclusion of the lemma is false, then $\phi^n(D) \subseteq D(0,1)$ for all $n \geq 0$. In particular, we can write $\phi^n(D) = D(0, s_n)$ for $n \geq 1$, where

$$r < s_1 < s_2 < \cdots .$$

Because ϕ has no poles in $\phi^{n-1}(D)$, we may write $\phi(z) = \sum_{j\geq 0} c_j z^j$ as a power series converging on each $D(0, s_n)$. Thus, by Theorem 3.15, we have $s_1 = |c_d| r^d$, for some $d \geq 1$.

We claim that $s_n \geq (s_1/r)^n r$ for all $n \geq 1$. Since $s_1/r > 1$, this claim implies that $s_n > 1$ for some n, which would contradict our assumption that $\phi^n(D) \subseteq D(0,1)$.

The $n = 1$ case of the claim is trivial. For $n \geq 2$, Theorem 3.15 again tells us that $s_n \geq |c_d| s_{n-1}^d$. Therefore,

$$s_n \geq s_1 s_{n-1}^d r^{-d} \geq s_1^{dn-d+1} r^{d-dn} = (s_1/r)^{dn-d+1} r \geq (s_1/r)^n r,$$

where the first inequality is because $s_1 = |c_d| r^d$, the second is by the inductive assumption that $s_{n-1} \geq (s_1/r)^{n-1} r$, and the last is because

$$dn - d - n + 1 = (d-1)(n-1) \geq 0$$

and $s_1/r > 1$. Thus, we have proven the claim and hence the lemma. □

Proof of Theorem 5.19. If there is a $\mathbb{P}^1(\mathbb{C}_v)$-disk D containing x such that the complement of $\bigcup_{n\geq 0} \phi^n(D)$ contains at least two points, then $\{\phi^n\}_{n\geq 0}$ is equicontinuous on D by Corollary 5.18.

Conversely, suppose that $x \in \mathcal{F}_{\phi,\mathrm{I}}$. Choose a $\mathbb{P}^1(\mathbb{C}_v)$-disk D as in Lemma 5.8, so that $\phi^n(D)$ is contained in a residue class for each $n \geq 0$. We consider three cases.

First, if there is some $m \geq 1$ for which $D \subsetneq \phi^m(D)$, then by Lemma 5.20 applied to ϕ^m, there is some $n \geq 0$ for which $\phi^{mn}(D)$ is not contained in a residue class, contradicting our choice of D.

Second, if $D \cap \phi^n(D) = \varnothing$ for all $n \geq 1$, let $D' \subsetneq D$ be a strictly smaller $\mathbb{P}^1(\mathbb{C}_v)$-disk containing x. Then $\mathbb{P}^1(\mathbb{C}_v) \smallsetminus \bigcup_{n\geq 0} \phi^n(D')$ contains the uncountable set $D \smallsetminus D'$, and we are done.

Finally, by Proposition 3.26, the only remaining possibility is that there is some integer $m \geq 1$ for which $\phi^m(D) \subseteq D$. Then for every $n \geq 0$, $\phi^n(D)$ is contained in one of the m $\mathbb{P}^1(\mathbb{C}_v)$-disks $\phi^i(D)$, for $i = 0, \ldots, m-1$. Each of those m disks is itself contained in a residue class, by our choice of D. Because there are infinitely many residue classes, it follows that $\mathbb{P}^1(\mathbb{C}_v) \smallsetminus \bigcup_{n\geq 0} \phi^n(D)$ contains a residue class and hence an uncountable set. □

Corollary 5.21. *Let $\phi(z) \in \mathbb{C}_v(z)$ be a rational function with classical Julia set $\mathcal{J}_{\phi,\mathrm{I}}$ and exceptional set E_ϕ. Suppose $\mathcal{J}_{\phi,\mathrm{I}}$ is nonempty. Then $\#E_\phi \leq 1$, and for any open set U intersecting $\mathcal{J}_{\phi,\mathrm{I}}$, we have*

$$\bigcup_{n\geq 0} \phi^n(U) \supseteq \mathbb{P}^1(\mathbb{C}_v) \smallsetminus E_\phi.$$

Proof. We have $\mathcal{J}_{\phi,\mathrm{I}} \neq \varnothing$ by hypothesis, and hence, by Theorem 1.19(a)–(b) and Proposition 5.11, we must have $\#E_\phi \leq 1$.

Without loss, U is a $\mathbb{P}^1(\mathbb{C}_v)$-disk containing a point $x \in \mathcal{J}_{\phi,\mathrm{I}}$. By Theorem 5.19, $\mathbb{P}^1(\mathbb{C}_v) \smallsetminus \bigcup_{n\geq 0} \phi^n(U)$ consists of at most one point. If this set is empty, we are done. Otherwise, the one point a in this set cannot have any preimages in $\bigcup_{n\geq 0} \phi^n(U)$, and therefore $\phi^{-1}(a) = \{a\}$. Thus, a is in the exceptional set, and we are done. □

Corollary 5.22. *Let $\phi(z) \in \mathbb{C}_v(z)$ be a rational function. Then the interior of the classical Julia set $\mathcal{J}_{\phi,\mathrm{I}}$ is empty.*

Proof. Pick $x \in \mathcal{J}_{\phi,\mathrm{I}}$ and an open set $U \subseteq \mathbb{P}^1(\mathbb{C}_v)$ containing x. Then by Corollary 5.21, $\bigcup_{n\geq 0} \phi^n(U)$ omits at most one point of $\mathbb{P}^1(\mathbb{C}_v)$. However, if $U \subseteq \mathcal{J}_{\phi,\mathrm{I}}$, then $\bigcup_{n\geq 0} \phi^n(U) \subseteq \mathcal{J}_{\phi,\mathrm{I}}$ by Proposition 5.10(b). Thus, $\mathcal{J}_{\phi,\mathrm{I}}$ is a closed set omitting at most one point of $\mathbb{P}^1(\mathbb{C}_v)$. Then $\mathcal{J}_{\phi,\mathrm{I}} = \mathbb{P}^1(\mathbb{C}_v)$, contradicting Corollary 5.15. Hence, $U \not\subseteq \mathcal{J}_{\phi,\mathrm{I}}$. □

Proposition 5.23. *Let $\phi(z) \in \mathbb{C}_v(z)$ be a rational function with classical Julia set $\mathcal{J}_{\phi,\mathrm{I}}$, and let $a \in \mathbb{P}^1(\mathbb{C}_v)$ be a nonexceptional point. Then the closure $\overline{\mathrm{Orb}^-(a)}$ of the backward orbit of a contains $\mathcal{J}_{\phi,\mathrm{I}}$, with equality if and only if $a \in \mathcal{J}_{\phi,\mathrm{I}}$.*

Proof. For the first statement, we need to show that for any point $x \in \mathcal{J}_{\phi,\mathrm{I}}$ and any open set $U \subseteq \mathbb{P}^1(\mathbb{C}_v)$ containing x, the backward orbit of a intersects

U. But given such x and U, Corollary 5.21 tells us that $a \in \phi^n(U)$ for some $n \geq 0$, as desired.

For the second statement, if $\overline{\mathrm{Orb}^-(a)} = \mathcal{J}_{\phi,\mathrm{I}}$, then clearly $a \in \mathcal{J}_{\phi,\mathrm{I}}$. Conversely, if $a \in \mathcal{J}_{\phi,\mathrm{I}}$, then $\mathrm{Orb}^-(a) \subseteq \mathcal{J}_{\phi,\mathrm{I}}$ by Proposition 5.10(b). Because $\mathcal{J}_{\phi,\mathrm{I}}$ is closed, we have

$$\overline{\mathrm{Orb}^-(a)} \subseteq \mathcal{J}_{\phi,\mathrm{I}}.$$

Finally, the reverse inclusion is the content of the first statement, proven in the previous paragraph. □

The next result says that the Julia set is the smallest closed and completely invariant set not contained in the exceptional set.

Corollary 5.24. *Let $\phi(z) \in \mathbb{C}_v(z)$ be a rational function with classical Julia set $\mathcal{J}_{\phi,\mathrm{I}}$. Let $S \subseteq \mathbb{P}^1(\mathbb{C}_v)$ be a closed subset of $\mathbb{P}^1(\mathbb{C}_v)$ that is not contained in the exceptional set and for which $\phi^{-1}(S) \subseteq S$. Then $\mathcal{J}_{\phi,\mathrm{I}} \subseteq S$.*

Proof. Pick any $a \in S$ not in the exceptional set. By hypothesis, the backward orbit $\mathrm{Orb}^-(a)$ is contained in S. Because S is closed, the closure $\overline{\mathrm{Orb}^-(a)}$ is also contained in S, and hence, by Proposition 5.23, $\mathcal{J}_{\phi,\mathrm{I}} \subseteq S$. □

Recall that a *perfect set* in a topological space is a closed subset with no isolated points. It is well known that a nonempty perfect subset of $\mathbb{R}^n$ is uncountable. This fact is true for subsets not just of $\mathbb{R}^n$, but of any complete metric space, including $\mathbb{P}^1(\mathbb{C}_v)$; see Exercise 5.7.

Corollary 5.25. *Let $\phi(z) \in \mathbb{C}_v(z)$ be a rational function. Then the classical Julia set $\mathcal{J}_{\phi,\mathrm{I}}$ is a perfect set. In particular, $\mathcal{J}_{\phi,\mathrm{I}}$ is either empty or uncountable.*

Proof. We already know from Proposition 5.10(a) that $\mathcal{J}_{\phi,\mathrm{I}}$ is a closed set. For any $a \in \mathcal{J}_{\phi,\mathrm{I}}$, we know that the backward orbit of ϕ is infinite, or else a would be exceptional and hence Fatou. Thus, there is some nonperiodic point b in the backward orbit of a, and hence in $\mathcal{J}_{\phi,\mathrm{I}}$. In particular, the backward orbit $\mathrm{Orb}^-(b)$ of b does not contain the point a. However, by Proposition 5.23, $\mathrm{Orb}^-(b)$, which is contained in $\mathcal{J}_{\phi,\mathrm{I}}$, has an accumulation point at a, as desired.

Finally, if $\mathcal{J}_{\phi,\mathrm{I}} \neq \varnothing$, then it is a nonempty perfect subset of the complete metric space $\mathbb{P}^1(\mathbb{C}_v)$. By Proposition 5.10(a), $\mathcal{J}_{\phi,\mathrm{I}}$ is also closed, and hence it is uncountable. □

We close this section with an alternative description of the Julia set in the case that ϕ is a polynomial.

Definition 5.26. Let $\phi(z) \in \mathbb{C}_v[z]$ be a polynomial of degree $d \geq 2$. The *(classical) filled Julia set* of ϕ is

$$\mathcal{K}_{\phi,\mathrm{I}} := \{x \in \mathbb{C}_v : \mathrm{Orb}_\phi^+(x) \text{ is bounded}\}.$$

Because ∞ is a superattracting fixed point, we can also describe $\mathcal{K}_{\phi,\mathrm{I}}$ as

$$\mathcal{K}_{\phi,\mathrm{I}} = \{x \in \mathbb{C}_v : \lim_{n\to\infty} \phi^n(x) \neq \infty\}.$$

Proposition 5.27. *Let $\phi(z) \in \mathbb{C}_v[z]$ be a polynomial of degree $d \geq 2$ with classical filled Julia set $\mathcal{K}_{\phi,\mathrm{I}}$ and classical Julia set $\mathcal{J}_{\phi,\mathrm{I}}$. Then $\mathcal{J}_{\phi,\mathrm{I}} = \partial\mathcal{K}_{\phi,\mathrm{I}}$, where ∂ denotes the topological boundary, as subsets of $\mathbb{P}^1(\mathbb{C}_v)$.*

Proof. Because the point at ∞ is superattracting and hence lies in the Fatou set, and because the Fatou set is open, there is some radius $R > 0$ such that $\mathbb{P}^1(\mathbb{C}_v) \smallsetminus \overline{D}(0, R) \subseteq \mathcal{F}_{\phi,\mathrm{I}}$, and such that $\lim_{n\to\infty} \phi^n(a) = \infty$ for all $|a| > R$. After a change of coordinates, we may assume that $R < 1$.

Clearly, $\mathcal{J}_{\phi,\mathrm{I}} \subseteq \mathcal{K}_{\phi,\mathrm{I}}$. Given $x \in \mathcal{J}_{\phi,\mathrm{I}}$ and any $\mathbb{P}^1(\mathbb{C}_v)$-disk U containing x, we wish to show that U intersects both $\mathcal{K}_{\phi,\mathrm{I}}$ and its complement. We already have $x \in \mathcal{J}_{\phi,\mathrm{I}} \subseteq \mathcal{K}_{\phi,\mathrm{I}}$. By Theorem 5.19, $\bigcup_{n\geq 0} \phi^n(U)$ omits at most two points of $\mathbb{P}^1(\mathbb{C}_v)$, and hence there is some $b \in U$ and $n \geq 0$ such that $|\phi^n(b)| > R$. Therefore, by our choice of R, $\lim_{n\to\infty} \phi^n(b) = \infty$, as desired.

For the reverse inclusion, given $x \in \partial\mathcal{K}_{\phi,\mathrm{I}}$, we wish to show that $x \in \mathcal{J}_{\phi,\mathrm{I}}$. Given any $\mathbb{P}^1(\mathbb{C}_v)$-disk U containing x, there are points $a, b \in U$ such that $a \in \mathcal{K}_{\phi,\mathrm{I}}$ but $b \notin \mathcal{K}_{\phi,\mathrm{I}}$. Hence, $|\phi^n(a)| \leq R < 1$ for all $n \geq 0$, but there is some $m \geq 0$ such that $|\phi^m(b)| > 1$. In particular, $\phi^m(U)$ is not contained in a single residue class. By Lemma 5.8, $x \in \mathcal{J}_{\phi,\mathrm{I}}$. □

Remark 5.28. Almost all of the results of this section are true in complex dynamics as well, and for essentially the same reasons. As noted, Hsia's theorem is an analogue of the Montel's theorem, with one fewer point involved. Similarly, the analogues of Corollary 5.18, Theorem 5.19, and Corollary 5.21 all hold over $\mathbb{C}$ but with one more point, because they are all based on the Montel–Carathéodory theorem instead of Hsia's theorem. In addition, Proposition 5.23, Corollary 5.24, and Proposition 5.27 all hold over $\mathbb{C}$. Corollary 5.22 and Corollary 5.25 must be modified only slightly—the former to allow the possibility that the Julia set is the whole Riemann sphere, and the latter to exclude the possibility that the Julia set is empty. For more details, see [**9**, Chapters 3–4], [**39**, Sections III.1,4], or [**104**, Sections 3–4].

5.4. Examples of non-archimedean Fatou and Julia sets

Throughout this section, K denotes a complete non-archimedean field, and $\mathbb{C}_v$ is the completion of an algebraic closure of K.

Example 5.29. Let $\phi(z) = Az^2+Bz+C \in K[z]$ be a quadratic polynomial. As we saw in Example 1.6, if char $K \neq 2$, then ϕ is conjugate over K to a polynomial $z^2 + c$, with $c \in K$. The dynamics of $z^2 + c$ over a complete non-archimedean field was first considered in [**136**] over $K = \mathbb{Q}_p$ for $p \neq 2$, and in [**12**, Section 3.3.1] over $K = \mathbb{C}_p$ for any p; see also [**47**, **48**]. A full description of the dynamics of a quadratic polynomial over any complete non-archimedean field (even of characteristic 2) appeared in [**23**], and we follow that strategy here.

Instead of conjugating ϕ to $z^2 + c$, we consider a fixed point α of ϕ. Thus, α is a root of the polynomial $Az^2 + (B - 1)z + C = 0$, and either $\alpha \in K$ or else the field $K(\alpha)$ is a quadratic extension of K.

Conjugating by translation by α and then by scaling by A, we obtain a map of the form $\psi(z) = z^2 + \lambda z$ for some $\lambda \in K(\alpha)$. If $|\lambda| \leq 1$, then ψ has explicit good reduction, and hence ϕ has potential good reduction.

On the other hand, if $|\lambda| > 1$, the reader can check the following.

- $z = 0$ is a repelling fixed point. Thus, $\mathcal{J}_{\psi,\mathrm{I}}$ and $\mathcal{J}_{\phi,\mathrm{I}}$ are nonempty, and ϕ does not have potential good reduction.
- Let $R := |\lambda|$ and $U_0 := \overline{D}(0, R)$. Then

  $$|\psi(z)| = |z|^2 \quad \text{and} \quad \lim_{n\to\infty} \psi^n(z) = \infty \quad \text{for } |z| > R.$$

  Thus, U_0 contains the filled Julia set $\mathcal{K}_{\phi,\mathrm{I}}$.
- For $n \geq 1$, let $U_n := \psi^{-n}(U_0)$. Then $U_1 = X \cup Y$, where

  $$X := \overline{D}(0, 1) \quad \text{and} \quad Y := \overline{D}(-\lambda, 1).$$

- $\psi : X \to U_0$ and $\psi : Y \to U_0$ are both bijective. Therefore, by Proposition 3.22, ψ has inverses

  $$h_X : U_0 \to X \quad \text{and} \quad h_Y : U_0 \to Y$$

  given by power series defined over $K(\alpha)$.
- For $n \geq 1$, U_n is a disjoint union of 2^n rational closed disks of radius R^{1-n}, each containing a point in $K(\alpha)$, with

  $$U_0 \supsetneq U_1 \supsetneq U_2 \supsetneq \cdots.$$

  In fact, each disk of U_{n-1} contains exactly two disks of U_n, and the distance between those two disks is R^{2-n}.
- $\mathcal{J}_{\psi,\mathrm{I}} = \mathcal{K}_{\psi,\mathrm{I}} = \bigcap_{n\geq 0} U_n$, which is homeomorphic to a Cantor set.

(See Exercise 5.8.) Thus, the Julia set is compact and contained in $K(\alpha)$, and every point of the Fatou set approaches ∞ under iteration.

The following example (first considered by Smart and Woodcock in [**133**] and then again by Hsia in [**77**, Example 4.1]) is of a higher-degree polynomial map with properties similar to the quadratic polynomials in Example 5.29. Although similar dynamics can be exhibited for higher-degree polynomials over any non-archimedean field, we present the following example (as in [**77, 133**]) only over p-adic fields, because of the particular form the Julia set takes in that case.

Example 5.30. Fix a prime p, let $K = \mathbb{Q}_p$, and let $\mathbb{C}_v = \mathbb{C}_p$. Let

$$\phi(z) = \frac{z^p - z}{p}.$$

Then $U_0 := \overline{D}(0,1)$ contains the filled Julia set and, as in Example 5.29, it is easy to show that the filled Julia set $\mathcal{K}_{\phi,\mathrm{I}}$ is

$$\mathcal{K}_{\phi,\mathrm{I}} = \bigcap_{n\geq 0} U_n, \quad \text{where } U_n := \phi^{-n}(U_0).$$

Each preimage U_n consists of p^n disks of radius $|p|^n$, with p disks in each disk of U_{n-1}, and each mapping bijectively onto its image. By the same techniques as in Example 5.29, it is not difficult to show that both the Julia set and the filled Julia set are exactly $\mathcal{J}_{\phi,\mathrm{I}} = \mathcal{K}_{\phi,\mathrm{I}} = \mathbb{Z}_p$, and that the whole Fatou set $\mathbb{P}^1(\mathbb{C}_p) \smallsetminus \mathbb{Z}_p$ is attracted to ∞ under iteration; see Exercise 5.9.

In complex dynamics, quadratic polynomials display a wide variety of dynamical properties, and the dynamics of many higher-degree maps look locally like dynamics found among quadratic polynomials. However, as Example 5.29 shows, non-archimedean quadratic polynomials are far simpler; either they have potential good reduction, or their Julia sets are Cantor sets arranged in a very understandable way, with all Fatou points attracted to the superattracting point at ∞.

By the same token, the complex Mandelbrot set, which is the set of complex c-values for which the critical point $z = 0$ of $\phi_c(z) := z^2 + c$ lies in the filled Julia set of ϕ_c, has a famously complicated structure, as it distinguishes those maps with one broad class of dynamics (connected Julia set) from those with another (totally disconnected Julia set). By constrast, the non-archimedean version of the Mandelbrot set is simply a disk. More specifically, if $|\lambda| \leq 1$, then $z^2 + \lambda z$ is in the Mandelbrot set; otherwise, it is not. For a description of the non-archimedean Mandelbrot set for the family $z^2 + c$, see Exercise 5.10.

However, even non-archimedean cubic polynomials display far more intricate dynamics than their quadratic counterparts, as the following example illustrates. It was first considered by Hsia in [**77**, Example 4.2]. We leave some claims for the reader in Exercises 5.11–5.13.

Example 5.31. Pick $a, b, c \in K$ with $0 < |a| < 1$ and $|b|, |c| \le 1$, and let

$$\phi(z) = az^3 + z^2 + bz + c.$$

Using the results of Chapter 3, we make the following observations.

- ϕ has one fixed point $\gamma \in \overline{D}(-1/a, 1)$, with $\gamma \in K$.
- $|\phi'(\gamma)| = |a|^{-1} > 1$, so γ is repelling, and hence $\gamma \in \mathcal{J}_{\phi,\mathrm{I}}$.
- ϕ has two nonrepelling fixed points in $\overline{D}(0,1)$, a superattracting fixed point at ∞, and no other fixed points.
- $\phi(\overline{D}(0,1)) = \overline{D}(0,1)$, and hence $\overline{D}(0,1) \subseteq \mathcal{F}_{\phi,\mathrm{I}}$.

See Exercise 5.11(a).

We can describe the filled Julia set of ϕ using the same methods as in Examples 5.29 and 5.30, as follows.

- $\lim_{n\to\infty} \phi^n(x) = \infty$ for all $x \in \mathbb{C}_v$ with $|x| > |a|^{-1}$.
- Let $U_0 := \overline{D}(0, |a|^{-1})$. Then the filled Julia set is
$$\mathcal{K}_{\phi,\mathrm{I}} = \bigcap_{n\ge 0} \phi^{-n}(U_0).$$
- $\phi^{-1}(U_0) = A \cup B$, where
$$A := \overline{D}(\gamma, 1) \quad \text{and} \quad B := \overline{D}(0, |a|^{-1/2}).$$
- $\phi : A \to U_0$ is bijective, but $\phi : B \to U_0$ has Weierstrass degree 2.

See Exercise 5.11(b).

The final bullet point above marks a contrast with Example 5.29, since the disk B does not map bijectively onto U_0. As a result, even though each successive preimage $\phi^{-n}(U_0)$ is a union of disks, it is far more complicated to give a precise description of their radii and arrangement. In addition, the Julia set $\mathcal{J}_{\phi,\mathrm{I}}$ is not compact, for the following reason. The point γ has a preimage γ_1 with $|\gamma_1| = |a|^{-1/2}$. (In fact, there are two such preimages, but we fix one.) Similarly, γ_1 has a preimage γ_2 with $|\gamma_2| = |a|^{-1/4}$. In general, γ has a preimage γ_n under ϕ^n with $|\gamma_n| = |a|^{-1/2^n}$; see Exercise 5.11(c). The sequence $\{\gamma_n\}_{n\ge 1}$ lies in $\mathcal{J}_{\phi,\mathrm{I}}$ but has no accumulation points in $\mathbb{P}^1(\mathbb{C}_v)$, and thus $\mathcal{J}_{\phi,\mathrm{I}}$ is not compact.

Another contrast is that $\mathcal{F}_{\phi,\mathrm{I}}$ includes many points not attracted to ∞ because $\overline{D}(0,1) \subseteq \mathcal{F}_{\phi,\mathrm{I}}$. Still, $\mathcal{J}_{\phi,\mathrm{I}}$ is the topological boundary of the filled Julia set, according to Proposition 5.27.

The dynamics on $\overline{D}(0,1)$ itself are given by the dynamics of the reduced map $\overline{\phi}(z) = z^2 + \overline{b}z + \overline{c}$, in the sense that residue classes in $\overline{D}(0,1)$ are mapped to each other as specified by $\overline{\phi}(z)$. According to Theorem 4.18, then, any periodic residue class V either is attracted to a single attracting periodic point inside or V maps bijectively onto itself with indifferent periodic points but no attracting periodic points.

In Example 5.31, ϕ acts on $\overline{D}(0,1)$ essentially like the quadratic polynomial z^2+bz+c, but for $|z| \ge |a|^{-1}$, the az^3-term is powerful enough to make other things happen. Similar phenomena occur for any polynomial of the form $aq(z)z^d + r(z)$, where $r \in \mathcal{O}_K[z]$ is a polynomial of explicit good reduction, and where

$$q(z) \in \mathcal{O}_K[z] \smallsetminus \{0\}, \quad 0 < |a| < 1, \quad \text{and } d \ge 1 + \deg(r).$$

Further intricacies can arise if any of the disks making up the preimages $\phi^{-n}(U_0)$ map to each other with Weierstrass degree divisible by the residue characteristic. Even the cubic map of Example 5.31 exhibits different behavior depending on the residue characteristic; see Exercises 5.12 and 5.13.

The rational function of the next example first appeared in [**16**, Example 1].

Example 5.32. Assume the residue characteristic of K is not 2, pick $a \in K$ with $0 < |a| < 1$, and let

$$\phi(z) = z^2 + \frac{az^2}{z+1} = \frac{z^3 + (a+1)z^2}{z+1}.$$

The reader can check the following in Exercise 5.14(a).

- ϕ has a superattracting fixed point at ∞, and
$$\lim_{n\to\infty} \phi^n(x) = \infty \quad \text{for } |x| > 1.$$
- ϕ has a superattracting fixed point at 0, and
$$\lim_{n\to\infty} \phi^n(x) = 0 \quad \text{for } |x| < 1.$$
- ϕ has an indifferent fixed point $\alpha \in D(1,1)$.
- ϕ has a repelling fixed point $\beta \in D(-1,1)$.
- $\phi\big(D(-1,1)\big) = \mathbb{P}^1(\mathbb{C}_v)$.
- For any $x \in \mathbb{C}_v$ with $|x+1| = 1$, we have $\phi\big(D(x,1)\big) = D(x^2,1)$.

The last two bullet points above say that the dynamics on $\overline{D}(0,1)$ are mostly determined by the reduced map $\overline{\phi}(z) = z^2$, *except* on the residue class $D(-1,1)$, which maps onto $\mathbb{P}^1(\mathbb{C}_v)$. Since the Julia set $\mathcal{J}_{\phi,\mathrm{I}}$ contains the repelling fixed point β and is invariant under ϕ, it follows that for any $x \in \mathbb{C}_v$ for which

$$\big|x^{2^n} + 1\big| < 1 \quad \text{for some } n \ge 0,$$

the residue class $D(x,1)$ intersects $\mathcal{J}_{\phi,\mathrm{I}}$. Conversely, since $D(-1,1)$ is the only problematic residue class, if $y \in \mathbb{C}_v$ satisfies

$$\big|y^{2^n} + 1\big| \ge 1 \quad \text{for all } n \ge 0,$$

then $y \in \mathcal{F}_{\phi,\mathrm{I}}$. That is, the Fatou set contains any class that avoids -1 under iteration of $z \mapsto z^2$. (If the residue field $\overline{k}$ of $\mathbb{C}_v$ is algebraic over a finite field, all such Fatou residue classes are preperiodic; otherwise, some are preperiodic and others are wandering.)

As in Example 5.31, the Julia set is not compact, but this time the reason is that it intersects infinitely many residue classes. At the same time, the Fatou set has infinitely many periodic residue classes, corresponding to the infinitely many periodic points of $z \mapsto z^2$ in $\mathbb{P}^1(\overline{k})$. It can be shown that any point in $\mathcal{F}_{\phi,\mathrm{I}}$ lands in a Fatou residue class $D(y,1)$ after finitely many iterations; see Exercise 5.14(e).

Example 5.33. Let $c \in \mathbb{C}_v$ with $0 < |c| < 1$, and let

$$\phi(z) = \frac{z^4 - 8c^4 z - c^4}{4z^3 + z^2 + 4c^4}.$$

This is the map of Example 4.15 with $a = c^4$. (We use c^4 in this example to avoid having to write fractional powers of a later.) Recall that ϕ is the Lattès map associated with the multiplication-by-2 map for a certain elliptic curve of multiplicative reduction.

As we saw in that example, ϕ has bad reduction. Nevertheless, ϕ still has empty Julia set. The reader can check the following.

- $\phi\big(\mathbb{P}^1(\mathbb{C}_v) \smallsetminus \overline{D}(0,1)\big) \subseteq \mathbb{P}^1(\mathbb{C}_v) \smallsetminus \overline{D}(0,1)$, and hence this open $\mathbb{P}^1(\mathbb{C}_v)$-disk is contained in $\mathcal{F}_{\phi,\mathrm{I}}$.
- For any $x \in \mathbb{C}_v$ with $|x| = 1$, either
$$|\phi(x)| = 1 \quad \text{and} \quad \phi\big(D(x,1)\big) \subseteq D\big(\phi(x),1\big),$$
or else
$$\phi\big(D(x,1)\big) \subseteq \mathbb{P}^1(\mathbb{C}_v) \smallsetminus \overline{D}(0,1) \subseteq \mathcal{F}_{\phi,\mathrm{I}}.$$
- For any $x \in \mathbb{C}_v$ with $|x| \le |c|^2$, either
$$|\phi(x)| = 1 \quad \text{and} \quad \phi\big(D(x,|c|^2)\big) \subseteq D\big(\phi(x),1\big),$$
or else
$$\phi\big(D(x,|c|^2)\big) \subseteq \mathbb{P}^1(\mathbb{C}_v) \smallsetminus \overline{D}(0,1) \subseteq \mathcal{F}_{\phi,\mathrm{I}}.$$
- For any $x \in \mathbb{C}_v$ with $|c| < |x| < 1$, we have
$$|\phi(x)| = |x|^2 \quad \text{and} \quad \phi\big(D(x,|x|)\big) \subseteq D\big(\phi(x),|\phi(x)|\big).$$
- For any $x \in \mathbb{C}_v$ with $|c|^2 < |x| < |c|$, we have
$$|\phi(x)| = |c|^4/|x|^2 \quad \text{and} \quad \phi\big(D(x,|x|)\big) \subseteq D\big(\phi(x),|\phi(x)|\big).$$
- For any $x \in \mathbb{C}_v$ with $|x| = |c|$, we have $\phi(x) \in \overline{D}(0,|c|^2)$.

The reader can also check that the above bullet points imply that the (classical) Julia set of ϕ is empty. (See Exercise 5.15.) Later, in Example 8.20, we will see how Berkovich spaces help us understand this dynamical system better.

Exercises for Chapter 5

5.1. Let $z, w \in \mathbb{P}^1(\mathbb{C}_v)$.

(a) Verify that equation (5.1) defining $d_{\mathrm{sph}}(z, w)$ is independent of the choices $[x, y]$ and $[u, v]$ of homogeneous coordinates for z and w.

(b) Verify that if $z, w \neq \infty$, then the two definitions of $d_{\mathrm{sph}}(z, w)$ in equations (5.1) and (5.2) agree.

5.2. Prove that the spherical metric d_{sph} satisfies the first two conditions to be a metric on $\mathbb{P}^1(\mathbb{C}_v)$. That is, for any $z, w \in \mathbb{P}^1(\mathbb{C}_v)$, prove that

(a) $d_{\mathrm{sph}}(z, w) \in [0, \infty)$, with $d_{\mathrm{sph}}(z, w) = 0$ if and only if $z = w$.

(b) $d_{\mathrm{sph}}(w, z) = d_{\mathrm{sph}}(z, w)$.

5.3. Prove that $\mathrm{PGL}(2, \mathcal{O})$ is generated by

- maps of the form $z \mapsto z + \alpha$ for $\alpha \in \mathcal{O}$,
- maps of the form $z \mapsto \beta z$ for $\beta \in \mathcal{O}^\times$, and
- $z \mapsto 1/z$.

(*Suggestion*: It's just Gaussian elimination.)

5.4. Given $z, w \in \mathbb{P}^1(\mathbb{C}_v)$, prove that there is some $\eta \in \mathrm{PGL}(2, \mathcal{O})$ such that $\eta(z) = w$.

5.5. Let $z, w \in \mathbb{P}^1(\mathbb{C}_v)$, let $\alpha \in \mathcal{O}$, and let $\beta \in \mathcal{O}^\times$. Prove that

$$d_{\mathrm{sph}}(z + \alpha, w + \alpha) = d_{\mathrm{sph}}(\beta z, \beta w) = d_{\mathrm{sph}}(1/z, 1/w) = d_{\mathrm{sph}}(z, w).$$

5.6. Let a, b, c be distinct points in $\mathbb{P}^1(\mathbb{C}_v)$, let D be a $\mathbb{P}^1(\mathbb{C}_v)$-disk, and let $\mathcal{G} \subseteq \mathbb{C}_v(z)$ be a set of rational functions. Suppose that for each $f \in \mathcal{G}$, the image $f(D)$ contains at most one of the points a, b, c. Prove that $\mathcal{G}$ is equicontinuous on D with respect to the spherical metric.

5.7. Let X be a complete metric space, and let $Y \subseteq X$ be a nonempty closed perfect subset. Prove that Y is uncountable. You may *not* assume that X or Y is compact.

(*Suggestion*: If Y were countably infinite, write $Y = \{y_1, y_2, \ldots\}$ and construct a nested sequence of open balls $\{V_n\}_{n\geq 1}$ intersecting Y, with radii decreasing to 0 and with $y_n \notin V_n$.)

5.8. Prove all the claims made in the six bullet points of Example 5.29.

5.9. Prove all the claims made in Example 5.30.

5.10. Assume that $\mathbb{C}_v$ is non-archimedean, and let $p \geq 0$ be the residue characteristic of $\mathbb{C}_v$. For each $c \in \mathbb{C}_v$, define

$$\phi_c(z) = z^2 + c.$$

Define the non-archimedean Mandelbrot set to be

$$\mathcal{M}(\mathbb{C}_v) := \{c \in \mathbb{C}_v : \phi_c \text{ has potential good reduction}\}.$$

(a) If $p \neq 2$, prove that $\mathcal{M}(\mathbb{C}_v) = \overline{D}(0,1)$.

(b) If $p = 2$ and $\operatorname{char} \mathbb{C}_v = 0$, prove that $\mathcal{M}(\mathbb{C}_v) = \overline{D}(0, |2|^{-2})$.

(c) If $p = \operatorname{char} \mathbb{C}_v = 2$, prove that $\mathcal{M}(\mathbb{C}_v) = \mathbb{C}_v$.
In fact, show that ϕ_c is conjugate to $z \mapsto z^2$ for every $c \in \mathbb{C}_v$; see Exercise 1.6(d). Thus, the family $\{\phi_c : c \in \mathbb{C}_v\}$ is not very interesting in characteristic 2.

5.11. In this exercise, you will prove some of the claims made in Example 5.31. Let $a, b, c \in \mathbb{C}_v$ with $0 < |a| < 1$ and $|b|, |c| \leq 1$, and let

$$\phi(z) := az^3 + z^2 + bz + c.$$

(a) Prove the first set of four bullet points of Example 5.31.

(b) Prove the second set of four bullet points of Example 5.31.

(c) Prove that for any $x \in \mathbb{C}_v$ with $1 < |x| \leq |a|^{-1}$, the inverse image $\phi^{-1}(x)$ consists of one point in $\overline{D}(\gamma, 1)$ and two points of absolute value $|x|^{1/2}$.

(d) Prove that for any $x \in \overline{D}(0,1)$, the reductions $\overline{x}$ and $\overline{\phi(x)}$ are related by

$$\overline{\phi(x)} = \overline{\phi}(\overline{x}), \quad \text{where } \overline{\phi(z)} = z^2 + \overline{b}z + \overline{c}.$$

5.12. Define a, b, c, ϕ as in Exercise 5.11. In addition, assume that $\mathbb{C}_v$ does not have residue characteristic 2.

(a) Prove that ϕ has a critical point in $D(-b/2, 1)$.

(b) Let $x \in \overline{D}(0,1)$ be a periodic point such that $|2\phi^n(x) + b| < 1$ for some $n \geq 0$. Prove that x is attracting.

(c) Let $x \in \overline{D}(0,1)$ be a periodic point such that $|2\phi^n(x) + b| = 1$ for all $n \geq 0$. Prove that x is indifferent.

(d) Prove that every periodic point in $\mathbb{C}_v \smallsetminus \overline{D}(0,1)$ is repelling.

(e) Prove that ϕ has at most two attracting periodic cycles (including the one at ∞), infinitely many indifferent cycles, and infinitely many repelling cycles.

(f) Prove that every element of the Fatou set either eventually maps into $\overline{D}(0,1)$ or is attracted to ∞ under iteration.

5.13. Define a, b, c, ϕ as in Exercise 5.11. In addition, assume that $\mathbb{C}_v$ has residue characteristic 2.

(a) If $|b| = 1$, prove that every periodic point in $\overline{D}(0,1)$ is indifferent.

(b) If $|b| < 1$, prove that every periodic point in $\overline{D}(0,1)$ is attracting.

(c) Prove that ϕ has infinitely many attracting periodic points in the $\mathbb{P}^1(\mathbb{C}_v)$-disk $\mathbb{C}_v \setminus \overline{D}(0,1)$.

(*Suggestion*: Look for n-periodic points x with $|x| = |a|^{-1/2^{(n-1)}}$.)

5.14. In this exercise, you will prove some of the claims made in Example 5.32. Assume that the residue characteristic of $\mathbb{C}_v$ is not 2, pick $a \in \mathbb{C}_v$ with $0 < |a| < 1$, and let

$$\phi(z) = z^2 + \frac{az^2}{z+1} = \frac{z^3 + (a+1)z^2}{z+1}.$$

(a) Prove all six bullet points of Example 5.32.

(b) Prove that all four critical points γ of ϕ have the property that $\phi(\gamma)$ lies in one of the fixed residue classes $\overline{0} = D(0,1)$, $\overline{1} = D(1,1)$, or $\overline{\infty} = \mathbb{P}^1(\mathbb{C}_v) \setminus \overline{D}(0,1)$.

(c) Let $x \in \mathbb{C}_v$ with $|x - \gamma| \geq 1$ for all critical points γ of ϕ. Prove that $|\phi'(x)| \geq 1$.

(d) Prove that for any $x \in D(-1,1)$, either $|\phi(x) - 1| < 1$ or $|\phi'(x)| > 1$. (Or both.)

(e) Let $x \in \mathcal{F}_{\phi,\mathrm{I}} \cap D(-1,1)$. Prove that there is some $N \geq 1$ such that

$$\phi^n(x) \notin D(-1,1) \quad \text{for all } n \geq N.$$

(*Suggestion*: Use parts (b)–(d) and consider the forward orbit of a small disk containing x.)

5.15. Prove each of the six bullet points of Example 5.33, and then show that they together imply that the Lattès map ϕ of that example has empty classical Julia set.

Part 3

The Berkovich Line

Chapter 6

The Berkovich Projective Line

In his thesis [**114**], Rivera-Letelier pioneered the use of the Berkovich projective line $\mathbb{P}^1_{\mathrm{an}}$ in the study of non-archimedean dynamics. Roughly speaking, $\mathbb{P}^1_{\mathrm{an}}$ is a topological space containing $\mathbb{P}^1(\mathbb{C}_v)$ but which has the advantages of being compact, Hausdorff, and path-connected.

In the interest of streamlining our presentation, we defer many of the proofs of results in this chapter to Chapter 15. Nevertheless, readers new to Berkovich's theory might find even this lighter version to be rather demanding. It may help to keep in mind throughout that the principal conclusions are the following.

(a) The Berkovich projective line $\mathbb{P}^1_{\mathrm{an}}$ is a path-connected, Hausdorff compactification of $\mathbb{P}^1(\mathbb{C}_v)$.

(b) The points of $\mathbb{P}^1(\mathbb{C}_v)$ in $\mathbb{P}^1_{\mathrm{an}}$ are said to be of Type I.

(c) Every closed subdisk $\overline{D}(a,r) \subseteq \mathbb{C}_v$ corresponds to a point in $\mathbb{P}^1_{\mathrm{an}}$ that we denote $\zeta(a,r)$. This point is said to be of Type II if $r \in |\mathbb{C}_v^\times|$, or of Type III otherwise.

(d) All other points in $\mathbb{P}^1_{\mathrm{an}}$ correspond to decreasing sequences of disks $D_1 \supsetneq D_2 \supsetneq \cdots$ with empty intersection. Such points are said to be of Type IV.

See Figure 6.3, which is a sketch of the Berkovich disk $\overline{D}_{\mathrm{an}}(a,r)$, and Figure 6.1, which illustrates a path between a Type I point $0 \in \mathbb{C}_v$ and a Type II point $\zeta(x,r)$. Try to keep all these goals in mind as you work your way through the details of this chapter.

Berkovich spaces were introduced by Vladimir Berkovich in the late 1980s as an improvement on the theory of rigid analytic spaces. Related seminorm constructions appeared previously in [**51**], but Berkovich's theory seems to be the proper setting for non-archimedean dynamics. For deeper background on Berkovich spaces, we refer the reader to Berkovich's monograph [**26**]. The first two chapters of [**8**] provide more accessible introduction, given that they focus on the projective line rather than arbitrary varieties as in [**26**]. Much of our presentation here is influenced by [**8**]. Finally, Rivera-Letelier's papers [**114, 115, 117**] construct the Berkovich projective line in a different way, motivated more directly by the dynamical applications; see Appendix B.4.

6.1. Seminorms as Berkovich points

Definition 6.1. Let A be a ring with unity. A *multiplicative seminorm* on A is a function $\|\cdot\| : A \to [0,\infty)$ such that

(a) $\|0\| = 0$ and $\|1\| = 1$;

(b) for all $f, g \in A$, $\|fg\| = \|f\| \cdot \|g\|$; and

(c) for all $f, g \in A$, $\|f+g\| \leq \|f\| + \|g\|$.

If $\|f+g\| \leq \max\{\|f\|, \|g\|\}$ for all $f, g \in A$, we say $\|\cdot\|$ is *non-archimedean*.

We say another multiplicative seminorm $\|\cdot\|_\zeta$ is *bounded* with respect to $\|\cdot\|$, and that $\|\cdot\|$ *bounds* $\|\cdot\|_\zeta$, if there is a constant $C > 0$ such that

$$\|f\|_\zeta \leq C\|f\| \quad \text{for all } f \in A.$$

Notably absent from Definition 6.1 is the requirement that $\|f\| = 0$ implies $f = 0$, whence the terminology *semi*norm.

We write most of our seminorms with a subscript, as we did with $\|\cdot\|_\zeta$ above. In that case, we often write simply ζ to indicate the seminorm $\|\cdot\|_\zeta$.

The standard constructions of the Berkovich projective line and affine line start with the Berkovich unit disk, which is the set of multiplicative seminorms on the power series ring $\overline{\mathcal{A}}(0,1)$ that are bounded by the sup-norm on $\overline{D}(0,1)$. However, since we do not need the full theory of Berkovich spaces, we begin with the following simpler but equivalent definition of the Berkovich affine line.

Definition 6.2. The *Berkovich affine line*, denoted $\mathbb{A}^1_{\text{an}}$, is the set of multiplicative seminorms $\zeta = \|\cdot\|_\zeta$ on the polynomial ring $\mathbb{C}_v[z]$ that extend the absolute value on $\mathbb{C}_v$. We equip $\mathbb{A}^1_{\text{an}}$ with the *Gel'fand topology*, which is the weakest topology such that for every $f \in \mathbb{C}_v[z]$, the function mapping $\mathbb{A}^1_{\text{an}}$ to $\mathbb{R}$ by $\zeta \mapsto \|f\|_\zeta$ is continuous.

The above definition of the Gel'fand topology means that for any open subset V of $\mathbb{R}$ and any $f \in \mathbb{C}_v[z]$, the set

$$\{\zeta \in \mathbb{A}^1_{\text{an}} : \|f\|_\zeta \in V\}$$

is an open subset of $\mathbb{A}^1_{\text{an}}$. More precisely, the Gel'fand topology is the topology generated by such sets. Later, in Theorem 6.17, we present an alternative description of the Gel'fand topology that may provide a better intuition for what open sets in $\mathbb{P}^1_{\text{an}}$ can be. Until then, simply understanding Berkovich *points* will occupy much of our attention through the end of Section 6.3.

As noted above, we abuse notation and use the symbols ζ and $\|\cdot\|_\zeta$ interchangeably. Generally, however, we write ζ when we want to emphasize that we are talking about a Berkovich point, and we write $\|\cdot\|_\zeta$ when we want to emphasize that we are talking about a seminorm.

6.1.1. The four types of Berkovich points. Before developing the theoretical properties of $\mathbb{A}^1_{\text{an}}$, we give some examples of its elements. We will formalize the following categorization of Berkovich points in Definition 6.8.

First, for any point $x \in \mathbb{C}_v$, define a function $\|\cdot\|_x : \mathbb{C}_v[z] \to [0, \infty)$ by

$$\|f\|_x := |f(x)|.$$

The reader may check that $\|\cdot\|_x$ is indeed a multiplicative seminorm extending $|\cdot|$, although it is important to note that it is *not* a norm, since $\|z - x\|_x = 0$; see Exercise 6.1. We will see in Theorem 6.21 that the map $x \mapsto \|\cdot\|_x$ is an embedding of $\mathbb{A}^1(\mathbb{C}_v)$ into $\mathbb{A}^1_{\text{an}}$. Berkovich points of the form $\|\cdot\|_x$ are called *Type I points*.

Next, for any (rational or irrational) closed disk $\overline{D}(a, r) \subseteq \mathbb{C}_v$, define $\zeta(a, r)$ to be

$$\|f\|_{\zeta(a,r)} := \sup\{|f(x)| : x \in \overline{D}(a, r)\},$$

which is the sup-norm of Definition 3.6. By Proposition 3.7, it is a multiplicative norm on $\overline{\mathcal{A}}(a, s)$ for any $s > r$. Hence, $\zeta(a, r)$ is a multiplicative norm on $\mathbb{C}_v[z] \subseteq \overline{\mathcal{A}}(a, s)$ extending $|\cdot|$. That is, $\zeta(a, r)$ is a point in $\mathbb{A}^1_{\text{an}}$. Such points are called *Type II points* if $r \in |\mathbb{C}_v^\times|$, or *Type III points* otherwise.

There is also a fourth type of Berkovich point, at least in the case that $\mathbb{C}_v$ is not spherically complete; see Definition 2.6. Consider a decreasing sequence of closed disks $D_1 \supsetneq D_2 \supsetneq \cdots$ with empty intersection. Denoting the sup-norm on $\mathbb{C}_v[z]$ associated to D_n by ζ_n, define a corresponding multiplicative norm $\|\cdot\| : \mathbb{C}_v[z] \to [0, \infty)$ by

$$\|f\| := \lim_{n\to\infty} \|f\|_{\zeta_n}.$$

We leave it to the reader to check, in Exercise 6.2, that this limit converges, that $\|\cdot\|$ is a multiplicative norm extending $|\cdot|$, and that any equivalent sequence of closed disks $\{E_n\}$ gives the same norm. Here, we say that two

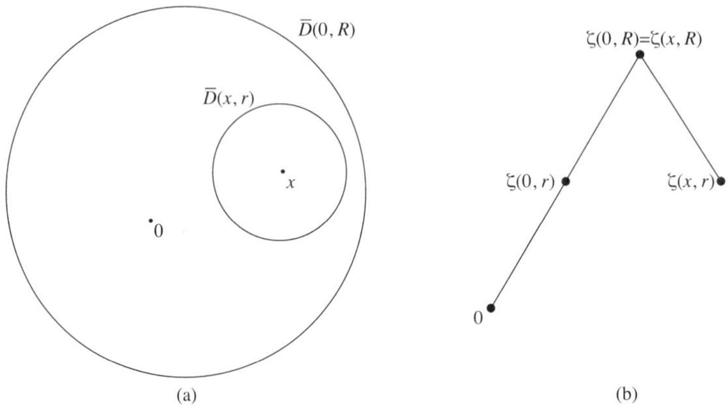

Figure 6.1. The path from 0 to $\zeta(x,r)$ in $\mathbb{A}^1_{\text{an}}$

decreasing sequences $\{D_n\}$, $\{E_n\}$ of closed disks with empty intersection are *equivalent* if for every $N \geq 1$, there is some $M \geq N$ such that $E_M \subseteq D_N$ and $D_M \subseteq E_N$. A point $\zeta \in \mathbb{A}^1_{\text{an}}$ is said to be of *Type IV* if it is associated in this way with an equivalence class of decreasing sequences of disks with empty intersection.

Even though we have not discussed the topology on $\mathbb{A}^1_{\text{an}}$ carefully, we can already begin to understand why the space is path-connected. For example, given $x \in \mathbb{C}_v$ and $r > 0$ such that $0 < r < |x|$, we can construct a path from the Type I point 0 to the Type II or III point $\zeta(x,r)$ as follows.

Set $R = |x|$, and consider the arrangement of the points 0 and x and the disks $\overline{D}(x,r)$ and $\overline{D}(0,R)$ in Figure 6.1(a). To draw the path in Berkovich space, start at the Type I point 0, which we can informally think of as a disk of radius $s = 0$. Increase the radius s through a path consisting of Type II and III points of the form $\zeta(0,s)$ until $s = R$. Then $\zeta(0,R) = \zeta(x,R)$, because $\overline{D}(0,R) = \overline{D}(x,R)$. Thus, we may now consider Type II and III points $\zeta(x,s)$ corresponding to disks $\overline{D}(x,s)$ centered at s. By decreasing the radius s, we eventually arrive at the desired point $\zeta(x,r)$, corresponding to the disk $\overline{D}(x,r)$; see Figure 6.1(b).

6.1.2. Diameter and absolute value.

Definition 6.3. Let $\zeta \in \mathbb{A}^1_{\text{an}}$. The *absolute value* of ζ is

$$|\zeta| := \|z\|_\zeta,$$

and the *diameter* of ζ is

$$\operatorname{diam}(\zeta) := \inf\big\{\|z-a\|_\zeta : a \in \mathbb{C}_v\big\}.$$

Clearly, the absolute value of a Type I point x coincides with the usual definition of $|x|$. On the other hand, the absolute value of a Type II or III point $\zeta(a,r)$ is $|\zeta(a,r)| = \max\{|a|, r\}$. If ζ is a Type IV point approximated by a decreasing sequence $\overline{D}(a_1,r_1) \supsetneq \overline{D}(a_2,r_2) \supsetneq \cdots$ of disks with empty intersection, then $|\zeta| = |\zeta(a_n,r_n)|$ for any n large enough that $0 \notin \overline{D}(a_n,r_n)$.

If $\zeta = \zeta(a,r)$ is the point of Type II or III corresponding to the disk $\overline{D}(a,r)$, then

$$\operatorname{diam}(\zeta) = r = \operatorname{diam}\left(\overline{D}(a,r)\right),$$

where the second use of diam above is the metric space diameter of the disk, justifying our use of the term "diameter" in this context. The diameter of a Type I point $x \in \mathbb{C}_v$ is 0, which is the same as the diameter of the set $\{x\}$. For a Type IV point ζ defined by a decreasing sequence $D_1 \supsetneq D_2 \supsetneq \cdots$, the reader can check that

$$\operatorname{diam}(\zeta) = \lim_{n\to\infty} \operatorname{diam}(D_n).$$

(See Exercise 6.3.) In particular, by Lemma 2.7(b), Type IV points have strictly positive diameter. Unlike points of the other three types, however, there is no corresponding subset of $\mathbb{C}_v$ whose diameter is $\operatorname{diam}(\zeta)$.

6.2. Disks in the Berkovich affine line

Just as $\mathbb{A}^1_{\mathrm{an}}$ is a Berkovich space version of $\mathbb{A}^1(\mathbb{C}_v)$, we also have Berkovich versions of disks, as follows.

Definition 6.4. Let $a \in \mathbb{C}_v$, and let $r > 0$.

(a) The *closed Berkovich disk* $\overline{D}_{\mathrm{an}}(a,r)$ is

$$\overline{D}_{\mathrm{an}}(a,r) := \left\{\zeta \in \mathbb{A}^1_{\mathrm{an}} : \|z-a\|_\zeta \le r\right\}.$$

(b) The *open Berkovich disk* $D_{\mathrm{an}}(a,r)$ is

$$D_{\mathrm{an}}(a,r) := \left\{\zeta \in \mathbb{A}^1_{\mathrm{an}} : \|z-a\|_\zeta < r\right\}.$$

Open and closed disks both inherit the Gel'fand topology from $\mathbb{A}^1_{\mathrm{an}}$. If $r \in |\mathbb{C}_v^\times|$, we say the Berkovich disks $\overline{D}_{\mathrm{an}}(a,r)$ and $D_{\mathrm{an}}(a,r)$ are *rational*; otherwise, they are *irrational*.

The names "open" and "closed" in Definition 6.4 provide some intuition about the Gel'fand topology, via the following proposition.

Proposition 6.5. *Let $a \in \mathbb{C}_v$ and $r > 0$. Then*

$$D_{\mathrm{an}}(a,r) \text{ is open, and } \overline{D}_{\mathrm{an}}(a,r) \text{ is closed in } \mathbb{A}^1_{\mathrm{an}}.$$

Proof. Let $f(z) = z - a \in \mathbb{C}_v[z]$, and let $U := (-\infty, r) \subseteq \mathbb{R}$. By definition of the Gel'fand topology, the map $F : \mathbb{A}^1_{\mathrm{an}} \to \mathbb{R}$ by $F(\zeta) := \|f\|_\zeta$ is continuous, and hence

$$D_{\mathrm{an}}(a,r) = \left\{\zeta \in \mathbb{A}^1_{\mathrm{an}} : \|z-a\|_\zeta < r\right\} = F^{-1}(U)$$

is an open subset of $\mathbb{A}^1_{\mathrm{an}}$.

Similarly, $V := (-\infty, r] \subseteq \mathbb{R}$ is closed in $\mathbb{R}$, and hence

$$\overline{D}_{\mathrm{an}}(a,r) = \left\{\zeta \in \mathbb{A}^1_{\mathrm{an}} : \|z-a\|_\zeta \le r\right\} = F^{-1}(V)$$

is closed in $\mathbb{A}^1_{\mathrm{an}}$. □

It is immediate from Definition 6.4 that $\overline{D}_{\mathrm{an}}(a,r)$ contains all Type I points of the form $\|\cdot\|_x$ for $x \in \overline{D}(a,r)$, that $D_{\mathrm{an}}(a,r)$ contains all Type I points of the form $\|\cdot\|_x$ for $x \in D(a,r)$, and that

$$D_{\mathrm{an}}(a,r) = \bigcup_{0<s<r} \overline{D}_{\mathrm{an}}(a,s).$$

We leave it to the reader to check that

$$\mathbb{A}^1_{\mathrm{an}} = \bigcup_{s>0} \overline{D}_{\mathrm{an}}(a,s) = \bigcup_{s>0} D_{\mathrm{an}}(a,s). \tag{6.1}$$

(See Exercise 6.4.) It is also easy to check that

- $\overline{D}(a,r) \subseteq \overline{D}(b,s)$ if and only if $\overline{D}_{\mathrm{an}}(a,r) \subseteq \overline{D}_{\mathrm{an}}(b,s)$,
- $D(a,r) \subseteq D(b,s)$ if and only if $D_{\mathrm{an}}(a,r) \subseteq D_{\mathrm{an}}(b,s)$,

and similarly if we replace the inclusions by equalities. In particular, the Berkovich disk $\overline{D}_{\mathrm{an}}(a,r)$ or $D_{\mathrm{an}}(a,r)$ is independent of the choice of center a of the corresponding closed disk $\overline{D}(a,r) \subseteq \mathbb{C}_v$ or, respectively, the corresponding open disk $D(a,r) \subseteq \mathbb{C}_v$.

Thus, there is a one-to-one correspondence between closed Berkovich disks and closed disks in $\mathbb{C}_v$, and between open Berkovich disks and open disks in $\mathbb{C}_v$. However, the reader should note that an important subtlety arises for irrational disks. Recall that if $r \notin |\mathbb{C}_v^\times|$, then the irrational closed and open disks $\overline{D}(a,r)$ and $D(a,r)$ in $\mathbb{C}_v$ coincide. However, the corresponding Berkovich disks $\overline{D}_{\mathrm{an}}(a,r)$ and $D_{\mathrm{an}}(a,r)$ differ, solely because the former contains the Type III point $\zeta(a,r)$ but the latter does not.

Of course, if $r \in |\mathbb{C}_v^\times|$, then the rational closed disk $\overline{D}_{\mathrm{an}}(a,r)$ contains *many* points not in $D_{\mathrm{an}}(a,r)$, including all the Type I points in the closed annulus $\{|z-a| = r\} := \overline{D}(a,r) \smallsetminus D(a,r)$. In fact, we have the following result.

Proposition 6.6. *Let $a \in \mathbb{C}_v$, and let $r > 0$. Then $\overline{D}_{\mathrm{an}}(a,r) \smallsetminus \{\zeta(a,r)\}$ is the disjoint union*

$$\overline{D}_{\mathrm{an}}(a,r) \smallsetminus \{\zeta(a,r)\} = \coprod_{c \in C} D_{\mathrm{an}}(c,r),$$

where C is a set consisting of one representative from each open disk $D(c,r)$ of radius r contained in $\overline{D}(a,r)$.

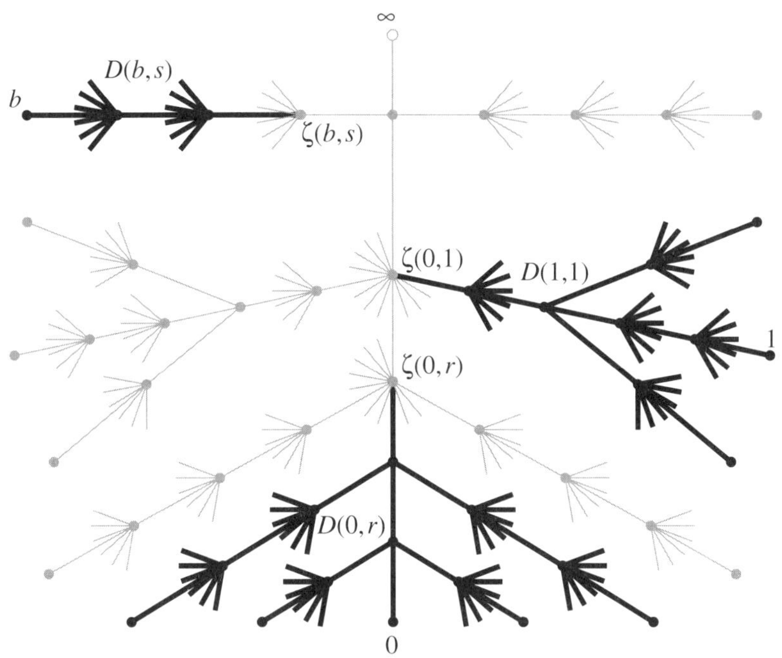

Figure 6.2. $\mathbb{A}^1_{\mathrm{an}}$ with three rational open disks highlighted

Proof. See Section 15.2. □

Figure 6.2 illustrates $\mathbb{A}^1_{\mathrm{an}}$, with three rational open disks highlighted: $D_{\mathrm{an}}(0,r)$ for some $0 < r < 1$; $D_{\mathrm{an}}(1,1)$; and $D_{\mathrm{an}}(b,s)$ for some $|b| > 1$ and $0 < s < |b|$. Note that each open disk has a unique boundary point (in this case, $\zeta(0,r)$, $\zeta(1,1) = \zeta(0,1)$, and $\zeta(b,s)$, respectively), and that the rational open disk consists of precisely one of the infinitely many branches emanating from that boundary point.

Remark 6.7. Proposition 6.6 implicitly distinguishes between points of Types II and III, as follows. Any closed Berkovich disk $\overline{D}_{\mathrm{an}}(a,r)$ has a corresponding point $\zeta(a,r)$ of Type II or III that we say *bounds* the disk. If $r \in |\mathbb{C}_v^\times|$, then there are infinitely many open disks $D(c,r)$ contained in $\overline{D}(a,r)$, corresponding to the infinitely many elements of the residue field. On the other hand, if $r \notin |\mathbb{C}_v^\times|$, then $D(a,r) = \overline{D}(a,r)$. Thus, a closed Berkovich disk bounded by a Type II point can be partitioned into infinitely many open subdisks of the same radius (along with the bounding point), whereas if the bounding point is Type III, there is only one open disk in the partition.

6.3. Berkovich's classification

We now formally define the four types of Berkovich points we described in Section 6.1.1. Recall our notation of the seminorms $\|\cdot\|_x$ and $\|\cdot\|_{\zeta(a,r)}$, given by $\|f\|_x := |f(x)|$ and

$$\|f\|_{\zeta(a,r)} := \sup\{|f(x)| : x \in D(a,r)\} = \sup\{|f(x)| : x \in \overline{D}(a,r)\}.$$

Definition 6.8. We say that a Berkovich point $\zeta \in \mathbb{A}^1_{\text{an}}$ is of

Type I if $\|\cdot\|_\zeta = \|\cdot\|_x$ for some $x \in \mathbb{C}_v$;

Type II if $\|\cdot\|_\zeta = \|\cdot\|_{\zeta(a,r)}$ for some rational closed disk $\overline{D}(a,r) \subseteq \mathbb{C}_v$;

Type III if $\|\cdot\|_\zeta = \|\cdot\|_{\zeta(a,r)}$ for some irrational disk $\overline{D}(a,r) \subseteq \mathbb{C}_v$; or

Type IV if $\|\cdot\|_\zeta = \lim_{n\to\infty} \|\cdot\|_{\zeta_n}$, where the norms $\{\zeta_n\}$ correspond to a decreasing sequence of closed disks in $\mathbb{C}_v$ with empty intersection.

Of the four types, note that only the Type I points fail to be norms.

Theorem 6.9 (Berkovich's classification). *Let $\zeta \in \mathbb{A}^1_{\text{an}}$. Then ζ is of exactly one of the four types in Definition* 6.8. *Moreover, ζ determines the corresponding geometric data up to equivalence, in the sense that*

(a) *for points of Type I, $\|\cdot\|_x = \|\cdot\|_y$ if and only if $x = y$;*

(b) *for points of Type II or III, $\zeta(a,r) = \zeta(b,s)$ if and only if $\overline{D}(a,r) = \overline{D}(b,s)$;*

(c) *for points of Type IV, $\lim_{n\to\infty} \zeta_n = \lim_{n\to\infty} \xi_n$ if and only if the sequences $\{D_n\}$ and $\{E_n\}$ of closed disks corresponding to $\{\zeta_n\}$ and $\{\xi_n\}$ are equivalent.*

Proof. See Section 15.2. □

Berkovich proved Theorem 6.9 in [**26**, Section 1.4.4]; another presentation appears in [**8**, Theorem 1.2].

With Theorem 6.9 in hand to help us understand what Berkovich points are, we have the language needed to describe the structure of Berkovich disks, at least informally. Intuitively, the closed Berkovich disk $\overline{D}_{\text{an}}(a,r)$ looks like a tree branching downward from the top-most point $\zeta(a,r)$, with infinitely many branches at every Type II point, as described by Proposition 6.6 and Remark 6.7. More precisely, the branches at a Type II point $\zeta(b,s)$ correspond to the infinitely many open subdisks $D(c,s)$ of $\overline{D}(b,s)$ of the same radius, as well as (if $s < r$) one more branch extending upward, corresponding to increasing the radius. On the other hand, by the same proposition and remark, the Type III points are interior points with no branching.

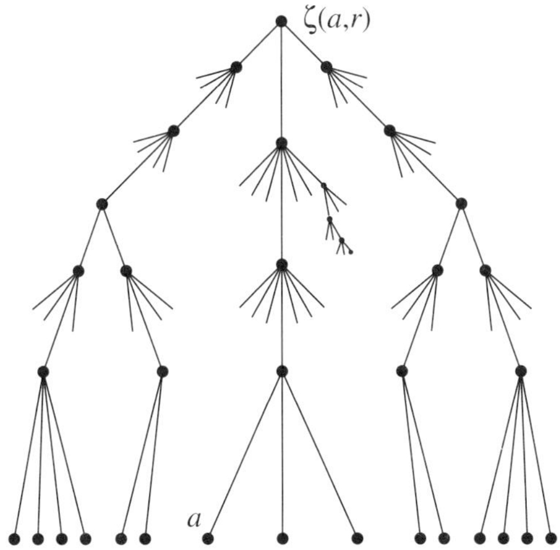

Figure 6.3. The closed Berkovich disk $\overline{D}_{\text{an}}(a, r)$

The limbs of the tree end at the Type I and Type IV points. Figure 6.3 illustrates the tree, with Berkovich points of larger diameter appearing higher up in the figure. Only a few of the (Type II) branch points appear in Figure 6.3; in reality, such points would be dense in each line segment, with an infinite number of branches at each one. The tips of the branches appearing at the bottom of the tree are Type I points. Figure 6.3 also attempts to illustrate one Type IV point, which appears as a small chain of branches near the middle of the tree; the full Berkovich disk has such points scattered throughout. The Type IV point in question would be at the end of an infinite chain of such branches, but still with positive diameter and hence strictly above the bottom of the figure. Figure 6.3 also contains a diagram of the open Berkovich disk $D_{\text{an}}(a, r)$, which is the single branch emanating from the point $\zeta(a, r)$ and containing the Type I point a.

Warning 6.10. The Gel'fand topology is weaker than the tree image might suggest at first. For example, the Type I points (at the tips of the branches) are dense in the full space, as we will see in Theorem 6.25. In particular, any open set containing a Type II point ζ must in fact contain all but finitely many of the branches emanating from ζ; see Proposition 6.19 and Lemma 15.6.

6.4. The Berkovich projective line

Just as the classical projective line in algebraic geometry can be defined by gluing two copies of the affine line $\mathbb{A}^1$ via the map $z \mapsto 1/z$, the Berkovich projective line can be defined by a similar gluing. First, however, we need

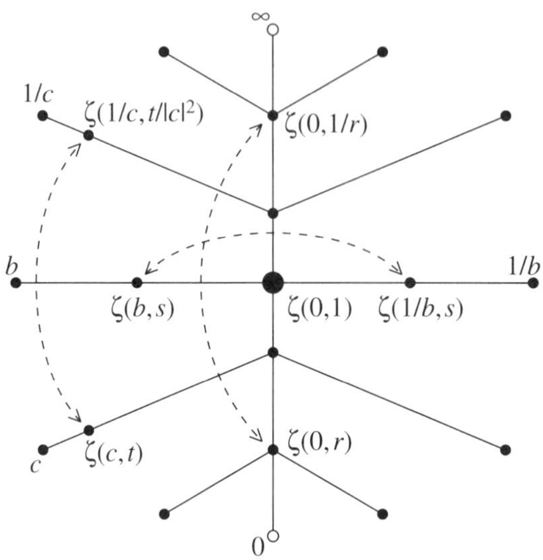

Figure 6.4. The map $\zeta \mapsto 1/\zeta$ on $\mathbb{A}^1_{\text{an}} \smallsetminus \{0\}$

to define the map $\zeta \mapsto 1/\zeta$, which we do as follows. This map is a function from $\mathbb{A}^1_{\text{an}} \smallsetminus \{0\}$ to itself, taking the seminorm $\|\cdot\|_\zeta$ on $\mathbb{C}_v[z]$ to the seminorm $\|\cdot\|_{1/\zeta}$ given by

$$\|f\|_{1/\zeta} := \frac{\|z^{\deg(f)} f(1/z)\|_\zeta}{\|z\|_\zeta^{\deg(f)}} \quad \text{for } f \in \mathbb{C}_v[z]. \tag{6.2}$$

We leave it to the reader to check that

- $\mathbb{A}^1_{\text{an}} \smallsetminus \{0\}$ is an open subset of $\mathbb{A}^1_{\text{an}}$;
- $1/\zeta$, as defined in (6.2), is indeed an element of $\mathbb{A}^1_{\text{an}} \smallsetminus \{0\}$;
- the map $\zeta \mapsto 1/\zeta$ is a homeomorphism of $\mathbb{A}^1_{\text{an}} \smallsetminus \{0\}$; and
- the map $\zeta \mapsto 1/\zeta$ preserves Berkovich types (I–IV).

See Exercises 6.5, 6.6, and 6.7. Figure 6.4 shows how $\zeta \mapsto 1/\zeta$ acts on some points in $\mathbb{A}^1_{\text{an}} \smallsetminus \{0\}$, as described in Exercise 6.7.

Having verified the above properties of this map, we are now prepared to define $\mathbb{P}^1_{\text{an}}$.

Definition 6.11. The Berkovich projective line $\mathbb{P}^1_{\text{an}}$ is the topological space formed by gluing two copies of $\mathbb{A}^1_{\text{an}}$ along their subsets $\mathbb{A}^1_{\text{an}} \smallsetminus \{0\}$ via the map $\zeta \mapsto 1/\zeta$ of equation (6.2).

As is common for the classical projective line $\mathbb{P}^1$, we often write

$$\mathbb{P}^1_{\text{an}} = \mathbb{A}^1_{\text{an}} \cup \{\infty\},$$

where ∞ is the point 0 on the second copy of $\mathbb{A}^1_{\mathrm{an}}$. Note that $\mathbb{A}^1_{\mathrm{an}}$ is an open subset of $\mathbb{P}^1_{\mathrm{an}}$, by the first bullet point above.

We now extend Definition 6.3 of the diameter and absolute value to every point $\zeta \in \mathbb{P}^1_{\mathrm{an}}$, by

$$|\zeta| := \begin{cases} |\zeta| & \text{if } \zeta \in \mathbb{A}^1_{\mathrm{an}}, \\ \infty & \text{if } \zeta = \infty, \end{cases} \quad \text{and} \quad \operatorname{diam}(\zeta) := \begin{cases} \operatorname{diam}(\zeta) & \text{if } \zeta \in \mathbb{A}^1_{\mathrm{an}}, \\ 0 & \text{if } \zeta = \infty. \end{cases}$$

In particular, a point $\zeta \in \mathbb{P}^1_{\mathrm{an}}$ is of Type I if and only if its diameter is zero.

Warning 6.12. The absolute value function $|\cdot| : \mathbb{P}^1_{\mathrm{an}} \to \mathbb{R} \cup \{\infty\}$ is continuous with respect to the Gel'fand topology, but the diameter function $\operatorname{diam} : \mathbb{P}^1_{\mathrm{an}} \to \mathbb{R}$ is *not*. The problem is not just the point at ∞; the diam function fails to be continuous on $\mathbb{A}^1_{\mathrm{an}}$ and on every affinoid. However, diam is upper semicontinuous on $\mathbb{A}^1_{\mathrm{an}}$; see Exercise 6.9.

Unlike the classical projective line from algebraic geometry, however, $\mathbb{P}^1_{\mathrm{an}}$ can also be constructed by gluing two closed unit *disks* via $\zeta \mapsto 1/\zeta$. The intuitive reason why this is possible in the non-archimedean setting is that the set $\{x \in \mathbb{P}^1(\mathbb{C}_v) : |x| = 1\}$ is an *open* subset of $\mathbb{P}^1(\mathbb{C}_v)$, unlike the unit circle $\{x \in \mathbb{P}^1(\mathbb{C}) : |x| = 1\}$ in the Riemann sphere.

Proposition 6.13. *The Berkovich projective line $\mathbb{P}^1_{\mathrm{an}}$ is homeomorphic to the space formed by gluing two copies of the Berkovich disk $\overline{D}_{\mathrm{an}}(0,1)$ along the Berkovich annulus*

$$\overline{D}_{\mathrm{an}}(0,1) \smallsetminus D_{\mathrm{an}}(0,1) = \{\zeta \in \mathbb{A}^1_{\mathrm{an}} : |\zeta| = 1\},$$

via the identification $\zeta \mapsto 1/\zeta$ of equation (6.2).

Proof. See Section 15.3. Incidentally, the function $\zeta \mapsto 1/\zeta$ maps this annulus $\{\zeta : |\zeta| = 1\}$ bijectively onto itself; see Exercise 6.10. □

Proposition 6.13 allows us to picture $\mathbb{P}^1_{\mathrm{an}}$ as $\overline{D}_{\mathrm{an}}(0,1)$ with an extra copy of the open tree $D_{\mathrm{an}}(0,1)$ attached to the top of the point $\zeta(0,1)$. The new top portion contains all points ζ of $\mathbb{P}^1_{\mathrm{an}}$ with $|\zeta| > 1$, including ∞. That is, it contains all Type I points $x \in \mathbb{P}^1(\mathbb{C}_v)$ with $|x| > 1$, all Type II and III points $\zeta(a,r)$ with $|a| > 1$ or $r > 1$, and all Type IV points corresponding to decreasing sequences of disks outside $\overline{D}(0,1)$. For a rough idea of the space, see Figure 6.5, which highlights the four Type I points $0, 1, \infty$, and a, for some $a \in \mathbb{C}_v$ with $0 < |a| < 1$. It also highlights several Type II points, including $\zeta(0,1)$ and $\zeta(0,|a|)$. As is the case with $\mathbb{A}^1_{\mathrm{an}}$, the space looks like a tree with infinite branching at a dense set of points (the Type II points) along every edge.

The infinite branching at each Type II point, as well as the lack of branching at Type III points, can be summed up more formally by the

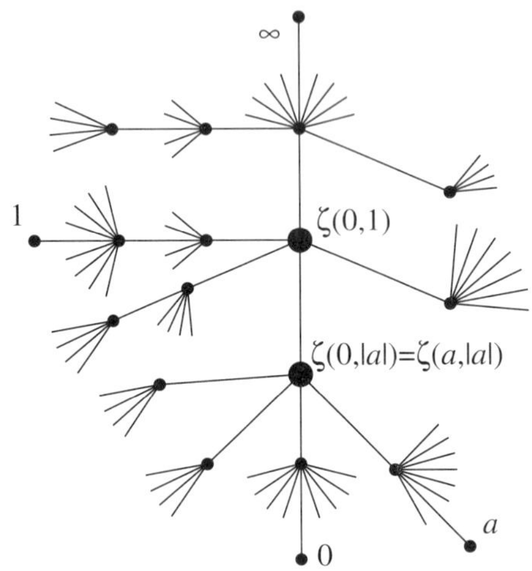

Figure 6.5. The Berkovich projective line $\mathbb{P}^1_{\text{an}}$

following extension of Proposition 6.6. To see the contrast between points of Type II and III, recall again that if $\zeta(a,r)$ is of Type III, then there is only one open disk of the form $D(c,r)$ contained in $\overline{D}(a,r)$; but if $\zeta(a,r)$ is of Type II, there are infinitely many.

Corollary 6.14. *Let $a \in \mathbb{C}_v$, and let $r > 0$. The set $\mathbb{P}^1_{\text{an}} \smallsetminus \{\zeta(a,r)\}$ formed by deleting the Type II or III point $\zeta(a,r)$ from the Berkovich projective line is the disjoint union*

$$\mathbb{P}^1_{\text{an}} \smallsetminus \{\zeta(a,r)\} = W_\infty \amalg \coprod_{c \in \mathcal{C}} D_{\text{an}}(c,r),$$

where $\mathcal{C}$ is a set consisting of one representative from each open disk $D(c,r)$ of radius r contained in $\overline{D}(a,r)$, and

$$W_\infty := \mathbb{P}^1_{\text{an}} \smallsetminus \overline{D}_{\text{an}}(a,r).$$

Proof. This is immediate from Proposition 6.6. □

Remark 6.15. The Type II point $\zeta(0,1)$ corresponding to the disk $\overline{D}(0,1)$ is not qualitatively different from any other Type II point. For example, we will see in Remark 7.11 that any Type II point may be moved to $\zeta(0,1)$ by a change of coordinates. Still, it is common to fix a choice of coordinate on $\mathbb{P}^1_{\text{an}}$, just as it is common to fix a choice of coordinate on the classical projective line. In that case, the point $\zeta(0,1)$ is often called the *Gauss point* (for example, in [**8**]) or the *canonical point* (for example, in [**114, 115, 117**]).

6.5. Disks and affinoids in $\mathbb{P}^1_{\mathrm{an}}$

We saw in Proposition 6.5 that open disks in $\mathbb{A}^1_{\mathrm{an}}$ are topologically open, and similarly for closed disks. That observation leads us to define the following Berkovich analogues of $\mathbb{P}^1(\mathbb{C}_v)$-disks and affinoids.

Definition 6.16.

(a) A *closed Berkovich disk* in $\mathbb{P}^1_{\mathrm{an}}$ is either a closed Berkovich disk $\overline{D}_{\mathrm{an}}(a,r)$ or the complement $\mathbb{P}^1_{\mathrm{an}} \smallsetminus D_{\mathrm{an}}(a,r)$ of an open Berkovich disk.

An *open Berkovich disk* in $\mathbb{P}^1_{\mathrm{an}}$ is either an open Berkovich disk $D_{\mathrm{an}}(a,r)$ or the complement $\mathbb{P}^1_{\mathrm{an}} \smallsetminus \overline{D}_{\mathrm{an}}(a,r)$ of a closed Berkovich disk.

If $r \in |\mathbb{C}_v^\times|$, we say the Berkovich disk is *rational*; otherwise, we say it is *irrational*.

(b) A *connected Berkovich affinoid* is a nonempty intersection of finitely many $\mathbb{P}^1_{\mathrm{an}}$-disks.

If $D_1, \ldots, D_n$ are closed (respectively, open, rational, irrational) Berkovich disks, we say that the connected Berkovich affinoid

$$D_1 \cap \cdots \cap D_n$$

is also *closed* (respectively, *open*, *rational*, *irrational*).

(c) A *Berkovich affinoid* is a nonempty finite union of connected Berkovich affinoids. If all of the components in the union are closed (respectively, open, rational, irrational) connected Berkovich affinoids, we say the resulting Berkovich affinoid is also *closed* (respectively, *open*, *rational*, *irrational*).

Figure 6.6 shows $\mathbb{P}^1_{\mathrm{an}} \smallsetminus (D_{\mathrm{an}}(1,1) \cup D_{\mathrm{an}}(b,s) \cup D_{\mathrm{an}}(0,r))$, which is a closed connected Berkovich affinoid obtained by removing three open disks from $\mathbb{P}^1_{\mathrm{an}}$; in fact, the disks removed are precisely the ones shown earlier in Figure 6.2. On the other hand, Figure 6.7 shows an open connected Berkovich affinoid formed by removing two closed disks from $\mathbb{P}^1_{\mathrm{an}}$. (*Note*: The radii r and s are different in the latter figure.) One of the two disks removed contains ∞, and thus the affinoid in question is the (Berkovich) annulus $U = D_{\mathrm{an}}(0,s) \smallsetminus \overline{D}_{\mathrm{an}}(a,r)$. Indeed, it is appropriate to call U an annulus, because it can also be written as

$$U = D_{\mathrm{an}}(a,s) \smallsetminus \overline{D}_{\mathrm{an}}(a,r) = \big\{\zeta \in \mathbb{P}^1_{\mathrm{an}} : r < \|z-a\|_\zeta < s\big\}.$$

Note that according to Definition 6.16, the full set $\mathbb{P}^1_{\mathrm{an}}$ is considered a connected Berkovich affinoid, since it is the intersection of zero disks. In addition, as noted above, all open Berkovich disks are open subsets of $\mathbb{P}^1_{\mathrm{an}}$, and closed Berkovich disks are closed. Hence, open Berkovich affinoids are

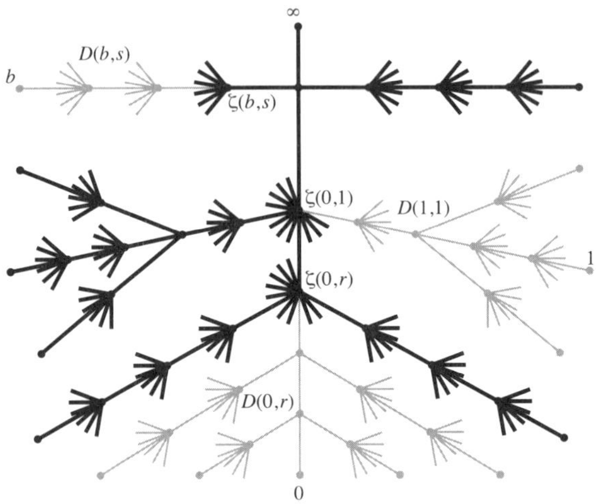

Figure 6.6. The closed connected Berkovich affinoid $\mathbb{P}^1_{\mathrm{an}} \setminus (D_{\mathrm{an}}(1,1) \cup D_{\mathrm{an}}(b,s) \cup D_{\mathrm{an}}(0,r))$

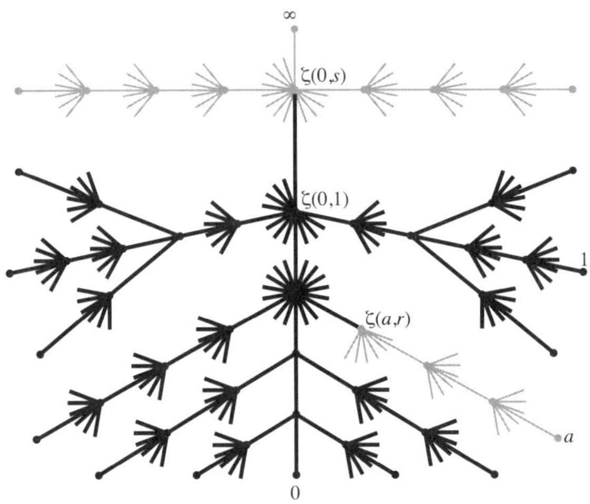

Figure 6.7. The open connected Berkovich affinoid $D_{\mathrm{an}}(0,s) \setminus \overline{D}_{\mathrm{an}}(a,r)$

open subsets, and closed Berkovich affinoids are closed. In addition, as we will see in Theorem 6.32, connected Berkovich affinoids are connected in the topological sense. For the moment, however, the following result shows that Berkovich affinoids provide us with a very useful characterization of the Gel'fand topology. Recall that a *basis* $\mathcal{B}$ for a topology $\mathcal{U}$ is a subset of $\mathcal{U}$ such that every element of $\mathcal{U}$ is a union of elements of $\mathcal{B}$.

Theorem 6.17. *The set of open connected Berkovich affinoids in* $\mathbb{P}^1_{\mathrm{an}}$ *forms a basis for the Gel'fand topology.*

Proof. See Section 15.3. □

Warning 6.18.

(a) The set of open Berkovich *disks* does *not* form a basis for the Gel'fand topology; see Exercise 6.12.

(b) There is a strong parallel between the $\mathbb{P}^1(\mathbb{C}_v)$ and the $\mathbb{P}^1_{\mathrm{an}}$ versions of affinoids, including disks; but this parallel has its limitations. In particular, unlike affinoids in $\mathbb{P}^1(\mathbb{C}_v)$, closed Berkovich affinoids are *not* topologically open, and open Berkovich affinoids are *not* topologically closed. (That's good, because we will show in Theorem 6.32 that $\mathbb{P}^1_{\mathrm{an}}$ is path-connected.)

(c) The closure of a rational open Berkovich disk is *not* the corresponding closed disk. Instead, the closure of the rational open disk $D_{\mathrm{an}}(a,r)$ is $D_{\mathrm{an}}(a,r) \cup \{\zeta(a,r)\}$. Since $r \in |\mathbb{C}_v^\times|$, the closure is *not* the closed disk $\overline{D}_{\mathrm{an}}(a,r)$, which also contains infinitely many other open disks $D_{\mathrm{an}}(c,r)$, for $c \in \{|z-a| = r\}$; see Proposition 6.6.

In light of Theorem 6.17, it is important to understand open connected Berkovich affinoids.

Proposition 6.19. *Let $U \subseteq \mathbb{P}^1_{\mathrm{an}}$ be an open connected Berkovich affinoid.*

(a) *There is a finite set of pairwise disjoint closed Berkovich disks $\{V_1, \ldots, V_n\}$ such that*

$$U = \mathbb{P}^1_{\mathrm{an}} \smallsetminus \bigl(V_1 \cup \cdots \cup V_n\bigr).$$

(b) *Let $\zeta = \zeta(a,r) \in U$ be a Type II point. Then U contains all but finitely many of the branches emanating from ζ. That is, U contains all but finitely many of the disks W_∞ and $D_{\mathrm{an}}(c,r)$ for $c \in \overline{D}(a,r)$ from Corollary 6.14.*

In addition, for all *of the branches W emanating from ζ, we have $U \cap W \neq \varnothing$.*

Proof. This is left to the reader; see Exercise 6.11. □

Proposition 6.19(b) says that an open connected Berkovich affinoid U is obtained from the tree of Figure 6.5 by choosing a point ζ in the tree, making cuts at finitely many other points of Type II or III, and letting U be the portion still connected to ζ via some path through the tree. Note that when the cuts are made, the point ξ of the cut itself is removed, along with any branches off ξ, besides the one containing ζ.

Example 6.20. Choose $a \in \mathbb{C}_v$ with $0 < |a| < 1$, as in Figure 6.5. Let

$$U := \{|a| < |\zeta| < 1\} = D_{\mathrm{an}}(0,1) \smallsetminus \overline{D}_{\mathrm{an}}(0,|a|),$$

which is an open Berkovich annulus. (That is, it is the open connected Berkovich affinoid formed by intersecting the two open disks $D_{\text{an}}(0,1)$ and $\mathbb{P}^1_{\text{an}} \smallsetminus \overline{D}_{\text{an}}(0,|a|)$.) Let $\zeta = \zeta(0,|a|^{1/2}) \in U$, which lies on the line segment between $\zeta(0,|a|)$ and $\zeta(0,1)$ in Figure 6.5. Then U is formed by cutting the tree at $\zeta(0,|a|)$ and $\zeta(0,1)$, and only keeping the portion that is still connected to our chosen point ζ.

Thus, U appears in Figure 6.5 as the interval from $\zeta(0,|a|)$ to $\zeta(0,1)$, *not* including the two endpoints. Of course, there are infinitely branches (not shown in Figure 6.5) emanating from each of the infinitely many Type II points in that interval. However, $\zeta(0,1)$ and all of its branches have been removed, with the exception of the one pointing downward toward 0. (This does not contradict Proposition 6.19(b), because $\zeta(0,1) \notin U$.) Similarly, $\zeta(0,|a|)$ has also been removed along with all of its branches other than the one pointing upward toward ∞.

On the other hand, $\zeta = \zeta(0,|a|^{1/2})$ is a Type II point in U, and we now confirm that it satisfies the conclusions of Proposition 6.19(b). Although U does *not* contain the branch $W_\infty := \mathbb{P}^1_{\text{an}} \smallsetminus \overline{D}_{\text{an}}(0,|a|^{1/2})$, it *does* intersect it, for example, at $\zeta(0,|a|^{1/4})$ and at any Type I point x with $|a|^{1/2} < |x| < 1$. Similarly, U does not contain the branch $D_{\text{an}}(0,|a|^{-1/2})$, but it intersects it at points of absolute value strictly between $|a|$ and $|a|^{1/2}$. Every other branch off ζ is an open Berkovich disk of the form $V = D_{\text{an}}(c,|a|^{1/2})$ with $|c| = |a|^{1/2}$, and we have $V \subseteq U$, as predicted by Proposition 6.19(b).

The topological boundary of a Berkovich disk is a single point of Type II or III. Similarly, the boundary of a Berkovich affinoid is a finite set of points of Types II and III; see Exercise 6.19. For instance, in Example 6.20, the boundary of the annulus $U = \{|a| < |\zeta| < 1\}$ consists of the two points $\zeta(0,|a|)$ and $\zeta(0,1)$. The line segment between these two points is called the *skeleton* of U; see Exercise 6.29 for the definition of skeleta of arbitrary connected affinoids.

Among its many uses, Theorem 6.17 allows us to prove the next two results, which together say that $\mathbb{P}^1_{\text{an}}$ is a Hausdorff compactification of $\mathbb{P}^1(\mathbb{C}_v)$. In particular, in light of the following theorem, we hereafter consider $\mathbb{P}^1(\mathbb{C}_v)$ to be simply the set of Type I points in $\mathbb{P}^1_{\text{an}}$.

Theorem 6.21. *The function $i : \mathbb{P}^1(\mathbb{C}_v) \to \mathbb{P}^1_{\text{an}}$ given by*

$$i(x) = \begin{cases} \|\cdot\|_x & \text{if } x \in \mathbb{C}_v, \\ \infty & \text{if } x = \infty, \end{cases}$$

is an embedding of topological spaces. Moreover, the image $i(\mathbb{P}^1(\mathbb{C}_v))$ is precisely the set of Type I points of $\mathbb{P}^1_{\text{an}}$.

Proof. See Section 15.4. □

Theorem 6.22. *The Berkovich projective line $\mathbb{P}^1_{\mathrm{an}}$ is compact and Hausdorff.*

Proof. See Section 15.4. □

As a compact Hausdorff space, $\mathbb{P}^1_{\mathrm{an}}$ is also normal, meaning that any two disjoint closed subsets can be separated by open sets. Incidentally, it follows fairly quickly from the definitions that any two disjoint closed Berkovich affinoids can be separated by open Berkovich affinoids; see Exercise 6.20.

Remark 6.23. The reader can easily check the following relationship between Berkovich disks and their classical counterparts in $\mathbb{P}^1(\mathbb{C}_v)$; see Exercise 6.15. For any $a \in \mathbb{C}_v$ and any $r > 0$,

$$(6.3)\qquad \begin{aligned} \overline{D}_{\mathrm{an}}(a,r) \cap \mathbb{P}^1(\mathbb{C}_v) &= \overline{D}(a,r), \\ D_{\mathrm{an}}(a,r) \cap \mathbb{P}^1(\mathbb{C}_v) &= D(a,r), \end{aligned}$$

and similarly for the disks $\mathbb{P}^1_{\mathrm{an}} \smallsetminus \overline{D}_{\mathrm{an}}(a,r)$ and $\mathbb{P}^1_{\mathrm{an}} \smallsetminus D_{\mathrm{an}}(a,r)$. Since every disk in $\mathbb{P}^1_{\mathrm{an}}$ or $\mathbb{P}^1(\mathbb{C}_v)$ is of this form, every Berkovich disk gives us a classical disk, and every classical disk comes from a Berkovich disk.

This relationship is almost a one-to-one correspondence between Berkovich disks and classical disks, but not quite. The same subtlety we noted about irrational disks in Section 6.2 (just before Proposition 6.6) is again the issue. That is, if $r \notin |\mathbb{C}_v^\times|$, then

$$\overline{D}_{\mathrm{an}}(a,r) \cap \mathbb{P}^1(\mathbb{C}_v) = D_{\mathrm{an}}(a,r) \cap \mathbb{P}^1(\mathbb{C}_v) = \overline{D}(a,r) = D(a,r),$$

even though $\overline{D}_{\mathrm{an}}(a,r) \supsetneq D_{\mathrm{an}}(a,r)$. The same thing happens for the disks $\mathbb{P}^1_{\mathrm{an}} \smallsetminus \overline{D}_{\mathrm{an}}(a,r)$ and $\mathbb{P}^1_{\mathrm{an}} \smallsetminus D_{\mathrm{an}}(a,r)$. Fortunately, this hiccup for irrational disks is the only obstacle to the correspondence being one-to-one; see Exercise 6.15(b).

Similarly, if $W \subseteq \mathbb{P}^1_{\mathrm{an}}$ is a connected Berkovich affinoid, then its set of Type I points $W \cap \mathbb{P}^1(\mathbb{C}_v)$ is a classical connected affinoid, as can be easily deduced by writing W as an intersection of disks and applying equation (6.3) to each disk. Of course, if some of the disks are irrational, then the hiccup above makes things a little more complicated. Fortunately, if we specify in advance that the disks involved are either all open or all closed, then the ambiguity vanishes; see, for example, Exercise 6.28. Thus, we now set the convention that when switching back and forth between Berkovich affinoids and their classical analogues, then unless otherwise specified, **we consider each irrational disk in $\mathbb{C}_v$ to be open**.

Remark 6.24. The reader can easily check that a point $\zeta \in \mathbb{A}^1_{\mathrm{an}}$ belongs to a given rational Berkovich disk $D \subseteq \mathbb{A}^1_{\mathrm{an}}$ if and only if the geometric data corresponding to ζ is contained in the associated classical disk $D_{\mathrm{I}} := D \cap \mathbb{C}_v$;

see Exercise 6.16. That is,

- if $\zeta = x \in \mathbb{C}_v$ is of Type I, we need $x \in D_{\mathrm{I}}$;
- if $\zeta = \zeta(b,s)$ is of Type II or III, we need $\overline{D}(b,s) \subseteq D_{\mathrm{I}}$; and
- if ζ is of Type IV, we need all but finitely many of the disks approximating ζ to lie in D_{I}.

The condition for ζ to belong to an irrational Berkovich disk $D \subseteq \mathbb{A}^1_{\mathrm{an}}$ is the same, with the one exception that if $D = D_{\mathrm{an}}(a,r)$ is an irrational *open* Berkovich disk, then $\zeta(a,r) \notin D$ even though $\overline{D}(a,r) \subseteq D \cap \mathbb{C}_v$.

For a Berkovich affinoid W, the situation is a bit more complicated. If $\zeta \in \mathbb{P}^1_{\mathrm{an}}$ is of Type I or IV, then it turns out that again, $\zeta \in W$ if and only if the geometric data corresponding to ζ is contained in the associated classical affinoid $W_{\mathrm{I}} := W \cap \mathbb{P}^1(\mathbb{C}_v)$; see Exercise 6.17. However, if ζ is of Type II or III, it is entirely possible that $\zeta \in W$ but the disk corresponding to ζ does *not* sit inside W. For example, if $0 < r < s < t$ and W is the annulus $W = \{r < |\zeta| < t\}$, then $\zeta(0,s) \in W$, even though $\overline{D}(0,s) \not\subseteq W$.

Nevertheless, it is easy to check whether a given point ζ lies in W, simply by checking whether ζ lies in each Berkovich disk defining W.

The following result states that each of the four types of points forms a dense subset of a Berkovich disk. We leave the proof as an exercise.

Theorem 6.25.

(a) *The set of Type I points in $\mathbb{P}^1_{\mathrm{an}}$ forms a dense subset of $\mathbb{P}^1_{\mathrm{an}}$.*

(b) *The set of Type II points in $\mathbb{P}^1_{\mathrm{an}}$ forms a dense subset of $\mathbb{P}^1_{\mathrm{an}}$.*

(c) *If $|\mathbb{C}_v^\times| \neq (0,\infty)$, then the set of Type III points in $\mathbb{P}^1_{\mathrm{an}}$ forms a dense subset of $\mathbb{P}^1_{\mathrm{an}}$.*

(d) *If $\mathbb{C}_v$ is not spherically complete, then the set of Type IV points in $\mathbb{P}^1_{\mathrm{an}}$ forms a dense subset of $\mathbb{P}^1_{\mathrm{an}}$.*

Proof. See Exercise 15.7. □

6.6. Paths and path-connectedness

In this section we discuss a key property of connected Berkovich affinoids, including disks and $\mathbb{P}^1_{\mathrm{an}}$ itself, that we have stated only informally so far: these spaces are path-connected.

Definition 6.26. Let $\zeta, \xi \in \mathbb{A}^1_{\mathrm{an}}$. We say that ζ *lies above* ξ, and we write $\xi \preceq \zeta$, if $\|\cdot\|_\xi$ is bounded by $\|\cdot\|_\zeta$; that is, if

$$\|f\|_\xi \le \|f\|_\zeta \quad \text{for all } f \in \mathbb{C}_v[z].$$

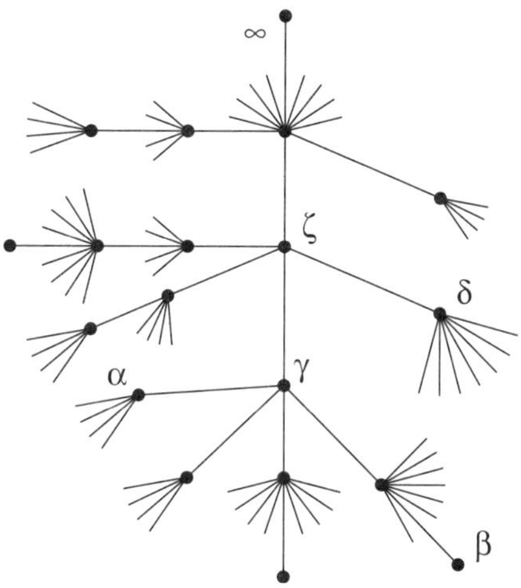

Figure 6.8. Lying above: $\alpha, \beta \preceq \gamma$ and $\alpha, \beta, \gamma, \delta \preceq \zeta$.

Lying above a seminorm is thus exactly the same as bounding it. We only use the different notation because "bounding" places an emphasis on the ring and the seminorm, where as "lying above" places the emphasis on ζ as a geometric point in a tree. In particular, the relation $\xi \preceq \zeta$ corresponds to the point ζ literally lying *above* ξ in our the tree image of $\mathbb{P}^1_{\text{an}}$. For example, in Figure 6.8, we have $\alpha, \beta \preceq \gamma$ and $\alpha, \beta, \gamma, \delta \preceq \zeta$.

Proposition 6.27. *Let $\preceq$ denote the relation on $\mathbb{A}^1_{\text{an}}$ from Definition* 6.26.

(a) *The relation $\preceq$ defines a partial order on $\mathbb{A}^1_{\text{an}}$.*

(b) *All Type I and IV points are minimal with respect to $\preceq$.*

(c) *For any $\xi \in \mathbb{A}^1_{\text{an}}$ and any closed Berkovich disk $\overline{D}_{\text{an}}(a,r)$, we have $\xi \in \overline{D}_{\text{an}}(a,r)$ if and only if $\xi \preceq \zeta(a,r)$.*

(d) *For any $\xi, \zeta \in \mathbb{A}^1_{\text{an}}$ with $\xi \preceq \zeta$, we have* $\operatorname{diam}(\xi) \le \operatorname{diam}(\zeta)$, *with equality if and only if $\xi = \zeta$.*

Proof. See Section 15.5. □

Among other things, Proposition 6.27 confirms our intuition about the locations of the various types of points in the tree. Specifically, the Type I and IV points in $\mathbb{P}^1_{\text{an}}$ are at the endpoints of their branches, while points of Type II and III have some points lying above them *and* some points lying below them.

The following definition is useful for finding the unique path between two Berkovich points.

Definition 6.28. Let $\zeta_0, \zeta_1 \in \mathbb{A}^1_{\mathrm{an}}$. The *least upper bound* of ζ_0 and ζ_1, denoted $\zeta_0 \vee \zeta_1$, is the unique element of $\mathbb{A}^1_{\mathrm{an}}$ such that

(a) $\zeta_0 \preceq \zeta_0 \vee \zeta_1$ and $\zeta_1 \preceq \zeta_0 \vee \zeta_1$;

(b) if $\xi \in \mathbb{A}^1_{\mathrm{an}}$ satisfies $\zeta_0 \preceq \xi$ and $\zeta_1 \preceq \xi$, then $\zeta_0 \vee \zeta_1 \preceq \xi$.

For example, in Figure 6.8, we have $\alpha \vee \beta = \gamma$. The point ζ also satisfies $\alpha \preceq \zeta$ and $\beta \preceq \zeta$ and hence is an upper bound for α and β, but $\gamma \preceq \zeta$ is the least upper bound. Similarly, $\alpha \vee \delta = \gamma \vee \delta = \beta \vee \delta = \zeta$.

Proposition 6.29. *Let $\zeta_0, \zeta_1 \in \mathbb{A}^1_{\mathrm{an}}$. Then the least upper bound $\zeta_0 \vee \zeta_1$ exists and is unique.*

Proof. See Section 15.5. □

Proposition 6.29 provides us with the key idea for showing that $\mathbb{P}^1_{\mathrm{an}}$ is path-connected. Given two points $\zeta, \zeta' \in \mathbb{A}^1_{\mathrm{an}}$ with $\zeta \preceq \zeta'$, the set

$$I := \{\xi \in \mathbb{A}^1_{\mathrm{an}} : \zeta \preceq \xi \preceq \zeta'\}$$

is homeomorphic to a compact real interval, as the tree image suggests; see Exercise 6.24. (That is, I is either a single point or else homeomorphic to $[0,1]$.) Thus, given any two points $\zeta_0, \zeta_1 \in \mathbb{A}^1_{\mathrm{an}}$, there is an interval I_0 from ζ_0 up to $\zeta_0 \vee \zeta_1$, and an interval I_1 from ζ_0 up to $\zeta_0 \vee \zeta_1$. Their intersection $I_0 \cap I_1$ consists solely of the point $\zeta_0 \vee \zeta_1$; the union $I_0 \cup I_1$ is then a path from ζ_0 to ζ_1.

Example 6.30. Let $x \in \mathbb{C}_v \smallsetminus \{0\}$, set $R := |x|$, and let $r \in (0, R)$. Then Figure 6.1(b) on page 124 shows the path from the Type I point 0 to the Type II or III point $\zeta(x, r)$, as we described informally in Section 6.1. More precisely, the least upper bound of the two points is

$$0 \vee \zeta(x, r) = \zeta(0, R). \tag{6.4}$$

Proposition 6.27 allows us to confirm equation (6.4), as follows. First, $\zeta(0, R)$ lies above both points, as $0 \in \overline{D}(0, R)$ and $\overline{D}(x, r) \subseteq \overline{D}(0, R)$. Second, any point ζ lying above both 0 and $\zeta(x, r)$ must be of Type II or III and correspond to a disk containing both 0 and $\overline{D}(x, r)$. That is, ζ must be of the form $\zeta(0, s)$, with $s \geq R$. Clearly, $\zeta(0, R)$ is the minimal such point, verifying equation (6.4). Thus, the path shown in Figure 6.1(b) is indeed the path described more generally above: I_0 is the leg running from 0 up to $0 \vee \zeta(x, r) = \zeta(0, R)$, and I_1 is the leg running from $\zeta(0, R)$ back down to $\zeta(x, r)$.

Of course, there are details to check: that each of I_0 and I_1 is indeed homeomorphic to a real interval, that the path lies inside any connected Berkovich affinoid that contains both ζ_0 and ζ_1, and so on. In addition, because the path is in fact unique, we need one more definition before we state the result formally.

Definition 6.31. Let X be a topological space. We say X is *uniquely path-connected* if for any two distinct points $x_0, x_1 \in X$, there is a unique subset $I \subseteq X$ containing x_0 and x_1 such that I is homeomorphic to the real closed interval $[0, 1]$, with the homeomorphism mapping x_0 to 0 and x_1 to 1.

Theorem 6.32. *Let $U \subseteq \mathbb{P}^1_{\text{an}}$ be a connected Berkovich affinoid. Then U is uniquely path-connected. Moreover, for any $\zeta_0, \zeta_1 \in U$, all points on the unique arc connecting them, other than ζ_0 and ζ_1 themselves, are of Type II or III.*

Proof. See Section 15.5. □

Corollary 6.33. *$\mathbb{P}^1_{\text{an}}$ is locally path-connected.*

Proof. Given $\zeta \in \mathbb{P}^1_{\text{an}}$ and an open set U containing ζ, there is an open connected affinoid V such that $\zeta \in V \subseteq U$, by Theorem 6.17. Then V is a path-connected neighborhood of ζ contained in U, by Theorem 6.32. □

Definition 6.34. Let $\zeta_0, \zeta_1 \in \mathbb{P}^1_{\text{an}}$. The unique arc from ζ_0 to ζ_1 is called the (*closed*) *interval* from ζ_0 to ζ_1 and is denoted $[\zeta_0, \zeta_1] \subseteq \mathbb{P}^1_{\text{an}}$. We say that any point $\xi \in [\zeta_0, \zeta_1]$ *lies between* ζ_0 and ζ_1. If $\xi \in [\zeta_0, \zeta_1]$, but $\xi \neq \zeta_0, \zeta_1$, we say that ξ *lies strictly between* ζ_0 and ζ_1.

If $\zeta_0 = \zeta_1$ in Definition 6.34, then the interval $[\zeta_0, \zeta_1]$ is the singleton $\{\zeta_0\}$. Otherwise, if $\zeta_0 \neq \zeta_1$, we can of course define the half-open and open intervals $[\zeta_0, \zeta_1)$, $(\zeta_0, \zeta_1]$, and (ζ_0, ζ_1) by removing one or both endpoints in the obvious way. The reader should be cautious about the use of the word "open" here, however. On the one hand, the closed interval $[\zeta_0, \zeta_1]$ is indeed closed in Gel'fand topology, since it is a compact subset of the Hausdorff space $\mathbb{P}^1_{\text{an}}$. On the other hand, none of $[\zeta_0, \zeta_1)$, $(\zeta_0, \zeta_1]$, or (ζ_0, ζ_1) is either open or closed in the Gel'fand topology; see Exercise 6.35.

Note that the sets $[\zeta_0, \zeta_1]$ and $[\zeta_1, \zeta_0]$ both make sense and coincide, because in contrast to $\mathbb{R}$, there is no canonical positive or negative direction on $\mathbb{P}^1_{\text{an}}$. In addition, for any distinct points $\zeta_0, \zeta_1 \in \mathbb{P}^1_{\text{an}}$, it is immediate from Definitions 6.31 and 6.34 that $[\zeta_0, \zeta_1]$ is homeomorphic to the interval $[0, 1] \subseteq \mathbb{R}$. If ξ lies strictly between ζ_0 and ζ_1, then by the final statement of Theorem 6.32, ξ must be a point of Type II or III.

Proposition 6.35. *Let $\zeta_0, \zeta_1, \zeta_2 \in \mathbb{P}^1_{\text{an}}$. Then*

$$[\zeta_0, \zeta_1] \cap [\zeta_0, \zeta_2] \cap [\zeta_1, \zeta_2]$$

consists of a single point. Moreover, if none of ζ_0, ζ_1, or ζ_2 lies between the other two, then this single point is of Type II.

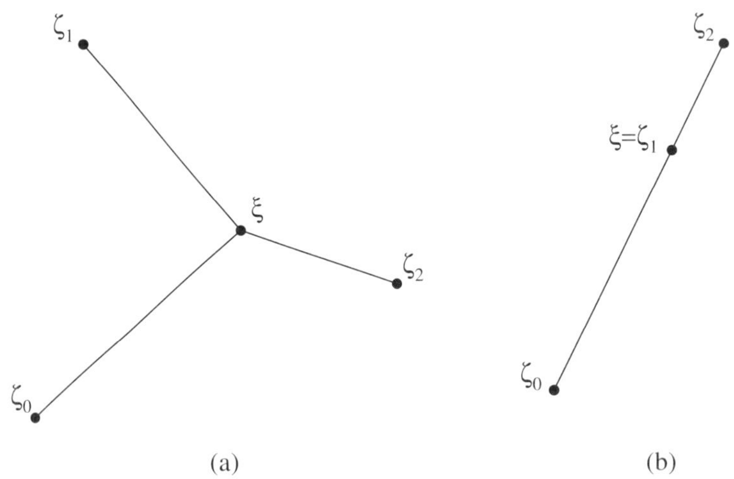

Figure 6.9. ξ lies between ζ_0, ζ_1, and ζ_2.

Proof. See Section 15.6. □

Equivalently, Proposition 6.35 says that for any $\zeta_0, \zeta_1, \zeta_2 \in \mathbb{P}^1_{\mathrm{an}}$, there is a unique point $\xi \in \mathbb{P}^1_{\mathrm{an}}$ that simultaneously

- lies between ζ_0 and ζ_1,
- lies between ζ_0 and ζ_2,
- lies between ζ_1 and ζ_2.

For example, in Figure 6.9(a), the point ξ is the unique point lying between any two of ζ_0, ζ_1, and ζ_2. The same is true in Figure 6.9(b), only this time the point ξ coincides with ζ_1, since ζ_1 already lies between ζ_0 and ζ_2.

Proposition 6.35 generalizes the least upper bound $\zeta_0 \vee \zeta_1$ of Definition 6.28. After all, if $\zeta_0, \zeta_1 \in \mathbb{A}^1_{\mathrm{an}}$, then $\zeta_0 \vee \zeta_1$ is the unique point of Proposition 6.35 that lies between all three of ζ_0, ζ_1, and ∞; see Exercise 6.26.

The notion of lying between also allows us to state the following useful definition.

Definition 6.36. Let $S \subseteq \mathbb{P}^1_{\mathrm{an}}$ be any subset of the Berkovich projective line. The *convex hull* of S is the set of all points $\xi \in \mathbb{P}^1_{\mathrm{an}}$ for which there exist $\zeta_0, \zeta_1 \in S$ such that ξ lies between ζ_0 and ζ_1.

Equivalently, the convex hull of S is the union of all arcs that start and end at points of S. It is immediate, then, that the convex hull of S is path-connected. In fact, it is the smallest connected subset of $\mathbb{P}^1_{\text{an}}$ containing S. That is, it is the intersection of all connected subsets of $\mathbb{P}^1_{\text{an}}$ containing S.

Remark 6.37. The spaces $\mathbb{P}^1_{\text{an}}$ and $\mathbb{A}^1_{\text{an}}$, as well as every Berkovich disk and affinoid, are examples of $\mathbb{R}$-*trees*, which are certain topological inverse limits of the better-known finite trees from graph theory. Like finite trees, $\mathbb{R}$-trees are always uniquely path-connected; but like $\mathbb{P}^1_{\text{an}}$, they allow infinite branching and dense sets of branch points. For more on $\mathbb{R}$-trees, especially in the context of Berkovich disks, see [**8**, Section 1.4, Appendix B] or [**58**].

6.7. Directions at Berkovich points

The following result generalizes Corollary 6.14. It emphasizes that in the tree structure on $\mathbb{P}^1_{\text{an}}$, the points of Types I and IV are endpoints, the points of Type II are branch points with infinite branching, and the points of Type III are unbranched interior points.

Proposition 6.38. *Let $\zeta \in \mathbb{P}^1_{\text{an}}$. The number of path-connected components of $\mathbb{P}^1_{\text{an}} \smallsetminus \{\zeta\}$ is*

(a) *one, if ζ is of Type I or IV;*

(b) *two, if ζ is of Type III; or*

(c) *infinite, if ζ is of Type II.*

Proof.

(a) If ζ is of Type I or IV, then by the final statement of Theorem 6.32, ζ does not lie along the unique arc between any two points $\zeta_0, \zeta_1 \in \mathbb{P}^1_{\text{an}} \smallsetminus \{\zeta\}$. Thus, $\mathbb{P}^1_{\text{an}} \smallsetminus \{\zeta\}$ is path-connected.

(b) By Corollary 6.14, if ζ is of Type III, $\mathbb{P}^1_{\text{an}} \smallsetminus \{\zeta\}$ consists of two irrational open Berkovich disks, both of which are path-connected, by Theorem 6.32.

(c) By Corollary 6.14, if ζ is of Type II, $\mathbb{P}^1_{\text{an}} \smallsetminus \{\zeta\}$ consists of infinitely many rational open Berkovich disks, each of which is path-connected, by Theorem 6.32. □

Proposition 6.38 inspires the following definition.

Definition 6.39. Let $\zeta \in \mathbb{P}^1_{\text{an}}$. The path-connected components of $\mathbb{P}^1_{\text{an}} \smallsetminus \{\zeta\}$ are called the *residue classes* of ζ or the *directions* at ζ.

If $x \in \mathbb{P}^1_{\text{an}} \smallsetminus \{\zeta\}$, we denote the direction at ζ containing x by $\vec{v}_\zeta(x)$, or simply $\vec{v}(x)$ if the point ζ is clear from context.

The term "residue class" comes from the fact that if $\zeta = \zeta(0,1)$ is the Gauss point, then the components of $\mathbb{P}^1_{\text{an}} \smallsetminus \{\zeta\}$ consist of open disks $D_{\text{an}}(c,1)$ with $|c| \leq 1$, as well as the disk at infinity, $\mathbb{P}^1_{\text{an}} \smallsetminus \overline{D}_{\text{an}}(0,1)$. The corresponding $\mathbb{P}^1(\mathbb{C}_v)$-disks are precisely the residue classes arising from the reduction map $\mathbb{P}^1(\mathbb{C}_v) \mapsto \mathbb{P}^1(\overline{k})$.

On the other hand, the term "direction" is meant to evoke the tree image: starting from ζ, one can move in any of the directions emanating from ζ. Indeed, the notation $\vec{v}_\zeta(x)$ is intended to emphasize that even though the object in question is a subset of $\mathbb{P}^1_{\text{an}}$, we should think of it as a vector, pointing along an edge of the tree, from ζ toward x.

Recall from Theorem 6.17 that the open connected Berkovich affinoids form a basis for the topology on $\mathbb{P}^1_{\text{an}}$. In particular, if $U \subseteq \mathbb{P}^1_{\text{an}}$ is an open set and if $\zeta \in U$, then U contains an open connected affinoid containing ζ, and hence U intersects all of the residue classes at ζ. Moreover, U contains all but finitely many of the residue classes at ζ.

Remark 6.40. In [**8**], the directions (or *tangent vectors*, as Baker and Rumely call them) at a point $\zeta \in \mathbb{P}^1_{\text{an}}$ are defined in a different way than in our Definition 6.39. Specifically, a tangent vector at ζ is an equivalence class of intervals of the form $[\zeta, \xi]$, where $\xi \in \mathbb{P}^1_{\text{an}} \smallsetminus \{\zeta\}$, and where two intervals $[\zeta, \xi_1]$ and $[\zeta, \xi_2]$ are said to be equivalent if their intersection consists of more than just the point ζ.

It is not difficult to show that there is a natural one-to-one correspondence between our directions at ζ and the tangent vectors of Baker and Rumely at ζ: the tangent vector represented by $[\zeta, \xi]$ corresponds to the direction containing ξ; see Exercise 6.33. Our definition is perhaps a little simpler, but Baker and Rumely's definition has the advantage that it suggests the intuition of direction on a graph more explicitly.

6.8. The hyperbolic metric

In [**114, 115, 117**], Rivera-Letelier noted a parallel between the set of Berkovich points of Types II–IV and the three-dimensional real hyperbolic space that has the Riemann sphere $\mathbb{P}^1(\mathbb{C})$ as its boundary. He therefore dubbed this subset of $\mathbb{P}^1_{\text{an}}$ to be a non-archimedean *hyperbolic* space, and he noted that it has a natural metric, as follows.

Definition 6.41. The set

$$\mathbb{H} := \mathbb{P}^1_{\text{an}} \smallsetminus \mathbb{P}^1(\mathbb{C}_v)$$

is the *hyperbolic space* over $\mathbb{C}_v$. The function $d_{\mathbb{H}} : \mathbb{H} \times \mathbb{H} \to [0, \infty)$ given by

$$d_{\mathbb{H}}(\zeta, \xi) := 2 \log\big(\operatorname{diam}(\zeta \vee \xi)\big) - \log\big(\operatorname{diam}(\zeta)\big) - \log\big(\operatorname{diam}(\xi)\big)$$

is the *hyperbolic metric*.

The hyperbolic metric $d_{\mathbb{H}}$ is also called the "big model metric" in [**8**], and simply "la distance" in [**115, 117**].

Theorem 6.42. *The function* $d_{\mathbb{H}} : \mathbb{H} \times \mathbb{H} \to [0, \infty)$ *is a metric on* $\mathbb{H}$.

Proof. See Section 15.6. □

Warning 6.43. Of course, the metric $d_{\mathbb{H}}$ induces a topology on $\mathbb{H}$. However, this metric topology is much stronger than the Gel'fand topology on $\mathbb{H}$; see Exercise 6.34. In the literature, therefore, the metric topology is sometimes called the "strong topology," in which case the Gel'fand topology is called the "weak topology."

To help explain the formula of Definition 6.41, first consider the special case that $\zeta = \zeta(a, R)$ and $\xi = \zeta(a, r)$ for some point $a \in \mathbb{C}_v$ and real numbers $0 < r < R$. Then ζ lies above ξ, and hence $\zeta \vee \xi = \zeta$. Since $\operatorname{diam}(\zeta) = R$ and $\operatorname{diam}(\xi) = r$, Definition 6.41 for $d_{\mathbb{H}}$ gives

$$d_{\mathbb{H}}(\zeta, \xi) = 2 \log R - \log R - \log r = \log R - \log r.$$

Thus, $d_{\mathbb{H}}(\zeta, \xi)$ is the logarithm of the ratio R/r which, as we saw in Remark 3.35, is invariant under the action of bijective Laurent series. That is, the distance from ζ to ξ is the (logarithmic) conformal invariant of the unique open Berkovich annulus bounded by ζ and ξ.

More generally, *any* two distinct points $\zeta, \xi \in \mathbb{H}$ not of Type IV define an open annulus, as follows. Let U_1 be the component of $\mathbb{P}^1_{\text{an}} \smallsetminus \{\zeta\}$ containing ξ, and let U_2 be the component of $\mathbb{P}^1_{\text{an}} \smallsetminus \{\xi\}$ containing ζ. Then U_1 and U_2 are both open Berkovich disks in $\mathbb{P}^1_{\text{an}}$, and their intersection is an open Berkovich annulus. As we will see in Proposition 7.17, $d_{\mathbb{H}}$ is invariant under the application of linear fractional maps; hence, $d_{\mathbb{H}}(\zeta, \xi)$ is again the logarithm of the conformal invariant of this annulus.

For the moment, however, note that the formula of Definition 6.41 makes intuitive sense, even when $\zeta = \zeta(a, r)$ and $\xi = \zeta(b, s)$, but neither point lies above the other. In that case, their least upper bound is

$$\zeta \vee \xi = \zeta(a, R) = \zeta(b, R),$$

where $R := |a - b|$. We have that $R > r, s$, because of our specification that neither of ζ nor ξ lies above the other. Since the unique path from ζ to ξ passes through $\zeta \vee \xi$, it is only appropriate to define

$$\begin{aligned} d_{\mathbb{H}}(\zeta, \xi) &= d_{\mathbb{H}}(\zeta \vee \xi, \zeta) + d_{\mathbb{H}}(\zeta \vee \xi, \xi) = \log R - \log r + \log R - \log s \\ &= 2 \log \big(\operatorname{diam}(\zeta \vee \xi) \big) - \log \big(\operatorname{diam}(\zeta) \big) - \log \big(\operatorname{diam}(\xi) \big), \end{aligned}$$

thus giving precisely the formula in Definition 6.41.

Exercises for Chapter 6

6.1. For any $x \in \mathbb{C}_v$, prove that the function $\|\cdot\|_x : \mathbb{C}_v[z] \to [0, \infty)$ given by $\|f\|_x = |f(x)|$ is a multiplicative seminorm extending the absolute value $|\cdot|$ on $\mathbb{C}_v$.

6.2. Let $D_1 \supsetneq D_2 \supsetneq \cdots$ be a decreasing sequence of closed disks with empty intersection, and define $\|\cdot\| : \mathbb{C}_v[z] \to [0, \infty)$ by

$$\|f\| := \lim_{n\to\infty} \|f\|_{\zeta_n}, \tag{6.5}$$

where ζ_n is the sup-norm on D_n.

(a) Prove that for each $f \in \mathbb{C}_v[z]$, the limit in equation (6.5) converges.

(b) Prove that $\|\cdot\|$ is a multiplicative norm (not just a seminorm) extending the absolute value $|\cdot|$ on $\mathbb{C}_v$.

(c) Let $E_1 \supsetneq E_2 \supsetneq \cdots$ be an equivalent sequence of disks. That is, for every $N \geq 1$, there is some $M \geq N$ such that $E_M \subseteq D_N$ and $D_M \subseteq E_N$. Prove that the norm on $\mathbb{C}_v[z]$ corresponding to $\{E_n\}_{n\geq 1}$ equals the norm $\|\cdot\|$ corresponding to $\{D_n\}_{n\geq 1}$.

6.3. In this exercise, you will confirm some claims made after Definition 6.3.

(a) Let $\zeta = x$ be a point of Type I. Prove that

$$|\zeta| = |x| \quad \text{and} \quad \operatorname{diam}(\zeta) = 0.$$

(b) Let $\zeta = \zeta(a, r)$ be a point of Type II or III. Prove that

$$|\zeta| = \max\{|a|, r\} \quad \text{and} \quad \operatorname{diam}(\zeta) = r.$$

(c) If ζ is a Type IV point determined by a decreasing sequence

$$\overline{D}(a_1, r_1) \supsetneq \overline{D}(a_2, r_2) \supsetneq \cdots$$

of disks with empty intersection, prove that

(i) $|\zeta| = |\zeta(a_n, r_n)|$ for any n such that $0 \notin \overline{D}(a_n, r_n)$, and

(ii) $\operatorname{diam}(\zeta) = \lim_{n\to\infty} r_n$.

6.4. For any $a \in \mathbb{C}_v$, prove equation (6.1).

6.5. Prove that for any Type I point $x \in \mathbb{C}_v$, the set $\{x\}$ is closed in $\mathbb{A}^1_{\text{an}}$. That is, prove that $\mathbb{A}^1_{\text{an}} \smallsetminus \{x\}$ is open.

(Do *not* use results from Section 6.4 and beyond. Some of those results rely on this exercise, including Theorem 6.22, that $\mathbb{P}^1_{\text{an}}$ is Hausdorff.)

6.6. Given $\zeta \in \mathbb{A}^1_{\text{an}} \smallsetminus \{0\}$, define a function $\|\cdot\|_{1/\zeta} : \mathbb{C}_v[z] \to [0, \infty)$ by equation (6.2), which we reproduce here:

$$\|f\|_{1/\zeta} := \left\|z^{\deg(f)} f(1/z)\right\|_\zeta \Big/ \|z\|_\zeta^{\deg(f)} \quad \text{for } f \in \mathbb{C}_v[z].$$

(a) Prove that $\|\cdot\|_{1/\zeta}$ is a multiplicative seminorm on $\mathbb{C}_v[z]$ extending $|\cdot|$ and that $\|\cdot\|_{1/\zeta} \neq \|\cdot\|_0$. In other words, prove that $1/\zeta \in \mathbb{A}^1_{\text{an}} \smallsetminus \{0\}$.

(b) Prove that $\zeta \mapsto 1/\zeta$ is a homeomorphism from $\mathbb{A}^1_{\mathrm{an}} \smallsetminus \{0\}$ to itself.

(Do *not* use results from Section 6.4 and beyond, as some of those results rely on this exercise.)

6.7. Consider the map $\zeta \mapsto 1/\zeta$ on $\mathbb{A}^1_{\mathrm{an}} \smallsetminus \{0\}$ given by equation (6.2).

(a) Prove that $\zeta \mapsto 1/\zeta$ maps a Type I point $x \in \mathbb{P}^1(\mathbb{C}_v)$ to $1/x$.

(b) Prove that $\zeta \mapsto 1/\zeta$ maps a Type II or III point $\zeta(a,r)$ to

$$\begin{cases} \zeta(0,1/r) & \text{if } 0 \in \overline{D}(a,r), \\ \zeta(1/a, r/|a|^2) & \text{otherwise.} \end{cases}$$

(c) Describe the action of $\zeta \mapsto 1/\zeta$ on a Type IV point corresponding to a chain of disks $D_1 \supsetneq D_2 \supsetneq \cdots$.

(d) Use parts (a)–(c) to confirm that $\zeta \mapsto 1/\zeta$ preserves the type of the Berkovich point ζ.

(e) Prove that $|1/\zeta| = 1/|\zeta|$ for all $\zeta \in \mathbb{A}^1_{\mathrm{an}} \smallsetminus \{0\}$.

(Do *not* use results from Section 6.4 and beyond, as some of those results rely on this exercise.)

6.8. Let $D \subseteq \mathbb{P}^1_{\mathrm{an}}$ be an open Berkovich disk. Prove that D can be written as a nested union $\bigcup_{n\geq 1} D_n$ of closed Berkovich disks

$$D_1 \subseteq D_2 \subseteq \cdots \subseteq \mathbb{P}^1_{\mathrm{an}}.$$

6.9. In this exercise, you will confirm several claims made in Warning 6.12.

(a) Prove that the absolute value function $|\cdot| : \mathbb{P}^1_{\mathrm{an}} \to \mathbb{R}$ is continuous.

(b) Prove that the diameter function $\operatorname{diam} : \mathbb{P}^1_{\mathrm{an}} \to \mathbb{R}$ is *not* continuous even when restricted to $\mathbb{A}^1_{\mathrm{an}}$ or to any Berkovich affinoid. (*Suggestion*: Use Theorem 6.17.)

(c) Prove that $\operatorname{diam} : \mathbb{A}^1_{\mathrm{an}} \to \mathbb{R}$ is upper semicontinuous. That is, for any $t \in \mathbb{R}$, prove that the set $\{\zeta \in \mathbb{A}^1_{\mathrm{an}} : \operatorname{diam}(\zeta) < t\}$ is open.

(d) Prove that $\operatorname{diam} : \mathbb{P}^1_{\mathrm{an}} \to \mathbb{R}$ is *not* upper semicontinuous.

6.10. Let $U = \{\zeta \in \mathbb{A}^1_{\mathrm{an}} : |\zeta| = 1\}$ be the annulus of Proposition 6.13. Prove that the function $\zeta \mapsto 1/\zeta$ of Section 6.4 maps U bijectively onto itself.

6.11. Prove Proposition 6.19.

6.12. Verify Warning 6.18(a): the set of open Berkovich disks does *not* form a basis for the Gel'fand topology for either $\mathbb{A}^1_{\mathrm{an}}$ or $\mathbb{P}^1_{\mathrm{an}}$.

(*Suggestion*: Prove that an annulus $\{\zeta \in \mathbb{P}^1_{\mathrm{an}} : r < |\zeta| < s\}$ is open in $\mathbb{P}^1_{\mathrm{an}}$ but cannot be written as a union of Berkovich disks.)

6.13. Let X be a dense subset of $\mathbb{C}_v$, and let Y be a dense subset of $(0, \infty)$. Let $\mathcal{D}$ be the set of open Berkovich disks of the form

$$D_{\mathrm{an}}(a, r) \quad \text{or} \quad \mathbb{P}^1_{\mathrm{an}} \smallsetminus \overline{D}_{\mathrm{an}}(a, r) \quad \text{with } a \in X \text{ and } r \in Y.$$

Let $\mathcal{B}$ be the set of nonempty intersections of finitely many elements of $\mathcal{D}$. Prove that $\mathcal{B}$ is a basis for the Gel'fand topology on $\mathbb{P}^1_{\mathrm{an}}$.

In particular, if $\mathbb{C}_v$ is separable (i.e., has a countable dense subset), then $\mathbb{P}^1_{\mathrm{an}}$ is second countable (i.e., has a countable basis).

6.14. Let $a \in \mathbb{C}_v$ and $r > 0$.

(a) If $r \notin |\mathbb{C}_v^\times|$, prove that the interior of $\overline{D}_{\mathrm{an}}(a, r)$ is $D_{\mathrm{an}}(a, r)$.

(b) If $r \in |\mathbb{C}_v^\times|$, prove that the interior of $\overline{D}_{\mathrm{an}}(a, r)$ is the union of the infinitely many disjoint open disks $D_{\mathrm{an}}(b, r)$, for $b \in \overline{D}(a, r)$.

6.15. In this exercise, you will confirm several claims made in Remark 6.23.

(a) Prove equation (6.3): for any $a \in \mathbb{C}_v$ and $r > 0$,

$$\overline{D}_{\mathrm{an}}(a, r) \cap \mathbb{P}^1(\mathbb{C}_v) = \overline{D}(a, r), \qquad \text{and}$$
$$D_{\mathrm{an}}(a, r) \cap \mathbb{P}^1(\mathbb{C}_v) = D(a, r).$$

(b) Let $D_1 \neq D_2 \subseteq \mathbb{P}^1_{\mathrm{an}}$ be Berkovich disks that are different but for which $D_1 \cap \mathbb{P}^1(\mathbb{C}_v) = D_2 \cap \mathbb{P}^1(\mathbb{C}_v)$. Prove that there is some $a \in \mathbb{C}_v$ and $r \in (0, \infty) \smallsetminus |\mathbb{C}_v^\times|$ such that D_1 and D_2 are either

$$\begin{cases} \overline{D}_{\mathrm{an}}(a, r) \quad \text{and} \quad D_{\mathrm{an}}(a, r), & \text{or} \\ \mathbb{P}^1_{\mathrm{an}} \smallsetminus \overline{D}_{\mathrm{an}}(a, r) \quad \text{and} \quad \mathbb{P}^1_{\mathrm{an}} \smallsetminus D_{\mathrm{an}}(a, r). \end{cases}$$

6.16. Let $D \subseteq \mathbb{A}^1_{\mathrm{an}}$ be a Berkovich disk, and let $D_{\mathrm{I}} := D \cap \mathbb{C}_v$ be the associated classical disk.

(a) For any Type I point $x \in \mathbb{C}_v$, prove that $x \in D$ if and only if $x \in D_{\mathrm{I}}$.

(b) For any Type II point $\zeta = \zeta(a, r)$, prove that $\zeta \in D$ if and only if $\overline{D}(a, r) \subseteq D_{\mathrm{I}}$.

(c) For any Type III point $\zeta = \zeta(a, r)$, prove that $\zeta \in D$ if and only if $\overline{D}(a, r) \subseteq D_{\mathrm{I}}$, *unless* $D = D_{\mathrm{an}}(a, r)$.

(d) For any Type IV point ζ corresponding to $D_1 \supsetneq D_2 \supsetneq \cdots$ in $\mathbb{C}_v$, prove that $\zeta \in D$ if and only if $D_n \subseteq D_{\mathrm{I}}$ for some $n \geq 1$.

6.17. Let $W \subseteq \mathbb{P}^1_{\mathrm{an}}$ be a Berkovich connected affinoid, and let W_{I} be the associated classical affinoid $W_{\mathrm{I}} := W \cap \mathbb{P}^1(\mathbb{C}_v)$.

(a) For any Type I point $x \in \mathbb{C}_v$, prove that $x \in W$ if and only if $x \in W_{\mathrm{I}}$.

(b) For any Type IV point ζ corresponding to a decreasing sequence of disks $D_1 \supsetneq D_2 \supsetneq \cdots$ in $\mathbb{C}_v$, $\zeta \in D$ if and only if $D_n \subseteq W_{\mathrm{I}}$ for some $n \geq 1$.

(c) If W is not a disk contained in $\mathbb{A}^1_{\mathrm{an}}$, prove that there are points $\zeta = \zeta(a, r)$ of Type II such that $\zeta \in W$ but $\overline{D}(a, r) \not\subseteq W_{\mathrm{I}}$.

6.18. Let $\zeta \in \mathbb{C}_v$, and let S be a set of disks in $\mathbb{P}^1_{\mathrm{an}}$ that each contain ζ. Define

$$U := \bigcup_{D \in S} D.$$

Prove that U is

(a) $\mathbb{P}^1_{\mathrm{an}} \smallsetminus \{\xi\}$ for some point $\xi \in \mathbb{P}^1_{\mathrm{an}}$ of Type I or IV,

(b) a disk D of S,

(c) an open Berkovich disk, or

(d) all of $\mathbb{P}^1_{\mathrm{an}}$.

(See also Exercise 2.14.)

6.19. This exercise concerns boundaries of Berkovich affinoids.

(a) Let $D \subseteq \mathbb{P}^1_{\mathrm{an}}$ be either the open Berkovich disk $D_{\mathrm{an}}(a, r)$ or the closed Berkovich disk $\overline{D}_{\mathrm{an}}(a, r)$. Prove that the boundary ∂D consists of the single point $\zeta(a, r)$. Note that this single boundary point is of Type II if D is rational and of Type III if D is irrational.

(b) Let $W \subseteq \mathbb{P}^1_{\mathrm{an}}$ be a connected Berkovich affinoid. Prove that its boundary ∂W is a finite set of points of Type II and III.

(c) Let $S \subseteq \mathbb{P}^1_{\mathrm{an}}$ be a finite set of points of Type II and III, and let $\zeta \in \mathbb{P}^1_{\mathrm{an}} \smallsetminus S$. Prove that the connected component of $\mathbb{P}^1_{\mathrm{an}} \smallsetminus S$ containing ζ is a connected Berkovich affinoid.

6.20. This exercise concerns separation properties of Berkovich affinoids.

(a) Let $U, V \subseteq \mathbb{P}^1_{\mathrm{an}}$ be disjoint closed Berkovich affinoids. Prove that there are disjoint open Berkovich affinoids $U', V' \subseteq \mathbb{P}^1_{\mathrm{an}}$ such that $U \subseteq U'$ and $V \subseteq V'$.

(b) Let $U \subseteq \mathbb{P}^1_{\mathrm{an}}$ be an open connected Berkovich affinoid, and let $V \subseteq U$ be a closed connected Berkovich affinoid that is neither a disk nor all of $\mathbb{P}^1_{\mathrm{an}}$. Prove that $U \smallsetminus V$ is not connected.

6.21. Let $\zeta, \xi \in \mathbb{P}^1_{\mathrm{an}}$. Prove that $\zeta \vee \xi = \zeta$ if and only if $\xi \preceq \zeta$.

6.22. Let $\xi \in \mathbb{P}^1_{\mathrm{an}}$, $a \in \mathbb{C}_v$, and $r > 0$. Prove that

$$\operatorname{diam}\big(\zeta(a, r) \vee \xi\big) = \max\big\{r, \|z - a\|_\xi\big\}.$$

6.23. Let $\zeta, \xi \in \mathbb{P}^1_{\mathrm{an}}$.

(a) Prove that for all $a \in \mathbb{C}_v$, we have

$$\|z-a\|_{\zeta\vee\xi} = \max\left\{\|z-a\|_\zeta, \|z-a\|_\xi\right\}.$$

(b) Let $\overline{D}_{\mathrm{an}}(a,r)$ be the smallest disk containing both ζ and ξ. Prove that $\mathrm{diam}(\zeta \vee \xi) = r$.

(c) If neither ζ nor ξ lies above the other, prove that there is some $f \in \mathbb{C}_v[z]$ such that

$$\|f\|_{\zeta\vee\xi} \neq \max\left\{\|f\|_\zeta, \|f\|_\xi\right\}.$$

6.24. Let $\zeta, \zeta' \in \mathbb{A}^1_{\mathrm{an}}$ with $\zeta \preceq \zeta'$, and let

$$I := \left\{\xi \in \mathbb{A}^1_{\mathrm{an}} : \zeta \preceq \xi \preceq \zeta'\right\}.$$

(a) If $\zeta \neq \zeta'$, prove that I is homeomorphic to the real interval $[0,1]$.

(b) If $\zeta = \zeta'$, prove that $I = \{\zeta\}$ is a single point.

6.25. Let S be a connected subset of $\mathbb{P}^1_{\mathrm{an}}$. Prove that the following two conditions are equivalent.

(a) S is an interval of the form

$$[\zeta_0, \zeta_1], \quad [\zeta_0, \zeta_1), \quad \text{or} \quad (\zeta_0, \zeta_1)$$

for some $\zeta_0, \zeta_1 \in \mathbb{P}^1_{\mathrm{an}}$.

(b) For any three points in S, one lies between the other two.

6.26. In this exercise, you will verify a claim made shortly after Proposition 6.35. Let $\zeta_0, \zeta_1 \in \mathbb{A}^1_{\mathrm{an}}$. Prove that the least upper bound $\zeta_0 \vee \zeta_1$ lies between any two of ζ_0, ζ_1, and ∞.

6.27. Let S be any subset of $\mathbb{P}^1_{\mathrm{an}}$. Prove that the convex hull of S is connected; see Definition 6.36.

6.28. Let W be a connected Berkovich affinoid, and let

$$V := W \cap \mathbb{P}^1(\mathbb{C}_v).$$

(a) If W is closed, prove that W is the closure of V in $\mathbb{P}^1_{\mathrm{an}}$.

(b) If W is open, prove that W is the interior of the closure of V in $\mathbb{P}^1_{\mathrm{an}}$.

(c) If W is open or rational closed, prove that W is the convex hull of V.

(d) If W is irrational closed, prove that the interior of W is the convex hull of V.

6.29. In this exercise, we define *skeleta* of Berkovich affinoids.

(a) Let $U \subset \mathbb{P}^1_{\mathrm{an}}$ be an open connected Berkovich affinoid. Let T be the convex hull of the boundary ∂U. (By Exercise 6.19, ∂U is a finite set of points of Type II and III.) Prove that T is a finite tree with endpoints at the points of ∂U.
(*Note*: We allow $T = \varnothing$ to be empty or $T = \{\zeta\}$ to be a single point.)
The finite tree $T \subseteq \mathbb{P}^1_{\mathrm{an}}$ is called the *skeleton* of U.

(b) Show that the mapping from

$$\{\text{open connected Berkovich affinoids } U \subsetneq \mathbb{P}^1_{\mathrm{an}}\}$$

to

$$\{\text{finite trees } T \subseteq \mathbb{P}^1_{\mathrm{an}} \text{ with endpoints of Type II or III}\}$$

given in part (a) is bijective.

6.30. Let S be any subset of $\mathbb{P}^1(\mathbb{C}_v)$, and let $\overline{S}$ be its closure in $\mathbb{P}^1_{\mathrm{an}}$. For any Type II point $\zeta = \zeta(a, r) \in \mathbb{P}^1_{\mathrm{an}}$, prove that $\zeta \in \overline{S}$ if and only if at least one of the following two conditions holds:

(a) S intersects infinitely many residue classes of ζ; or
(b) there is some $b \in \overline{D}(a, r)$ such that the set

$$\bigl\{|x - b| \,:\, x \in S,\ |x - b| \neq r\bigr\} \subseteq \mathbb{R}$$

has an accumulation point at r.

6.31. Let S be any subset of $\mathbb{P}^1(\mathbb{C}_v)$, and let $\overline{S}$ be its closure in $\mathbb{P}^1_{\mathrm{an}}$. For any Type III point $\zeta = \zeta(a, r) \in \mathbb{P}^1_{\mathrm{an}}$, prove that $\zeta \in \overline{S}$ if and only if the set

$$\bigl\{|x - a| \,:\, x \in S\bigr\} \subseteq \mathbb{R}$$

has an accumulation point at r.

6.32. Let S be any subset of $\mathbb{P}^1(\mathbb{C}_v)$, and let $\overline{S}$ be its closure in $\mathbb{P}^1_{\mathrm{an}}$. Let $\zeta \in \mathbb{P}^1_{\mathrm{an}}$ be a Type IV point represented by a chain of disks $D_1 \supsetneq D_2 \supsetneq \cdots$. Prove that $\zeta \in \overline{S}$ if and only if $D_n \cap S \neq \varnothing$ for all $n \geq 1$.

6.33. Fix a point $\zeta \in \mathbb{P}^1_{\mathrm{an}}$. In Remark 6.40, we noted that in [**8**], Baker and Rumely define a tangent vector at ζ to be an equivalence class of intervals of the form $[\zeta, \xi]$, where $\xi \in \mathbb{P}^1_{\mathrm{an}} \smallsetminus \{\zeta\}$, and where two intervals $[\zeta, \xi_1]$ and $[\zeta, \xi_2]$ are said to be equivalent if their intersection consists of more than just the point ζ.

Define a map from the set of Baker–Rumely tangent vectors at ζ to the set of directions at ζ by sending the equivalence class of the interval $[\zeta, \xi]$ to the unique direction at ζ containing ξ. Prove that this map is well-defined, one-to-one, and onto.

6.34. Show that every subset $U \subseteq \mathbb{H}$ that is open in the Gel'fand (weak) topology on $\mathbb{H}$ is also open in the metric (strong) topology on $\mathbb{H}$, as mentioned in Warning 6.43.

On the other hand, show that no open ball in the metric topology is open in the Gel'fand topology.

6.35. Let ζ_0, ζ_1 be distinct points in $\mathbb{P}^1_{\mathrm{an}}$.

(a) Prove that none of $[\zeta_0, \zeta_1)$, $(\zeta_0, \zeta_1]$, or (ζ_0, ζ_1) is open in the Gel'fand topology.

(b) Prove that none of $[\zeta_0, \zeta_1)$, $(\zeta_0, \zeta_1]$, or (ζ_0, ζ_1) is closed in the Gel'fand topology.

(c) If ζ_0 and ζ_1 both lie in $\mathbb{H}$, prove that none of the three intervals above is either open or closed in the metric topology.

Chapter 7

Rational Functions and Berkovich Space

Having discussed $\mathbb{P}^1_{\mathrm{an}}$ as a space in Chapter 6, we now study the action of a rational function $\phi \in \mathbb{C}_v(z)$ on $\mathbb{P}^1_{\mathrm{an}}$, which we define in Theorem 7.1 and Definition 7.2 below. (More generally, any morphism between $\mathbb{C}_v$-varieties extends to a continuous map between the corresponding Berkovich spaces; see [**26**, Section 3.4].) We also discuss local degrees at Berkovich points, generalizing the multiplicity of ϕ at a point of $\mathbb{P}^1(\mathbb{C}_v)$.

7.1. The action of rational functions

Recall from Definition 6.2 that a point in $\mathbb{A}^1_{\mathrm{an}}$ is a multiplicative seminorm $\|\cdot\|_\zeta$ on the polynomial ring $\mathbb{C}_v[z]$ which restricts to $|\cdot|$ on $\mathbb{C}_v$. We may extend $\|\cdot\|_\zeta$ to most or all rational functions $h \in \mathbb{C}_v(z)$ as follows. Write $h = f/g$ with $f, g \in \mathbb{C}_v[z]$. If $\|g\|_\zeta \neq 0$, then define $\|h\|_\zeta := \|f\|_\zeta / \|g\|_\zeta$. See Exercise 7.1 for details.

Theorem 7.1. *Let $\phi \in \mathbb{C}_v(z)$ be a rational function, and let $\zeta \in \mathbb{A}^1_{\mathrm{an}}$. Unless ζ is a Type I point $\zeta = a \in \mathbb{C}_v$ for which $\phi(a) = \infty$, then the function*

$$\|\cdot\|_{\phi(\zeta)} : \mathbb{C}_v[z] \to [0, \infty) \quad \text{by} \quad \|f\|_{\phi(\zeta)} := \|f \circ \phi\|_\zeta \tag{7.1}$$

is a multiplicative seminorm on $\mathbb{C}_v[z]$ extending $|\cdot|$ on $\mathbb{C}_v$.

Proof. See Section 16.1. □

We can now define the action of a rational function $\phi \in \mathbb{C}_v(z)$ on $\mathbb{P}^1_{\mathrm{an}}$. We also denote this extended map ϕ.

Definition 7.2. Let $\phi \in \mathbb{C}_v(z)$ be a rational function, and let $\zeta \in \mathbb{P}^1_{\text{an}}$. Define $\phi(\zeta) \in \mathbb{P}^1_{\text{an}}$ to be

(a) the Type I point $\phi(\infty)$ if $\zeta = \infty$,

(b) the Type I point ∞ if $\zeta = a$ is of Type I and $\phi(a) = \infty$, or

(c) the point $\phi(\zeta) \in \mathbb{A}^1_{\text{an}}$ of equation (7.1) otherwise.

Remark 7.3. The reader can check that for the map $\eta(z) = 1/z$ and for any $\zeta \in \mathbb{A}^1_{\text{an}} \setminus \{0\}$, the point $\eta(\zeta)$ given by Definition 7.2 coincides with the point $1/\zeta$ defined in equation (6.2). In fact, viewing $\mathbb{P}^1_{\text{an}}$ as two copies of $\mathbb{A}^1_{\text{an}}$ glued via $1/z$, one can alternatively define the action of $\phi \in \mathbb{C}_v(z)$ on $\mathbb{P}^1_{\text{an}}$, by gluing appropriately chosen affine patches; see Exercise 7.2. In a general treatment of Berkovich space morphisms, such a definition would be more appropriate than the ad hoc Definition 7.2.

Theorem 7.4. *Let $\phi \in \mathbb{C}_v(z)$ be a rational function. Then the function $\phi : \mathbb{P}^1_{\text{an}} \to \mathbb{P}^1_{\text{an}}$ of Definition 7.2 is the unique continuous extension of the function $\phi : \mathbb{P}^1(\mathbb{C}_v) \to \mathbb{P}^1(\mathbb{C}_v)$ to $\mathbb{P}^1_{\text{an}}$.*

Proof. See Section 16.1. □

Corollary 7.5. *Let $\phi, \psi \in \mathbb{C}_v(z)$ be rational functions, and let $\zeta \in \mathbb{P}^1_{\text{an}}$. Then $\psi(\phi(\zeta)) = (\psi \circ \phi)(\zeta)$.*

Proof. Recall from Theorem 6.25 that the set of Type I points is dense in $\mathbb{P}^1_{\text{an}}$. Since the desired equality holds for Type I points, the result is immediate from Theorem 7.4. □

Theorem 7.4 describes the action of ϕ on Type I points. For Type II and some Type III points, we have the following preliminary result.

Proposition 7.6. *Let $\phi \in \mathbb{C}_v(z)$ be a nonconstant rational function, let $\zeta = \zeta(a,r) \in \mathbb{P}^1_{\text{an}}$ be a point of Type II or III, and suppose that ϕ has no poles in the open disk $D(a,r)$. Then there exist $b \in \mathbb{C}_v$ and $s > 0$ such that*

$$\phi\big(D(a,r)\big) = D(b,s),$$

with $s \in |\mathbb{C}_v^\times|$ if and only if $r \in |\mathbb{C}_v^\times|$. Moreover, $\phi(\zeta) = \zeta(b,s)$.

Proof. See Section 16.1. □

If ζ is of Type II, then we can always find a disk $D(a,r)$ satisfying the hypotheses of Proposition 7.6. After all, ζ has infinitely many residue classes (see Proposition 6.38), but ϕ has only finitely many poles. Thus, ζ must have some residue class $D(a,r)$ containing no poles of ϕ.

Since Type IV points are approximated by disks, Proposition 7.6 leads to the following result, describing the images of Type IV points.

Proposition 7.7. *Let $\phi \in \mathbb{C}_v(z)$ be a nonconstant rational function, and let $\zeta \in \mathbb{P}^1_{\mathrm{an}}$ be a point of Type IV, corresponding to a decreasing sequence of disks*

$$D_1 \supseteq D_2 \supseteq \cdots$$

with empty intersection. Then there are positive integers $d, N \geq 1$ such that $\phi(D_n)$ is a disk in $\mathbb{C}_v$ for all $n \geq N$, and $\phi : D_n \to \phi(D_n)$ is a d-to-1 map. Moreover, $\phi(\zeta)$ is the Type IV point corresponding to the decreasing sequence of disks

$$\phi(D_N) \supseteq \phi(D_{N+1}) \supseteq \cdots .$$

Proof. See Section 16.1. □

Recall from Remark 6.23 that an open or closed affinoid in $\mathbb{P}^1(\mathbb{C}_v)$ corresponds to a unique open or closed Berkovich affinoid, respectively. The following result shows that this correspondence is respected by nonconstant rational functions acting on $\mathbb{P}^1_{\mathrm{an}}$.

Theorem 7.8. *Let $\phi \in \mathbb{C}_v(z)$ be a nonconstant rational function, let $W \subseteq \mathbb{P}^1_{\mathrm{an}}$ be a connected Berkovich affinoid, and let*

$$W_{\mathrm{I}} := W \cap \mathbb{P}^1(\mathbb{C}_v)$$

be the corresponding connected affinoid in $\mathbb{P}^1(\mathbb{C}_v)$. Then $\phi(W)$ is the Berkovich connected affinoid of the same type (if any) corresponding to $\phi(W_{\mathrm{I}})$, and $\phi^{-1}(W)$ is the Berkovich affinoid of the same type (if any) corresponding to $\phi^{-1}(W_{\mathrm{I}})$. Moreover, the following hold.

(a) $\partial(\phi(W)) \subseteq \phi(\partial W)$.

(b) *Each of the connected components $V_1, \ldots, V_m$ of $\phi^{-1}(W)$ is a connected Berkovich affinoid mapping onto W.*

(c) *For each $i = 1, \ldots, m$,*

$$\phi(\partial V_i) = \partial W \quad \text{and} \quad \phi(\operatorname{int} V_i) = \operatorname{int} W,$$

where $\operatorname{int} X$ denotes the interior of the set X.

(d) *If W is open, then $\phi(\partial V_i) \cap W = \varnothing$.*

Proof. See Section 16.1. □

Theorem 7.8 is only a partial analogue of Proposition 3.29, because it fails to provide any degree information for components of $\phi^{-1}(W)$. After we have defined local degrees appropriately, however, Proposition 7.33 will supply the missing degree formula.

Corollary 7.9. *Let $\phi \in \mathbb{C}_v(z)$ be a nonconstant rational function. Then $\phi : \mathbb{P}^1_{\mathrm{an}} \to \mathbb{P}^1_{\mathrm{an}}$ is an open map. That is, for every open set $U \subseteq \mathbb{P}^1_{\mathrm{an}}$, the image $\phi(U)$ is also open.*

Proof. This is immediate from Theorem 6.17, that the open connected Berkovich affinoids form a basis for the Gel'fand topology, and Theorem 7.8. □

Corollary 7.10. *Let $\phi \in \mathbb{C}_v(z)$ be a nonconstant rational function, and let $U \subseteq \mathbb{P}^1_{\text{an}}$ be an open connected Berkovich affinoid. Then the following are equivalent.*

(a) $\phi(U) \cap \phi(\partial U) = \varnothing$.

(b) *U is a connected component of $\phi^{-1}(\phi(U))$.*

Proof. See Section 16.1. □

Remark 7.11. Any Type II point can be moved to the Gauss point $\zeta(0,1)$ by a linear fractional transformation. More precisely, if $b, c \in \mathbb{C}_v$ are distinct and $r := |b-c| \in |\mathbb{C}_v^\times|$, then

$$\eta(z) := (c-b)^{-1}(z-b)$$

takes $b \mapsto 0$, $c \mapsto 1$, and $\infty \mapsto \infty$. Thus, $\eta(D(b,r)) = D(0,1)$, and hence $\eta(\zeta(b,r)) = \zeta(0,1)$ by Proposition 7.6.

The connected components of $\mathbb{P}^1_{\text{an}} \smallsetminus \zeta(0,1)$ correspond naturally to the elements $\mathbb{P}^1(\overline{k})$, where $\overline{k}$ is the residue field of $\mathbb{C}_v$. Thus, because η is a homeomorphism (as η and its inverse η^{-1} are both continuous), it follows that the connected components of $\mathbb{P}^1_{\text{an}} \smallsetminus \zeta(b,s)$ also correspond to the elements of $\mathbb{P}^1(\overline{k})$, although the bijection depends on η and hence on the choice of b and c. This correspondence motivates our use of the term *residue classes* for the connected components of $\mathbb{P}^1_{\text{an}} \smallsetminus \{\zeta\}$.

In addition, the elements of $\mathrm{PGL}(2,\mathbb{C}_v)$ that fix $\zeta(0,1)$ are precisely the elements of $\mathrm{PGL}(2,\mathcal{O})$, where $\mathcal{O}$ denotes the ring of integers of $\mathbb{C}_v$; see Exercise 7.3. Thus, the Type II points of $\mathbb{P}^1_{\text{an}}$ correspond to the left cosets of $\mathrm{PGL}(2,\mathcal{O})$ in $\mathrm{PGL}(2,\mathbb{C}_v)$.

7.2. Images of points of Types II and III

Proposition 7.6 does not apply to all points of Type III, and even for points of Type II it requires choosing a residue class containing no poles of ϕ. The following result remedies those omissions, at the expense of requiring a slightly more complicated statement.

Theorem 7.12. *Let $\phi \in \mathbb{C}_v(z)$ be a nonconstant rational function, and let $\zeta = \zeta(a,r) \in \mathbb{P}^1_{\text{an}}$ be a point of Type II or III. Then ϕ has a Laurent series expansion $\phi(z) = \sum_{n\in\mathbb{Z}} c_n(z-a)^n$ on the annulus*

$$U_\lambda := \{\lambda r < |z-a| < r\}$$

for all $\lambda < 1$ *sufficiently close to* 1. *Moreover, for sufficiently large* $\lambda < 1$, *the map* $\phi - c_0$ *has inner and outer Weierstrass degrees on* U_λ *both equal to some integer* $d \neq 0$ *between* $-\deg(\phi)$ *and* $\deg(\phi)$. *Setting* $s := |c_d| r^d$, *we have*

$$\phi(U_\lambda) = \begin{cases} \{\lambda^d s < |z - c_0| < s\} & \text{if } d > 0, \\ \{s < |z - c_0| < \lambda^d s\} & \text{if } d < 0. \end{cases}$$

Finally, $\phi(\zeta) = \zeta(c_0, s)$; *and* $s \in |\mathbb{C}_v^\times|$ *if and only if* $r \in |\mathbb{C}_v^\times|$.

Proof. See Section 16.1. □

Remark 7.13. Recall that we defined a point $\zeta(a, r)$ of Type II or III to be the sup norm for polynomials on $\overline{D}(a, r)$. A more refined definition, illustrated by Theorem 7.12, is as a limsup norm for *rational* functions on thin annuli approaching the boundary of the disk. Even for polynomials, Theorem 7.12 parallels the complex analytic fact that the supremum of an analytic function f on disk is attained on the boundary. Thus, to understand the limiting behavior of f at the boundary, it suffices to understand the behavior of f on a thin annulus U_λ.

Remark 7.14. Theorem 7.12 describes an annulus extending from the point $\zeta(a, r)$ in the direction of a. Alternatively, to represent the direction containing ∞, one could use an annulus of the form

$$U = \{r < |z - a| < \lambda r\} \quad \text{with } \lambda > 1.$$

For $\lambda > 1$ sufficiently close to 1, the image $\phi(U)$ would again be of the form either

$$\{s < |z - b| < \lambda^d s\} \quad \text{or} \quad \{\lambda^d s < |z - b| < s\},$$

depending on whether the Weierstrass degree d is positive or negative, respectively. To see this fact, we need only apply Theorem 7.12 to the map $\psi := \phi \circ h$ and the point $\zeta(0, 1/r)$, where $h(z) = a + 1/z$, and then note that $h(\zeta(0, 1/r)) = \zeta(a, r)$.

Example 7.15. Pick $a \in \mathbb{C}_v$ with $0 < |a| < 1$, pick an integer $m \geq 1$, and let

$$\phi(z) = az^{m+1} + az^m + \frac{a^2}{z - a} = \frac{az^{m+2} + a(1-a)z^{m+1} - a^2 z^m + a^2}{z - a}.$$

We now find the image $\phi(\zeta(0, 1))$ of the Gauss point via Theorem 7.12. (The reader can check that $\phi(D(0, 1)) = \mathbb{P}^1(\mathbb{C}_v)$, preventing us from applying Proposition 7.6 to the open disk $D(0, 1)$.)

Choose $\lambda \in (|a|^{1/(m+1)}, 1)$, and let $U := \{\lambda < |z| < 1\}$. Note that ϕ has no poles in the open annulus U. Moreover,

$$|ax^m| > \max\left\{ |ax^{m+1}|, \left|\frac{a^2}{x - a}\right| \right\} \quad \text{for any } x \in U, \tag{7.2}$$

and hence ϕ has inner and outer Weierstrass degree both equal to m on U. (There is no need to expand the power series for $a^2/(x-a)$, as Proposition 3.17 and inequality (7.2) guarantee that ax^m dominates all of its terms.) Let $s := |a| \cdot 1^m = |a|$, and note that the constant term c_0 of the Laurent series satisfies $|c_0| < s$. Thus, by Theorem 7.12,

$$\phi(U) = \{\lambda^m |a| < |z - c_0| < |a|\},$$

and $\phi(\zeta(0,1)) = \zeta(0,|a|)$.

Of course, what has happened here is that the az^m-term in the original definition of ϕ is dominant on the annulus U, and $z \mapsto az^m$ clearly maps $D(0,1)$ m-to-1 onto $D(0,|a|)$. Using the annulus U rather than the disk $D(0,1)$ allowed us to avoid the pole at a.

Alternatively, as we noted just after Proposition 7.6, we could determine $\phi(\zeta(0,1))$ by computing the image of a different residue class, like $D(1,1)$, that contains no poles. On the one hand, this strategy obviates the need to consider annuli. On the other hand, it would require more work, because we would need to expand ϕ as a power series centered at 1 rather than as a Laurent series centered at 0. If we had the patience to compute that power series, though, a quick application of Theorem 3.15 would show that $\phi(D(1,1)) = D(2a,|a|)$, and hence Proposition 7.6 would again show that $\phi(\zeta(0,1)) = \zeta(0,|a|)$.

Combined with the results of Section 7.1, Theorem 7.12 allows us to conclude that the type of a Berkovich point is preserved by nonconstant rational functions.

Corollary 7.16. *Let $\phi \in \mathbb{C}_v(z)$ be a nonconstant rational function, and let $\zeta \in \mathbb{P}^1_{\mathrm{an}}$. Then $\phi(\zeta)$ is a point of the same type as ζ.*

Proof. If ζ is of Type I, we are done by the second statement of Theorem 7.4. If ζ is of Type II or III, then Theorem 7.12 suffices. And if ζ is of Type IV, we are done by Proposition 7.7. □

The following result is useful whenever we want to change coordinates while discussing the metric $d_{\mathbb{H}}$ on $\mathbb{H}$.

Proposition 7.17. *Let $\phi \in \mathbb{C}_v(z)$ be a rational function. Then ϕ is an isometry on $\mathbb{H}$ with respect to $d_{\mathbb{H}}$ if and only if* $\deg(\phi) = 1$.

Proof. See Exercise 7.4. □

7.3. Local degrees in directions

Theorem 7.12 not only identifies the image of a given Type II or Type III point ζ under a rational function ϕ, but it also shows how each direction $\boldsymbol{v}$ at ζ is mapped to a direction $\phi_{\#}(\boldsymbol{v})$ at $\phi(\zeta)$. We make this mapping more precise now.

Definition 7.18. Let $\phi \in \mathbb{C}_v(z)$ be a nonconstant rational function, let $\zeta \in \mathbb{P}^1_{\text{an}}$, and let $\boldsymbol{v}$ be a direction at ζ. We define a positive integer $\deg_{\zeta,\boldsymbol{v}}(\phi)$ and a direction $\phi_{\#}(\boldsymbol{v})$ at $\phi(\zeta)$ as follows.

(a) If $\zeta = x$ is of Type I, then $\phi_{\#}(\boldsymbol{v})$ is the unique direction at $\phi(\zeta)$, and $\deg_{\zeta,\boldsymbol{v}}(\phi) := \deg_x(\phi)$ is the algebraic multiplicity of the map $\phi : \mathbb{P}^1(\mathbb{C}_v) \to \mathbb{P}^1(\mathbb{C}_v)$ at x.

(b) If $\zeta = \zeta(a, r)$ is of Type II or III and $\boldsymbol{v} = \vec{v}_\zeta(a)$ is the direction at ζ containing a, then $\phi_{\#}(\boldsymbol{v})$ is the direction at $\phi(\zeta)$ containing $\phi(U)$, where $U := \{\lambda r < |z - a| < r\}$ is the annulus of Theorem 7.12, and $\deg_{\zeta,\boldsymbol{v}}(\phi)$ is the multiplicity of the mapping $\phi : U \to \phi(U)$. That is, $\deg_{\zeta,\boldsymbol{v}}(\phi) := \pm d \geq 1$, where d is the common inner and outer Weierstrass degree of ϕ on U.

(c) If $\zeta = \zeta(a, r)$ is of Type II or III and $\boldsymbol{v}$ is the direction at ζ containing ∞, then $\phi_{\#}(\boldsymbol{v})$ is the direction at $\phi(\zeta)$ containing $\phi(U)$, where $U := \{r < |z - a| < \lambda r\}$ is the annulus of Remark 7.14, and $\deg_{\zeta,\boldsymbol{v}}(\phi)$ is the multiplicity of the mapping $\phi : U \to \phi(U)$. That is, $\deg_{\zeta,\boldsymbol{v}}(\phi) := \pm d \geq 1$, where d is the common inner and outer Weierstrass degree of ϕ on U.

(d) If ζ is of Type IV, then $\phi_{\#}(\boldsymbol{v})$ is the unique direction at $\phi(\zeta)$, and $\deg_{\zeta,\boldsymbol{v}}(\phi)$ is the multiplicity d of Proposition 7.7.

Warning 7.19. Recall that we defined a direction $\boldsymbol{v}$ at $\zeta \in \mathbb{P}^1_{\text{an}}$ to be a certain subset of $\mathbb{P}^1_{\text{an}}$. However, the image direction $\phi_{\#}(\boldsymbol{v})$ *may or may not* coincide with the image $\phi(\boldsymbol{v})$ of the subset $\boldsymbol{v} \subseteq \mathbb{P}^1_{\text{an}}$. In fact, the image $\phi(\boldsymbol{v})$ of the set may be the entire space $\mathbb{P}^1_{\text{an}}$; see, for example, Theorem 7.34(b). On the other hand, there are other cases in which $\phi_{\#}(\boldsymbol{v}) = \phi(\boldsymbol{v})$; see Theorem 7.34(c).

Example 7.20. Consider again the function

$$\phi(z) = az^{m+1} + az^m + \frac{a^2}{z - a}$$

of Example 7.15, where $0 < |a| < 1$ and $m \geq 1$. Let $\boldsymbol{v} := \vec{v}(0)$ be the direction at $\zeta(0, 1)$ containing 0. Since the image of the annulus

$$U = \{\lambda < |z| < 1\}$$

is $\phi(U) = \{\lambda^m < |z| < 1\}$, Definition 7.18(b) says that $\phi_\#(\boldsymbol{v})$ is the direction at $\zeta(0,|a|)$ containing 0, and $\deg_{\zeta(0,1),\boldsymbol{v}}(\phi) = m$.

The following result shows that directions at Berkovich points behave well under composition of functions, and that their local degrees behave in the same way that multiplicities do.

Proposition 7.21. *Let $\phi, \psi \in \mathbb{C}_v(z)$ be nonconstant rational functions, let $\zeta \in \mathbb{P}^1_{\text{an}}$, and let $\boldsymbol{v}$ be a direction at ζ. Then*

(a) $1 \le \deg_{\zeta,\boldsymbol{v}}(\phi) \le \deg(\phi)$.

(b) $\psi_\#\big(\phi_\#(\boldsymbol{v})\big) = \big(\psi \circ \phi\big)_\#(\boldsymbol{v})$.

(c) $\deg_{\zeta,\boldsymbol{v}}(\psi \circ \phi) = \deg_{\zeta,\boldsymbol{v}}(\phi) \cdot \deg_{\phi(\zeta),\phi_\#(\boldsymbol{v})}(\psi)$.

Proof. This is left to the reader; see Exercise 7.5. □

Local degrees in directions at points in $\mathbb{H}$ can also be understood as repulsion factors with respect to the hyperbolic metric $d_{\mathbb{H}}$, as the following result shows.

Theorem 7.22. *Let $\phi \in \mathbb{C}_v(z)$ be a nonconstant rational function, let $\zeta \in \mathbb{P}^1_{\text{an}}$, let $\boldsymbol{v}$ be a direction at ζ, and let $m := \deg_{\zeta,\boldsymbol{v}}(\phi)$. Then there is a point $\xi \in \boldsymbol{v}$ such that ϕ maps $[\zeta, \xi]$ homeomorphically onto $[\phi(\zeta), \phi(\xi)]$, and*

$$d_{\mathbb{H}}\big(\phi(\xi_1), \phi(\xi_2)\big) = m \cdot d_{\mathbb{H}}(\xi_1, \xi_2) \quad \text{for all } \xi_1, \xi_2 \in [\zeta, \xi] \cap \mathbb{H}. \tag{7.3}$$

Proof. See Section 16.2. □

Example 7.23. Fix a prime $p \ge 2$, pick $m \ge 2$, and set $\phi(z) = z^m$. Then for any $r > 0$, we have $\phi(D(0,r)) = D(0,r^m)$, and therefore

$$\phi\big(\zeta(0,r)\big) = \zeta(0,r^m)$$

by Proposition 7.6. In particular, ϕ maps the interval $[\zeta(0,r), \zeta(0,s)]$ homeomorphically onto $[\zeta(0,r^m), \zeta(0,s^m)]$, while stretching the hyperbolic metric by a factor of exactly m all along the interval, since

$$d_{\mathbb{H}}\big(\zeta(0,t_1), \zeta(0,t_2)\big) = \log t_2 - \log t_1 \quad \text{for } t_1 \le t_2.$$

Applying Theorem 7.12 to annuli emanating from $\zeta(0,r)$ toward 0, we see that $\deg_{\zeta(0,r),\vec{v}(0)}(\phi) = m$. Similarly, applying Remark 7.14 to annuli emanating toward ∞, we have $\deg_{\zeta(0,r),\vec{v}(\infty)}(\phi) = m$ as well. Thus, we have confirmed Theorem 7.22 for $\phi(z) = z^m$ at $\zeta(0,r)$ in the direction of 0 or ∞.

But what happens in other directions? For example, at the Gauss point $\zeta(0,1)$, consider the direction $\vec{v}(a)$, for some $a \in \mathbb{C}_v$ with $|a| = 1$. Expanding ϕ as a power series centered at a and noting that $\phi(a) = a^m$, we see

$$\phi(z) - a^m = \big((z-a) + a\big)^m - a^m = \sum_{n=1}^{m} \binom{m}{n} a^{m-n} (z-a)^n. \tag{7.4}$$

Write $m = p^e j$, with $p \nmid j$ and $e \geq 0$. Then by Exercise 2.15,

$$\left|\binom{m}{p^e}\right| = 1, \quad \text{but} \quad \left|\binom{m}{n}\right| < 1 \quad \text{for all } 1 \leq n < p^e.$$

Thus, for $\lambda < 1$ sufficiently close to 1, the $(z-a)^{p^e}$-term of equation (7.4) is uniquely dominant on $U_\lambda := \{\lambda < |z-a| < 1\}$. As a result,

- $\deg_{\zeta(0,1),\vec{v}(a)}(\phi) = p^e$,
- $\phi(U_\lambda) = \left\{\lambda^{p^e} < \left|z - a^m\right| < 1\right\}$, and
- for any $0 < r < 1$, we have $\phi(\zeta(a,r)) = \zeta(a^m, r^{p^e})$.

In particular, by the last bullet point,

$$\phi : \left[\zeta(0,1), \zeta(a,\lambda)\right] \to \left[\zeta(0,1), \zeta\left(a^m, \lambda^{p^e}\right)\right]$$

is a homeomorphism stretching all distances by a factor of exactly p^e.

If $e = 0$, i.e., if $p \nmid m$, then ϕ is an isometry on that interval. In fact, in that case, ϕ is an isometry from the infinite-length interval $[\zeta(0,1), a)$ to $[\zeta(0,1), a^m)$. On the other hand, if $e \geq 1$, then ϕ only stretches by a constant factor of p^e on the interval

$$\left[\zeta(0,1), \zeta\left(a, |p|^{1/(p^e - p^{(e-1)})}\right)\right].$$

After all,

$$\left|\binom{m}{p^{(e-1)}}\right| = |p|,$$

and hence the $z^{p^{(e-1)}}$-term becomes dominant for $\lambda < |p|^{1/(p^e - p^{(e-1)})}$.

Theorem 7.22 also allows us to draw some broader conclusions about the action of rational functions on the metric space $(\mathbb{H}, d_{\mathbb{H}})$.

Corollary 7.24. *Let $\phi \in \mathbb{C}_v(z)$ be a rational function of degree $d \geq 1$. Then ϕ is Lipschitz on $\mathbb{H}$ with respect to the hyperbolic metric $d_{\mathbb{H}}$, with Lipschitz constant at most d.*

Proof. Given $\zeta_1, \zeta_2 \in \mathbb{H}$, we may assume without loss that $\zeta_1 \neq \zeta_2$. At each point ζ in the interval $I := [\zeta_1, \zeta_2]$, consider the two directions $\boldsymbol{v}$ and $\boldsymbol{w}$ at ζ containing ζ_1 or ζ_2, with local degrees m and n, respectively. (Of course, there is only one such direction if ζ is itself one of those two endpoints.) By Theorem 7.22, ϕ stretches distances by a factor of exactly m in a small interval extending from ζ toward ζ_1, and by exactly n in a small interval extending from ζ toward ζ_2. Since $1 \leq m, n \leq d$ by Proposition 7.21(a), there is an open subset I_ζ of I containing ζ such that

$$d_{\mathbb{H}}\big(\phi(\xi_1), \phi(\xi_2)\big) \leq d \cdot d_{\mathbb{H}}(\xi_1, \xi_2) \qquad \text{for all } \xi_1, \xi_2 \in I_\zeta.$$

Covering the compact interval I by these open subsets I_ζ and taking a finite subcover, it follows that

$$d_{\mathbb{H}}\big(\phi(\zeta_1), \phi(\zeta_2)\big) \le d \cdot d_{\mathbb{H}}(\zeta_1, \zeta_2). \qquad \square$$

We close this section with the following relationship among poles, zeros, and local degrees in directions.

Proposition 7.25. *Let $\phi \in \mathbb{C}_v(z)$ be a nonconstant rational function, let $\zeta = \zeta(a, r) \in \mathbb{P}^1_{\text{an}}$ be a point of Type II or III, and let $\boldsymbol{v}$ be a direction at ζ. Suppose that 0 and ∞ lie in different directions at $\phi(\zeta)$.*

Let M_0 and M_∞, respectively, be the number of zeros and poles, counted with multiplicity, of ϕ in $\boldsymbol{v}$; and let N_0 and N_∞, respectively, be the number of zeros and poles, counted with multiplicity, of ϕ in $\overline{D}(a, r)$. Then

$$\deg_{\zeta,\boldsymbol{v}}(\phi) = \begin{cases} M_0 - M_\infty & \text{if } \infty \notin \boldsymbol{v} \text{ and } 0 \in \phi_{\#}(\boldsymbol{v}), \\ M_\infty - M_0 & \text{if } \infty \notin \boldsymbol{v} \text{ and } \infty \in \phi_{\#}(\boldsymbol{v}), \\ N_\infty - N_0 & \text{if } \infty \in \boldsymbol{v} \text{ and } 0 \in \phi_{\#}(\boldsymbol{v}), \\ N_0 - N_\infty & \text{if } \infty \in \boldsymbol{v} \text{ and } \infty \in \phi_{\#}(\boldsymbol{v}). \end{cases}$$

Proof. This is left to the reader; see Exercise 7.8. $\square$

7.4. Local degrees at Berkovich points

We wish to define the local degree $\deg_\zeta(\phi)$ of a rational function $\phi \in \mathbb{C}_v(z)$ at a point $\zeta \in \mathbb{P}^1_{\text{an}}$ itself. At a Type I point $\zeta = x \in \mathbb{P}^1(\mathbb{C}_v)$, the local degree $\deg_x(\phi)$ should be the algebraic multiplicity of ϕ at x, i.e., the multiplicity of x as a root of the equation $\phi(z) = \phi(x)$. However, the function $x \mapsto \deg_x(\phi)$ is not continuous on $\mathbb{P}^1(\mathbb{C}_v)$. (Consider $\phi(z) = z^d$ at $x = 0$, for example.) Thus, we cannot define $\deg_\zeta(\phi)$ as a continuous extension of $\deg_x(\phi)$ to $\mathbb{P}^1_{\text{an}}$.

Instead, note that $\deg_x(\phi)$ is the smallest integer $d \ge 1$ with the following property: x has an open neighborhood $U \subseteq \mathbb{P}^1(\mathbb{C}_v)$ such that every point in $\phi(U)$ has at most d preimages in U, counting multiplicity. Therefore, it makes sense to define $\deg_\zeta(\phi)$ as follows.

Definition 7.26. Let $\phi \in \mathbb{C}_v(z)$ be a nonconstant rational function, and let $\zeta \in \mathbb{P}^1_{\text{an}}$. The *local degree*, or *multiplicity*, of ϕ at ζ is

$$\deg_\zeta(\phi) := \inf_U \left(\sup_y \left(\sum_{x \in U \cap \phi^{-1}(y)} \deg_x(\phi) \right) \right), \tag{7.5}$$

where the infimum is over all open $U \subseteq \mathbb{P}^1_{\text{an}}$ containing ζ, the supremum is over all $y \in \phi(U) \cap \mathbb{P}^1(\mathbb{C}_v)$, and $\deg_x(\phi)$ denotes the algebraic multiplicity of x as a root of the equation $\phi(z) = y$.

The quantity $\sum_{x\in U\cap\phi^{-1}(y)}\deg_x(\phi)$ of equation (7.5) is an integer between 1 and $\deg(\phi)$, since it is the number of preimages of y in U, counting multiplicity. Thus, $\deg_\zeta(\phi)$ is also an integer lying between 1 and $\deg(\phi)$.

Lemma 7.27. *Let $\phi \in \mathbb{C}_v(z)$ be a nonconstant rational function, let $\zeta \in \mathbb{P}^1_{\mathrm{an}}$, and let $U \subseteq \mathbb{P}^1_{\mathrm{an}}$ be an open set containing ζ. Then there is an open connected Berkovich affinoid W such that*

(a) $\zeta \in W \subseteq U$,

(b) $\phi(\partial W) \subseteq \partial(\phi(W))$, *and*

(c) $\sum_{x\in W\cap\phi^{-1}(y)}\deg_x(\phi) = \deg_\zeta(\phi)$ *for every Type I point y in $\phi(W)$.*

Proof. See Section 16.2. □

Example 7.28. To illustrate Lemma 7.27, consider the function

$$\phi(z) = z^2 - az \in \mathbb{C}_v[z], \quad \text{where} \quad |a| = 1.$$

Fix $r > 1$, and let U be the open neighborhood

$$U := \left\{\zeta \in \mathbb{P}^1_{\mathrm{an}} : \frac{1}{r} < |\zeta| < r\right\},$$

of the Gauss point $\zeta(0,1)$. Let $U_{\mathrm{I}} := U \cap \mathbb{P}^1(\mathbb{C}_v)$ be the corresponding open annulus of Type I points. By applying Theorem 3.33 to smaller annuli contained in U_I, the reader can check that $\phi(U_{\mathrm{I}}) = D(0, r^2)$, and hence that $\phi(U) = D_{\mathrm{an}}(0, r^2)$, by Theorem 7.8. However, while Type I points in $\{1/r < |z| < r^2\}$ have two preimages in U, those in $\overline{D}(0, 1/r)$ have only one.

The problem, of course, is that both of the disks $\overline{D}(a, 1/r)$ and $\overline{D}(0, 1/r)$ map bijectively onto $\overline{D}(0, 1/r)$, but the former is contained in U_{I}, while the latter does not intersect U_{I}. Thus, if we define

$$W := D_{\mathrm{an}}(0, r) \smallsetminus \left(\overline{D}_{\mathrm{an}}(0, 1/r) \cup \overline{D}_{\mathrm{an}}(a, 1/r)\right) \subsetneq U,$$

then $\phi(W) = D_{\mathrm{an}}(0, r^2) \smallsetminus \overline{D}_{\mathrm{an}}(0, 1/r)$, and

$$\begin{aligned}\phi(\partial W) &= \phi\big(\{\zeta(0, r), \zeta(0, 1/r), \zeta(a, 1/r)\}\big)\\ &= \big\{\zeta(0, r^2), \zeta(0, 1/r)\big\} = \partial\big(\phi(W)\big).\end{aligned}$$

In addition, every Type I point in $\phi(W)$ has exactly two preimages in W, counting multiplicity. It can also be shown that $\deg_{\zeta(0,1)}(\phi) = 2$; see, for example, Theorem 7.34. Hence, W fits all three conclusions of Lemma 7.27.

In fact, the affinoid W above is precisely the one constructed in the proof of Lemma 7.27 in Section 16.2. Specifically, the original annulus U has boundary $\partial U = \{\zeta(0, 1/r), \zeta(0, r)\}$, which in turn has image

$$\phi(\partial U) = \big\{\zeta(0, 1/r), \zeta(0, r^2)\big\}.$$

The proof of Lemma 7.27 defines

$$T := \phi^{-1}\big(\phi(\partial U)\big) = \big\{\zeta(0,1/r), \zeta(a,1/r), \zeta(0,r)\big\}$$

and then defines W to be the connected component of $\mathbb{P}^1_{\text{an}} \smallsetminus T$ that contains the original point ζ.

The following result gives some basic properties of local degrees. It is essentially the same as [**63**, Proposition 2.2]; it also appears as a number of separate results in [**8**, Section 9.1].

Theorem 7.29. *Let $\phi, \psi \in \mathbb{C}_v(z)$ be nonconstant rational functions, and let $\zeta \in \mathbb{P}^1_{\text{an}}$. Then*

(a) *The function $\mathbb{P}^1_{\text{an}} \to \{1, \ldots, \deg(\phi)\}$ given by $\xi \mapsto \deg_\xi(\phi)$ is upper semicontinuous. That is, for every $m \geq 1$, the set*

$$\big\{\xi \in \mathbb{P}^1_{\text{an}} : \deg_\xi(\phi) \geq m\big\}$$

is closed in the Gel'fand topology.

(b) $\deg_\zeta(\psi \circ \phi) = \deg_\zeta(\phi) \cdot \deg_{\phi(\zeta)}(\psi)$.

(c) $\sum_{\phi(\xi)=\zeta} \deg_\xi(\phi) = \deg(\phi)$.

Proof. See Section 16.2. □

Recall that $1 \leq \deg_\zeta(\phi) \leq \deg(\phi)$. In particular, if $\deg(\phi) = 1$, then $\deg_\zeta(\phi) = 1$ for all $\zeta \in \mathbb{P}^1_{\text{an}}$. Thus, Theorem 7.29(b) implies that local degrees are invariant under coordinate changes on either the domain or range.

The following theorem relates local degrees to local degrees in directions. The summation formula (7.6) was essentially the definition of local degrees in [**115**, Section 4.2]; Baker and Rumely refer to it as Rivera-Letelier's "Repulsion Formula" in [**8**, Theorem 9.19].

Theorem 7.30. *Let $\zeta \in \mathbb{P}^1_{\text{an}}$, let $\phi \in \mathbb{C}_v(z)$ be a nonconstant rational function, and let $\boldsymbol{w}$ be a direction at $\phi(\zeta)$. Then*

$$\deg_\zeta(\phi) = \sum_{\phi_\#(\boldsymbol{v})=\boldsymbol{w}} \deg_{\zeta,\boldsymbol{v}}(\phi), \tag{7.6}$$

where the sum is over all directions $\boldsymbol{v}$ at ζ for which $\phi_\#(\boldsymbol{v}) = \boldsymbol{w}$.

Proof. See Section 16.2. □

Example 7.31. Let $c \in \mathbb{C}_v$ with $0 < |c| < 1$, and let

$$\phi(z) = \frac{z^4 - 8c^4 z - c^4}{4z^3 + z^2 + 4c^4} \in \mathbb{C}_v(z)$$

be the Lattès map of Example 5.33. Consider $\zeta = \zeta(0, r)$ with $|c| \le r \le 1$. Note that ϕ has no zeros or poles in the annulus

$$U_r := \{|c| < |x| < r\} = D(0, r) \setminus \overline{D}(0, |c|).$$

By Proposition 3.32, the inner and outer Weierstrass degrees of ϕ coincide on U_r. In fact, we may write

$$\phi(z) = z^2 \cdot \left(\frac{1 - 8c^4 z^{-1} - c^4 z^{-4}}{4z + 1 + 4c^4 z^{-2}} \right),$$

and because the expression in parentheses is clearly in $D(1, 1)$ for all $z \in U_r$, that common Weierstrass degree must be 2. (Put another way, the Laurent series on U_r for the expression in parentheses is dominated by its constant term. See also Exercise 3.20(b), since the numerator of ϕ has Weierstrass degree 4 on U_r, and the denominator has Weierstrass degree 2.) In fact, we have $|\phi(x)| = |x|^2$ for all $x \in U_r$.

In light of Theorems 3.33 and 7.12, then, we have

$$\phi\big(\zeta(0, r)\big) = \zeta(0, r^2).$$

By Definition 7.18, we also have

$$\phi_\#\big(\vec{v}(0)\big) = \vec{w}(0), \quad \text{with } \deg_{\zeta(0,r),\vec{v}(0)}(\phi) = 2$$

and

$$\phi_\#\big(\vec{v}(\infty)\big) = \vec{w}(\infty), \quad \text{with } \deg_{\zeta(0,r),\vec{v}(\infty)}(\phi) = 2,$$

where $\vec{v}(a)$ denotes the direction at $\zeta(0, r)$ containing a, and where $\vec{w}(a)$ denotes the direction at $\zeta(0, r^2)$ containing a.

On the other hand, $|\phi(b)| = |b|^2$ for any $b \in \mathbb{C}_v$ with $|b| = r$, and so

$$\phi_\#\big(\vec{v}(b)\big) \neq \vec{w}(0), \vec{w}(\infty).$$

Thus, the only direction at $\zeta(0, r)$ that maps to $\vec{w}(0)$ is $\vec{v}(0)$, and the only direction at $\zeta(0, r)$ that maps to $\vec{w}(\infty)$ is $\vec{v}(\infty)$. Using either $\boldsymbol{w} := \vec{w}(0)$ or $\boldsymbol{w} := \vec{w}(\infty)$, then, Theorem 7.30 tells us that $\deg_{\zeta(0,r)}(\phi) = 2$.

We leave it to the reader to check that for $|c|^2 < r < |c|$, we have $\phi(\zeta(0, r)) = \zeta(0, |c|^4/r^2)$, with $\deg_{\zeta(0,r)}(\phi) = 2$. This time, however,

$$\phi_\#\big(\vec{v}(0)\big) = \vec{w}(\infty) \quad \text{and} \quad \phi_\#\big(\vec{v}(\infty)\big) = \vec{w}(0),$$

with degree $\deg_{\zeta(0,r),\boldsymbol{v}}(\phi) = 2$ for both $\boldsymbol{v} = \vec{v}(0)$ and $\boldsymbol{v} = \vec{v}(\infty)$. Thus, ϕ maps each of the intervals

$$\big(\zeta(0, 1), \zeta(0, |c|)\big) \quad \text{and} \quad \big(\zeta(0, |c|), \zeta(0, |c|^2)\big)$$

bijectively onto the interval $(\zeta(0, 1), \zeta(0, |c|^2))$ in $\mathbb{P}^1_{\text{an}}$, but with local degree 2 at each point. By Theorem 7.29(c) and the fact that $\deg(\phi) = 4$, then, these two intervals make up the full preimage of $(\zeta(0, 1), \zeta(0, |c|^2))$; see Exercise 7.10.

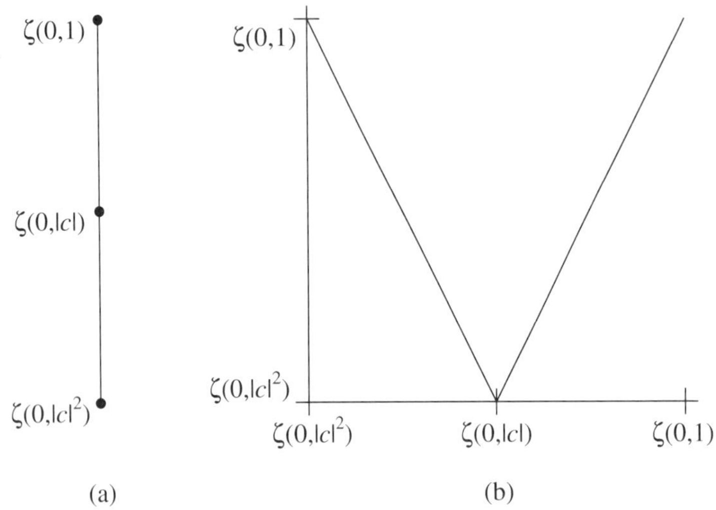

Figure 7.1. The Lattès map of Example 7.31 acting on the interval $J = [\zeta(0,1), \zeta(0,|c|^2)]$.

The reader can also check that

$$\phi\bigl(\zeta(0,1)\bigr) = \phi\bigl(\zeta(0,|c|^2)\bigr) = \zeta(0,1),$$

with

$$\deg_{\zeta(0,1)}(\phi) = \deg_{\zeta(0,|c|^2)}(\phi) = 2,$$

and that

$$\phi\bigl(\zeta(0,|c|)\bigr) = \zeta\bigl(0,|c|^2\bigr), \quad \text{with} \quad \deg_{\zeta(0,|c|)}(\phi) = 4.$$

(See Exercise 7.11.) Thus, we have accounted for the preimages of all points in the closed interval

$$J := \bigl[\zeta(0,1), \zeta(0,|c|^2)\bigr].$$

In addition, our computations above show that ϕ maps J onto itself by stretching it and then folding it in half over itself, as illustrated in Figure 7.1. The interval J, with its midpoint $\zeta(0,|c|)$, appears in Figure 7.1(a). The graph of $\phi|_J$ appears in Figure 7.1(b); the two halves

$$\bigl[\zeta(0,|c|^2), \zeta(0,|c|)\bigr] \quad \text{and} \quad \bigl[\zeta(0,|c|), \zeta(0,1)\bigr]$$

are each stretched by a factor of 2 and mapped onto J, with the lower half mapping from top to bottom, and the upper half from bottom to top.

Finally, because we accounted for all four preimages of each point $\zeta \in J$, we have $\phi^{-1}(J) = J$. We will see in Example 8.20 that J is precisely the Berkovich Julia set of ϕ.

Corollary 7.32. *Let $\phi \in \mathbb{C}_v(z)$ be a nonconstant rational function, and let $\zeta \in \mathbb{P}^1_{\mathrm{an}}$.*

(a) *If ζ is of Type I, then $\deg_\zeta(\phi)$ is the algebraic multiplicity of ϕ at ζ.*

(b) *If ζ is of Type III, then*

$$\deg_\zeta(\phi) = \deg_{\zeta,\boldsymbol{u}}(\phi) = \deg_{\zeta,\boldsymbol{v}}(\phi),$$

where $\boldsymbol{u}$ and $\boldsymbol{v}$ are the two directions at ζ.

(c) *If ζ is of Type IV, then $\deg_\zeta(\phi) = \mathrm{wdeg}_U(\phi)$ for any sufficiently small disk U containing ζ.*

Proof.

(a) and (c) This is immediate from Theorem 7.30 and Definition 7.18, since points of Types I and IV each have only one direction.

(b) We claim that $\phi_\#(\boldsymbol{u}) \neq \phi_\#(\boldsymbol{v})$. After all, $\phi(\zeta)$ is also a point of Type III, and hence if $\phi_\#(\boldsymbol{u}) = \phi_\#(\boldsymbol{v})$, then one of the two directions $\boldsymbol{w}$ at $\phi(\zeta)$ would not be the image under $\phi_\#$ of either direction at ζ. Using this direction $\boldsymbol{w}$ in Theorem 7.30, then, we would have $\deg_\zeta(\phi) = 0$, contradicting the fact that $\deg_\zeta(\phi) \geq 1$ and proving the claim.

Thus, each of the two directions at $\phi(\zeta)$ is the image under $\phi_\#$ of exactly one of the two directions at ζ. Applying Theorem 7.30 to these two directions $\phi_\#(\boldsymbol{v})$ and $\phi_\#(\boldsymbol{w})$, then, the desired equality is immediate. □

We close this section with the degree formula promised when we stated Theorem 7.8.

Proposition 7.33. *Let $\phi \in \mathbb{C}_v(z)$ be a nonconstant rational function of degree $d \geq 2$, and let $W \subseteq \mathbb{P}^1_{\mathrm{an}}$ be a connected Berkovich affinoid. Write*

$$\phi^{-1}(W) = V_1 \cup \cdots \cup V_m$$

as a disjoint union of connected Berkovich affinoids according to Theorem 7.8. Then for each $i = 1, \ldots, m$, there is an integer $1 \leq d_i \leq d$ such that every point in W has exactly d_i preimages in V_i, counting multiplicity. Moreover, $d_1 + \cdots + d_m = d$.

Proof. This is left to the reader; see Exercise 7.12. □

7.5. Computing local degrees

Corollary 7.32 provides an easy way to compute $\deg_\zeta(\phi)$ if the point ζ is of Type I, III, or IV. Theorem 7.34 below does the same for the remaining case where ζ is of Type II. This result, originally proven by Rivera-Letelier over $\mathbb{C}_p$, first appeared as [**114**, Corollaire 2.2, Proposition 2.4]; see also [**115**, Proposition 4.4] and [**8**, Theorem 9.26].

Recall from the discussion following Definition 6.39 that the directions at $\zeta(0,1)$ are open disks, with the direction containing $x \in \mathbb{P}^1(\mathbb{C}_v)$ denoted by $\vec{v}(x)$. Furthermore, recall that the set of directions is in one-to-one correspondence with $\mathbb{P}^1(\overline{k})$. In addition, as we did in Definition 4.6, we may write any rational function $\phi(z) \in \mathbb{C}_v(z)$ as $f(z)/g(z)$ with $f, g \in \mathcal{O}[z]$, and with at least one coefficient of f or g having absolute value 1. We can therefore define the reduction of ϕ to be the rational function

$$\overline{\phi}(z) = \frac{\overline{f}(z)}{\overline{g}(z)} \in k(z) \cup \{\infty\}.$$

We already made use of this idea in Examples 5.31 and 5.32 to understand the action of ϕ on residue classes in $\overline{D}(0,1)$. Note that if ϕ does not have explicit good reduction, then $\overline{\phi}$ has degree strictly smaller than $\deg(\phi)$; in fact, $\overline{\phi}$ may be constant.

Theorem 7.34. *Let $\phi \in \mathbb{C}_v(z)$ be a rational function of degree $d \geq 1$. Write $\phi = f/g$, where $f, g \in \mathcal{O}[z]$, and where at least one coefficient of f or g has absolute value 1. Define $\overline{\phi}(z) := \overline{f}(z)/\overline{g}(z)$. Then $\overline{\phi}$ is nonconstant if and only if $\phi(\zeta(0,1)) = \zeta(0,1)$. In that case,*

$$\deg_{\zeta(0,1)}(\phi) = \deg\left(\overline{\phi}\right). \tag{7.7}$$

Also in that case, define a direction $\boldsymbol{v}$ at $\zeta(0,1)$ to be bad *if $\boldsymbol{v}$ contains* both *a zero* and *a pole of ϕ, and define T to be the set of bad directions $\boldsymbol{v}$. Then*

(a) *T is a finite set,*

(b) *$\phi(\boldsymbol{v}) = \mathbb{P}^1_{\mathrm{an}}$ for each direction $\boldsymbol{v} \in T$, and*

(c) *$\phi_{\#}(\boldsymbol{v}) = \phi(\boldsymbol{v}) = \vec{v}(\phi(a))$ for all directions $\boldsymbol{v}$ at $\zeta(0,1)$ not in T and all $a \in \boldsymbol{v}$. For such directions $\boldsymbol{v}$, we have*

$$\deg_{\zeta(0,1),\boldsymbol{v}}(\phi) = \mathrm{wdeg}_{\boldsymbol{v}}(\phi) \tag{7.8}$$

where $\mathrm{wdeg}_{\boldsymbol{v}}(\phi)$ is the Weierstrass degree of ϕ on the disk $\boldsymbol{v}$, and $\deg_{\zeta(0,1),\boldsymbol{v}}(\phi)$ is the local degree in the direction $\boldsymbol{v}$ at $\zeta(0,1)$.

Finally, ϕ has explicit good reduction, i.e., $\deg(\overline{\phi}) = d$, if and only if $\overline{\phi}$ is nonconstant and $T = \varnothing$.

Theorem 7.34 does even more than provide a way to compute local degrees at Type II points. In particular, if ϕ fixes the Gauss point, then it *almost* has explicit good reduction, in the following sense. Outside of the finitely many bad directions in T, ϕ maps every other residue class onto the residue class dictated by the reduced map $\overline{\phi}$.

The proof of Theorem 7.34 relies on the following lemma, which is of interest in its own right. The proofs of both results appear in Section 16.3.

Lemma 7.35. *Let $\phi \in \mathbb{C}_v(z)$ be a rational function such that $\overline{\phi}$ is nonconstant. Then*

(a) $\phi(\zeta(0,1)) = \zeta(0,1)$.

(b) *For any $a, b \in \mathbb{P}^1(\mathbb{C}_v)$, we have*

$$\overline{\phi}(\overline{a}) = \overline{b} \quad \textit{if and only if} \quad \phi_{\#}(\vec{v}(a)) = \vec{v}(b),$$

where $\overline{a}, \overline{b}$ are the reductions of a and b in $\mathbb{P}^1(k)$, and where $\vec{v}(a)$ and $\vec{v}(b)$ are the directions at $\zeta(0,1)$ containing a and b, respectively.

(c) *For any $a \in \mathbb{P}^1(\mathbb{C}_v)$, we have*

$$\deg_{\zeta(0,1),\vec{v}(a)}(\phi) = \deg_{\overline{a}}(\overline{\phi}),$$

where $\vec{v}(a)$ is the direction of a at $\zeta(0,1)$, and $\deg_{\overline{a}}(\overline{\phi})$ is the multiplicity of $\overline{\phi} \in k(z)$ at the point $\overline{a} \in \mathbb{P}^1(k)$.

Warning 7.36. Because ϕ may not have explicit good reduction, many points $a \in \mathbb{P}^1(\mathbb{C}_v)$ do *not* satisfy $\overline{\phi}(\overline{a}) = \overline{\phi(a)}$. After all, according to Theorem 7.34(b), for each of the the finitely many bad directions $\boldsymbol{v} \in T$, we have $\phi(\boldsymbol{v}) = \mathbb{P}^1_{\text{an}}$. Thus, for Type I points $a \in \boldsymbol{v}$, $\phi(a)$ could be any point in $\mathbb{P}^1(\mathbb{C}_v)$, but $\overline{\phi}(\overline{a})$ is only a single residue class in $\mathbb{P}^1(\mathbb{C}_v)$.

Of course, Theorem 7.34 computes $\deg_\zeta(\phi)$ only in the case that ζ and its image $\phi(\zeta)$ are both the Gauss point. To do the same computation for a more general type II point ζ, we need to make coordinate changes on both the domain and range to move ζ and $\phi(\zeta)$ to $\zeta(0,1)$. We illustrate this technique with the following example.

Example 7.37. Fix a prime number p, let $\mathbb{C}_v = \mathbb{C}_p$, and define

$$\phi(z) = z^2 - z + 1 + \frac{p^2}{z} = \frac{z^3 - z^2 + z + p^2}{z}.$$

Let's use Theorem 7.34 to compute $\phi(\zeta)$ and $\deg_\zeta(\phi)$, where $\zeta = \zeta(0, |p|)$.

First, we can move $\zeta(0,1)$ to ζ with the linear fractional transformation $\eta(z) = pz$. Thus, $\phi(\zeta) = \phi(\eta(\zeta(0,1)))$. We have

$$\phi \circ \eta(z) = \phi(pz) = \frac{p^2z^3 - pz^2 + z + p}{z}.$$

Note that $\overline{\phi \circ \eta}(z) = z/z = 1$ is constant; by Theorem 7.34, $\phi \circ \eta$ does not map $\zeta(0,1)$ to itself.

In fact, since the constant value of $\overline{\phi \circ \eta}$ is 1, the constant value of $\overline{\phi \circ \eta - 1}$ is 0, and so we can scale by some power of p, as follows:

$$\phi \circ \eta(z) - 1 = \frac{p^2 z^3 - pz^2 + p}{z} = p\left(\frac{pz^3 - z^2 + 1}{z}\right).$$

Thus, setting $\theta(z) = (z-1)/p$, we have

$$\theta \circ \phi \circ \eta(z) = \frac{pz^3 - z^2 + 1}{z},$$

and hence

$$\overline{\theta \circ \phi \circ \eta}(z) = \frac{1 - z^2}{z},$$

which is nonconstant and, in fact, of degree 2. By Theorem 7.34, then, $\theta \circ \phi \circ \eta$ maps the Gauss point $\zeta(0,1)$ to itself, with multiplicity 2. Thus, ϕ maps the original point $\zeta = \zeta(0, |p|) = \eta(\zeta(0,1))$ to the point $\theta^{-1}(\zeta(0,1)) = \zeta(1, |p|)$, and $\deg_\zeta(\phi) = 2$.

Remark 7.38. Our approach to the action of $\phi(z) \in \mathbb{C}_v(z)$ on $\mathbb{P}^1_{\text{an}}$ in this chapter, and especially our treatment of multiplicities, is a hybrid of various other equivalent definitions. Our initial definition in Section 7.1 of this action via homogeneous coordinates roughly follows the strategy laid out in [**8**, Section 2.3]. On the other hand, our emphasis in Section 7.3 on local degrees in directions at Berkovich points comes from Rivera-Letelier's constructions in [**114**, Section 2.1] and [**115**, Section 4.1], although this concept also appears in [**8**, Section 9.1] with different terminology and notation.

As for local degrees themselves, our Definition 7.26 is inspired by [**8**, Proposition 9.15 and Corollary 9.17]. However, there are other equivalent ways of defining $\deg_\zeta(\phi)$. Baker and Rumely's version, which appears as [**8**, Definition 9.7], is measure-theoretic, while Favre and Rivera-Letelier's version, embedded in the proof of [**63**, Proposition-Définition 2.1], is algebraic. In his earlier papers, [**114**, Section 2.2] and [**115**, Section 4.2], Rivera-Letelier defined the local degree on a type-by-type basis, in line with our Corollary 7.32 and Theorem 7.34, but his definition essentially amounts to the sum of directional local degrees found in our Theorem 7.30.

7.6. The injectivity and ramification loci

Definition 7.39. Let $\phi \in \mathbb{C}_v(z)$ be a nonconstant rational function. The *ramification locus* of ϕ is the set

$$\operatorname{Ram}(\phi) := \{\zeta \in \mathbb{P}^1_{\text{an}} : \deg_\zeta(\phi) \geq 2\},$$

and the *injectivity locus* of ϕ is the complement

$$\operatorname{Inj}(\phi) := \mathbb{P}^1_{\text{an}} \smallsetminus \operatorname{Ram}(\phi) = \{\zeta \in \mathbb{P}^1_{\text{an}} : \deg_\zeta(\phi) = 1\}.$$

By Theorem 7.29(a), $\mathrm{Inj}(\phi)$ is an open set and $\mathrm{Ram}(\phi)$ is closed. In complex analysis, of course, the ramification locus of $\phi \in \mathbb{C}(z)$ is the finite set of critical points of ϕ. However, in our setting, the ramification locus is always infinite if $\deg(\phi) \geq 2$. Indeed, by the Riemann–Hurwitz formula (Theorem 1.9), ϕ has at least one critical point $a \in \mathbb{P}^1(\mathbb{C}_v)$, with some local degree $m \geq 2$. After a change of coordinates, we may assume that $a = 0$, and that $\phi(z) = c_0 + c_m z^m + \cdots$ on some disk $D(0,r)$. Thus, $\deg_{\zeta(0,s)}(\phi) = m$ for all sufficiently small s. In particular, $\mathrm{Ram}(\phi)$ contains a line segment extending from a.

Example 7.40. Let $\phi(z) = z^d$ with $d \geq 2$. Then $\deg_{\zeta(0,r)}(\phi) = d$ for all $r > 0$, and hence the Berkovich interval $[0, \infty]$ is contained in $\mathrm{Ram}(\phi)$.

If $|d| = 1$, i.e., if d is not divisible by the residue characteristic $\operatorname{char} k$, then $\mathrm{Ram}(\phi) = [0, \infty]$. Indeed, on any disk of the form $D(a,r)$ with $r < |a|$, we have

$$\phi(z) = a^d + da^{d-1}(z-a) + \sum_{i=2}^{d} \binom{d}{i} a^{d-i}(z-a)^i, \tag{7.9}$$

which has Weierstrass degree 1 on $D(a,r)$. Hence, every point of $D_{\mathrm{an}}(a,r)$ belongs to $\mathrm{Inj}(\phi)$.

If $|d| = 0$, i.e., if d is divisible by $p := \operatorname{char} \mathbb{C}_v$, then $\mathrm{Ram}(\phi) = \mathbb{P}^1_{\mathrm{an}}$. After all, the expansion of equation (7.9) has Weierstrass degree at least p on any disk $D(a,r)$.

If $0 < |d| < 1$, i.e., if $\mathbb{C}_v$ has mixed characteristic, with d divisible by the residue characteristic $p := \operatorname{char} k$, then

$$\mathrm{Ram}(\phi) = \left\{ \zeta \in \mathbb{A}^1_{\mathrm{an}} \; : \; \operatorname{diam}(\zeta) \geq |p|^{1/(p-1)} |\zeta| \right\} \cup \{\infty\},$$

as we leave the reader to check in Exercise 7.24. This set properly contains the interval $[0, \infty]$, with various extra points of Types II, III, and IV. In particular, the Type II and III points in $\mathrm{Ram}(\phi)$ are precisely those of the form $\zeta(a,r)$ with $r \geq |p|^{1/(p-1)} |a|$.

In Example 7.40, we see that even for the simple map z^d, the injectivity locus $\mathrm{Inj}(\phi)$ can consist of infinitely many connected components. In that example, all such components are disks; more generally, the components of $\mathrm{Inj}(\phi)$ are at worst affinoids, as the next result shows.

Theorem 7.41. *Let $\phi \in \mathbb{C}_v(z)$ be a rational function of degree $d \geq 2$. Then every connected component of the injectivity locus* $\mathrm{Inj}(\phi)$ *is a rational open connected affinoid.*

Proof. By Theorem 7.29(a), $\mathrm{Inj}(\phi)$ is an open set. Thus, a connected component $U \subseteq \mathrm{Inj}(\phi)$ is an open connected subset of $\mathbb{P}^1_{\mathrm{an}}$. We claim that every point of ∂U is of Type II and is isolated in ∂U.

Pick $\xi \in U$. Given $\zeta \in \partial U$, we have $\zeta \notin U$, since U is open, and hence $\deg_\zeta(\phi) = m \geq 2$. However, the half-open interval $(\zeta, \xi]$ must be contained in U, and hence in $\mathrm{Inj}(\phi)$.

Suppose first that ζ is of Type I. Change coordinates on the domain and the range separately, so that $\zeta = \phi(\zeta) = 0$. The path $(0, \xi] \subseteq U$ must contain all points of the form $\zeta(0, r)$ for $r > 0$ small enough. However, by definition of $\deg_0(\phi)$, the power series expansion of ϕ is of the form $\phi(z) = a_m z^m + \cdots$ with $a_m \neq 0$. For all $r > 0$ small enough, then, $\deg_{\zeta(0,r)}(\phi) = m \geq 2$, contradicting the fact that $\zeta(0, r) \in U \subseteq \mathrm{Inj}(\phi)$.

Suppose next that ζ is of Type III. After a change of coordinates, we may assume that $\zeta = \zeta(0, s)$ for some $s \in (0, \infty) \smallsetminus |\mathbb{C}_v^\times|$, and that $\xi \in D_{\mathrm{an}}(0, s)$. The interval $(\zeta, \xi]$ must therefore contain all points of the form $\zeta(0, r)$ for $r < s$ large enough. By definition of $\deg_\zeta(\phi)$, however, the Laurent series expansion of ϕ on a thin enough open annulus $\{\lambda s < |\zeta| < s\}$ has inner and outer Weierstrass degree equal to $\pm m$. Thus,

$$\deg_{\zeta(0,r)}(\phi) = m \geq 2$$

for all $r < s$ large enough, contradicting the fact that $\zeta(0, r) \in U \subseteq \mathrm{Inj}(\phi)$.

Similarly, if ζ is of Type IV, represented by a decreasing sequence of disks $D(a_1, r_1) \supsetneq D(a_2, r_r) \supsetneq \cdots$, then for all sufficiently large n, the point $\zeta(a_n, r_n)$ must lie in the path $(\zeta, \xi] \subseteq U$. By definition, however, we have $\deg_{\zeta(a_n,r_n)}(\phi) = m \geq 2$ for all sufficiently large n, contradicting the fact that $\zeta(a_n, r_n) \in U \subseteq \mathrm{Inj}(\phi)$.

Thus, ζ is indeed of Type II. Let $\boldsymbol{v}$ be the direction at ζ containing ξ. Then $U \subseteq \boldsymbol{v}$, since $\zeta \notin \mathrm{Inj}(\phi)$ and U is a a connected component of $\mathrm{Inj}(\phi)$. After a change of coordinates, we may assume that $\zeta = \zeta(0, 1)$, and that $\boldsymbol{v} = \vec{v}_\zeta(0)$. Hence, there is some $\lambda < 1$ sufficiently large that ϕ has inner and outer Weiestrass degree equal to some $m \in \mathbb{Z}$ on the annulus

$$V := \{\lambda < |\zeta| < 1\}.$$

The interval $(\zeta, \xi] \subseteq U$ contains every point of the form $\zeta(0, r)$ for $r < 1$ sufficiently large. Because $\deg_{\zeta(0,r)}(\phi) = 1$ for all such r, we must have $m = \pm 1$. Thus, $\deg_\eta(\phi) = 1$ for all $\eta \in V$, and therefore $V \subseteq \mathrm{Inj}(\phi)$. Because V is connected, in fact we have $V \subseteq U$. As a result,

$$(\partial U) \cap \left(\mathbb{P}^1_{\mathrm{an}} \smallsetminus \overline{D}_{\mathrm{an}}(0, \lambda)\right) = \{\zeta\},$$

and hence ζ is an isolated point of ∂U, proving our claim.

On the other hand, ∂U is a closed subset of $\mathbb{P}^1_{\mathrm{an}}$, and therefore it is compact. Since every point is isolated, it follows that ∂U is finite. Thus, U is a connected component of $\mathbb{P}^1_{\mathrm{an}} \smallsetminus \partial U$, where ∂U is a finite set of Type II points. Hence, U is a rational open connected affinoid. $\square$

Any rational open connected affinoid can be realized as a component of $\mathrm{Inj}(\phi)$ for some $\phi \in \mathbb{C}_v(z)$; see Exercises 7.28 and 7.29. For more on the ramification locus $\mathrm{Ram}(\phi)$, see [**52, 53**].

Exercises for Chapter 7

7.1. Let $\zeta \in \mathbb{A}^1_{\mathrm{an}}$, define $I := \{g \in \mathbb{C}_v[z] : \|g\|_\zeta = 0\}$, and let

$$A := \mathbb{C}_v[z]_I = \big\{f/g \,:\, f, g \in \mathbb{C}_v[z] \text{ and } g \notin I\big\} \subseteq \mathbb{C}_v(z)$$

be the associated localization. Extend $\|\cdot\|_\zeta$ to A by

$$\left\|\frac{f}{g}\right\|_\zeta := \frac{\|f\|_\zeta}{\|g\|_\zeta} \quad \text{for} \quad f \in \mathbb{C}_v[z] \text{ and } g \in \mathbb{C}_v[z] \smallsetminus I. \tag{7.10}$$

(a) Prove that I is a prime ideal of $\mathbb{C}_v[z]$.

(b) If $\zeta = a \in \mathbb{C}_v$ is of Type I, prove that $I = \langle z - a \rangle$.

(c) If ζ is not of Type I, prove that $I = \{0\}$, and hence $A = \mathbb{C}_v(z)$.

(d) Prove that equation (7.10) is a well-defined multiplicative seminorm on A extending the original seminorm $\|\cdot\|_\zeta$ on $\mathbb{C}_v[z]$.

7.2. In this exercise, we flesh out the construction hinted at in Remark 7.3. Write $\mathbb{P}^1_{\mathrm{an}}$ as the gluing of two copies $\mathbb{A}^1_{\mathrm{an},1}$ and $\mathbb{A}^1_{\mathrm{an},2}$ of the affine line, as in Definition 6.11. Let $\phi \in \mathbb{C}_v[z]$, write $\phi = g/h$ for relatively prime polynomials $g, h \in \mathbb{C}_v[z]$, and let

$$d := \deg(\phi) = \max\big\{\deg(g), \deg(h)\big\}.$$

(a) Define the following four subsets of $\mathbb{A}^1_{\mathrm{an},1} \amalg \mathbb{A}^1_{\mathrm{an},2}$:
- $U_{11} := \{\zeta \in \mathbb{A}^1_{\mathrm{an},1} : \|h\|_\zeta > 0\}$,
- $U_{12} := \{\zeta \in \mathbb{A}^1_{\mathrm{an},1} : \|g\|_\zeta > 0\}$,
- $U_{21} := \{\zeta \in \mathbb{A}^1_{\mathrm{an},2} : \|z^d h(1/z)\|_\zeta > 0\}$,
- $U_{22} := \{\zeta \in \mathbb{A}^1_{\mathrm{an},2} : \|z^d g(1/z)\|_\zeta > 0\}$.

Prove that each U_{ij} is an open subset of $\mathbb{A}^1_{\mathrm{an},i}$ and that together the four sets cover $\mathbb{A}^1_{\mathrm{an},1} \amalg \mathbb{A}^1_{\mathrm{an},2}$.

(b) Define maps $\phi_{ij} : U_{ij} \to \mathbb{A}^1_{\mathrm{an},j}$ by
- $\|f(z)\|_{\phi_{11}(\zeta)} := \|f(\phi(z))\|_\zeta$,
- $\|f(z)\|_{\phi_{12}(\zeta)} := \|f(1/\phi(z))\|_\zeta$,
- $\|f(z)\|_{\phi_{21}(\zeta)} := \|f(\phi(1/z))\|_\zeta$,
- $\|f(z)\|_{\phi_{22}(\zeta)} := \|f(1/\phi(1/z))\|_\zeta$,

for $f \in \mathbb{C}_v[z]$. Prove that for each $i, j \in \{1, 2\}$ and any $\zeta \in U_{ij}$, the expression $\phi_{ij}(\zeta)$ above is indeed defined and is a seminorm on $\mathbb{C}_v[z]$ extending $|\cdot|$, and hence an element of $\mathbb{A}^1_{\mathrm{an}}$.

(*Suggestion*: For ϕ_{21} and ϕ_{22}, write $\phi(1/z) = [z^d g(1/z)]/[z^d h(1/z)]$.)

(c) Prove that the maps ϕ_{ij} induce a well-defined map from $\mathbb{P}^1_{\mathrm{an}}$ to $\mathbb{P}^1_{\mathrm{an}}$. That is, prove that the four maps agree on their overlaps, after gluing $\mathbb{A}_{\mathrm{an},1}$ to $\mathbb{A}_{\mathrm{an},2}$ via $\zeta \mapsto 1/\zeta$ on both the domain and range.

(d) Prove that the map from part (c) above coincides with the description of $\phi : \mathbb{P}^1_{\mathrm{an}} \to \mathbb{P}^1_{\mathrm{an}}$ of Definition 7.2.

7.3. Prove the following claim made in Remark 7.11: The set of linear fractional transformations $(az+b)/(cz+d) \in \mathrm{PGL}(2, \mathbb{C}_v)$ that fix $\zeta(0,1)$ is precisely $\mathrm{PGL}(2, \mathcal{O})$.

7.4. Prove Proposition 7.17. (Use nothing beyond Section 7.2, as Proposition 7.17 was used in the proof of Theorem 7.22 and subsequent results.)

7.5. Prove Proposition 7.21.

7.6. Prove that local degrees in directions are invariant under coordinate changes. That is, if $\phi \in \mathbb{C}_v(z)$ and $\eta, \theta \in \mathrm{PGL}(2, \mathbb{C}_v)$, then for any $\zeta \in \mathbb{P}^1_{\mathrm{an}}$ and any direction $\boldsymbol{v}$ at ζ, if we set $\boldsymbol{w}$ to be the direction $\eta_\#(\boldsymbol{v})$ at $\eta(\zeta)$, then

$$\deg_{\eta(\zeta),\boldsymbol{w}}(\phi) = \deg_{\zeta,\boldsymbol{v}}(\theta \circ \phi \circ \eta).$$

7.7. Prove the following strengthening of Theorem 7.22 for points of Type II. Let $\phi \in \mathbb{C}_v(z)$ be a nonconstant rational function, and let $\zeta \in \mathbb{P}^1_{\mathrm{an}}$ be a point of Type II. Prove that there exists $\delta > 0$ such that for *any* direction $\boldsymbol{v}$ at ζ, there is an integer $m \geq 1$ so that for any point $\xi \in \boldsymbol{v}$ with $d_{\mathbb{H}}(\zeta, \xi) \leq \delta$,

- ϕ maps $[\zeta, \xi]$ homeomorphically onto $[\phi(\zeta), \phi(\xi)]$, and
- $d_{\mathbb{H}}(\phi(\xi_1), \phi(\xi_2)) = m \cdot d_{\mathbb{H}}(\xi_1, \xi_2)$ for all $\xi_1, \xi_2 \in [\zeta, \xi]$.

That is, show that the quantity λ of Theorem 7.12 can be chosen independently of the direction $\boldsymbol{v}$.

7.8. Prove Proposition 7.25.

(*Suggestion*: Use Exercise 3.21.)

7.9. Let $\phi \in \mathbb{C}_v(z)$ be a nonconstant rational function, and let $\zeta \in \mathbb{P}^1_{\mathrm{an}}$ be a point of Type III. Prove that $\deg_\zeta(\phi)$ is the Weierstrass degree of ϕ on any small enough annulus containing ζ.

More precisely, prove that for any $\xi_1, \xi_2 \in \mathbb{P}^1_{\mathrm{an}}$ with $\zeta \in (\xi_1, \xi_2)$ and $d_{\mathbb{H}}(\zeta, \xi_i)$ small enough, if we define U to be the annulus that is the connected component of $\mathbb{P}^1_{\mathrm{an}} \smallsetminus \{\xi_1, \xi_2\}$ containing ζ, then the inner and outer Weierstrass degrees of ϕ on U both equal $\deg_\zeta(\phi)$.

7.10. Fix $c \in \mathbb{C}_v$ with $0 < |c| < 1$, and let

$$\phi(z) = \frac{z^4 - 8c^4 z - c^4}{4z^3 + z^2 + 4c^4} \in \mathbb{C}_v(z)$$

as in Example 7.31. Prove the following claims made in that example.

For $|c|^2 < r < |c|$,

(a) $\deg_{\zeta(0,r)}(\phi) = 2$;

(b) $\phi(\zeta(0,r)) = \zeta(0,|c|^4/r^2)$;

(c) $\phi_\#$ maps the direction at $\zeta(0,r)$ containing 0 to the direction at $\zeta(0,|c|^4/r^2)$ containing ∞; and

(d) $\phi_\#$ maps the direction at $\zeta(0,r)$ containing ∞ to the direction at $\zeta(0,|c|^4/r^2)$ containing 0.

7.11. For the same function ϕ as in Exercise 7.10, prove the following claims made in Example 7.31:

(a) $\phi(\zeta(0,1)) = \phi(\zeta(0,|c|^2)) = \zeta(0,1)$;

(b) $\deg_{\zeta(0,1)}(\phi) = \deg_{\zeta(0,|c|^2)}(\phi) = 2$;

(c) $\phi(\zeta(0,|c|)) = \zeta(0,|c|^2)$; and

(d) $\deg_{\zeta(0,|c|)}(\phi) = 4$.

7.12. Prove Proposition 7.33.

(*Suggestion*: Use Lemma 7.27, Theorem 7.29(c), and Theorem 7.30.)

7.13. Let $\phi \in \mathbb{C}_v(z)$ be a nonconstant rational function, and let $\zeta_1, \zeta_2 \in \mathbb{P}^1_{\text{an}}$. Prove that the image $\phi([\zeta_1,\zeta_2])$ of the interval $[\zeta_1,\zeta_2] \subseteq \mathbb{P}^1_{\text{an}}$ is a finite union of closed intervals.

7.14. Let $\phi \in \mathbb{C}_v(z)$ be a rational function of degree $d \geq 1$, and let $\zeta_1, \zeta_2 \in \mathbb{H}$. Suppose that ϕ is injective on the interval $[\zeta_1,\zeta_2]$. Let $b := d_{\mathbb{H}}(\zeta_1,\zeta_2)$, and for each real number $t \in [0,b]$, define $\zeta(t)$ to be the unique point in $[\zeta_1,\zeta_2]$ for which $d_{\mathbb{H}}(\zeta_1,\zeta(t)) = t$. Prove that

$$d_{\mathbb{H}}\big(\phi(\zeta_1),\phi(\zeta_2)\big) = \int_0^b \deg_{\zeta(t)}(\phi)\,dt.$$

In particular, $d_{\mathbb{H}}(\phi(\zeta_1),\phi(\zeta_2)) \geq d_{\mathbb{H}}(\zeta_1,\zeta_2)$.

7.15. Fix a prime $p \geq 2$, and let $\mathbb{C}_v = \mathbb{C}_p$. Let $\phi(z) = z^p$. Fix $r > 0$.

(a) Let $\zeta = \zeta(0,r)$. Prove that $\phi(\zeta) = \zeta(0,r^p)$ and that $\deg_\zeta(\phi) = p$.

(b) Suppose that $r \geq |p|^{1/(p-1)}$ and let $\zeta = \zeta(1,r)$. Prove that $\phi(\zeta) = \zeta(1,r^p)$ and that $\deg_\zeta(\phi) = p$.

(c) Suppose that $r < |p|^{1/(p-1)}$, and let $\zeta = \zeta(1,r)$. Prove that $\phi(\zeta) = \zeta(1,|p|r)$ and that $\deg_\zeta(\phi) = 1$.

(d) Let $D_1 \supsetneq D_2 \supsetneq \cdots$ be a decreasing sequence of closed disks in $\mathbb{C}_p$ with empty intersection. Let $\zeta \in \mathbb{P}^1_{\text{an}}$ be the corresponding Type IV point. Writing $D_i = \overline{D}(a_i,r_i)$, assume that $|a_i| = 1$ and $|p|^{1/(p-1)} < r_i < 1$ for all i. Prove that $\phi(\zeta)$ is the Type IV point corresponding to the sequences of disks

$$\overline{D}(a_1^p, r_1^p) \supsetneq \overline{D}(a_2^p, r_2^p) \supsetneq \cdots$$

and that $\deg_\zeta(\phi) = p$.

7.16. Let $\phi \in \mathbb{C}_v(z)$ be nonconstant, let $d = \deg(\phi)$, and let p be the residue characteristic of $\mathbb{C}_v$. Suppose either that $p = 0$ or that $d < p$. Prove that for any Type IV point $\zeta \in \mathbb{P}^1_{\text{an}}$, we have $\deg_\zeta(\phi) = 1$.

(*Suggestion*: Use Proposition 3.9.)

7.17. Suppose $|3| = 1$. Fix $a \in \mathbb{C}_v$ with $0 < |a| < 1$, and let

$$\phi(z) = \frac{z+2}{z^3 + 3z^2 + a^2 z - 5}.$$

Let $\xi_0 := \zeta(-1, |a|)$ and $\xi_1 := \zeta(0, 1)$. Prove that

$$\phi([\xi_0, \xi_1]) = \left[\zeta\left(-\frac{1}{3}, |a|^3\right), \zeta(0,1)\right],$$

with $\deg_\zeta(\phi) = 3$ for all $\zeta \in [\xi_0, \xi_1]$.

7.18. Suppose that $\operatorname{char} \mathbb{C}_v = p > 0$, let $\phi(z) \in \mathbb{C}_v(z)$ be nonconstant, and define $\psi(z) := (\phi(z))^p$. Prove that

$$\deg_\zeta(\psi) = p \deg_\zeta(\phi) \quad \text{and} \quad \deg_{\zeta,\boldsymbol{v}}(\psi) = p \deg_{\zeta,\boldsymbol{v}}(\phi)$$

for all $\zeta \in \mathbb{P}^1_{\text{an}}$ and all directions $\boldsymbol{v}$ at ζ.

7.19. Fix $a \in \mathbb{C}_v$ with $0 < |a| < 1$, and let

$$\phi(z) = \frac{z-1}{az^3 + z^2 + 1}.$$

Find all $\xi \in \mathbb{P}^1_{\text{an}}$ for which $\phi(\xi) = \zeta(0,1)$, and confirm Theorem 7.29(c) for ϕ and $\zeta(0,1)$ by computing $\deg_\xi(\phi)$ for each such ξ. Do this in each of the following cases.

(a) Assume the residue characteristic of $\mathbb{C}_v$ is not 2.

(b) Assume the residue characteristic of $\mathbb{C}_v$ *is* 2.
(*Note*: There are a number of different possibilities in this case, depending on the exact value of a.)

7.20. Let $\psi(z) := (pz^3 - z^2 + 1)/z \in \mathbb{C}_p(z)$. We saw in Example 7.37 that $\psi(\zeta(0,1)) = \zeta(0,1)$.

(a) Find the absolute values of all zeros of ψ, and show that the associated set T from Theorem 7.34 of bad directions at $\zeta(0,1)$ is $T = \{\vec{v}(\infty)\}$.

(b) Show directly that $\psi\big(\mathbb{P}^1(\mathbb{C}_v) \smallsetminus \overline{D}(0,1)\big) = \mathbb{P}^1(\mathbb{C}_v)$, but that

$$\psi_\#\big(\vec{v}(a)\big) = \vec{v}\big(\psi(a)\big) \quad \text{for all} \quad a \in \overline{D}(0,1).$$

(c) Find all ramification points of the reduction $\overline{\psi}(z) = (1-z^2)/z$ in $\mathbb{P}^1(k)$.
(*Note*: The answer is different if the residue characteristic is $p = 2$.)

(d) Let $a \in \overline{D}(0,1)$. Confirm that the mapping

$$\psi : D(a,1) \to D(\psi(a),1)$$

is two-to-one if $\overline{a}$ is one of the ramification points of $\overline{\psi}$ from part (c), and is one-to-one otherwise.

(e) Show that ψ maps the annulus $\{1 < |z| < r\}$ bijectively onto itself for $r > 1$ sufficiently small. Find the largest such value of r.

7.21. Fix $a \in \mathbb{C}_v$ with $0 < |a| < 1$, and let

$$\phi(z) = \frac{az^3 + z^2 - a^2}{z^2 + a}.$$

For each of the following Berkovich points ζ, use Theorem 7.34 and the method of Example 7.37 to compute $\phi(\zeta)$ and $\deg_\zeta(\phi)$. In addition, compute $\deg_{\zeta,\boldsymbol{v}}(\phi)$ for each direction $\boldsymbol{v}$ at ζ.

(a) $\zeta = \zeta(0,1)$
(b) $\zeta = \zeta(0,|a|)$
(c) $\zeta = \zeta(\frac{-1}{a},1)$
(d) $\zeta = \zeta(1,|a|)$, assuming the residue characteristic of $\mathbb{C}_v$ is not 3.
(e) $\zeta = \zeta(1,|a|)$, assuming the residue characteristic of $\mathbb{C}_v$ is 3, and $|a| > |3|$.
(f) $\zeta = \zeta(1,|a|)$, assuming the residue characteristic of $\mathbb{C}_v$ is 3, and $|a| < |3|$.
(g) $\zeta = \zeta(1,|a|)$, assuming the residue characteristic of $\mathbb{C}_v$ is 3, and $|a| = |3|$.
(*Caution*: There are *many* further subcases in part (g).)

7.22. Fix $a \in \mathbb{C}_v$ with $0 < |a| < 1$, and let

$$\phi(z) = \frac{z-1}{az^3 + z^2 + a}.$$

Let $\zeta := \zeta(0,1)$, and for any $x \in \mathbb{P}^1(\mathbb{C}_v)$, let $\vec{v}(x)$ denote the direction at ζ containing x.

(a) Confirm that $\phi(\zeta) = \zeta$ in two ways: once by Theorem 7.34, and once by Theorem 7.12.

(b) Compute $\deg_{\zeta,\vec{v}(x)}(\phi)$ for $x = 0, 1, \infty$. In each case, do the computation in in two different ways: once by finding an annulus as in Theorem 7.12 or Remark 7.14, and once by Lemma 7.35(c).

(c) Let $\boldsymbol{w} := \vec{v}(0)$. Find all directions $\boldsymbol{v}$ at ζ for which $\phi_{\#}(\boldsymbol{v}) = \boldsymbol{w}$. For each such $\boldsymbol{v}$, compute $\deg_{\zeta,\boldsymbol{v}}(\phi)$ in two different ways: once by finding an annulus as in Theorem 7.12 or Remark 7.14, and once by Lemma 7.35(c). Then confirm that Theorem 7.30 holds for ϕ, ζ, and $\boldsymbol{w}$.

(d) Repeat part (c) for $\boldsymbol{w} := \vec{v}(\infty)$.

7.23. Let $a \in \mathbb{C}_v$, $\phi(z) \in \mathbb{C}_v(z)$, and ζ be as in Exercise 7.22.

(a) Find all $\xi \in \mathbb{P}^1_{\text{an}}$ for which $\phi(\xi) = \zeta$, and compute $\deg_\xi(\phi)$ at each by Theorem 7.34. Then confirm that Theorem 7.29(c) holds for ϕ and ζ.

(b) Fix $r \in \mathbb{R}$ with $1 < r < 1/|a|$, and let $U = D_{\text{an}}(0, r)$. Compute $\phi^{-1}(U)$.

(c) With r as in part (b), let $V = \{1/r < |z| < r\}$. Compute $\phi^{-1}(V)$.

(d) With V as in part (c), let W be the connected component of $\phi^{-1}(V)$ containing ζ. Determine how many connected components $\phi^{-1}(W)$ has.

(*Note*: You will get slightly different answers in parts (b) and (c) depending on whether or not the residue characteristic is 2.)

7.24. Suppose that $\operatorname{char} \mathbb{C}_v = 0$ but $\operatorname{char} k = p > 0$, and let $\phi(z) = z^d$ for some integer $d \geq 2$ divisible by p. Prove that the ramification locus of ϕ is

$$\operatorname{Ram}(\phi) = \left\{\zeta \in \mathbb{A}^1_{\text{an}} : \operatorname{diam}(\zeta) \geq |p|^{1/(p-1)}|\zeta|\right\} \cup \{\infty\},$$

as claimed in Example 7.40.

(*Suggestion*: Use Exercise 7.15.)

7.25. Let $\phi(z) \in \mathbb{C}_v(z)$ be a rational function of degree $d \geq 2$ and of explicit good reduction. Suppose either that $\operatorname{char} k = 0$ or that $d < \operatorname{char} k$.

(a) Prove that $\operatorname{Ram}(\phi)$ is the union of all line segments of the form $[a, \zeta(0,1)]$, where $a \in \mathbb{P}^1(\mathbb{C}_v)$ is a (Type I) critical point of ϕ.

(b) Prove, equivalently, that $\operatorname{Ram}(\phi)$ is the convex hull of the set of Type I critical points of ϕ.

7.26. Let $\phi(z) \in \mathbb{C}_v(z)$ be a rational function of degree 2, and suppose that $\operatorname{char} k \neq 2$. Then ϕ has two distinct Type I critical points $a, b \in \mathbb{P}^1(\mathbb{C}_v)$. Prove that $\operatorname{Ram}(\phi) = [a, b]$.

7.27. Let $\phi(z) \in \mathbb{C}_v(z)$ be a rational function of degree $d \geq 2$, and suppose either that $\operatorname{char} k = 0$ or that $d < \operatorname{char} k$. Let S be the (finite) set of Type I critical points of ϕ. Prove that $\operatorname{Ram}(\phi)$ is contained in the convex hull of S.

(*Note*: This is a weak version of [**52**] Corollary 7.13.)

7.28. Fix $a \in \mathbb{C}_v$ with $0 < |a| < 1$, and let

$$\phi(z) = \frac{a^2}{z} + z + z^2.$$

Prove that the open affinoid $\{|a| < |\zeta| < 1\}$ is a connected component of the injectivity locus $\mathrm{Inj}(\phi)$.

7.29. Let $D_1, \ldots, D_m$ be rational closed Berkovich disks, all contained in $D_{\mathrm{an}}(0,1)$. Let $U := D_{\mathrm{an}}(0,1) \smallsetminus (D_1 \cup \cdots \cup D_m)$. Find a rational function $\phi(z) \in \mathbb{C}_v(z)$ such that U is a connected component of $\mathrm{Inj}(\phi)$.

Part 4

Dynamics on the Berkovich Line

Chapter 8

Introduction to Dynamics on Berkovich Space

Having studied the fundamental properties of the Berkovich projective line, we are prepared to return to dynamics. In particular, we can now define Berkovich analogues of the Fatou and Julia sets of Chapter 5, admitting a rich theory of the connected components of the Fatou set.

8.1. Berkovich Fatou and Julia sets

It is possible to define Fatou sets in $\mathbb{P}^1_{\text{an}}$ in terms of appropriate notions of equicontinuity or normality; see [**57**, Theorems C and 5.4]. However, it is more convenient for our purposes to propose the following equivalent definition, inspired by Theorem 5.19.

Definition 8.1. Let $\phi \in \mathbb{C}_v(z)$. We say an open set $U \subseteq \mathbb{P}^1_{\text{an}}$ is *dynamically stable* under ϕ if $\bigcup_{n\geq 0} \phi^n(U)$ omits infinitely many points of $\mathbb{P}^1_{\text{an}}$.

The (*Berkovich*) *Fatou set* of ϕ, denoted $\mathcal{F}_{\phi,\text{an}}$, is the subset of $\mathbb{P}^1_{\text{an}}$ consisting of all points $x \in \mathbb{P}^1_{\text{an}}$ having a dynamically stable neighborhood.

The (*Berkovich*) *Julia set* of ϕ is the complement $\mathcal{J}_{\phi,\text{an}} := \mathbb{P}^1_{\text{an}} \smallsetminus \mathcal{F}_{\phi,\text{an}}$ of the Berkovich Fatou set.

As we will see in Theorem 8.3, the classical Fatou set $\mathcal{F}_{\phi,\text{I}}$ is precisely the set of Type I points in the Berkovich Fatou set $\mathcal{F}_{\phi,\text{an}}$, and similarly for the Julia sets. First, however, we note the Berkovich Fatou and Julia sets have the same basic properties described in Proposition 5.10, as follows.

Proposition 8.2. *Let $\phi \in \mathbb{C}_v(z)$ be a rational function with Berkovich Fatou set $\mathcal{F}_{\phi,\mathrm{an}}$ and Berkovich Julia set $\mathcal{J}_{\phi,\mathrm{an}}$.*

(a) *$\mathcal{F}_{\phi,\mathrm{an}}$ is an open subset of $\mathbb{P}^1_{\mathrm{an}}$, and $\mathcal{J}_{\phi,\mathrm{an}}$ is a closed subset of $\mathbb{P}^1_{\mathrm{an}}$.*

(b) *$\phi^{-1}(\mathcal{F}_{\phi,\mathrm{an}}) = \phi(\mathcal{F}_{\phi,\mathrm{an}}) = \mathcal{F}_{\phi,\mathrm{an}}$, and $\phi^{-1}(\mathcal{J}_{\phi,\mathrm{an}}) = \phi(\mathcal{J}_{\phi,\mathrm{an}}) = \mathcal{J}_{\phi,\mathrm{an}}$.*

(c) *For every integer $m \geq 1$, $\mathcal{F}_{\phi^m,\mathrm{an}} = \mathcal{F}_{\phi,\mathrm{an}}$, and $\mathcal{J}_{\phi^m,\mathrm{an}} = \mathcal{J}_{\phi,\mathrm{an}}$.*

(d) *For any $\eta \in \mathrm{PGL}(2, \mathbb{C}_v)$, if we set $\psi = \eta \circ \phi \circ \eta^{-1}$, then*

$$\mathcal{F}_{\psi,\mathrm{an}} = \eta(\mathcal{F}_{\phi,\mathrm{an}}) \quad \textit{and} \quad \mathcal{J}_{\psi,\mathrm{an}} = \eta(\mathcal{J}_{\phi,\mathrm{an}}).$$

Proof.

(a),(c),(d) This is left to the reader; see Exercise 8.2.

(b) It suffices to show that $\phi^{-1}(\mathcal{F}_{\phi,\mathrm{an}}) = \mathcal{F}_{\phi,\mathrm{an}}$ for nonconstant ϕ.

($\supseteq$): Given $y \in \mathcal{F}_{\phi,\mathrm{an}}$, let $x := \phi(y)$; we must show $x \in \mathcal{F}_{\phi,\mathrm{an}}$. Let U be an open neighborhood of y that is dynamically stable, and let $V := \phi(U)$, which is an open neighborhood of x by Corollary 7.9. Then

$$\bigcup_{n\geq 0} \phi^n(V) = \bigcup_{n\geq 1} \phi^n(U) \subseteq \bigcup_{n\geq 0} \phi^n(U)$$

omits infinitely many points of $\mathbb{P}^1_{\mathrm{an}}$. Thus, V is also dynamically stable, and hence $x \in \mathcal{F}_{\phi,\mathrm{an}}$, as desired.

($\subseteq$): Given $x \in \mathcal{F}_{\phi,\mathrm{an}}$ and $y \in \phi^{-1}(x)$, we must show that $y \in \mathcal{F}_{\phi,\mathrm{an}}$. Let U be an open neighborhood of x that is dynamically stable. If there is some $n \geq 0$ such that $y \in \phi^n(U)$, then as in the ($\supseteq$) direction, the set $V := \phi^n(U)$ is a dynamically stable open neighborhood of y, and we are done.

Otherwise, let $W := \phi^{-1}(U)$, which is an open neighborhood of y, and pick $t \in \phi^{-1}(y)$. Since $y \notin \phi^n(U)$ for any $n \geq 0$, the backward orbit $\mathrm{Orb}^-_\phi(t)$ must be disjoint from $\phi^n(W)$ for all $n \geq 0$. Moreover, y cannot be periodic or else we would have $y \in \phi^n(U)$ for some $n \geq 1$ after all. Thus, $\mathrm{Orb}^-_\phi(t)$ must be infinite. That is, $\bigcup_{n\geq 0} \phi^n(W)$ omits the infinite set $\mathrm{Orb}^-_\phi(t)$, and hence W is a dynamically stable open neighborhood of y, as desired. □

As promised above, the Type I and Berkovich Fatou and Julia sets are related to each other in the following unsurprising way.

Theorem 8.3. *Let $\phi \in \mathbb{C}_v(z)$. Then*

$$\mathcal{F}_{\phi,\mathrm{I}} = \mathcal{F}_{\phi,\mathrm{an}} \cap \mathbb{P}^1(\mathbb{C}_v) \quad \textit{and} \quad \mathcal{J}_{\phi,\mathrm{I}} = \mathcal{J}_{\phi,\mathrm{an}} \cap \mathbb{P}^1(\mathbb{C}_v).$$

Proof. Given $x \in \mathcal{F}_{\phi,\mathrm{I}}$, then by the final statement of Theorem 5.19, there is a $\mathbb{P}^1(\mathbb{C}_v)$-disk D containing x such that the iterates $\phi^n(D)$ omit uncountably many points of $\mathbb{P}^1(\mathbb{C}_v)$. Shrinking D if necessary, we may assume that D is rational open. Let D_{an} be the corresponding Berkovich disk. By

Theorem 7.8, the iterates $\phi^n(D_{\mathrm{an}})$ omit the same $\mathbb{P}^1(\mathbb{C}_v)$-points that the iterates $\phi^n(D)$ do. Since this omitted set is uncountable, we have $x \in \mathcal{F}_{\phi,\mathrm{an}}$.

Conversely, given $x \in \mathcal{F}_{\phi,\mathrm{an}} \cap \mathbb{P}^1(\mathbb{C}_v)$, there is an open connected Berkovich affinoid U containing x whose iterates $\phi^n(U)$ omit infinitely many points of $\mathbb{P}^1_{\mathrm{an}}$. Shrinking U if necessary, we may assume that U is a disk. (Note that this is possible because x is a Type I point.) Each iterate $\phi^n(U)$ must be either a disk or all of $\mathbb{P}^1_{\mathrm{an}}$, by Proposition 3.25 and Theorem 7.8, but it cannot be all of $\mathbb{P}^1_{\mathrm{an}}$, since it omits points.

Choose two distinct points $\zeta_0, \zeta_1 \in \mathbb{P}^1_{\mathrm{an}}$ omitted by $\bigcup_{n\geq 0} \phi^n(U)$, and choose a Type II point ξ lying between them. By a change of coordinates, we may assume that $\xi = \zeta(0,1)$.

We claim that $\xi \notin \bigcup_{n\geq 0} \phi^n(U)$. Indeed, if $\xi \in \phi^n(U)$ for some $n \geq 0$, then the Berkovich disk $\phi^n(U)$ contains all but at most one residue class of ξ. Since ζ_0 and ζ_1 lie in different residue classes of ξ, it follows that at least one of ζ_0 or ζ_1 lies in $\phi^n(U)$, giving a contradiction and proving the claim.

Thus, for each $n \geq 0$, $\phi^n(U)$ lies in a residue class of ξ. Let U_{I} be the $\mathbb{P}^1(\mathbb{C}_v)$-disk $U_{\mathrm{I}} := U \cap \mathbb{P}^1(\mathbb{C}_v)$ corresponding to U. Then for each $n \geq 0$, the image $\phi^n(U_{\mathrm{I}}) \subseteq \phi^n(U)$ is also contained in a residue class of the Gauss point ξ. Hence, by Lemma 5.8, we have $x \in \mathcal{F}_{\phi,\mathrm{I}}$. □

8.2. Classifying Berkovich periodic points

Just as all periodic points in $\mathbb{P}^1(\mathbb{C}_v)$ can easily be classified as Fatou or Julia based on the absolute value of their multipliers, one can usually make a similar determination for periodic points in Berkovich space.

Definition 8.4. Let $\phi \in \mathbb{C}_v(z)$, and let $\zeta \in \mathbb{P}^1_{\mathrm{an}} \smallsetminus \mathbb{P}^1(\mathbb{C}_v)$ be a point of Type II, III, or IV. Suppose that ζ is a periodic point of ϕ of minimal period $n \geq 1$. Then the *multiplier* of ζ is the local degree

$$\lambda := \deg_\zeta(\phi^n)$$

of ϕ^n at ζ. We say that ζ is *repelling* if $\lambda \geq 2$; otherwise, if $\lambda = 1$, we say that ζ is *indifferent*.

Warning 8.5.

(a) Although the multiplier of a Type I periodic point is an element of $\mathbb{C}_v$, the multiplier of a periodic point of Type II–IV is a positive integer, which we choose to *not* consider as an element of $\mathbb{C}_v$. In particular, we do not want to take the $\mathbb{C}_v$-absolute value of λ.

(b) There is no such thing as an attracting point of Type II, III, or IV. After all, there are no positive integers strictly less than 1.

(c) The name "repelling" is a bit misleading here, because repelling periodic points do not actually repel most nearby points! Instead, a repelling periodic point ζ has repelling directions, but usually only very few, as the following simple example illustrates.

Example 8.6. Let $m \geq 2$. The function $\phi(z) = z^m$ has a repelling fixed point of multiplier m at the Gauss point $\zeta(0,1)$. The directions (i.e., residue classes) $\overline{0} = \vec{v}(0)$ and $\overline{\infty} = \vec{v}(\infty)$ are repelling directions, in that they stretch distances, as in Theorem 7.22. More precisely, for any $r > 0$, we have

$$\phi(\zeta(0,r)) = \zeta(0,r^m),$$

and hence the hyperbolic metric $d_{\mathbb{H}}$ satisfies

$$d_{\mathbb{H}}\big(\phi(\zeta_1), \phi(\zeta_2)\big) = m \cdot d_{\mathbb{H}}(\zeta_1, \zeta_2)$$

for all ζ_1, ζ_2 in the Berkovich interval $[0, \infty]$. Therefore,

$$\deg_{\zeta(0,1),\vec{v}(0)}(\phi) = \deg_{\zeta(0,1),\vec{v}(\infty)}(\phi) = m.$$

However, if m is not divisible by the residue characteristic of $\mathbb{C}_v$, then for any $a \in \mathbb{C}_v$ with $|a| = 1$, and for any $0 < r < 1$, the reader can check that

$$\phi\big(\zeta(a,r)\big) = \zeta(a^m, r),$$

and hence

$$d_{\mathbb{H}}\Big(\phi\big(\zeta(0,1)\big), \phi\big(\zeta(a,r)\big)\Big) = \log r = d_{\mathbb{H}}\big(\zeta(0,1), \zeta(a,r)\big).$$

Thus, ϕ moves points $\zeta(a,r)$ from one residue class to another, but it does not actually change their hyperbolic distance from the Gauss point. In terms of local degrees in directions, we have $\deg_{\zeta(0,1),\vec{v}(a)}(\phi) = 1$.

On the other hand, if the residue characteristic p *does* divide m—specifically, if $p^e | m$ but $p^{e+1} \nmid m$, for some $e \geq 1$—then the reader can check (see Exercise 8.3) that

$$\phi\big(\zeta(a,r)\big) = \zeta(a^m, r^{p^e})$$

for all $0 < r < 1$ sufficiently close to 1, yielding

$$d_{\mathbb{H}}\Big(\phi\big(\zeta(0,1)\big), \phi\big(\zeta(a,r)\big)\Big) = p^e d_{\mathbb{H}}\big(\zeta(0,1), \zeta(a,r)\big)$$

for such r. In terms of local degrees in directions, then, we have

$$\deg_{\zeta(0,1),\vec{v}(a)}(\phi) = p^e.$$

Thus, in this very special case, ϕ *does* stretch distances in all directions at $\zeta(0,1)$, by a factor of either p^e or m in any given direction. But it is important to note that the point $\zeta(0,1)$ is considered a repelling fixed point regardless of whether $p|m$.

Even though Example 8.6 shows that repelling periodic points may not actually repel all nearby points in a geometric sense, the following result shows that the names "repelling" and "indifferent" are appropriate, from the standpoint of the Fatou and Julia sets.

Theorem 8.7. *Let $\phi \in \mathbb{C}_v(z)$, and let $\zeta \in \mathbb{P}^1_{\mathrm{an}}$ be a periodic point of Type II, III, or IV of period $n \geq 1$.*

(a) *If ζ is of Type III or IV, then ζ is indifferent, $\zeta \in \mathcal{F}_{\phi,\mathrm{an}}$, and every direction at ζ is fixed by $\phi^n_{\#}$.*

(b) *If ζ is of Type II and indifferent, let T be the finite set of bad directions $\boldsymbol{v}$ at ζ, as in Theorem 7.34. If every $\boldsymbol{v} \in T$ is periodic under the action of $\phi^n_{\#}$ at ζ, then there is an open Berkovich affinoid U containing ζ such that $\phi^n : U \to U$ is bijective, and hence $\zeta \in \mathcal{F}_{\phi,\mathrm{an}}$. Otherwise, $\zeta \in \mathcal{J}_{\phi,\mathrm{an}}$.*

(c) *If ζ is repelling, then $\zeta \in \mathcal{J}_{\phi,\mathrm{an}}$.*

Warning 8.8. Points of Type III or IV *can* be in the Julia set; Theorem 8.7(a) only excludes them if they are *periodic*. See [**115**, Lemmes 5.3–5.4].

Remark 8.9. If the residue field k of $\mathbb{C}_v$ is algebraic over a finite field (as is the case for the p-adic field $\mathbb{C}_p$, for example), then the technical condition in Theorem 8.7(b), that all bad directions $\boldsymbol{v} \in T$ at ζ are periodic under $\phi_{\#}$, is vacuous. After all, in the case of an indifferent Type II point—without loss, the Gauss point $\zeta(0,1)$—Theorem 7.34 says that the reduction $\overline{\phi}$ has degree 1 and hence belongs to $\mathrm{PGL}(2,k)$. If the residue field is indeed algebraic over a finite field, then every element of $\mathrm{PGL}(2,k)$ has finite order, and hence *every* direction at ζ is periodic.

On the other hand, if the residue field is not of this type (i.e., if $\operatorname{char} k = 0$, or if $\operatorname{char} k = p > 0$ but k is not algebraic over $\mathbb{F}_p$), then $\mathrm{PGL}(2,k)$ contains elements of infinite order. If $\overline{\phi}$ is such an element, then our indifferent Type II periodic point may or may not be in the Berkovich Julia set, depending on whether or not there are any nonperiodic bad directions.

Example 8.10. To illustrate Theorem 8.7(b), pick $\pi \in \mathbb{C}_v$ such that $0 < |\pi| < 1$, pick $a \in \mathbb{C}_v$ with $|a| = 1$, and define

$$\phi(z) = az + \frac{\pi}{z-1} = \frac{az^2 - az + \pi}{z-1}.$$

Since $\overline{\phi}(z) = \overline{a}z$ is a map of degree 1, Theorem 7.34 tells us that the Gauss point $\zeta = \zeta(0,1)$ is an indifferent fixed point of ϕ. Moreover, since ϕ has poles at $1, \infty$, and (noting the Newton polygon of the numerator) one zero in each of $D(0,1)$ and $D(1,1)$, the only bad direction at ζ is $\vec{v}_\zeta(1) = D_{\mathrm{an}}(1,1)$.

Again by Theorem 7.34, we have $\phi^n_{\#}(\vec{v}(1)) = \vec{v}_\zeta(a^n)$ for every $n \geq 0$. Thus, the bad direction $\vec{v}_\zeta(1)$ is periodic under $\phi_{\#}$ if and only if $\overline{a}^n = 1$ for some $n \geq 1$. That is, by Theorem 8.7(b), the point $\zeta(0,1)$ is Fatou if and only if $\overline{a}$ is of finite multiplicative order in the residue field k.

If k is algebraic over a finite field, then every nonzero element of k has finite multiplicative order, and hence every choice of a as above forces $\zeta(0,1)$ to lie in $\mathcal{F}_{\phi,\mathrm{an}}$. On the other hand, if k is *not* algebraic over a finite field, then $k^\times$ must contain an element $\overline{a}$ that is *not* of finite multiplicative order. If the coefficient a were chosen so that its reduction $\overline{a}$ has this property, then we would have $\zeta(0,1) \in \mathcal{J}_{\phi,\mathrm{an}}$. On the other hand, if we chose $a = 1$ (or any other choice for which $\overline{a}$ has finite multiplicative order), then we would still have $\zeta(0,1) \in \mathcal{F}_{\phi,\mathrm{an}}$.

To prove Theorem 8.7, we need the following lemma.

Lemma 8.11. *Let $\phi \in \mathbb{C}_v(z)$ be a rational function fixing the Gauss point, and suppose that ϕ also fixes the direction $\vec{v}(0)$ at $\zeta(0,1)$, with multiplicity $\deg_{\zeta(0,1),\vec{v}(0)}(\phi) \geq 2$. Let U be a Berkovich annulus of the form*

$$U = \{r < |\zeta| < 1\} = D_{\mathrm{an}}(0,1) \smallsetminus \overline{D}_{\mathrm{an}}(0,r),$$

for some radius $0 < r < 1$. Then $\bigcup_{n\geq 0} \phi^n(U)$ contains all but at most one point of the open Berkovich disk $D_{\mathrm{an}}(0,1)$. Moreover,

(a) *If there is one such point $a \in D_{\mathrm{an}}(0,1)$, then it is of Type I, and for all $0 < s < 1$, there is some $N \geq 1$ such that*

$$\{s < |\zeta - a| < 1\} \subseteq \phi^N(U).$$

(b) *If $\bigcup_{n\geq 0} \phi^n(U)$ contains $D_{\mathrm{an}}(0,1)$, then there is some $N \geq 1$ such that $D_{\mathrm{an}}(0,1) \subseteq \phi^N(U)$.*

Here, we have continued our abuse of notation, writing $\{s < |\zeta - a| < 1\}$ to denote the annulus

$$\{s < |\zeta - a| < 1\} = D_{\mathrm{an}}(a,1) \smallsetminus \overline{D}_{\mathrm{an}}(a,s) = \{\zeta \in \mathbb{P}^1_{\mathrm{an}} : s < \|z - a\|_\zeta < 1\}.$$

Proof of Lemma 8.11. Increasing $r < 1$ if necessary, we may assume that ϕ has no zeros or poles in U, and hence that ϕ has inner and outer Weierstrass degrees both equal to $m := \deg_{\zeta(0,1),\vec{v}(0)}(\phi) \geq 2$ on U, by Proposition 3.32. Define $U_n := \phi^n(U)$ for each $n \geq 0$. By Theorem 3.33(b) and the fact that $\phi(\zeta(0,1)) = \zeta(0,1)$, the image $U_1 = \phi(U)$ must be the annulus

$$U_1 = \{r^m < |\zeta| < 1\} \supseteq U_0.$$

It follows immediately that $U = U_0 \subseteq U_1 \subseteq U_2 \subseteq \cdots$.

If $U_n \supseteq D(0,1)$ for some n, then $U_n \supseteq D_{\text{an}}(0,1)$ as well, and we are done. Thus, we assume for the remainder of the proof that for each $n \geq 0$, there exists $a_n \in D(0,1) \smallsetminus U_n$.

We claim that for each $n \geq 0$, the set $D_{\text{an}}(0,1) \smallsetminus U_n$ is a closed disk $\overline{D}_{\text{an}}(a_n, s_n)$, where $0 < s_n \leq r^{m^n}$. We prove the claim by induction on n; we already know it for $n = 0, 1$. Given its truth for n, observe that because $a_{n+1} \notin U_{n+1}$, the function $\psi(z) := 1/(\phi(z) - a_{n+1})$ has no poles in U_n. By Proposition 3.32(a), ψ has a Laurent series expansion on U_n. Note that $\psi(\zeta(0,1)) = \zeta(0,1)$, and that ψ has outer Weierstrass degree $-m$ on U_n. If ψ also has inner Weierstrass degree $-m$ on U_n, then by Theorem 3.33(c), we have

$$\psi(U_n) = \{1 < |\zeta| < s_n^{-m}\}.$$

Otherwise, by Theorem 3.33(a), we have $\psi(U_n) \supseteq D_{\text{an}}(0, s_n^{-m})$. Either way,

$$D_{\text{an}}(0,1) \smallsetminus U_{n+1} \subseteq \overline{D}_{\text{an}}(a_n, s_n^m).$$

Thus, $0 < s_{n+1} \leq s_n^m \leq r^{m^{n+1}}$, proving the claim.

The disks $\overline{D}_{\text{an}}(a_0, s_0) \supseteq \overline{D}_{\text{an}}(a_1, s_1) \supseteq \cdots$ are nested, with radii decreasing to zero. By Lemma 2.7(b), their intersection is a singleton set $\{a\}$, where $a \in D(0,1)$ is a point of Type I. □

In most instances of Lemma 8.11, the union $\bigcup_{n\geq 0} \phi^n(U)$ contains the whole disk $D_{\text{an}}(0,1)$, but the omission of a single point of Type I can indeed occur. A full analysis of all such possibilities appears in Exercises 8.5–8.8.

Proof of Theorem 8.7. Replacing ϕ by ϕ^n, we may assume without loss that $\phi(\zeta) = \zeta$.

(a) First, we consider the case that ζ is of Type III. Change coordinates so that $\zeta = \zeta(0,r)$ for some $r \notin |\mathbb{C}_v|$. Pick $\lambda < 1$ sufficiently small according to Theorem 7.12 for the annulus $U = \{\lambda r < |z| < r\}$, write $\phi(z) = \sum_{n\in\mathbb{Z}} c_n z^n$ as a Laurent series on U, and let $d \in \mathbb{Z} \smallsetminus \{0\}$ be the associated Weierstrass degree. By the same theorem, we have

$$\zeta(0,r) = \zeta = \phi(\zeta) = \zeta(c_0, |c_d| r^d).$$

In particular, then, $r = |c_d| r^d$. Since $r \notin |\mathbb{C}_v|$, it follows that $d = 1$, and also $|c_d| = 1$. Thus, by Corollary 7.32(b),

$$\deg_\zeta(\phi) = \deg_{\zeta,\vec{v}(0)}(\phi) = d = 1.$$

We also see that $\phi_\#$ fixes both directions at ζ, and that $|\phi(x)| = |x|$ for all $x \in U$. It remains to show that ζ is in the Fatou set.

Applying a similar argument to the annulus

$$U' := \{r < |z| < \lambda^{-1} r\},$$

and bearing in mind Remark 7.14, we have $|\phi(x)| = |x|$ for all $x \in U'$ also. Thus, setting $V = \{\lambda r < |z| < \lambda^{-1} r\}$, we have $\phi(V) \subseteq V$. By Theorem 7.8, then, we also have $\phi(V_{\text{an}}) \subseteq V_{\text{an}}$. Hence, V_{an} is open, is dynamically stable under ϕ, and contains ζ. That is, $\zeta \in \mathcal{F}_{\phi,\text{an}}$.

Next, suppose ζ is of Type IV. Then the one direction at ζ is obviously fixed by $\phi_\#$. Choose a sequence of closed disks $D_1 \supseteq D_2 \supseteq \cdots$ representing ζ. By Proposition 7.7, there is an integer $N \geq 1$ such that for all $n \geq N$, the map $\phi : D_n \to \phi(D_n)$ is d-to-1, where $d := \deg_\zeta(\phi)$.

Since $\bigcap D_n = \varnothing$, we may choose $m \geq N$ large enough that D_m contains no Type I fixed points of ϕ. On the other hand, because ζ itself is fixed, the sequences of disks $\{D_n\}_{n\geq 1}$ and $\{\phi(D_n)\}_{n\geq 1}$ must be equivalent. Thus, the two disks D_m and $\phi(D_m)$ intersect each other. If $d \geq 2$ or if $\phi(D_m) \neq D_m$, then by Proposition 4.17, D_m contains a (Type I) fixed point, contradicting our choice of m at the start of this paragraph. Hence, $d = 1$, proving that ζ is indeed indifferent, and in addition, $\phi(D_m) = D_m$. The corresponding Berkovich disk $(D_m)_{\text{an}}$ therefore contains ζ and is dynamically stable, since

$$\phi\big((D_m)_{\text{an}}\big) = \big(\phi(D_m)\big)_{\text{an}} = (D_m)_{\text{an}}.$$

Once again, it follows that $\zeta \in \mathcal{F}_{\phi,\text{an}}$, proving part (a).

(b) Since ζ is of Type II, we may assume by a change of coordinates that $\zeta = \zeta(0,1)$. By Theorem 7.34, the assumption that ζ is indifferent shows that the reduction $\overline{\phi}$ is a linear fractional transformation in $\mathrm{PGL}(2,k)$. Let T be the finite set of bad directions from the same theorem.

First, suppose that some $\boldsymbol{v} \in T$ is not periodic under $\phi_\#$, and pick $a \in \boldsymbol{v}$, so that $\boldsymbol{v} = \vec{v}(a)$. By Lemma 7.35(b), the action of $\phi_\#$ on the directions at ζ is exactly given by the action of $\overline{\phi}$ on $\mathbb{P}^1(k)$. Let $\overline{\psi} \in \mathrm{PGL}(2,k)$ be the inverse of $\overline{\phi}$. Then $\overline{\psi}^m(\overline{a}) \neq \overline{\psi}^n(\overline{a})$ for all distinct nonnegative integers $m, n \geq 0$, or else $\overline{a}$ would be periodic under $\overline{\phi}$.

Given any open set $U \subseteq \mathbb{P}^1_{\text{an}}$ containing ζ, there is an open connected Berkovich affinoid V such that $\zeta \in V \subseteq U$. Then V, and hence U, contains all but finitely many residue classes at ζ. In particular, U must contain the residue class $\boldsymbol{w}$ corresponding to $\overline{\psi}^n(\overline{a})$ for some $n \geq 0$. By Theorem 7.34, we have

$$\phi^{n+1}(U) \supseteq \phi^{n+1}(\boldsymbol{w}) \supseteq \phi(\boldsymbol{v}) = \mathbb{P}^1_{\text{an}},$$

proving that U is not dynamically stable under ϕ. Thus, $\zeta \in \mathcal{J}_{\phi,\text{an}}$.

Second, suppose that every $\boldsymbol{v} \in T$ is periodic. Replacing ϕ by ϕ^m for some appropriate $m \geq 1$, we may assume that $\phi_\#(\boldsymbol{v}) = \boldsymbol{v}$ for every $\boldsymbol{v} \in T$.

Write $T = \{\boldsymbol{v}_1, \ldots, \boldsymbol{v}_M\}$. Because $\deg_\zeta(\phi) = 1$, Theorem 7.30 implies that $\deg_{\zeta,\boldsymbol{v}}(\phi) = 1$ for all directions $\boldsymbol{v}$ at ζ. Therefore, by Theorem 7.12, for each $i = 1, \ldots, M$, there is a (Berkovich) annulus U_i extending from ζ

in the direction $\boldsymbol{v}_i$ such that ϕ maps U_i bijectively onto itself. On the other hand, by Theorem 7.34(c), we have $\phi(\boldsymbol{v}) = \phi_{\#}(\boldsymbol{v})$ for each residue class $\boldsymbol{v}$ at ζ that does not belong to T. In particular, $\phi(\boldsymbol{v}) \cap \boldsymbol{v}_i = \varnothing$ for all such $\boldsymbol{v}$ and all $i = 1, \ldots, M$.

Thus, for each $i = 1, \ldots, M$, we may write $U_i = \boldsymbol{v}_i \smallsetminus V_i$, where V_i is a closed (Berkovich) disk properly contained in the residue class $\boldsymbol{v}_i$. Define

$$U := \mathbb{P}^1_{\text{an}} \smallsetminus (V_1 \cup \cdots \cup V_M),$$

which is an open connected Berkovich affinoid containing ζ. By the previous paragraph, we have $\phi(U) \subseteq U$, and hence U is dynamically stable under ϕ. Therefore, ζ lies in the Fatou set.

In addition, because $\deg_\zeta(\phi) = 1$, the mapping $\phi : U \to U$ is one-to-one. Furthermore, the reduction $\overline{\phi}$ permutes the directions at ζ; since we also have $\phi(U_i) = U_i$ for each $i = 1, \ldots, M$, it follows that $\phi(U) = U$. Thus, the mapping $\phi : U \to U$ is bijective, as desired.

(c) By part (a), the fact that ζ is repelling implies that it is of Type II, and therefore we can again change coordinates to assume that $\zeta = \zeta(0,1)$. As usual, denote the reduction of ϕ at ζ by $\overline{\phi}$, which is a rational function in $\overline{k}(z)$ of degree at least 2, by Theorem 7.34.

Given an open set $U \subseteq \mathbb{P}^1_{\text{an}}$ containing ζ, there is an open connected Berkovich affinoid V such that $\zeta \in V \subseteq U$. Hence, there are only finitely many residue classes at ζ that are not contained in U. Thus, there are finitely many (in fact, probably zero) residue classes $\boldsymbol{v}_1, \ldots, \boldsymbol{v}_\ell$ at ζ that are not contained in *any* iterate $\phi^n(U)$. Let $S = \{\overline{x}_1, \ldots, \overline{x}_\ell\}$ be the corresponding set of points in $\mathbb{P}^1(\overline{k})$. By Theorem 7.34, we have $\overline{\phi}^{-1}(S) \subseteq S$. Thus, S is contained in the exceptional set of $\overline{\phi}$, and therefore, by Theorem 1.19, each $\overline{x}_i$ is periodic.

Replacing ϕ by an iterate, we may further assume that each point of S is fixed by $\overline{\phi}$, and indeed that $\phi_{\#}^{-1}(\boldsymbol{v}_i) = \{\boldsymbol{v}_i\}$ for each $i = 1, \ldots, \ell$. In particular, $\deg_{\zeta(0,1),\boldsymbol{v}_i}(\phi) = \deg(\overline{\phi}) \geq 2$.

Therefore, for each $i = 1, \ldots, \ell$, we can make a $\mathrm{PGL}(2, \mathcal{O})$-change of coordinates to move $\boldsymbol{v}_i$ to $\overline{0}$. Then the affinoid V and, hence, our open set U, contains an annulus of the form $\{r < |\zeta| < 1\}$ for some $0 < r < 1$. Thus, $\bigcup_{n \geq 0} \phi^n(U)$ contains all but at most one point of $D_{\text{an}}(0,1)$, by Lemma 8.11. Changing coordinates back, $\bigcup_{n \geq 0} \phi^n(U)$ contains all but at most one point y_i of the residue class $\boldsymbol{v}_i$.

Combining the preceding paragraphs, we have shown that

$$\bigcup_{n \geq 0} \phi^n(U) \supseteq \mathbb{P}^1_{\text{an}} \smallsetminus \{y_1, \ldots, y_\ell\},$$

and hence U is *not* dynamically stable under ϕ. Since this is true for all such U, we have $\zeta \in \mathcal{J}_{\phi,\text{an}}$. ☐

8.3. Good reduction in Berkovich space

Among its many applications in non-archimedean dynamics, the use of the Berkovich projective line $\mathbb{P}^1_{\text{an}}$ allows the following elegant characterization of explicit good reduction.

Proposition 8.12. *Let $\phi \in \mathbb{C}_v(z)$ be a rational function of degree $d \geq 2$. Then the following statements are equivalent.*

(a) *ϕ has explicit good reduction.*

(b) *The Gauss point $\zeta(0,1)$ is a fixed point of multiplier d.*

(c) $\phi^{-1}(\zeta(0,1)) = \{\zeta(0,1)\}$.

(d) *The Berkovich Julia set $\mathcal{J}_{\phi,\text{an}}$ is the singleton set*

$$\mathcal{J}_{\phi,\text{an}} = \{\zeta(0,1)\}$$

consisting solely of the Gauss point.

Proof.

(a)$\Leftrightarrow$(b): By definition, ϕ has explicit good reduction if and only if its reduction $\overline{\phi} \in k(z)$ has degree d. By Theorem 7.34 and Definition 8.4, we have $\deg(\overline{\phi}) = d$ if and only if $\zeta(0,1)$ is a fixed point of multiplier d, as desired.

(b)$\Rightarrow$(d): Because $\zeta(0,1)$ is a repelling fixed point, it belongs to $\mathcal{J}_{\phi,\text{an}}$, by Theorem 8.7(c). Thus, it suffices to show that every other point of $\mathbb{P}^1_{\text{an}}$ lies in $\mathcal{F}_{\phi,\text{an}}$.

The set T of Theorem 7.34 is empty by the final statement of that theorem. Therefore, each direction $\boldsymbol{v}$ at $\zeta(0,1)$ satisfies $\phi(\boldsymbol{v}) = \phi_{\#}(\boldsymbol{v})$. That is, each residue class maps onto a residue class.

If a given residue class $\boldsymbol{v}$ is periodic under ϕ, then clearly it is dynamically stable, and hence it is contained in $\mathcal{F}_{\phi,\text{an}}$. On the other hand, if $\boldsymbol{v}$ is not periodic, then no forward iterate of $\phi(\boldsymbol{v})$ can intersect $\boldsymbol{v}$. Thus, the residue class $\phi(\boldsymbol{v})$ is dynamically stable and hence contained in $\mathcal{F}_{\phi,\text{an}}$. Therefore, by Proposition 8.2(b), $\boldsymbol{v} \subseteq \mathcal{F}_{\phi,\text{an}}$ as well, finishing the proof that $\mathbb{P}^1_{\text{an}} \smallsetminus \{\zeta(0,1)\} \subseteq \mathcal{F}_{\phi,\text{an}}$.

(d)$\Rightarrow$(c): Since $\mathcal{J}_{\phi,\text{an}} = \{\zeta(0,1)\}$ and because the Julia set is invariant by Proposition 8.2(b), we have $\phi^{-1}(\zeta(0,1)) = \{\zeta(0,1)\}$.

(c)$\Rightarrow$(b): Since $\phi^{-1}(\zeta(0,1)) = \{\zeta(0,1)\}$, ϕ must map $\zeta(0,1)$ to itself with multiplicity $\deg_{\zeta(0,1)}(\phi) = d$ by Theorem 7.29(c). ☐

Recall that Definition 1.13 defined the exceptional set in $\mathbb{P}^1(\mathbb{C}_v)$ to be the set of (Type I) points with finite grand orbit under ϕ. We can analogously define the exceptional set in $\mathbb{P}^1_{\text{an}}$ to be the set of points of *any* type that have finite grand orbit. Our next result says that ϕ has explicit good reduction if and only if $\zeta(0,1)$ is an exceptional point of ϕ. A variant of [**117**, Théorème 4] and [**15**, Proposition 3.1], it says that maps of potential good reduction are the *only* maps with exceptional points outside of $\mathbb{P}^1(\mathbb{C}_v)$. Moreover, for such maps there is only one such exceptional point.

Proposition 8.13. *Let $\phi \in \mathbb{C}_v(z)$ be a rational function of degree $d \geq 2$. Suppose there is a point $\zeta \in \mathbb{P}^1_{\text{an}} \smallsetminus \mathbb{P}^1(\mathbb{C}_v)$ of Type II, III, or IV with finite grand orbit under ϕ. Then there is only one such point, and it is a repelling fixed point of Type II, with multiplier d.*

In that case, $\mathcal{J}_{\phi,\text{an}} = \{\zeta\}$, and there exists $\eta \in \mathrm{PGL}(2, \mathbb{C}_v)$ such that $\eta(\zeta) = \zeta(0,1)$. Moreover, for any such η, the function $\eta \circ \phi \circ \eta^{-1} \in \mathbb{C}_v(z)$ has explicit good reduction.

Proof. A finite grant orbit cannot contain any wandering or even strictly preperiodic points, because every point in $\mathbb{P}^1_{\text{an}}$ has at least one preimage under ϕ. Thus, given a point ζ with finite grand orbit, there is some $n \geq 1$ such that $\phi^n(\zeta) = \zeta$. In fact, we must have $\phi^{-n}(\zeta) = \{\zeta\}$. By Theorem 7.29(c), then, we have

$$\deg_\zeta(\phi^n) = d^n \geq 2,$$

so that ζ is repelling. Therefore, by Theorem 8.7(a), ζ must be of Type II. After a $\mathrm{PGL}(2, \mathbb{C}_v)$-change of coordinates η, then, we may assume that $\zeta = \zeta(0,1)$. In addition, we have

$$\mathcal{J}_{\phi,\text{an}} = \mathcal{J}_{\phi^n,\text{an}} = \{\zeta(0,1)\}$$

by Propositions 8.2 and 8.12.

If $\xi \in \mathbb{P}^1_{\text{an}} \smallsetminus \mathbb{P}^1(\mathbb{C}_v)$ also has finite grant orbit, then by the same logic, we have $\xi \in \mathcal{J}_{\phi,\text{an}} = \{\zeta\}$, proving that ζ is indeed unique. In addition, for any other $\eta' \in \mathrm{PGL}(2, \mathbb{C}_v)$ that moves ζ to $\zeta(0,1)$, the Berkovich Julia set of $\eta' \circ \phi \circ (\eta')^{-1}$ is again $\{\zeta(0,1)\}$. By Proposition 8.12, then, $\eta' \circ \phi \circ (\eta')^{-1}$ has explicit good reduction. □

Corollary 8.14. *Let $\phi \in \mathbb{C}_v(z)$ be a rational function of degree $d \geq 2$, and suppose that ϕ^n has explicit good reduction for some $n \geq 2$. Then ϕ itself has explicit good reduction.*

Proof. By Proposition 8.12, the Gauss point $\zeta := \zeta(0,1)$ is fixed by ϕ^n. Hence, for each $j \geq 0$, writing $j = i + mn$ with $0 \leq i \leq n-1$, we have

$\phi^{-j}(\zeta) = \phi^{-i}(\zeta)$. Thus, the grand orbit of ζ under ϕ is the finite set

$$\bigcup_{i=0}^{n-1} \phi^{-i}(\zeta).$$

By Proposition 8.13, then, ϕ has explicit good reduction. $\square$

8.4. More basic properties of Berkovich Julia sets

We have the following Berkovich analogues of several of the results from Section 5.3, with only occasional modifications to deal with the case of potential good reduction.

Theorem 8.15. *Let $\phi \in \mathbb{C}_v(z)$ be a rational function of degree at least 2 and with Berkovich Julia set $\mathcal{J}_{\phi,\mathrm{an}}$. Let E_ϕ be the exceptional set of ϕ in $\mathbb{P}^1(\mathbb{C}_v)$. Then*

(a) *Let $U \subseteq \mathbb{P}^1_{\mathrm{an}}$ be an open set intersecting $\mathcal{J}_{\phi,\mathrm{an}}$. Then*

$$\mathbb{P}^1_{\mathrm{an}} \smallsetminus \bigcup_{n\geq 0} \phi^n(U)$$

is a finite set contained in E_ϕ.

(b) *$\mathcal{J}_{\phi,\mathrm{an}}$ has empty interior.*

(c) *For any point $\zeta \in \mathbb{P}^1_{\mathrm{an}} \smallsetminus E_\phi$, the closure of the backward orbit of ζ contains $\mathcal{J}_{\phi,\mathrm{an}}$, with equality if and only if $\zeta \in \mathcal{J}_{\phi,\mathrm{an}}$.*

(d) *Let $S \subseteq \mathbb{P}^1_{\mathrm{an}}$ be a closed subset of $\mathbb{P}^1_{\mathrm{an}}$ that is not contained in E_ϕ and for which $\phi^{-1}(S) \subseteq S$. Then $\mathcal{J}_{\phi,\mathrm{an}} \subseteq S$.*

(e) *Assuming $\mathcal{J}_{\phi,\mathrm{an}}$ is nonempty, either*
   (i) *$\mathcal{J}_{\phi,\mathrm{an}}$ consists of a single Type II point, and ϕ has potential good reduction; or*
   (ii) *$\mathcal{J}_{\phi,\mathrm{an}}$ is a perfect set, and it is uncountable.*

(f) *If ϕ is a polynomial and we define*

$$\mathcal{K}_{\phi,\mathrm{an}} := \left\{ \zeta \in \mathbb{P}^1_{\mathrm{an}} : \lim_{n\to\infty} \phi^n(\zeta) \neq \infty \right\}$$

to be the Berkovich filled Julia set *of ϕ, then $\mathcal{J}_{\phi,\mathrm{an}} = \partial\mathcal{K}_{\phi,\mathrm{an}}$.*

The assumption that $\mathcal{J}_{\phi,\mathrm{an}} \neq \varnothing$ in part (e) turns out to be unnecessary; we will see in Corollary 12.6 that the Berkovich Julia set is always nonempty.

Proof.

(a) By definition, $W := \bigcup_{n\geq 0} \phi^n(U)$ must contain all but finitely many points of $\mathbb{P}^1_{\mathrm{an}}$. All preimages of each such point must also lie outside W, and hence all such points have finite grand orbit. Any such point that is Type I is therefore in E_ϕ. By Proposition 8.13, there can be at most one such point ζ that is not of Type I, and if there is one, then $\mathcal{J}_{\phi,\mathrm{an}} = \{\zeta\}$. In that case, then, we have $\zeta \in U \subseteq W$. Thus, the only points that could be in $\mathbb{P}^1_{\mathrm{an}} \smallsetminus W$ are exceptional points of Type I, as desired.

(b) This is similar to the proof of Corollary 5.22.

(c) This is similar to the proof of Proposition 5.23.

(d) This is similar to the proof of Corollary 5.24.

(e) This is similar to the proof of Corollary 5.25, except that we need to consider the possibility that the backward orbit of the given point $a \in \mathcal{J}_{\phi,\mathrm{an}}$ is *finite*. In that case, Proposition 8.13 says that $\mathcal{J}_{\phi,\mathrm{an}}$ consists of a single Type II point and that ϕ has potential good reduction.

In addition, because $\mathbb{P}^1_{\mathrm{an}}$ is in general not a metric space, our uncountability argument from Corollary 5.25 does not apply here. Fortunately, it is also true that any nonempty perfect subset of a compact Hausdoff space (such as $\mathbb{P}^1_{\mathrm{an}}$) is uncountable. We leave the proof of this well-known topological statement to the reader; see Exercise 8.15.

(f) This is similar to the proof of Proposition 5.27, except that we cannot assume the open set U is a disk when proving that $\partial\mathcal{K}_{\phi,\mathrm{an}} \subseteq \mathcal{J}_{\phi,\mathrm{an}}$. A slightly different argument, which we leave to the reader, is required here; see Exercise 8.16. □

8.5. Examples

We now revisit some of our earlier examples of dynamical systems on $\mathbb{P}^1(\mathbb{C}_v)$, to understand their action on the larger space $\mathbb{P}^1_{\mathrm{an}}$.

Example 8.16. Let $\phi(z) = z^2 + \lambda z \in \mathbb{C}_v[z]$, as in Example 5.29.

If $|\lambda| \le 1$, then ϕ has explicit good reduction, and hence by Proposition 8.12, its Berkovich Julia set is $\mathcal{J}_{\phi,\mathrm{an}} = \{\zeta(0,1)\}$, and the Fatou set is therefore $\mathcal{F}_{\phi,\mathrm{an}} = \mathbb{P}^1_{\mathrm{an}} \smallsetminus \{\zeta(0,1)\}$.

If $|\lambda| > 1$, recall from Example 5.29 that the classical filled Julia set $\mathcal{K}_{\phi,\mathrm{I}}$ is a certain Cantor set contained in the disk $U_0 = \overline{D}(0,|\lambda|)$. Define

$$W := \mathbb{P}^1_{\mathrm{an}} \smallsetminus \mathcal{K}_{\phi,\mathrm{I}},$$

which is an open subset of $\mathbb{P}^1_{\mathrm{an}}$ containing all points of Types II–IV. By similar reasoning as in Example 5.29, every point $\zeta \in W$ has

$$\lim_{n\to\infty} \phi^n(\zeta) = \infty,$$

and hence ζ does not belong to the Berkovich filled Julia set $\mathcal{K}_{\phi,\mathrm{an}}$; see Exercise 8.17. Therefore, by Theorem 8.15(f), the Berkovich Julia set and filled Julia set are the same Cantor set

$$\mathcal{K}_{\phi,\text{an}} = \mathcal{J}_{\phi,\text{an}} = \mathcal{K}_{\phi,\text{I}} = \mathcal{J}_{\phi,\text{I}}.$$

The Fatou set $\mathcal{F}_{\phi,\text{an}}$ is the set W above, and every Fatou point is attracted to ∞ under iteration.

Example 8.17. Fix a prime p, and let

$$\phi(z) = \frac{z^p - z}{p} \in \mathbb{C}_p[z]$$

as in Example 5.30. By similar reasoning as in Example 8.16, the Berkovich Julia set and filled Julia set of ϕ are the Cantor set

$$\mathcal{K}_{\phi,\text{an}} = \mathcal{J}_{\phi,\text{an}} = \mathcal{K}_{\phi,\text{I}} = \mathcal{J}_{\phi,\text{I}} = \mathbb{Z}_p,$$

and all Fatou points are attracted to ∞; see Exercise 8.18.

The dynamics on the Julia sets of Examples 8.16 and 8.17 can also be described in terms of symbolic dynamics, which was introduced in Exercise 1.26. See Exercise 8.19 for more on this connection, which occurs when there are $d = \deg(\phi)$ disks mapping onto the same disk U_0 that contains the filled Julia set. For more general maps, however, the dynamics can be more intricate.

Example 8.18. Pick $a, b, c \in \mathbb{C}_v$ with $0 < |a| < 1$ and $|b|, |c| \leq 1$, and let

$$\phi(z) = az^3 + z^2 + bz + c,$$

as in Example 5.31. This time, the Berkovich Julia set and filled Julia set do *not* coincide, as $\mathcal{J}_{\phi,\text{an}}$ includes points that are not of Type I. In particular, since $\zeta(0,1)$ is a repelling fixed point—as noted in Example 5.31, the reduction $\overline{\phi}(z) = z^2 + \overline{b}z + \overline{c}$ has degree 2—we have $\zeta(0,1) \in \mathcal{J}_{\phi,\text{an}}$. On the other hand, just as in Example 5.31, setting $U_0 := \overline{D}_{\text{an}}(0, |a|^{-1})$, the Berkovich filled Julia set is

$$\mathcal{K}_{\phi,\text{an}} = \bigcap_{n \geq 0} \phi^{-n}(U_0).$$

(See Exercise 8.20.) Moreover, by Theorem 8.15(f), its boundary is the Berkovich Julia set $\mathcal{J}_{\phi,\text{an}}$.

As a side note, recall from Example 5.31 that the classical Julia set $\mathcal{J}_{\phi,\text{I}}$ is not compact, as illustrated by the sequence of points

$$\{\gamma_n\}_{n \geq 1} \subseteq \mathcal{J}_{\phi,\text{I}} \quad \text{with} \quad |\gamma_n| = |a|^{-1/2^n}$$

from that example. While this sequence of Type I points has no accumulation points in $\mathbb{P}^1(\mathbb{C}_v)$, observe that we have

$$\lim_{n \to \infty} \gamma_n = \zeta(0,1) \in \mathcal{J}_{\phi,\text{an}}.$$

See Exercise 8.22 for further properties of $\mathcal{J}_{\phi,\text{an}}$ for this map.

Example 8.19. Assume the residue characteristic of $\mathbb{C}_v$ is not 2, pick $a \in \mathbb{C}_v$ with $0 < |a| < 1$, and let

$$\phi(z) = z^2 + \frac{az^2}{z+1} = \frac{z^3 + (a+1)z^2}{z+1},$$

as in Example 5.32. Because $\overline{\phi}(z) = z^2$ has degree 2, the Gauss point $\zeta(0,1)$ is a repelling fixed point and hence belongs to the Julia set.

Recall that the residue class $\vec{v}(-1) = D_{\rm an}(-1,1)$ contains a Type I repelling fixed point β. In addition, $\vec{v}(-1)$ is the only bad direction at $\zeta(0,1)$ in the sense of Theorem 7.34, and hence every other residue class $\vec{v}(x)$ at $\zeta(0,1)$ maps onto $\vec{v}(x^2)$. The conclusions of Example 5.32 apply to the Berkovich Fatou and Julia sets as well: if $|x^{2^n} + 1| \geq 1$ for all $n \geq 0$, then the residue class $\vec{v}(x)$ is contained in $\mathcal{F}_{\phi,\rm an}$; otherwise, it intersects the Julia set.

Moreover, as in Example 8.18, the noncompactness of the classical Julia set $\mathcal{J}_{\phi,\rm I}$ is again rectified by passing to $\mathcal{J}_{\phi,\rm an}$. In particular, any sequence of Type I Julia points $\{\gamma_n\}_{n\geq 1}$ lying in different residue classes has no accumulation points in $\mathbb{P}^1(\mathbb{C}_v)$, but in $\mathbb{P}^1_{\rm an}$ it approaches the Type II point $\zeta(0,1) \in \mathcal{J}_{\phi,\rm an}$. See Exercise 8.23 for further properties of $\mathcal{J}_{\phi,\rm an}$ for this map.

Example 8.20. Let $c \in \mathbb{C}_v$ with $0 < |c| < 1$, and consider the Lattès map

$$\phi(z) = \frac{z^4 - 8c^4 z - c^4}{4z^3 + z^2 + 4c^4}$$

of Examples 5.33 and 7.31. We now show that $\mathcal{J}_{\phi,\rm an}$ is the interval

$$\mathcal{J}_{\phi,\rm an} = J := \big[\zeta(0,1), \zeta(0,|c|^2)\big],$$

as promised in Example 7.31.

To see that $J \subseteq \mathcal{J}_{\phi,\rm an}$, note that $\zeta(0,1)$ is a repelling fixed point, since the reduction

$$\overline{\phi}(z) = \frac{z^4}{4z^3 + z^2} = \frac{z^2}{4z+1}$$

has degree 2. By considering the action of ϕ on J as described in Example 7.31 and Figure 7.1, the reader can check that the backward orbit $\mathrm{Orb}^-_\phi(\zeta(0,1))$ is dense in J; see Exercise 8.24. Thus, since the Julia set is closed and contains all preimages of repelling periodic points, we do indeed have $J \subseteq \mathcal{J}_{\phi,\rm an}$.

Conversely, we claim that $\mathbb{P}^1_{\rm an} \smallsetminus J \subseteq \mathcal{F}_{\phi,\rm an}$. To see this, we consider the various bullet points of Example 5.33. For instance, since

$$\phi\big(\mathbb{P}^1(\mathbb{C}_v) \smallsetminus \overline{D}(0,1)\big) \subseteq \mathbb{P}^1(\mathbb{C}_v) \smallsetminus \overline{D}(0,1),$$

we have

$$\phi\big(\mathbb{P}^1_{\mathrm{an}} \smallsetminus \overline{D}_{\mathrm{an}}(0,1)\big) \subseteq \mathbb{P}^1_{\mathrm{an}} \smallsetminus \overline{D}_{\mathrm{an}}(0,1)$$

by Theorem 7.8, and hence the disk $\mathbb{P}^1_{\mathrm{an}} \smallsetminus \overline{D}_{\mathrm{an}}(0,1)$ is contained in $\mathcal{F}_{\phi,\mathrm{an}}$. Applying Theorem 7.8 to the other bullet points of Example 5.33, our claim follows, and hence the Julia set is indeed $\mathcal{J}_{\phi,\mathrm{an}} = J$.

Exercises for Chapter 8

8.1. Let $\phi \in \mathbb{C}_v(z)$, and let $U \subseteq \mathbb{P}^1_{\mathrm{an}}$ be an open set. Suppose that it is *not* the case that $\operatorname{char} \mathbb{C}_v = p > 0$ and ϕ is conjuate to $z \mapsto z^{p^m}$ for some $m \geq 1$. If $\bigcup \phi^n(U)$ omits at least three points of $\mathbb{P}^1_{\mathrm{an}}$, prove that U is dynamically stable under ϕ.

(*Suggestion*: Use Theorem 1.19 and Proposition 8.13.)

8.2. Prove parts (a), (c), and (d) of Proposition 8.2.

8.3. In this exercise, you will verify some of the claims made in Example 8.6 concerning the map $\phi(z) = z^m \in \mathbb{C}_v(z)$, where $m \geq 2$. Let $p \geq 0$ be the residue characteristic of $\mathbb{C}_v$.

(a) If $p \nmid m$, then for any $a \in \mathbb{C}_v$ with $|a| = 1$ and any $r \in (0,1)$, prove that $\phi(\zeta(a,r)) = \zeta(a^m, r)$.

(b) If $p^e | m$ but $p^{e+1} \nmid m$ for some integer $e \geq 1$, write

$$\phi\big(\zeta(a,r)\big) = \zeta(a^m, s),$$

and give a formula for s in terms of $r \in (0,1)$. (The correct formula is a piecewise-defined function of r.)

In particular, if r is close enough to 1, show that $s = r^{p^e}$, as claimed in Example 8.6.

8.4. Let k be a field. Show that k is algebraic over a finite field if and only if every element of $\mathrm{PGL}(2,k)$ has finite order; cf. Remark 8.9.

8.5. Let $d \geq m \geq 2$, and let $\phi(z) \in \mathbb{C}_v(z)$ be a rational function of degree d with a superattacting fixed point at 0, with $\phi(D_{\mathrm{an}}(0,1)) = D_{\mathrm{an}}(0,1)$, and with

$$\deg_0(\phi) = \mathrm{wdeg}_{D(0,1)}(\phi) = m.$$

Let $0 < r < 1$, and let $U = \{r < |\zeta| < 1\}$. Prove that

$$\bigcup_{n \geq 0} \phi^n(U) = D_{\mathrm{an}}(0,1) \smallsetminus \{0\}.$$

8.6. Prove that $\phi \in \mathbb{C}_v(z)$ satisfies the hypotheses of Exercise 8.5 if and only if ϕ is of the form

$$\phi(z) = \frac{a_m z^m + \cdots + a_d z^d}{b_0 + \cdots + b_d z^d}$$

with $|a_m| = |b_0| = 1$, with either a_d or b_d nonzero (or both), and with $|a_j|, |b_j| \leq 1$ for all j.

8.7. Given ϕ and U as in Lemma 8.11, suppose there is some $a \in D(0,1)$ such that $a \notin \phi^n(U)$ for any $n \geq 0$. Prove that either

(a) $\phi(D_{\text{an}}(0,1)) = D_{\text{an}}(0,1)$, and a is a superattracting fixed point of ϕ with

$$\deg_a(\phi) = \deg_{\zeta(0,1),\vec{v}(0)}(\phi) = \text{wdeg}_{D(0,1)}(\phi);$$

or

(b) ϕ has at least one pole in $D(0,1)$, but a is a totally invariant fixed point of ϕ, so that

$$\psi(z) := \frac{1}{\phi\left(\dfrac{1}{z} + a\right) - a} \quad \text{is a polynomial.}$$

8.8. Fix $d \geq m \geq 1$, and let $\psi(z) = a_0 + a_1 z + \cdots + a_d z^d \in \mathbb{C}_v[z]$.

(a) Prove that ψ fixes $\zeta(0,1)$ with multiplier m if and only if $|a_j| \leq 1$ for all $j \leq m$, $|a_m| = 1$, and $|a_j| < 1$ for all $j > m$.

(b) Let $\phi(z) := 1/\psi(1/z)$. Prove that $\deg_{\zeta(0,1),\vec{v}(0)}(\phi) = m$.

(c) Assume that $2 \leq m < d$. Let $U := \{r < |\zeta| < 1\}$, where $0 < r < 1$. Prove that $\bigcup_{n\geq 0} \phi^n(U) = \mathbb{P}^1_{\text{an}} \smallsetminus \{0\}$.

8.9. In the proof of Theorem 8.7(b), we constructed a certain open connected Berkovich affinoid W and proved that $\phi(W) \subseteq W$. Prove that in fact, $\phi(W) = W$.

More precisely, suppose that $\zeta := \zeta(0,1)$ is an indifferent fixed point of $\phi \in \mathbb{C}_v(z)$, and that each of the finitely many bad directions $\boldsymbol{v}_1, \ldots, \boldsymbol{v}_M$ at ζ is fixed by $\phi_\#$. For each $i = 1, \ldots, M$, let $V_i \subsetneq \boldsymbol{v}_i$ be a closed disk such that ϕ maps the annulus $\boldsymbol{v}_i \smallsetminus V_i$ bijectively onto itself. Prove that ϕ maps the Berkovich affinoid $W := \mathbb{P}^1_{\text{an}} \smallsetminus (V_1 \cup \cdots \cup V_M)$ bijectively onto itself.

8.10. Let $\phi \in \mathbb{C}_v(z)$, and let ζ be a Type II Julia periodic point of ϕ.

(a) If ζ is repelling, prove that infinitely many of the residue classes of ζ are periodic components of the Fatou set of ϕ.

(b) If ζ is indifferent, prove that at most two of the residue classes of ζ are periodic components of the Fatou set of ϕ.
(*Suggestion*: Reduce to the case $\zeta = \zeta(0,1)$, and diagonalize the reduction $\overline{\phi} \in \text{PGL}(2,k)$, if possible.)

8.11. In Exercise 8.10(b), prove that infinitely many residue classes of ζ are wandering Fatou components.

8.12. Let $\phi \in \mathbb{C}_v(z)$ be a rational function of degree $d = 1$.

(a) Prove that statements (a)–(c) of Proposition 8.12 are equivalent for ϕ.

(b) Prove that the Berkovich Julia set $\mathcal{J}_{\phi,\mathrm{an}}$ is either empty or else consists of a single Type I point. In particular, statement (d) of Proposition 8.12 can never hold for ϕ.

8.13. Let $x_1, x_2, x_3 \in \mathbb{P}^1(\mathbb{C}_v)$ be three points whose reductions in $\mathbb{P}^1(k)$ are distinct. Let $y_1, y_2, y_3 \in \mathbb{P}^1(\mathbb{C}_v)$ also be three points whose reductions in $\mathbb{P}^1(k)$ are distinct. Let $f \in \mathrm{PGL}(2, \mathbb{C}_v)$ be the unique linear transformation satisfying $f(x_i) = y_i$ for $i = 1, 2, 3$. Prove that f has explicit good reduction.

(*Suggestion*: Consider the unique point lying between all the x_i's, and similarly for the y_i's. Then use Exercise 8.12(a).)

8.14. Let $\phi \in \mathbb{C}_v(z)$ be a rational function that fixes $\zeta := \zeta(0,1)$, and let U be a residue class at ζ such that ϕ maps U m-to-1 onto itself, for some $m \geq 2$. If ϕ has no critical points in U, prove that the reduction $\overline{\phi}$ is inseparable.

8.15 (See Theorem 8.15(e))**.** Let X be a compact Hausdoff topological space, and let $Y \subseteq X$ be a nonempty closed perfect subset. Prove that Y is uncountable. (Do *not* assume that X or Y is a metric space.)

8.16. Prove Theorem 8.15(f).

8.17. Let $\phi(z) = z^2 + \lambda z \in \mathbb{C}_v(z)$ with $|\lambda| > 1$, and let

$$U_0 := \overline{D}_{\mathrm{an}}(0, |\lambda|) \quad \text{and} \quad W := \mathbb{P}^1_{\mathrm{an}} \smallsetminus \mathcal{K}_{\phi,\mathrm{I}}$$

as in Example 8.16.

(a) Prove that $\lim_{n\to\infty} \phi^n(\zeta) = \infty$ for any point ζ in the open disk $\mathbb{P}^1_{\mathrm{an}} \smallsetminus U_0$.

(b) Prove that $\lim_{n\to\infty} \phi^n(\zeta) = \infty$ for all $\zeta \in W$.

(c) Observe that $\mathbb{H} \subseteq W$, and conclude that all periodic points of ϕ are of Type I, and that $\mathcal{J}_{\phi,\mathrm{an}} = \mathcal{J}_{\phi,\mathrm{I}}$.

8.18. Fix a prime p, and let $\phi(z) = (z^p - z)/p \in \mathbb{C}_p[z]$ as in Example 8.17. Let

$$U_0 := \overline{D}_{\mathrm{an}}(0, 1) \quad \text{and} \quad W := \mathbb{P}^1_{\mathrm{an}} \smallsetminus \mathcal{K}_{\phi,\mathrm{I}}.$$

(a) Prove that $\lim_{n\to\infty} \phi^n(\zeta) = \infty$ for any point ζ in the open disk $\mathbb{P}^1_{\mathrm{an}} \smallsetminus U_0$.

(b) Prove that $\lim_{n\to\infty} \phi^n(\zeta) = \infty$ for all $\zeta \in W$.

8.19. Let $\phi \in \mathbb{C}_v[z]$ be a polynomial of degree $d \geq 2$, let $U_0 \subseteq \mathbb{A}^1_{\mathrm{an}}$ be a closed disk, and let $Y_1, \ldots, Y_d \subseteq U_0$ be pairwise disjoint closed disks. Suppose that ϕ maps Y_i bijectively onto U_0 for each $i = 1, \ldots, d$.

(a) Prove that $U_0 \supseteq \phi^{-1}(U_0) \supseteq \phi^{-2}(U_0) \supseteq \cdots$.

(b) Prove that the Julia set and filled Julia set of ϕ are

$$\mathcal{J}_{\phi,\mathrm{an}} = \mathcal{K}_{\phi,\mathrm{an}} = \bigcap_{n\ge 0} \phi^{-n}(U_0),$$

and that this set consists entirely of Type I points.

(c) In the terminology of symbolic dynamics (see Exercise 1.26), let

$$\Sigma := \prod_{n\ge 0}\{1,\dots d\}$$

be the set of symbol sequences on d symbols, and define

$$f : \mathcal{J}_{\phi,\mathrm{an}} \to \Sigma$$

by setting $f(x)$ to be the symbol sequence $(\alpha_0, \alpha_1, \alpha_2, \dots) \in \Sigma$ with $\phi^n(x) \in Y_{\alpha_n}$ for all $n \ge 0$. Prove that $f : \mathcal{J}_{\phi,\mathrm{an}} \to \Sigma$ is a bijective map, and that $f \circ \phi = T \circ f$, where $T : \Sigma \to \Sigma$ is the shift map

$$T(\alpha_0, \alpha_1, \alpha_2, \dots) = (\alpha_1, \alpha_2, \alpha_3, \dots). \tag{8.1}$$

8.20. Let $\phi(z) = az^3 + z^2 + bz + c \in \mathbb{C}_v[z]$, with $0 < |a| < 1$ and $|b|, |c| \le 1$, and let $U_0 := \overline{D}_{\mathrm{an}}(0, |a|^{-1})$. Prove that the Berkovich filled Julia set is

$$\mathcal{K}_{\phi,\mathrm{an}} = \bigcap_{n\ge 0} \phi^{-n}(U_0),$$

as claimed in Example 8.18.

8.21. Let ϕ and U_0 be the map and disk of Exercise 8.20.

(a) Show that $\phi^{-1}(U_0)$ is the disjoint union of two closed disks Y_1 and Y_2, where Y_1 maps bijectively onto U_0, and Y_2 maps two-to-one onto U_0.

(b) Let $\Sigma := \prod_{n\ge 0}\{1, 2\}$ be the set of symbol sequences on two symbols, and define $f : \mathcal{K}_{\phi,\mathrm{an}} \to \Sigma$ as in Exercise 8.19(c). Prove that

$$f \circ \phi = T \circ f,$$

where $T : \Sigma \to \Sigma$ is the shift map of equation (8.1), but that f is *not* injective.

(c) Prove that f is surjective, and for an arbitrary symbol sequence $A = (\alpha_0, \alpha_1, \dots)$, describe the inverse image $f^{-1}(A)$.

8.22. Assume the residue characteristic of $\mathbb{C}_v$ is not 2, and let ϕ be the map of Exercise 8.20.

(a) Prove that for every $x \in \mathbb{C}_v$ with $|x| > 1$, either $\phi^n(x) \to \infty$ as $n \to \infty$, or ϕ is one-to-one on the disk $D_{\mathrm{an}}(x, |x|)$. (Or both.)

(b) Prove that every point of $\mathcal{J}_{\phi,\mathrm{an}}$ is either a Type II point in the backward orbit of $\zeta(0, 1)$ or else a Type I point.

8.23. Assume the residue characteristic of $\mathbb{C}_v$ is not 2, pick $a \in \mathbb{C}_v$ with $0 < |a| < 1$, and let

$$\phi(z) = z^2 + \frac{az^2}{z+1} = \frac{z^3 + (a+1)z^2}{z+1},$$

as in Examples 5.32 and 8.19.

(a) Prove that ϕ is one-to-one on $D_{\mathrm{an}}(x,1)$ for every $x \in \overline{D}(0,1)$ with $|x| = |x+1| = 1$.

(b) Prove that for every $x \in D(-1,1)$,
   (i) $|\phi(x)| > 1$ if $|x+1| < |a|$;
   (ii) $\phi(x) \in D(1,1)$ if $|x+1| > |a|$; and
   (iii) ϕ maps $D_{\mathrm{an}}(x,|a|)$ bijectively onto $D_{\mathrm{an}}\big(1 + \frac{a}{x+1}, 1\big)$ if $|x+1| = |a|$.

(c) Prove that every point of $\mathcal{J}_{\phi,\mathrm{an}}$ is either a Type II point in the backward orbit of $\zeta(0,1)$ or else a Type I point.

8.24. Let $0 < |c| < 1$, and let

$$\phi(z) = \frac{z^4 - 8c^4 z - c^4}{4z^3 + z^2 + 4c^4}$$

be the Lattès map of Example 8.20. Let

$$J := \big[\zeta(0,1), \zeta(0,|c|^2)\big].$$

Use Example 7.31 and Figure 7.1 to prove that $\mathrm{Orb}_\phi^-(\zeta(0,1))$ is dense in J.

8.25. Let $\phi(z) \in \mathbb{C}_v[z]$ be a polynomial of degree $d \geq 2$ with Fatou set $\mathcal{F}_{\phi,\mathrm{an}}$. Let $U \subseteq \mathbb{A}^1_{\mathrm{an}}$ be a connected component of $\mathcal{F}_{\phi,\mathrm{an}}$ not containing ∞. Prove that U is an open disk.

Chapter 9

Classifying Berkovich Fatou Components

In this chapter, we discuss dynamics on periodic components of the Berkovich Fatou set. We follow the work of Rivera-Letelier, who classified dynamics on such components into two types: attracting and indifferent. In his thesis [**114**], he worked over the field $\mathbb{C}_p$. Most of the modifications required to extend the classification to arbitrary non-archimedean fields appeared in his later papers [**115, 117**], and we fill in the remaining details here.

9.1. Berkovich Fatou components

As in the complex setting, a rational function $\phi \in \mathbb{C}_v(z)$ acts on the set of connected components of its Berkovich Fatou set according to the following result.

Theorem 9.1. *Let $\phi \in \mathbb{C}_v(z)$ be a rational function of degree $d \geq 2$ with Berkovich Fatou set $\mathcal{F}_{\phi,\mathrm{an}}$, and let U be a connected component of $\mathcal{F}_{\phi,\mathrm{an}}$. Then the following hold.*

(a) *$\phi(U)$ is also a connected component of $\mathcal{F}_{\phi,\mathrm{an}}$.*

(b) *There is an integer $1 \leq m \leq d$ such that every point in $\phi(U)$ has exactly m preimages in U, counting multiplicity.*

(c) *The inverse image $\phi^{-1}(U)$ is a finite union of components of $\mathcal{F}_{\phi,\mathrm{an}}$, each of which is mapped onto U by ϕ.*

To prove Theorem 9.1, we need the following lemma.

Lemma 9.2. *Let $\phi \in \mathbb{C}_v(z)$ be a rational function of degree $d \geq 2$ with Berkovich Fatou set $\mathcal{F}_{\phi,\mathrm{an}}$, let U be a connected component of $\mathcal{F}_{\phi,\mathrm{an}}$, and let $x \in U$. Then U is the union of all open connected Berkovich affinoids containing x and contained in $\mathcal{F}_{\phi,\mathrm{an}}$.*

Proof. By Theorem 6.32, any open connected Berkovich affinoid $W \subseteq \mathcal{F}_{\phi,\mathrm{an}}$ containing x is connected and, hence, contained in U. It remains to show that U is contained in the union of such sets W.

By Corollary 6.33, $\mathbb{P}^1_{\mathrm{an}}$ is locally path-connected, and hence U, as a connected component of an open subset of $\mathbb{P}^1_{\mathrm{an}}$, is in fact path-connected. Therefore, given $y \in U \smallsetminus \{x\}$, there is a subset $I \subseteq U$ homeomorphic to $[0,1]$ that contains both x and y.

For each point $\zeta \in I \subseteq U$, there is an open connected Berkovich affinoid W_ζ such that $\zeta \in W_\zeta \subseteq U$, by Theorem 6.17. Since $I \cong [0,1]$ is compact, there is in fact a finite set of such open connected Berkovich affinoids $W_1, \ldots, W_m$ covering I. Because I is connected, we may re-index so that $x \in W_1$, $y \in W_\ell$, and

$$W_i \cap (W_1 \cup \cdots \cup W_{i-1}) \neq \varnothing \quad \text{for each } i = 1, \ldots, \ell - 1.$$

By Proposition 3.28 and Exercise 6.28(b), it follows that $W_1 \cup \cdots \cup W_\ell$ is a connected affinoid contained in U, and containing both x and y. □

Proof of Theorem 9.1.

(a) Let U be a connected component of $\mathcal{F}_{\phi,\mathrm{an}}$, and let $x \in U$. Then $\phi(U)$ is also a connected subset of $\mathcal{F}_{\phi,\mathrm{an}}$, by Proposition 8.2(b) and the continuity of ϕ. By Lemma 9.2, then, to show that $\phi(U)$ is a connected component of $\mathcal{F}_{\phi,\mathrm{an}}$, we need to show that $\phi(U)$ contains every open connected Berkovich affinoid containing $\phi(x)$ and contained in $\mathcal{F}_{\phi,\mathrm{an}}$.

Let $W \subseteq \mathcal{F}_{\phi,\mathrm{an}}$ be an open connected Berkovich affinoid containing $\phi(x)$. By Theorem 7.8, $\phi^{-1}(W)$ is the disjoint union of finitely many open connected Berkovich affinoids $V_1, \ldots, V_\ell$, each of which maps onto W. Since $\phi(x) \in W$, one such affinoid, without loss of V_1, contains x. By Proposition 8.2(b), we have $V_1 \subseteq \mathcal{F}_{\phi,\mathrm{an}}$, and by Lemma 9.2, we have $V_1 \subseteq U$. Thus,

$$\phi(U) \supseteq \phi(V_1) = W,$$

as desired.

(b) Given $y, y' \in \phi(U)$, each has d preimages in $\mathbb{P}^1_{\mathrm{an}}$. Call their preimages that lie in U

$$x_1, \ldots, x_m \quad \text{and} \quad x'_1, \ldots, x'_\ell,$$

respectively, repeated according to multiplicity. We need to show that $m = \ell$. We begin by claiming that there is an open connected Berkovich affinoid $X \subseteq U$ that contains every x_i and x'_i.

To see that X exists, observe that for every index i, there is an open connected Berkovich affinoid $X_i \subseteq U$ containing both x_1 and x_i, and an open connected Berkovich affinoid $X'_i \subseteq U$ containing both x_1 and x'_i, by Lemma 9.2. Then

$$X := \bigcup_{i=1}^{m} X_i \cup \bigcup_{i=1}^{\ell} X'_i \subseteq U$$

is also an open connected Berkovich affinoid, by Proposition 3.28 and Exercise 6.28(b). The claim is proven.

By Theorem 7.8, $\phi(X)$ is also an open connected Berkovich affinoid, and $\phi^{-1}(\phi(X))$ is a finite disjoint union of open connected Berkovich affinoids. Let W be the one open connected Berkovich affinoid in this union that contains x. Then ϕ maps W onto $\phi(X)$ with some multiplicity $M \geq 1$. Because X is connected and

$$x \in X \subseteq \phi^{-1}(\phi(X)),$$

we must have $X \subseteq W$. In particular, y and y' both have exactly M preimages in W, counting multiplicity, since $y, y' \in \phi(X)$.

However, by Proposition 8.2(b), $W \subseteq \mathcal{F}_{\phi,\mathrm{an}}$, and hence

$$X \subseteq W \subseteq U.$$

Thus, y has exactly m preimages in W, and y' has exactly ℓ preimages in W. By the previous paragraph, then, $m = M = \ell$.

(c) Given $x \in U$, let $t_1, \ldots, t_d \in \mathbb{P}^1_{\mathrm{an}}$ be the preimages of x under ϕ, repeated according to multiplicity; each lies in $\mathcal{F}_{\phi,\mathrm{an}}$, by Proposition 8.2(b). Let $V_1, \ldots, V_\ell$ be the (finitely many) distinct Berkovich Fatou components that together contain all of $t_1, \ldots, t_d$. Then for each i, $\phi(V_i) \cap U$ contains x and hence is nonempty. However, by part (a), both $\phi(V_i)$ and U are connected components of $\mathcal{F}_{\phi,\mathrm{an}}$, and hence we must have $\phi(V_i) = U$.

By part (b), then, each V_i maps m_i-to-1 onto U, for some integer $m_i \geq 1$. Moreover, counting the preimages $t_1, \ldots, t_d$ in $V_1 \cup \cdots \cup V_\ell$, we must have

$$m_1 + \cdots + m_\ell = d.$$

Therefore, all d preimages of *every* point in U lie in $V_1 \cup \cdots \cup V_\ell$. Thus,

$$\phi^{-1}(U) = V_1 \cup \cdots \cup V_\ell. \qquad \square$$

9.2. Attracting components

In this section, we discuss the first of the two types of fixed components possible in non-archimedean dynamics.

Definition 9.3. Let $\phi \in \mathbb{C}_v(z)$ be a rational function of degree $d \geq 2$, and let $a \in \mathbb{P}^1(\mathbb{C}_v)$ be an attracting periodic point of minimal period $m \geq 1$. Let $V \subseteq \mathbb{P}^1_{\text{an}}$ be an open Berkovich disk containing a and small enough that

$$\lim_{n\to\infty} \phi^{mn}(x) = a \quad \text{for all} \quad x \in V \cap \mathbb{P}^1(\mathbb{C}_v).$$

(a) The *attracting basin* of a is

$$\mathfrak{A}_\phi(a) := \bigcup_{n\geq 0} \phi^{-n}(V),$$

i.e., the set of all $\zeta \in \mathbb{P}^1_{\text{an}}$ such that $\phi^n(\zeta) \in V$ for some $n \geq 0$.

(b) The *immediate attracting basin* of a is the connected component U of $\mathfrak{A}_\phi(a)$ containing a. We call U an *attracting component* for ϕ.

It is easy to see that the definition of the (immediate) attracting basin is independent of the choice of the small disk V.

For example, if $\phi(z) = z^2$, then the attracting basin of the fixed point 0 is precisely $D_{\text{an}}(0,1)$, which is also the immediate attracting basin.

As a slightly more complicated example, if $\phi(z) = z^3 - z^2$, then the immediate attracting basin of 0 is again $D_{\text{an}}(0,1)$. This time, however, the attracting basin consists of infinitely many open disks, including $D_{\text{an}}(1,1)$, all of which are residue classes at $\zeta(0,1)$, and all of which eventually map to $D_{\text{an}}(0,1)$.

On the other hand, if $\phi(z) = z^2 + \lambda z$ with $|\lambda| > 1$, consider the attracting fixed point at ∞. Recall from Example 8.16 that the Julia set $\mathcal{J}_{\phi,\text{I}} = \mathcal{J}_{\phi,\text{an}}$ is a Cantor set contained in $\mathbb{P}^1(\mathbb{C}_v)$, and that all points of the Fatou set are attracted to ∞ under iteration. Thus, the Fatou set is both the attracting basin and the immediate attracting basin of ∞. This more complicated basin inspires the following definition.

Definition 9.4. Let $U \subseteq \mathbb{P}^1_{\text{an}}$ be an open set. We say that U is a *domain of Cantor type* if there is an increasing sequence of open connected Berkovich affinoids $U_1 \subsetneq U_2 \subsetneq U_3 \subsetneq \cdots$ such that

(a) for every $n \geq 1$, each connected component of $\mathbb{P}^1_{\text{an}} \smallsetminus U_n$ contains at least two points of ∂U_{n+1}; and

(b) $U = \bigcup_{n\geq 1} U_n$.

A domain U of Cantor type is not itself anything like the Cantor set; rather, as in the $z^2 + \lambda z$ example, the *boundary* ∂U is homeomorphic to the Cantor set; see Theorem 9.5(d) and Exercise 9.2.

Note that $(\partial U_{n+1}) \cap (\partial U_n) = \varnothing$ in Definition 9.4. After all, if ζ were a boundary point of both U_{n+1} and U_n, then the component of $\mathbb{P}^1_{\text{an}} \smallsetminus U_n$ containing ζ could not contain a second point of ∂U_{n+1}. Indeed, U_{n+1} is connected and contains U_n. Thus, with $\overline{U}_n$ denoting the closure of U_n in $\mathbb{P}^1_{\text{an}}$, we must in fact have $\overline{U}_n \subsetneq U_{n+1}$.

Clearly, any attracting basin is an open subset of the Berkovich Fatou set. However, we can say much more. We have the following result, which first appeared for $\mathbb{C}_p$ as [**114**, Théorème 2], augmented by [**116**, Proposition 6.1].

Theorem 9.5 (Rivera-Letelier). *Let $\phi \in \mathbb{C}_v(z)$ be a rational function of degree $d \geq 2$ with Berkovich Fatou set $\mathcal{F}_{\phi,\text{an}}$, let $a \in \mathbb{P}^1(\mathbb{C}_v)$ be an attracting periodic point of minimal period $m \geq 1$, and let $U \subseteq \mathbb{P}^1_{\text{an}}$ be the immediate attracting basin of a. Then the following hold.*

(a) *U is a periodic connected component of $\mathcal{F}_{\phi,\text{an}}$, of minimal period m.*

(b) *U is either an open Berkovich disk or a domain of Cantor type.*

(c) *If U is a disk, then its unique boundary point is a Type II repelling periodic point, of minimal period dividing m.*

(d) *If U is of Cantor type, then its boundary is homeomorphic to the Cantor set and is contained in the Berkovich Julia set.*

(e) *The mapping $\phi^m : U \to U$ is ℓ-to-1, for some integer $\ell \geq 2$.*

Note that the attracting component U and the attracting periodic point a must have the same minimal period, but if the component is a disk, the boundary point might have strictly smaller period. This situation can arise if other iterates of U share the same boundary point.

For example, if $\phi(z) = z^2 - 1$, then the two points 0 and -1 form an attracting 2-cycle, and the corresponding components are the disks $D_{\text{an}}(0, 1)$ and $D_{\text{an}}(-1, 1)$. However, both disks share the same boundary point $\zeta(0, 1)$, which is fixed, as ϕ has explicit good reduction.

Proof of Theorem 9.5. First, note that $\phi^n(U)$ cannot intersect U for any integer $1 \leq n \leq m-1$. If it did, then under the iteration of ϕ^m, a point x in the intersection would have to approach both a and $\phi^n(a)$, a contradiction. Thus, if we can show that U is periodic, then its minimal period must be m. Replacing ϕ by ϕ^m, then, we may assume for the remainder of the proof that $m = 1$, i.e., that a is fixed.

Note that a is not a multiple fixed point, since its multiplier is not 1. Thus, there is at least one other fixed point b. If b is superattracting, then it, too, is not a multiple fixed point, and hence there is a third fixed point c. On the other hand, if b is not superattracting, then b has another distinct preimage c. Either way, then, there are at least two distinct points $b, c \in \mathbb{P}^1(\mathbb{C}_v)$ not lying in U.

By Proposition 4.3, we may choose a small open Berkovich disk V_0 containing a such that $\phi(V_0) \subseteq V_0$, and such that all points of $V_0 \cap \mathbb{P}^1(\mathbb{C}_v)$ are attracted to a under iteration of ϕ. For each $n \geq 1$, define V_n to be the connected component of $\phi^{-n}(V_0)$ containing a. Then each V_n is an open connected Berkovich affinoid. Moreover,

$$V_0 \subseteq V_1 \subseteq \cdots,$$

and ϕ maps each V_n onto V_{n-1}.

Define $\tilde{U} := \bigcup_{n\geq 0} V_n$. Then

$$\phi(\tilde{U}) = \tilde{U}, \quad \text{and} \quad \tilde{U} \subseteq U.$$

Our principal goal is to show that in fact, $\tilde{U} = U$. We consider two cases.

Case 1. Suppose every V_n is a disk. Because there are at least two points $b, c \in \mathbb{P}^1(\mathbb{C}_v) \smallsetminus U$, the union $\tilde{U} = \bigcup V_n$ must also be a an open disk; see Exercise 6.18. After a change of coordinates, we may assume

$$a = 0 \quad \text{and} \quad \tilde{U} = D_{\text{an}}(0, r) \quad \text{for some } r > 0.$$

Each disk V_n is of the form $D_{\text{an}}(0, r_n)$, with $r_n \nearrow r$. Writing ϕ as a power series

$$\phi(z) = \sum_{n\geq 1} c_n z^n,$$

we have $|c_1| < 1$ (since 0 is an attracting fixed point) but

$$\max_{n\geq 1}\{|c_n| r^n\} = r$$

(since $\phi(\tilde{U}) = \tilde{U}$). Thus, the Weierstrass degree of ϕ on $D(0, r)$ must be some integer $\ell \geq 2$. The point $\zeta := \zeta(0, r)$ is therefore fixed, with $\deg_{\zeta, \vec{v}(0)}(\phi) = \ell$.

It follows that $\phi(\zeta) = \zeta$, with $\deg_\zeta(\phi) \geq \ell \geq 2$. By definition, then, ζ is a repelling fixed point, and hence, by Theorem 8.7, it is of Type II and belongs to $\mathcal{J}_{\phi,\text{an}}$. The disk $\tilde{U}$ is therefore a full connected component of the Fatou set, proving that $\tilde{U} = U$. All of the desired conclusions follow immediately from what we have just shown about $\tilde{U}$.

Case 2. Otherwise, some V_n is not a disk. If n is the smallest integer for which V_n is not a disk, then putting V_{n-1} in the role of V_0, we may assume that V_1 is not a disk. Let ζ_0 be the unique boundary point of V_0. Note that

$\phi(\overline{V}_0) \subseteq V_0$, or else V_1 would be a disk. In particular, $\phi(\zeta_0) \in V_0$. It follows inductively, by Theorem 7.8(c), that $\phi(\partial V_n) \subseteq V_n$ for all $n \geq 0$.

Step 1. We begin by claiming that $\tilde{U}$ is of Cantor type. To do so, we must show that for all $n \geq 0$, each component of $\mathbb{P}^1_{\text{an}} \smallsetminus V_n$ contains at least two points of ∂V_{n+1}. For $n = 0$, we know that V_1 is a connected affinoid but not a disk, and hence it has at least two boundary points. All of these boundary points map to ζ_0, by Theorem 7.8(c), and hence they lie outside V_0, as desired.

To finish the proof of the claim, we proceed by induction. Given that the claim holds for $n-1$, let W be a component of $\mathbb{P}^1_{\text{an}} \smallsetminus V_n$. Since V_n is an open connected Berkovich affinoid, W is a closed disk with unique boundary point ξ. Let $\zeta := \phi(\xi)$, which belongs to ∂V_{n-1}, again by Theorem 7.8(c). Let X be the component of $\mathbb{P}^1_{\text{an}} \smallsetminus V_{n-1}$ bounded by ζ. Note that $\zeta \in X \cap V_n$ (since $\phi(\partial V_n) \subseteq V_n$), and hence

$$Y := X \cap V_n$$

is a nonempty connected Berkovich affinoid (although it is neither open nor closed). Moreover, ∂Y consists of ζ and at least two points of ∂V_n, by our inductive hypothesis.

In addition, by Theorem 7.8, $\phi^{-1}(Y)$ has a connected component Y' whose boundary $\partial Y'$ consists of ξ and at least two boundary points of V_{n+1}. Since $Y \cap V_{n-1} = \varnothing$, we also have $Y' \cap V_n = \varnothing$, and hence $Y' \subseteq W$. Thus, the closed disk W contains $\partial Y'$ and therefore at least two boundary points of V_{n+1}, proving the claim.

Step 2. Next, we show that $\partial\tilde{U} \subseteq \mathcal{J}_{\phi,\text{an}}$. Given $\zeta \in \partial\tilde{U}$ and an open set W containing ζ, we must show that W is not dynamically stable. Because W is open, if ζ is of Type I or IV, then W contains some disk W' containing ζ; and if ζ is of Type II or III, then W contains some thin annulus W' extending from ζ in the direction of a. Let η be the unique boundary point of W' in the Type I/IV case, or the boundary point other than ζ in the Type II/III case.

Because $\tilde{U}$ is open, we have $\zeta \notin \tilde{U}$. Therefore, for each $n \geq 0$, there is a unique $\zeta_n \in \partial V_n$ lying between a and ζ. Moreover, because $\zeta \in \partial\tilde{U}$, we have $\lim_{n\to\infty} \zeta_n = \zeta$. Thus, $\zeta_n \in W'$ for all sufficiently large n. Fix a large enough n, and as we showed is possible in Step 1, choose two distinct points $\xi_1, \xi_2 \in \partial V_{n+1}$ lying in the component of $\mathbb{P}^1_{\text{an}} \smallsetminus V_n$ containing ζ_n. Since

$$\xi_1, \xi_2 \in V_{n+2} \subseteq \tilde{U},$$

it follows that ξ_1 and ξ_2 both lie in the Berkovich annulus (or punctured disk) with boundary points ζ_n and ζ. Hence, $\xi_1, \xi_2 \in W'$.

Let X_i be the component of $\mathbb{P}^1_{\text{an}} \smallsetminus V_{n+1}$ containing ξ_i, for $i = 1, 2$. Then X_1, X_2 are disjoint closed disks, and neither contains η, since $\eta \in V_n \subseteq V_{n+1}$. Hence, at most one of these disks can contain ζ. The other, which without loss is X_1, must therefore be contained in W'.

Recall that X_1 is a closed disk with boundary point $\xi_1 \in \partial V_{n+1}$, and hence $\phi^{n+1}(\xi_1)$ is ζ_0, the unique boundary point of V_0. Moreover, if $\boldsymbol{v}$ denotes the direction at ξ_1 containing a (i.e., the one direction at ξ_1 not in X_1), then applying $\phi_{\#}$ inductively, it is easy to see that $\phi_{\#}^{n+1}(\boldsymbol{v})$ is the direction at ζ_0 containing a. Applying Theorem 7.34 to the directions at ξ_1 extending into X_1, then, $\phi^{n+1}(X_1)$ is either $\mathbb{P}^1_{\text{an}} \smallsetminus V_0$ or else all of $\mathbb{P}^1_{\text{an}}$. Either way, since

$$X_1 \subseteq W' \subseteq W,$$

we have $\phi^{n+1}(W) \supseteq \mathbb{P}^1_{\text{an}} \smallsetminus V_0$, and hence $\phi^{n+2}(W) \supseteq \mathbb{P}^1_{\text{an}} \smallsetminus \phi(V_0)$.

Recall that $\phi(V_0)$ is a strictly smaller disk contained in V_0 (as we noted early in Case 2), and that all points of V_0 approach a under iteration. Thus, $\{\phi^n(V_0)\}_{n\geq 0}$ is a decreasing sequence of disks with

$$\bigcap_{n\geq 0} \phi^n(V_0) = \{a\}.$$

Since $\phi : \mathbb{P}^1_{\text{an}} \to \mathbb{P}^1_{\text{an}}$ is onto, it follows that

$$\bigcup_{n\geq 0} \phi^n(W) \supseteq \bigcup_{n\geq 0} \left(\mathbb{P}^1_{\text{an}} \smallsetminus \phi^n(V_0)\right) = \mathbb{P}^1_{\text{an}} \smallsetminus \{a\},$$

and therefore W is not dynamically stable.

Step 3. Since $\partial\tilde{U} \subseteq \mathcal{J}_{\phi,\text{an}}$, we have shown that $\tilde{U}$ is indeed the full Fatou component containing a, and that $\tilde{U} = U$. We leave the fact that ∂U is homeomorphic to the Cantor set to the reader; see Exercise 9.2. It remains to show that $\phi : U \to U$ is ℓ-to-1, for some $\ell \geq 2$. By Theorem 9.1(b), we need only show that $\phi : U \to U$ is not one-to-one.

By Theorem 7.8(c), each of the (at least two) boundary points of V_1 maps to the unique boundary point ζ_0 of V_0. Therefore, since

$$\partial V_0 \cup \partial V_1 \subseteq V_2 \subseteq U,$$

ϕ is not injective on U, as desired. □

The boundary of an attracting domain of Cantor type generally includes both preperiodic and wandering points. It can include points of each Berkovich type as well. For example, the boundary of the attracting component at ∞ for the quadratic polynomial of Example 8.16 consists solely of Type I points. On the other hand, the boundary of the attracting component at ∞ for the cubic polynomial of Example 8.18 consists of both Types I and II points. There can also be Type IV points in the boundary of an attracting domain. This phenomenon was first noted by Rivera-Letelier in [**114**, Exemple 6.3]; see Exercise 9.1. The only known way for Type III points to appear in the boundary is in the presence of certain wandering domains; see Remark 11.18.

9.3. Indifferent components

We now turn to the second of the two types of periodic components.

Definition 9.6. Let $\phi \in \mathbb{C}_v(z)$ be a rational function of degree $d \geq 2$. We define the *indifference domain* of ϕ to be the set of points $\zeta \in \mathbb{P}^1_{\text{an}}$ for which there is an open connected affinoid W containing ζ and an integer $m \geq 1$ such that ϕ^m maps W bijectively onto itself.

A connected component of the indifference domain is called an *indifferent component* for ϕ.

The indifference domain is clearly an open subset of the Berkovich Fatou set, but again we can say much more. The following result first appeared for $\mathbb{C}_p$ as [**114**, Théorème 3] and [**116**, Proposition 6.1].

Theorem 9.7 (Rivera-Letelier). *Let $\phi \in \mathbb{C}_v(z)$ be a rational function of degree $d \geq 2$ with Berkovich Fatou set $\mathcal{F}_{\phi,\text{an}}$, and let $U \subseteq \mathbb{P}^1_{\text{an}}$ be an indifferent component for ϕ. Then the following hold.*

(a) *U is a periodic connected component of $\mathcal{F}_{\phi,\text{an}}$, of some minimal period $m \geq 1$.*

(b) *U is a rational open connected Berkovich affinoid.*

(c) *Each of the finitely many points of the boundary ∂U is a Type II point lying in the Berkovich Julia set.*

(d) *ϕ^m permutes the boundary points of U; in particular, each is periodic.*

(e) *The mapping $\phi^m : U \to U$ is one-to-one and onto.*

The proof requires two technical lemmas.

Lemma 9.8. *Let $a \in \mathbb{C}_v$, let $r > 0$, and let $\psi \in \mathbb{C}_v[[z-a]]$ be a convergent power series on $D(a,r)$ with Weierstrass degree 1 and image $\psi(D(a,r)) = D(a,r)$. For any $b \in D(a,r)$ and $0 < s < r$, either*

(a) *$\zeta(b,s)$ is periodic under ψ, or*

(b) *there is some $t \in |\mathbb{C}_v^\times| \cap (s,r)$ such that $\zeta(b,t)$ is periodic under ψ, and there are infinitely many directions at $\zeta(b,t)$ that contain an iterate $\psi^n(b)$.*

Remark 9.9. Conclusion (b) of Lemma 9.8 cannot occur if the residue field of $\mathbb{C}_v$ is algebraic over a finite field. In that case, in fact, conclusion (a) of the lemma holds for bijective self-maps of an arbitrary connected affinoid, not just of a disk; see Exercises 9.3 and 9.6.

Proof of Lemma 9.8.

Step 1. After changing coordinates by translating, we may assume that $a = b = 0$. Write

$$\psi(z) = \sum_{i\geq 0} c_i z^i \quad \text{with} \quad c_i \in \mathbb{C}_v.$$

Then by the hypothesis that $\psi : D(0,r) \to D(0,r)$ is bijective, we have

$$|c_0| < r, \quad |c_1| = 1, \quad \text{and} \quad |c_i| \leq r^{1-i} \quad \text{for all } i \geq 2.$$

If $|c_0| \leq s$, then $\psi(\zeta(0,s)) = \zeta(0,s)$, and we are done. Thus, we may assume for the remainder of the proof that $|c_0| > s$.

Writing $P(z) := \psi(z) - c_0 - c_1 z = c_2 z^2 + c_3 z^3 + \cdots$, we have

$$|P(x)| \leq \frac{|c_0|^2}{r} \quad \text{and} \quad |P'(x)| \leq \frac{|c_0|}{r} \quad \text{for all } x \in \overline{D}(0,|c_0|), \tag{9.1}$$

since $|c_0|/r \leq 1$. In addition, a simple induction shows that

$$\psi^n(z) = c_0\big(1 + c_1 + \cdots + c_1^{n-1}\big) + c_1^n z + \sum_{i=0}^{n-1} c_1^{n-1-i} P(\psi^i(z)) \tag{9.2}$$

for all $n \geq 0$. Combined with (9.1), then, it follows that $|\psi^n(0)| \leq |c_0|$, and furthermore that

$$\Big|\psi^n(0) - c_0\big(1 + c_1 + \cdots + c_1^{n-1}\big)\Big| \leq \frac{|c_0|^2}{r} \quad \text{for all } n \geq 0. \tag{9.3}$$

In addition, differentiating equation (9.2), and applying both inequality (9.1) and the fact that $|\psi'(x)| = 1$ for all $x \in \overline{D}(0,r)$, we have

$$\Big|(\psi^n)'(0) - c_1^n\Big| \leq \frac{|c_0|}{r}.$$

Therefore,

$$\Big|(\psi^n)'(0) - 1\Big| \leq \max\left\{\frac{|c_0|}{r}, |c_1^n - 1|\right\} \quad \text{for all } n \geq 0. \tag{9.4}$$

Step 2. If $|c_1^n - 1| = 1$ for all $n \geq 1$, then $|c_1^m - c_1^n| = 1$ for all $m \neq n$, and hence any two of the values

$$c_0\big(1 + c_1 + \cdots + c_1^{n-1}\big) = \frac{c_0(c_1^n - 1)}{c_1 - 1}$$

are at distance $|c_0|$ from one another. Thus, choosing $t = |c_0| \in (s,r)$, we have $\psi(\zeta(0,t)) = \zeta(0,t)$, and the points $\{\psi^n(0)\}_{n\geq 0}$ lie in infinitely many distinct directions at $\zeta(0,t)$, as desired.

Therefore, we may assume that $|c_1^m - 1| < 1$ for some $m \geq 1$. Replacing ψ by ψ^m, we may in fact assume that $|c_1 - 1| < 1$. In particular, by inequality (9.3), we have $|\psi^n(0) - nc_0| < |c_0|$ for all $n \geq 1$. If $\mathbb{C}_v$ has residue characteristic 0, then again the iterates of 0 lie in infinitely many distinct directions at the fixed point $\zeta(0, |c_0|)$, and we are done.

Step 3. We are left with one remaining case: that $|c_1 - 1| < 1$, and that $\mathbb{C}_v$ has positive residue characteristic $p \geq 2$. Define

$$\alpha := \max\left\{\frac{|c_0|}{r}, |c_1 - 1|\right\} < 1 \quad \text{and} \quad \beta := \max\left\{\frac{|c_0|}{r}, |p|, \alpha^{p-1}\right\} < 1.$$

Recall from Exercise 2.16(a) that

$$\left|1 + c_1 + \cdots + c_1^{p-1}\right| \leq \max\left\{|p|, |c_1 - 1|^{p-1}\right\}.$$

Thus, inequalities (9.3) and (9.4) yield

$$\left|\psi^p(0)\right| \leq \beta|c_0| = \beta|\psi(0)| \quad \text{and} \quad \left|\left(\psi^p\right)'(0) - 1\right| \leq \alpha.$$

In particular, note that the bounds α and β associated with ψ^p, and hence inductively for ψ^{p^j}, are no bigger than those associated with ψ. Thus, choosing j large enough that $\beta^j|c_0| \leq s$, we see that

$$\psi^{p^j}\left(\zeta(0, s)\right) = \zeta(0, s). \qquad \square$$

The next lemma is a variant of the Lemme d'Approximation of [**114**, Section 5]. Its conclusions are designed to mesh with the hypotheses of Lemma 9.8.

Lemma 9.10 (Rivera-Letelier's approximation lemma). *Let $b, c \in \mathbb{C}_v$ and $r > s > 0$. Let $\phi \in \mathbb{C}_v(z)$ be a rational function that maps the open Berkovich annulus*

$$\{s < |\zeta - b| < r\} \quad \textit{bijectively onto} \quad \{s < |\zeta - c| < r\}$$

with $\phi(\zeta(b, r)) = \zeta(c, r)$. Then there is a rational function $\psi \in \mathbb{C}_v(z)$ such that

(a) *ψ maps $D_{\text{an}}(b, r)$ onto $D_{\text{an}}(c, r)$ with Weierstrass degree 1, and*
(b) *$\psi(\xi) = \phi(\xi)$ for any $\xi \in D_{\text{an}}(b, r)$ with $\operatorname{diam}(\xi) \geq s$.*

Remark 9.11. Lemma 9.10 holds not just for annuli but for arbitrary open connected affinoids; see Exercise 9.5. To prove Theorem 9.7, however, annuli are enough.

Proof of Lemma 9.10. After changing coordinates on the domain and range, we may assume that $b = c = 0$. Let $U := \{s < |\zeta| < r\}$.

Write $\phi = f/g$ with $f, g \in \mathbb{C}_v[z]$. Since $\phi(U) = U$, neither f nor g has any zeros in U. In light of Theorem 3.33, ϕ maps U to itself with inner and outer Weierstrass degrees equal to 1; moreover, $|\phi(x)| = |x|$ for all $x \in U$, so that $|f(x)| = |x| \cdot |g(x)|$ for all $x \in U$. Thus, setting

$$m := \operatorname{wdeg}_{D(0,r)}(g) = \operatorname{wdeg}_{\overline{D}(0,s)}(g)$$

to be the number of zeros of g in $\overline{D}(0,s)$, we see that f has

$$\operatorname{wdeg}_{\overline{D}(0,s)}(f) = m+1$$

zeros in $\overline{D}(0,s)$, counting multiplicity. After another change of coordinates on the domain (specifically, translating by a root α of f with $|\alpha| \leq s$), we may assume that $f(0) = 0$.

Thus, we may write

$$f(z) = zf_1(z)f_2(z) \quad \text{and} \quad g(z) = g_1(z)g_2(z),$$

where $f_1, g_1 \in \mathbb{C}_v[z]$ are monic of degree m with all their roots in $\overline{D}(0,s)$, and $f_2, g_2 \in \mathbb{C}_v[z]$ have all their roots outside $D(0,r)$. More precisely, writing

$$f_1(z) = \prod_{i=1}^{m}(z-\alpha_i) \quad \text{and} \quad g_1(z) = \prod_{i=1}^{m}(z-\beta_i)$$

with $|\alpha_i|, |\beta_i| \leq s$, we have

$$\frac{g_1(z)}{f_1(z)} = \prod_{i=1}^{m} \frac{z-\beta_i}{z-\alpha_i} = \prod_{i=1}^{m}\left(1 + \frac{\alpha_i - \beta_i}{z-\alpha_i}\right),$$

and hence

$$\left|\frac{g_1(x)}{f_1(x)} - 1\right| \leq \frac{s}{|x|} \quad \text{for all } x \in U. \tag{9.5}$$

Define

$$\psi(z) := \frac{zf_2(z)}{g_2(z)} = \phi(z) \cdot \frac{g_1(z)}{f_1(z)},$$

which has no poles and exactly one zero in $D(0,r)$. Thus, ψ is a convergent power series on $D(0,r)$ of Weierstrass degree 1. In addition, since $|f_1(x)| = |g_1(x)|$ for all $x \in U$, we have

$$|\psi(x)| = |\phi(x)| = |x| \quad \text{for all } x \in U,$$

and hence ψ maps $D_{\mathrm{an}}(0,r)$ onto itself with Weierstrass degree 1, proving conclusion (a).

In fact, by inequality (9.5), we have

$$\big|\psi(x) - \phi(x)\big| = |x| \cdot \left|\frac{g_1(x)}{f_1(x)} - 1\right| \leq s \quad \text{for all } x \in U.$$

For any $\xi = \zeta(a,t)$ in $D_{\mathrm{an}}(0,r)$ with $t > s$, then, and for any annulus V of the form $V = \{\lambda t < |\zeta - a| < t\}$ with $\lambda < 1$ sufficiently close to 1, the images $\phi(V)$ and $\psi(V)$ must coincide. As a result, $\phi(\xi) = \psi(\xi)$. By continuity, this equality also extends to $\zeta(a,s)$ and to Type IV points of diameter at least s, proving conclusion (b). □

Proof of Theorem 9.7. It suffices to show parts (b), (c), and (e). Part (a) is then immediate from Definition 9.6, and part (d) is immediate from Theorem 7.8(c).

By hypothesis, there is an open connected Berkovich affinoid $U_0 \subseteq U$ such that $\phi^m : U_0 \to U_0$ is bijective. Replacing ϕ by ϕ^m, we may assume that $m = 1$. In addition, by Theorem 7.8(c), ϕ must permute the finitely many points of ∂U_0; replacing ϕ by an iterate again, we may assume that ϕ fixes all the boundary points of U_0. After a change of coordinates, we may assume that $\infty \in U_0$, so that

$$U_0 = \mathbb{P}^1_{\text{an}} \smallsetminus (D_1 \cup \cdots \cup D_q),$$

where $D_1, \dots, D_q \subseteq \mathbb{A}^1_{\text{an}}$ are disjoint closed disks.

For the remainder of the proof, fix a disk $D \in \{D_1, \dots, D_q\}$, and write it as $D = \overline{D}_{\text{an}}(a,r)$. By the previous paragraph, ϕ fixes the point $\zeta(a,r)$. Moreover, if $\boldsymbol{w}$ denotes the direction at $\zeta(a,r)$ containing ∞, then because $\phi(U_0) = U_0$, we have $\phi_{\#}(\boldsymbol{w}) = \boldsymbol{w}$. We need to show that $D \cap U$ is simply D with finitely many rational closed disks removed, that its boundary points, other than $\zeta(a,r)$, are Julia periodic points, and that ϕ maps $D \cap U$ bijectively onto itself. We do this below by inductively constructing connected affinoids

$$U_0 \subseteq U_1 \subseteq \cdots \subseteq U_N$$

extending further and further into D. Crucially, we set the number N at the start of the process, so that it necessarily stops after finitely many steps. After one last tweak to construct $U_N' \supseteq U_N$, we will show that $U_N' \cap D = U \cap D$.

Step 1. Let V be the connected component of the injectivity locus $\text{Inj}(\phi)$ containing ∞. Thus, $V \supseteq U \supseteq U_0$, with $\deg_\zeta(\phi) = 1$ for all $\zeta \in V$. Moreover, by Theorem 7.41, V is an open connected affinoid. Hence,

$$V \cap D = D \smallsetminus \big(B_1 \cup \cdots \cup B_\ell\big),$$

where $B_1, \dots, B_\ell \subseteq D$ are disjoint closed disks. Write $B_i = \overline{D}_{\text{an}}(b_i, s_i)$, and let

$$\mathcal{S} := \{r\} \cup \{s_i : 1 \le i \le \ell\} \cup \big\{|b_i - b_j| : 1 \le i < j \le \ell\big\},$$

which is a finite subset of $(0, \infty)$. Denote the elements of $\mathcal{S}$ by

$$r = S_0 > S_1 > S_2 > \cdots > S_N.$$

In the next step, we inductively construct the affinoids $U_0 \subseteq \cdots \subseteq U_N$ we promised above, satisfying the following properties:

(a) $U_i \subseteq V \cap (U_0 \cup D)$;

(b) ϕ maps U_i bijectively onto itself;

(c) the finitely many points $\zeta \in D \cap \partial U_i$ are permuted by ϕ, with each satisfying $\mathrm{diam}(\zeta) \geq S_i$; and

(d) any $\zeta \in D \cap \partial U_i$ with $\mathrm{diam}(\zeta) > S_i$ is a Julia periodic point.

Our affinoid U_0 obviously satisfies conditions (a) and (b). In addition, $D \cap \partial U_0 = \{\zeta(a, r)\}$, and $\mathrm{diam}(\zeta(a, r)) = r = S_0$. Thus, conditions (c) and (d) are also trivially satisfied.

Step 2. Suppose we already have U_{i-1} for some $1 \leq i \leq N$; let us construct U_i. Let Z be the finite set

$$Z := D \cap \mathcal{F}_{\phi,\mathrm{an}} \cap \partial U_{i-1}.$$

By properties (c) and (d) of Step 1 for U_{i-1}, each point $\zeta \in Z$ is periodic, with $\mathrm{diam}(\zeta) = S_{i-1}$. In fact, since ϕ maps $\mathcal{F}_{\phi,\mathrm{an}}$ into itself, ϕ must permute the points of Z. For each $\zeta \in Z$, let n be the corresponding minimal period. By Theorem 8.7 and the fact that $\zeta \in \mathcal{F}_{\phi,\mathrm{an}}$, we have $\deg_\zeta(\phi^n) = 1$, and hence $\deg_\zeta(\phi) = 1$. Thus, each $\zeta \in Z$ belongs to $\mathrm{Inj}(\phi)$, and therefore $Z \subseteq V$.

Moreover, if $\zeta \in Z$ is of Type II, then each of the bad directions of ϕ^n at ζ (in the sense of Theorem 7.34) is periodic under $\phi^n_\#$. Since $\deg_\zeta(\phi^n) = 1$, each good residue class maps bijectively onto a good residue class under ϕ^n. Therefore, each such residue class and all of its forward iterates under ϕ (not just under ϕ^n) are contained in the component V of $\mathrm{Inj}(\phi)$.

For each $\zeta \in Z$, every residue class $\boldsymbol{v}$ of ζ (other than the residue class $\boldsymbol{w}$ containing ∞) may be written as $D_{\mathrm{an}}(b_{\boldsymbol{v}}, S_{i-1})$ for some Type I point $b_{\boldsymbol{v}} \in \boldsymbol{v}$. If $\boldsymbol{v}$ is not contained in the component V of $\mathrm{Inj}(\phi)$, we may assume that $b_{\boldsymbol{v}} \notin V$; otherwise, our choice of $b_\zeta \in \boldsymbol{v}$ is arbitrary. By definition of our set $\mathcal{S}$ of radii, then, the annulus

$$\left\{S_i < \left|z - b_\zeta\right| < S_{i-1}\right\} = D_{\mathrm{an}}(b_\zeta, S_{i-1}) \smallsetminus \overline{D}_{\mathrm{an}}(b_\zeta, S_i)$$

is contained in V. In addition, $\phi(\zeta)$ also belongs to Z, and therefore

$$\phi_\#(\boldsymbol{v}) = \zeta\big(b_{\phi_\#(\boldsymbol{v})}, S_{i-1}\big).$$

By Lemma 9.10, there is a rational function $\psi_{\boldsymbol{v}} \in \mathbb{C}_v(z)$ such that $\psi_{\boldsymbol{v}}$ maps $\boldsymbol{v} = D_{\mathrm{an}}(b_\zeta, S_{i-1})$ onto $\phi_\#(\boldsymbol{v}) = D_{\mathrm{an}}(b_{\phi_\#(\boldsymbol{v})}, S_{i-1})$ with Weierstrass degree 1, and such that for any $\xi \in \boldsymbol{v}$ with $\mathrm{diam}(\xi) \geq S_i$, we have $\phi(\xi) = \psi_{\boldsymbol{v}}(\xi)$. Because $\mathrm{wdeg}_{\boldsymbol{v}}(\psi_{\boldsymbol{v}}) = 1$, we also have

$$\mathrm{diam}\left(\psi_{\boldsymbol{v}}(\xi)\right) = \mathrm{diam}(\xi) \quad \text{for any} \quad \xi \in \boldsymbol{v} = D_{\mathrm{an}}\big(b_{\boldsymbol{v}}, S_{i-1}\big).$$

Thus, for every $\zeta \in Z$ with minimal period n and each ϕ^n-bad direction $\boldsymbol{v}$ at ζ, let m be the minimal period of $\boldsymbol{v}$ under $\phi^n_{\#}$, and define

$$\psi := \psi_{\phi_{\#}^{mn-1}(\boldsymbol{v})} \circ \cdots \circ \psi_{\phi_{\#}(\boldsymbol{v})} \circ \psi_{\boldsymbol{v}} \in \mathbb{C}_v(z).$$

Then ψ maps $D_{\text{an}}(b_{\boldsymbol{v}}, S_{i-1})$ onto itself with Weierstrass degree 1, and

$$\psi(\xi) = \phi^{mn}(\xi) \quad \text{for any} \quad \xi \in D_{\text{an}}(b_{\boldsymbol{v}}, S_{i-1}) \quad \text{with} \quad \operatorname{diam}(\xi) \geq S_i.$$

By Lemma 9.8, there is a (minimal) radius $t_{\boldsymbol{v}} \in [S_i, S_{i-1})$ such that the point $\zeta(b_{\boldsymbol{v}}, t_{\boldsymbol{v}})$ is periodic under ψ, and hence under ϕ. Moreover, if $t_{\boldsymbol{v}} > S_i$, then $t_{\boldsymbol{v}} \in |\mathbb{C}_v^{\times}|$, and the direction $\boldsymbol{u}$ at $\zeta(b_{\boldsymbol{v}}, t_{\boldsymbol{v}})$ containing b_ζ has infinite forward orbit under $\phi^{mn}_{\#}$. In addition, each of the finitely many disks making up

$$(D \cap U_{i-1}) \smallsetminus V$$

is contained in a disk $\overline{D}_{\text{an}}(b_{\boldsymbol{v}}, t_{\boldsymbol{v}})$ for some ϕ^n-bad direction $\boldsymbol{v}$ at some $\zeta \in Z$.

Let Z' be the (finite) set of all points $\zeta(b_{\boldsymbol{v}}, t_{\boldsymbol{v}})$ constructed in this way, and define

$$Y_1 := \{\phi^j(\xi) : \xi \in Z' \text{ and } j \geq 0\}, \quad \text{and}$$
$$Y_2 := D \cap \mathcal{J}_{\phi,\text{an}} \cap \partial U_{i-1} = (D \cap \partial U_{i-1}) \smallsetminus Z.$$

Observe that Y_1 is a finite set, because each $\xi \in Z'$ is periodic under ϕ. We also know that Y_2 is finite, and that every $\zeta \in Y_2$ satisfies

$$\operatorname{diam}(\zeta) \geq S_{i-1} > S_i$$

and is a Julia periodic point. Finally, define $Y := Y_1 \cup Y_2$, and let U_i be the connected component of $(U_0 \cup D) \smallsetminus Y$ containing U_0.

By construction, $U_i \supseteq U_{i-1}$ satisfies properties (a) and (b) of Step 1. Its boundary points in D belong to the finite set Y, and hence U_i is indeed an open connected affinoid. In addition, the points of Y are all periodic and have diameter at least S_i, so that U_i also has property (c) of Step 1. It remains to show that U_i satisfies property (d) of Step 1.

Any point $\xi \in D \cap \partial U_i$ with $\operatorname{diam}(\xi) > S_i$ either belongs to Y_2, in which case we already know it is Julia, or it is a Type II point of the form $\phi^j(\zeta)$ for some $\zeta = \zeta(b_{\boldsymbol{v}}, t_{\boldsymbol{v}}) \in Z'$ with $t_{\boldsymbol{v}} > S_i$. In the latter case, we only need to show that $\zeta \in \mathcal{J}_{\phi,\text{an}}$; hence we may assume that $j = 0$, so that $\xi = \zeta(b_{\boldsymbol{v}}, t_{\boldsymbol{v}})$. Let M be the minimal period of ξ under ϕ. If $\deg_\xi(\phi^M) \geq 2$, then ξ is repelling and hence Julia, leaving us with the case that ξ is indifferent.

By definition of Z' and our choice of $b_{\boldsymbol{v}}$, we have $b_{\boldsymbol{v}} \notin V$. That is, setting $\boldsymbol{u}$ to be the direction at ξ containing $b_{\boldsymbol{v}}$, ϕ cannot be one-to-one on $\boldsymbol{u}$. Since $\deg_\xi(\phi^M) = 1$, $\boldsymbol{u}$ is one of the finitely many bad directions at ξ from Theorem 7.34. On the other hand, by construction, $\boldsymbol{u}$ has infinite forward orbit under $(\phi^M)_{\#}$. By Theorem 8.7(b), then, ξ is an indifferent

Julia periodic point, verifying property (d) of Step 1, and completing our induction.

Step 3. Stop the inductive process of Step 2 once we have constructed U_N. Each $\zeta \in D \cap \partial U_N$ either is a Julia periodic point or else is an indifferent Fatou periodic point $\zeta = \zeta(b, S_N)$ of some period n and of diameter exactly S_N. In the latter case, by Exercise 9.7, we must have $N = 0$ and $D \subseteq V$; thus, ϕ maps D bijectively onto itself. In that case, define $U_0' := U_0 \cup D$. Otherwise, define $U_N' := U_N$. Either way, U_N' is a connected affinoid, ϕ maps U_N' onto itself with degree 1, and all points of $D \cap \partial U_N'$ are periodic and in the Julia set.

Repeating this process for each of the disks D among $D_1, \ldots, D_q$, we are left with an open connected Berkovich affinoid U, with ϕ mapping U one-to-one onto itself and permuting the boundary points, all of which lie in the Julia set. Thus, U is a component of the Fatou set. Finally, by Theorem 8.7, the finitely many boundary points of U are not of Type III, because they are periodic and Julia. Hence, U is indeed a *rational* open connected Berkovich affinoid. □

One key feature of the proof of Theorem 9.7 is the definition in Step 1 of the *finite* set $\mathcal{S}$ of radii at which something interesting might happen. Thus, the inductive construction of Step 2 stops after finitely many steps, forcing U to be only a connected affinoid, and not, say, a domain of Cantor type.

Example 9.12. Fix $a, b \in \mathbb{C}_v$ with $0 < |b| < |a| = 1$, and define

$$\phi(z) = az + \frac{b^2}{z-1} = \frac{az^2 - az + b^2}{z-1}.$$

Note that ∞ is an indifferent fixed point, with multiplier $1/a$. Setting

$$U_0 := \mathbb{P}^1_{\text{an}} \smallsetminus \overline{D}_{\text{an}}(0,1),$$

note also that ϕ maps U_0 bijectively onto itself. We use the method of the proof of Theorem 9.7 to compute the indifferent component $U \supseteq U_0$ for ϕ containing ∞.

Observe that

$$b^{-1}\big[\phi(bz+1) - a\big] = az + \frac{1}{z} = \frac{az^2+1}{z},$$

which has reduction $(\overline{a}z^2 + 1)/z$ of degree 2. Hence

$$\phi\big(\zeta(1,|b|)\big) = \zeta\big(a,|b|\big) \quad \text{with} \quad \deg_{\zeta(1,|b|)}(\phi) = 2.$$

In addition, ϕ maps

$$V := \mathbb{P}^1_{\text{an}} \smallsetminus \overline{D}_{\text{an}}(1,|b|)$$

bijectively onto $\phi(V) = \mathbb{P}^1_{\text{an}} \smallsetminus \overline{D}_{\text{an}}(a,|b|)$. Thus, V is the component of the injectivity locus $\text{Inj}(\phi)$ containing ∞.

In the notation of the proof of Theorem 9.7, we have

$$U_0 = \mathbb{P}^1_{\mathrm{an}} \smallsetminus D, \quad \text{where} \quad D := \overline{D}_{\mathrm{an}}(0,1).$$

In addition,

$$V \cap D = D \smallsetminus B_1, \quad \text{where} \quad B_1 = \overline{D}_{\mathrm{an}}\big(1, |b|\big).$$

Thus, the set $\mathcal{S}$ of radii is $\mathcal{S} = \{1, |b|\}$. In particular, we have $N = 1$.

To construct U_1 from U_0, note that the one boundary point of U_0 is the Gauss point $\zeta(0,1)$. Since $\overline{\phi}(z) = az$, the Gauss point is an indifferent fixed point, with one bad residue class in the direction of 1. Moreover, the approximating rational function $\psi(z) = az$ satisfies the conclusions of Lemma 9.10 for the annulus $\{|b| < |\zeta - 1| < 1\}$. There are four possibilities.

First, if $\overline{a}$ has infinite order in the multiplicative group $k^\times$ of the residue field, then $\zeta(0,1) \in \mathcal{J}_{\phi,\mathrm{an}}$, and the inductive process stops immediately. We set $U := U_0$, which is the desired indifferent component.

Otherwise, we have $|a^m - 1| < 1$ for some minimal integer $m \geq 1$. In that case, the bad residue class at $\zeta(0,1)$ is periodic of period m, and $\phi^n(\zeta(1,|b|)) = \zeta(a^n, |b|)$ for all $n \geq 0$; see Exercise 9.8.

Second, then, suppose $|a^m - 1| \leq |b|$. In that case, $\zeta(1,|b|)$ is a repelling periodic point of period m, and

$$U := \mathbb{P}^1_{\mathrm{an}} \smallsetminus \bigcup_{i=0}^{m-1} \overline{D}_{\mathrm{an}}\big(a^i, |b|\big)$$

is the indifferent component.

Third, if $|b| < |a^m - 1|$ and the residue characteristic $\operatorname{char} k$ is zero, then as in Step 2 of the proof of Lemma 9.8, we have

$$\big|a^{mn} - a^{m\ell}\big| = \big|a^m - 1\big| \quad \text{for all } n > \ell \geq 0.$$

It follows that $\zeta(1, |a^m - 1|)$ is an indifferent Julia periodic point of period m, and hence

$$U := \mathbb{P}^1_{\mathrm{an}} \smallsetminus \bigcup_{i=0}^{m-1} \overline{D}_{\mathrm{an}}\big(a^i, |a^m - 1|\big)$$

is the indifferent component.

Fourth, if $|b| < |a^m - 1|$ and $\operatorname{char} k = p > 0$, then as in Step 3 of the proof of Lemma 9.8, there is some minimal $j \geq 1$ such that $\zeta(1,|b|)$ is a repelling periodic point of period mp^j; see Exercises 9.9 and 9.10 for further details. Thus,

$$U := \mathbb{P}^1_{\mathrm{an}} \smallsetminus \bigcup_{i=0}^{mp^j-1} \overline{D}_{\mathrm{an}}\big(a^i, |b|\big)$$

is the indifferent component.

Rivera-Letelier demonstrated in [**114**, Propositions 6.4,6.7] that *any* rational open connected affinoid (other than the trivial $\mathbb{P}^1_{\mathrm{an}}$ and $\varnothing$) can be an indifferent component; see Exercises 9.12–9.15. By contrast, over $\mathbb{C}$, an indifferent component is either a Siegel disk or a Herman ring (i.e., annulus). The non-archimedean setting is therefore a bit more flexible than the complex, allowing indifferent components with arbitrarily many holes.

9.4. Rivera-Letelier's classification

As their names suggest, attracting periodic points lie in attracting components of the Fatou set, and (Fatou) indifferent periodic points lie in indifferent components, as the following result makes precise.

Proposition 9.13. *Let $\phi \in \mathbb{C}_v(z)$ be a rational function of degree $d \geq 2$ with indifference domain $\mathfrak{I}$. Then $\mathfrak{I}$ contains*

- *every Type I indifferent periodic point,*
- *every Type II Fatou periodic point,*
- *every Type III periodic point, and*
- *every Type IV periodic point.*

Moreover, $\mathfrak{I}$ does not contain any other periodic points.

Proof. Let $\zeta \in \mathbb{P}^1_{\mathrm{an}}$ be a periodic point. Replacing ϕ by ϕ^m if necessary, we may assume that ζ is fixed.

If ζ is a Type I indifferent fixed point, then by Proposition 4.3(c), there is an open disk U containing ζ such that $\phi : U \to U$ is bijective.

If ζ is a Type II Fatou periodic point, then by Theorem 8.7(b), there is an open Berkovich affinoid U containing ζ such that $\phi : U \to U$ is bijective.

If ζ is a Type III or IV periodic point, then by Theorem 8.7(a), ζ is indifferent and Fatou. Indeed, since ϕ fixes each direction at ζ with degree 1, there is an open annulus (if ζ is of Type III) or disk (if ζ is of Type IV) containing ζ such that $\phi : U \to U$ is bijective.

If $\zeta \in \mathcal{J}_{\phi,\mathrm{an}}$, then $\zeta \notin \mathfrak{I}$, since $\mathfrak{I} \subseteq \mathcal{F}_{\phi,\mathrm{an}}$, as we have noted.

By Theorems 5.14 and 8.7, the only remaining case is that ζ is a Type I attracting fixed point. Let U be the component of $\mathcal{F}_{\phi,\mathrm{an}}$ containing ζ. By Theorem 9.5, U is the immediate attracting basin of ζ, and ϕ maps U onto itself with some degree $\ell \geq 2$. If ζ were also in $\mathfrak{I}$, then by Theorem 9.7, U would be the connected component of $\mathfrak{I}$ containing ζ, and ϕ maps U onto itself with degree 1, a contradiction. Thus, $\zeta \notin \mathfrak{I}$. □

One central result of Rivera-Letelier's thesis [**114**] and following work [**116**] is that the two types of dynamics we saw in Sections 9.2 and 9.3 are the only possibilities on a periodic Berkovich Fatou component.

Theorem 9.14 (Rivera-Letelier's classification). *Let $\phi \in \mathbb{C}_v(z)$ be a rational function of degree $d \geq 2$ with Berkovich Fatou set $\mathcal{F}_{\phi,\mathrm{an}}$, and let U be a periodic connected component of $\mathcal{F}_{\phi,\mathrm{an}}$. Then U is either an indifferent component or an attracting component, but not both.*

Theorem 9.14 parallels the situation in complex dynamics, where two kinds of domains are drawn to a periodic point (attracting and parabolic), and two more are analogous to our indifferent components (Siegel disks and Herman rings). Rivera-Letelier phrases Theorem 9.14 as the *trichotomy* of non-archimedean dynamics: every component of the Berkovich Fatou set is either preperiodic to an indifferent periodic component, preperiodic to an attracting periodic component, or wandering; and the three cases are mutually exclusive. We will discuss wandering domains in Chapter 11.

To prove Theorem 9.14, we need some lemmas.

Lemma 9.15. *Let $\phi \in \mathbb{C}_v(z)$, let $X \subseteq \mathbb{P}^1_{\mathrm{an}}$ be a connected set, let $a \in \mathbb{C}_v$, and let $\{s_n\}_{n\geq 1}$ be a strictly increasing sequence of positive real numbers with $s_n \nearrow s \in \mathbb{R}$. Suppose that for all $n \geq 1$, we have*

- $\zeta(a, s_n) \in X$, *and*
- $\phi(\zeta(a, s_n)) \in \{s_n \leq |\zeta - a| < s\} = D_{\mathrm{an}}(a, s) \smallsetminus D_{\mathrm{an}}(a, s_n)$.

Then X contains a Type II fixed point.

Proof. Clearly, $\lim_{n\to\infty} \zeta(a, s_n) = \zeta(a, s)$. Moreover, any open affinoid containing $\zeta(a, s)$ must also contain all but finitely many of the images $\phi(\zeta(a, s_n))$. Hence,

$$\lim_{n\to\infty} \phi\big(\zeta(a, s_n)\big) = \zeta(a, s).$$

By the continuity of ϕ, then, we have $\phi(\zeta(a, s)) = \zeta(a, s)$. In addition, because each $\zeta(a, s_n)$ belongs to X, and because X is connected, the interval $[\zeta(a, s_1), \zeta(a, s))$ is contained in X.

Let $\boldsymbol{v} = D_{\mathrm{an}}(a, s)$ be the direction at $\zeta(a, s)$ containing a. Note that $\phi_\#(\boldsymbol{v}) = \boldsymbol{v}$, since $\phi(\zeta(a, s_n)) \in D_{\mathrm{an}}(a, s)$ for all n. We claim that

$$\deg_{\zeta(a,s),\boldsymbol{v}}(\phi) = 1. \tag{9.6}$$

To prove the claim, let $m := \deg_{\zeta(a,s),\boldsymbol{v}}(\phi) \geq 1$. By Theorem 7.12, we have

$$\phi\big(\zeta(a, s_n)\big) = \zeta\big(a, s_n^m s^{1-m}\big) \quad \text{for all sufficiently large } n.$$

If $m \geq 2$, then $s_n^m s^{1-m} = (s_n/s)^{1-m} s_n < s_n$. Thus, $\phi(\zeta(a, s_n)) \in D_{\mathrm{an}}(a, s_n)$ for sufficiently large n, contradicting the hypotheses and hence proving equation (9.6), as claimed.

Therefore, we have $\phi(\zeta(a,t)) = \zeta(a,t)$ for all sufficiently large $t < s$. In particular, we may choose $t \in |\mathbb{C}_v^\times|$ with this property, and with $t > s_1$. The resulting point $\zeta(a,t)$ is a Type II fixed point in X. □

Lemma 9.16. *Let $\phi \in \mathbb{C}_v(z)$, and let $X \subseteq \mathbb{P}^1_{\mathrm{an}}$ be a connected set with at least two points. If X contains a Type III fixed point, then it also contains a Type II fixed point.*

Proof. Write the Type III fixed point as $\zeta(a,r)$. Since X has at least two points, one of the two directions $\boldsymbol{v}$ at $\zeta(a,s)$ contains some $\xi \in X$. After a coordinate change, we may assume without loss that $\boldsymbol{v}$ also contains the point ∞. By Theorem 8.7(a), $\deg_\zeta(\phi) = 1$, and $\phi_\#$ fixes $\boldsymbol{v}$. It follows that $\phi(\zeta(a,s)) = \zeta(a,s)$ for all $s > r$ sufficiently close to r. Some such $\zeta(a,s)$ is of Type II and belongs to the interval $[\xi, \zeta) \subseteq X$, and we are done. □

Lemmas 9.15 and 9.16 are needed to prove the following result, which is itself the heart of the proof of Theorem 9.14. It first appeared in [**116**, Section 8.3], and a more general version for $\mathbb{R}$-trees appears in [**59**, Theorem 4.5]; see also [**8**, Theorem 10.85].

Lemma 9.17 (Rivera-Letelier's stable fixed-point lemma)**.** *Let $\phi \in \mathbb{C}_v(z)$, and let $X \subseteq \mathbb{P}^1_{\mathrm{an}}$ be a connected set with at least two points. Suppose that $\phi(X) \subseteq X$. Then either X contains a Type II fixed point, or the closure of X in $\mathbb{P}^1_{\mathrm{an}}$ contains an attracting fixed point.*

Proof. Since X is connected and has at least two points, it must contain an interval, and in particular it must contain a Type II point $\zeta(a,s_0)$. Define

$$\mathcal{R} := \Big\{ r > 0 : \zeta(a,r) \in X \text{ and } \phi\big(\zeta(a,r)\big) \in \overline{D}_{\mathrm{an}}(a,r) \Big\}.$$

Case 1. Suppose that $\mathcal{R} = \varnothing$. Then for any $r > 0$ such that $\zeta(a,r) \in X$, we have $\|z-a\|_{\phi(\zeta(a,r))} > r$. Writing $\phi(\zeta(a,s_0)) = \zeta(b,t)$, we define

$$s_1 := \max\big\{ |b-a|, t \big\} = \|z-a\|_{\phi(\zeta(a,s_0))} > s_0.$$

Observe that $\zeta(a,s_1)$ lies between $\zeta(a,s_0)$ and $\phi(\zeta(a,s_0))$. (In fact, it is the unique point lying between all three of $\zeta(a,s_0)$, $\phi(\zeta(a,s_0))$, and ∞; see Figure 9.1.) Because both $\zeta(a,s_0)$ and its image lie in X, it follows that $\zeta(a,s_1) \in X$ also, since X is connected. Repeating this process, we can inductively define a strictly increasing sequence $\{s_n\}_{n\geq 0}$ by

$$s_{n+1} := \|z-a\|_{\phi(\zeta(a,s_n))} > s_n,$$

with $\zeta(a,s_n) \in X$.

If $s_n \nearrow \infty$, then $\phi(\infty) = \infty \in \overline{X}$. This (Type I) fixed point at ∞ cannot be indifferent or repelling, since $\phi(\zeta(a,s_n)) \notin \overline{D}_{\mathrm{an}}(a,s_n)$ even when s_n is very large. Hence, the fixed point at ∞ is attracting.

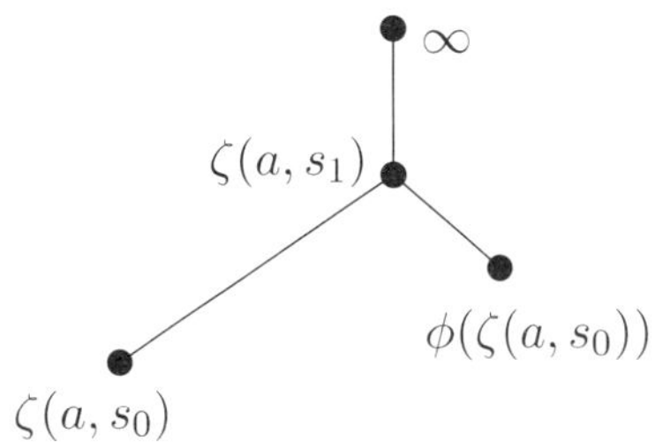

Figure 9.1. Proof of Lemma 9.17, Case 1.

Otherwise, we have $s_n \nearrow s$ for some $s \in (0,\infty)$, and the hypotheses of Lemma 9.15 hold, giving us our desired Type II fixed point in X.

Case 2. For each of the distinct poles $y_1,\ldots,y_\ell$ of ϕ in $\mathbb{C}_v$, define

$$\mathcal{R}_i := \Big\{ r>0 : \zeta(y_i,r)\in X \text{ and } \phi\big(\zeta(y_i,r)\big) \in \overline{D}_{\text{an}}(y_i,r) \Big\},$$

and let $r_i := \inf \mathcal{R}_i$, where we understand r_i to be ∞ if $\mathcal{R}_i = \varnothing$.

For this case, suppose that each $\mathcal{R}_i$ is empty, but that $\mathcal{R} \neq \varnothing$. Thus, there is some $r>0$ such that $\zeta(a,r)\in X$ and $\phi(\zeta(a,r)) \in \overline{D}_{\text{an}}(a,r)$. Since $r \notin \mathcal{R}_i$ for each i, there cannot be any poles in $\overline{D}_{\text{an}}(a,r)$. Therefore,

$$\phi\big(\overline{D}_{\text{an}}(a,r)\big) \subseteq \overline{D}_{\text{an}}(a,r).$$

If $\phi(\overline{D}_{\text{an}}(a,r)) = \overline{D}_{\text{an}}(a,r)$, then $\zeta(a,r)\in X$ is a fixed point of Type II or III; either way, by Lemma 9.16, X contains a fixed point of Type II, as desired.

Otherwise, we have $\phi(\overline{D}_{\text{an}}(a,r)) \subsetneq D_{\text{an}}(b,r)$ for some $b\in\overline{D}(a,r)$. In particular,

$$\phi\big(D(b,r)\big) \subsetneq D(b,r) \quad\text{and}\quad \phi\big(\zeta(a,r)\big) \in D_{\text{an}}(b,r).$$

By Theorem 4.18(a), there is an attracting fixed point c in $D(b,r)$, and

$$\lim_{n\to\infty} \phi^n\big(\zeta(a,r)\big) = c.$$

Since $\phi^n(\zeta(a,r)) \in X$ for each $n \geq 0$, then, we have $c\in\overline{X}$, as desired.

Case 3. Suppose that some $\mathcal{R}_i$ is nonempty. We may assume without loss that $r_1 < \infty$, and that $r_1 \leq r_2,\ldots,r_\ell$. For this case, suppose also that $r_1 \in \mathcal{R}_1$, so that

$$r_1 > 0, \quad \zeta(y_1,r_1)\in X, \quad\text{and}\quad \phi\big(\zeta(y_1,r_1)\big) \in \overline{D}_{\text{an}}(y_1,r_1).$$

If $\phi(\zeta(y_1,r_1)) = \zeta(y_1,r_1)$, then $\zeta(y_1,r_1)$ is a fixed point of Type II or III in X, and we are done by Lemma 9.16. Otherwise, we have

$$\phi\big(\zeta(y_1,r_1)\big) = \zeta(b,s) \quad\text{for some}\quad b\in\overline{D}(y_1,r_1) \quad\text{and}\quad 0<s<r_1.$$

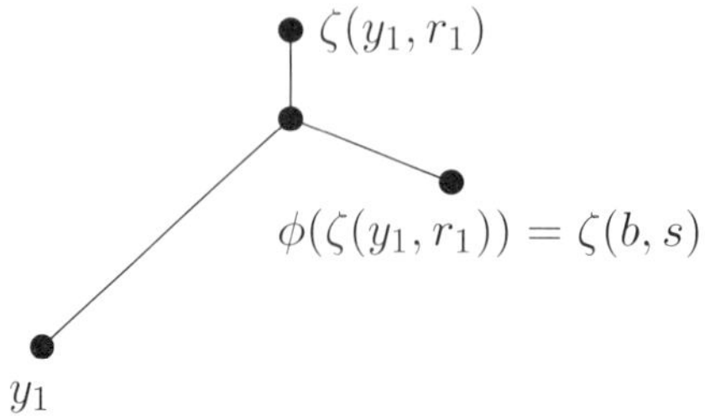

Figure 9.2. Proof of Lemma 9.17, Case 3.

We claim that $|b - y_1| = r_1$. If not, then $|b - y_1| < r_1$. Because the connected set X contains both $\zeta(b, s)$ and $\zeta(y_1, r_1)$, and by the continuity of ϕ, there is some $t < r_1$ sufficiently large such that $\zeta(y_1, t) \in X$ and $\phi(\zeta(y_1, t)) \in \overline{D}_{\mathrm{an}}(y_1, t)$; see Figure 9.2. That is, $t \in \mathcal{R}_1$, contradicting the fact that $r_1 = \inf \mathcal{R}_1$, and proving our claim.

Thus, the open disk $D(b, r_1)$ contains no poles, and hence

$$\phi\big(D(b, r_1)\big) \subseteq \overline{D}(b, s) \subsetneq D(b, r_1).$$

In addition, $\phi(\zeta(y_1, r_1)) \in D_{\mathrm{an}}(b, r_1)$. As in Case 2, Theorem 4.18(a) shows that $D(b, r_1)$ contains an attracting fixed point c such that

$$\lim_{n\to\infty} \phi^n\big(\zeta(y_1, r_1)\big) = c.$$

Since $\zeta(y_1, r_1) \in X$ and $\phi(X) \subseteq X$, we have $c \in \overline{X}$, as desired.

Case 4. Finally, still assuming that

$$\mathcal{R}_1 \neq \varnothing \quad \text{and} \quad r_1 \leq r_2, \ldots, r_\ell,$$

suppose that $r_1 \notin \mathcal{R}_1$. We claim that $r_1 > 0$. Because $\phi(y_1) = \infty$, we may choose $0 < s < 1$ small enough that

$$\phi\big(D_{\mathrm{an}}(y_1, s)\big) \subseteq \mathbb{P}^1_{\mathrm{an}} \smallsetminus \overline{D}_{\mathrm{an}}(y_1, 1),$$

by the continuity of ϕ. Then $r > s$ for all $r \in \mathcal{R}_1$. Hence $r_1 \geq s > 0$, proving the claim.

Since $r_1 = \inf \mathcal{R}_1$, there is a sequence $\{s_n\}_{n\geq 1} \subseteq \mathcal{R}_1$ that is strictly decreasing, with $s_n \searrow r_1$. Clearly,

$$\lim_{n\to\infty} \zeta(y_1, s_n) = \zeta(y_1, r_1).$$

In addition, by definition of $\mathcal{R}_1$, any accumulation point of the sequence $\{\phi(\zeta(y_1, s_n))\}_{n\geq 1}$ must lie in $\overline{D}_{\mathrm{an}}(y_1, r_1)$. Since ϕ is continuous, then, we have $\phi(\zeta(y_1, r_1)) \in \overline{D}_{\mathrm{an}}(y_1, r_1)$.

If $\phi(\zeta(y_1, r_1)) \neq \zeta(y_1, r_1)$, then we must also have

$$\phi(\zeta(y_1, s_n)) \in \overline{D}_{\mathrm{an}}(y_1, r_1) \smallsetminus \{\zeta(y_1, r_1)\}$$

for all sufficiently large n. Since $\zeta(y_1, r_1)$ lies between $\zeta(y_1, s_n) \in X$ and $\phi(\zeta(y_1, s_n)) \in X$, we have $\zeta(y_1, r_1) \in X$ as well. Hence, $r_1 \in \mathcal{R}_1$, contrary to our assumption.

Thus, we must have $\phi(\zeta(y_1, r_1)) = \zeta(y_1, r_1)$. Define $\eta \in \mathrm{PGL}(2, \mathbb{C}_v)$ by $\eta(z) = 1/(z - y_1)$, which moves y_1 to ∞ and ∞ to 0. Also define $s'_n := 1/s_n$ and $s := 1/r_1$. Then $s'_n \nearrow s$, and the points $\zeta(0, s'_n)$ satisfy the hypotheses of Lemma 9.15 for $\psi := \eta \circ \phi \circ \eta^{-1}$ with respect to the connected set $\eta(X)$. Therefore, $\eta(X)$ contains a Type II fixed point of ψ, and hence X contains a Type II fixed point of ϕ. □

Remark 9.18. The above proof of Lemma 9.17 is inspired by Rivera-Letelier's proof in [**116**, Section 8], which Baker and Rumely presented more formally as [**8**, Theorem 10.85]. The idea is as follows: starting from an arbitrary point $\zeta \in X$, gradually build a segment extending from ζ that will ultimately terminate at a fixed point. To do so, we proceed in stages. At each stage, we creep a short distance from the current endpoint ξ of the segment in the direction of $\phi(\xi)$. (But not too far, lest we take a wrong turn at a Type II point where ξ, $\phi(\xi)$, and the desired fixed point all lie in different directions.) Baker and Rumely make explicit what Rivera-Letelier does not: that the infinitely many choices inherent in this argument require transfinite induction.

Our proof above is a little simpler, because in reality, there are only a few turns in the road from our starting point in X to the fixed point. Instead of creeping along, we can jump to the nearest point $\zeta = \zeta(y_1, r_1) \in X$ at which the direction $\boldsymbol{v}$ containing $\phi(\zeta)$ does not contain ∞. If ζ itself is not already fixed, then Cases 2–4 show that the direction $\boldsymbol{v}$ immediately leads to an attracting basin. On the other hand, if no such starting point ζ exists, then Case 1 jumps us as high as we can go, hitting a fixed point.

Proof of Theorem 9.14. Replacing ϕ by ϕ^n if necessary, we may assume that $\phi(U) = U$. By Lemma 9.17, either U contains a Type II fixed point, or else $\overline{U}$ contains an attracting fixed point.

If there is a Type II fixed point $\zeta \in U \subseteq \mathcal{F}_{\phi,\mathrm{an}}$, then by Proposition 9.13, ζ belongs to the indifference domain $\mathfrak{I}$ of ϕ. By Theorem 9.7, U must be a component of $\mathfrak{I}$ and hence an indifferent component.

If $\overline{U}$ contains an attracting fixed point a, then by Theorem 5.14, we have $a \in \mathcal{F}_{\phi,\mathrm{an}}$, and therefore $a \in U$. By Theorem 9.5, U must be the immediate attracting basin of a and hence an attracting component.

Finally, by Proposition 9.13 again, no indifferent component can contain an attracting periodic point. Thus, U cannot be both an attracting and an indifferent component. □

Exercises for Chapter 9

9.1. Fix a prime $p \geq 2$, and let

$$\phi(z) = p^{-1}(z^p - z^{p^2}) \in \mathbb{C}_p[z].$$

Prove that the attracting basin U of ∞ is an attracting Fatou component of Cantor type, and that its boundary ∂U includes some points of Type IV. (This observation first appeared in [**114**, Exemple 6.3].)

9.2. Let $U_1 \subsetneq U_2 \subsetneq U_3 \subsetneq \cdots$ be open connected Berkovich affinoids satisfying the conditions of Definition 9.4, so that $U := \bigcup_{n\geq 0} U_n$ is a domain of Cantor type. In this exercise, you will prove the claim from Theorem 9.5(d), that ∂U is homeomorphic to the Cantor set.

For each $n \geq 0$, define $f_n : \partial U_{n+1} \to \partial U_n$ by setting $f_n(\zeta)$ to be the unique $\xi \in \partial U_n$ lying in the same component of $\mathbb{P}^1_{\text{an}} \smallsetminus U_n$ as ζ.

Define X_U to be the inverse limit $\varprojlim \partial U_n$. That is, an element of X_U is a sequence $\{\zeta_n\}_{n\geq 0}$ such that $\zeta_n \in \partial U_n$, and $f_n(\zeta_{n+1}) = \zeta_n$. The set X_U is equipped with the inverse limit topology inherited from the discrete sets ∂U_n.

(a) Show that every sequence $\{\zeta_n\}_{n\geq 0} \in X_U$ has a unique limit point and that limit point lies in ∂U.

(b) Show that the function $F : X_U \to \partial U$ given by

$$F\big(\{\zeta_n\}_{n\geq 0}\big) := \lim_{n\to\infty} \zeta_n$$

is continuous.

(c) Show that the function F of part (b) is one-to-one and onto; thus, since X_U is compact, it is a homeomorphism.

(d) Show that X_U is also homeomorphic to the (middle thirds) Cantor set.
(*Suggestion*: For each $n \geq 0$, partition the Cantor set into subsets corresponding to the elements of ∂U_n, in such a way as to be consistent with the maps f_n.)

9.3. Let $\psi \in \mathbb{C}_v[[z-a]]$ be a power series as in Lemma 9.8, i.e., mapping $D(a,r)$ bijectively onto itself. Let k be the residue field of $\mathbb{C}_v$.

(a) Suppose that k is an algebraic extension of a finite field. Prove that the first case of the lemma always holds: every Type II or III point $\zeta(b,s)$ in $D_{\text{an}}(a,r)$ is periodic.

(b) Suppose that k is *not* an algebraic extension of a finite field. Prove that there is a choice of ψ and the point $\zeta(b,s)$ fitting the hypotheses of Lemma 9.8 for which $\zeta(b,s)$ is not periodic.

9.4. Let $r > s > 0$, let $\phi \in \mathbb{C}_v(z)$, and let W be an open connected Berkovich affinoid such that $\zeta(0,r), \zeta(0,s) \in \partial W$. Suppose that ϕ is one-to-one on W, with

$$\phi(W) \subseteq D_{\text{an}}(0,r) \quad \text{and} \quad \phi\big(\zeta(0,r)\big) = \zeta(0,r).$$

Prove that there is some $a \in D(0,r)$ such that

$$\phi\big(\zeta(0,s)\big) = \zeta(a,s) \quad \text{and} \quad \big|\phi(x) - a\big| = |x| \quad \text{for all } x \in W.$$

(*Note*: This fact will be useful in Exercise 9.5.)

9.5. Prove the following generalization of Rivera-Letelier's approximation lemma, Lemma 9.10. Let $b, c \in \mathbb{C}_v$ and $r > s > 0$. Let $D_1, \ldots, D_\ell$ be closed Berkovich disks contained in $D_{\text{an}}(b,r)$, each of diameter at most s, and let

$$W := D_{\text{an}}(b,r) \smallsetminus \big(D_1 \cup \cdots \cup D_\ell\big).$$

Let $\phi \in \mathbb{C}_v(z)$ be a rational function that is one-to-one on W, with

$$\phi(W) \subseteq D_{\text{an}}(c,r) \quad \text{and} \quad \phi\big(\zeta(b,r)\big) = \zeta(c,r).$$

Then there is a rational function $\psi \in \mathbb{C}_v(z)$ such that

(a) ψ maps $D_{\text{an}}(b,r)$ onto $D_{\text{an}}(c,r)$ with Weierstrass degree 1, and

(b) $\psi(\xi) = \phi(\xi)$ for any $\xi \in D_{\text{an}}(b,r)$ with $\text{diam}(\xi) \geq s$.

(*Suggestion*: Mimic the proof of Lemma 9.10, knocking out one disk D_i at a time, and using Exercise 9.4 when needed.)

9.6. Suppose that the residue field k of $\mathbb{C}_v$ is algebraic over a finite field. Let W be an open connected Berkovich affinoid, and let $\phi \in \mathbb{C}_v(z)$ be a rational function that maps W bijectively onto itself. Prove that every Type II or III point in W is periodic under ϕ.

(*Suggestion*: Use Exercises 9.3 and 9.5.)

9.7. In this exercise, we use the notation of the proof of Theorem 9.7.

(a) In the notation of Step 1 of that proof, prove that either $N = 0$ or else the smallest element S_N of $\mathcal{S}$ is $S_N = s_i$ for some $i = 1, \ldots, \ell$.

(b) In Step 3 of that proof, if $\zeta \in D \cap \partial U_N$ lies in the Fatou set $\mathcal{F}_{\phi,\text{an}}$, prove that $N = 0$, that $\zeta = \zeta(a,r)$, and that $D \subseteq \text{Inj}(\phi)$.

9.8. Let $\phi(z) = az + \frac{b^2}{z-1} \in \mathbb{C}_v(z)$ be the function of Example 9.12, with $0 < |b| < |a| = 1$. Let $c \in \overline{D}(0,1)$, and let $\zeta := \zeta(c, |b|)$.

(a) Prove that $\phi(\zeta) = \zeta(ac, |b|)$.

(b) If $|c - 1| \leq |b|$, prove that $\deg_\zeta(\phi) = 2$.

(c) If $|b| < |c - 1| \leq 1$, prove that $\deg_\zeta(\phi) = 1$.

9.9. Consider a, b, ϕ as in Exercise 9.8 and Example 9.12. Suppose we are in the fourth case of that example, that $\operatorname{char} k = p > 0$, and $|b| < |a^m - 1|$, where $m \geq 1$ is the smallest positive integer such that $|a^m - 1| < 1$. Prove that the minimal period of the repelling periodic point $\zeta(1, |b|)$ is of the form $p^j m$ for some $j \geq 1$.

9.10. Consider a, b, ϕ as in Exercise 9.8 and j as in Exercise 9.9.

(a) Find a formula for j in terms of a, b, and p. (There will be many cases.)

(b) How are the iterates $\phi^i(\zeta(1, |b|))$ arranged relative to one another?

9.11. Fix $a, b, c \in \mathbb{C}_v$. Each of the following maps has an indifferent fixed point at ∞. Compute the corresponding indifferent component in each case.

(a) $\phi(z) = az + \frac{b^2}{z-1} + \frac{b^2}{z-(1+c)}$ with $0 < |b| < |c| < |a| = 1$.

(b) $\phi(z) = az + \frac{b^2}{z-1} + \frac{b^2}{z-c}$ with $0 < |b| < |c| = |c - 1| = |a| = 1$.

(c) $\phi(z) = az + \frac{b^2}{z-1} + \frac{b^2}{z}$ with $0 < |b| < |a| = 1$.

9.12. Let $a_1, \ldots, a_n \in \overline{D}(0, 1)$, and let $U_1, \ldots, U_n$ be disjoint open connected Berkovich affinoids with $U_i \subseteq D_{\mathrm{an}}(a_i, 1)$. (*Note*: The points a_i need not lie in separate residue classes, nor even be distinct.) Prove that there are integers $m_i \geq 2$ such that the polynomial

$$\phi(z) := z + \prod_{i=1}^{n} (z - a_i)^{m_i} :$$

(a) maps U_i bijectively onto U_i, for each $i = 1, \ldots, n$;

(b) fixes each point of ∂U_i, for each $i = 1, \ldots, n$; and

(c) has a repelling fixed point at $\zeta(0, 1)$.

(*Note*: This is a version of [**114**, Lemme 6.5], modified to apply to arbitrary $\mathbb{C}_v$ rather than just $\mathbb{C}_p$.)

9.13. Let $U_1, \ldots, U_n$ be disjoint rational open connected Berkovich affinoids; if $n = 1$, assume $U_1 \neq \mathbb{P}^1_{\mathrm{an}}$. Prove that there is a rational function $\phi \in \mathbb{C}_v(z)$ for which each U_i is a fixed indifferent component of ϕ.

(*Suggestion*: For each boundary point ξ of each U_i, make a change of coordinates to move ξ to $\zeta(0, 1)$, and apply Exercise 9.12. Let ϕ be the composition of all the resulting maps.)

(*Note*: This is essentially [**114**, Proposition 6.4].)

9.14. Let $\psi \in \mathbb{C}_v(z)$ fix the Gauss point, and let U be an open subset of the indifference domain of ψ satisfying

$$\psi(U) \subseteq U \subseteq \overline{D}_{\mathrm{an}}(0, 1).$$

Let $b \in \mathbb{C}_v$ with $0 < |b| < 1$, and assume that

$$\overline{D}_{\mathrm{an}}\big(x, |b|\big) \subseteq U \quad \text{for every } x \in U \cap \mathbb{P}^1(\mathbb{C}_v).$$

Prove that the indifference domain of

$$\phi(z) := \frac{z}{z-b}\psi(z)$$

contains $U \smallsetminus \overline{D}_{\mathrm{an}}(0, |b|)$.

(*Note*: This fact will be useful in Exercise 9.15.)

9.15. Let $\psi \in \mathbb{C}_v(z)$ have explicit good reduction, with indifference domain W. Let $r \in |\mathbb{C}_v^\times|$ with $0 < r < 1$, and let $a_1, \ldots, a_n \in \overline{D}(0,1)$. Prove that there exists $\phi \in \mathbb{C}_v(z)$ of degree at most $n + \deg(\psi)$ such that for every connected component U of W,

$$U \smallsetminus \big[\overline{D}_{\mathrm{an}}(a_1, r) \cup \cdots \cup \overline{D}_{\mathrm{an}}(a_n, r)\big]$$

is an indifferent component for ϕ.

(*Suggestion*: Reduce to the case that the disks $\overline{D}_{\mathrm{an}}(a_i, r)$ are disjoint and have forward orbits contained in $\overline{D}_{\mathrm{an}}(0,1)$. Then inductively apply Exercise 9.14.)

(*Note*: This is essentially [**114**, Proposition 6.7].)

Chapter 10

Further Results on Periodic Components

The results of Chapter 9 describe the geometric structure of periodic Fatou components, as well as the dynamics on attracting components. In this chapter, we study dynamics on indifferent components, and we discuss the possible number of cycles of Fatou components. As in Chapter 9, many of these results first appeared for $\mathbb{C}_p$ in Rivera-Letelier's thesis [**114**].

10.1. Periodic points in the indifference domain

Recall that a complex indifferent component contains at most one periodic point. Specifically, a fixed Siegel disk contains a unique fixed point and no other periodic points, and a Herman ring contains no periodic points at all. By contrast, a non-archimedean indifferent component often contains many Type I periodic points of many different periods. To state the relevant result, which appears as [**114**, Proposition 5.10], we need the following definition.

Definition 10.1. Let $\phi \in \mathbb{C}_v(z)$ be a rational function, and let $\zeta \in \mathcal{J}_{\phi,\mathrm{an}}$ be a Type II Julia fixed point. Let $\boldsymbol{v}$ be a direction at ζ that is fixed by $\phi_\#$. Define $N(\phi, \boldsymbol{v})$ to be the *multiplicity with which* ϕ *fixes* $\boldsymbol{v}$.

That is, choose $\theta \in \mathrm{PGL}(2, \mathbb{C}_v)$ with

$$\theta(\zeta) = \zeta(0,1) \quad \text{and} \quad \theta_\#(\boldsymbol{v}) = \vec{v}_{\zeta(0,1)}(0),$$

and define $\Phi := \theta \circ \phi \circ \theta^{-1}$, so that $\overline{\Phi}(z) \in k(z)$ fixes $0 \in \mathbb{P}^1(k)$. Then

$$N(\phi, \boldsymbol{v}) := \operatorname{ord}_0 \left(\overline{\Phi}(z) - z\right)$$

is the order of vanishing of $\overline{\Phi}(z) - z$ at $0 \in \mathbb{P}^1(k)$.

The reduced map $\overline{\Phi}$ of Definition 10.1 cannot be the identity function $z \mapsto z$. If it were, then $\phi_\#$ would fix all the bad directions at ζ, and hence ζ would be Fatou, by Theorem 8.7(b). Thus, $\overline{\Phi}(z) - z$ is a nonconstant rational function of degree at most $1 + \deg_\zeta(\phi)$, and hence

$$1 \leq N(\phi, \boldsymbol{v}) \leq 1 + \deg_\zeta(\phi) \leq 1 + \deg(\phi).$$

Theorem 10.2 (Rivera-Letelier). *Let $\phi \in \mathbb{C}_v(z)$ be a rational function of degree $d \geq 2$, let $U \subseteq \mathcal{F}_{\phi,\mathrm{an}}$ be an indifferent component fixed by ϕ, and let $\zeta_1, \ldots, \zeta_m$ be the points of ∂U. For each $i = 1, \ldots, m$, let $\boldsymbol{v}_i$ be the direction at ζ_i containing U. Suppose that*

$$\phi(\zeta_i) = \zeta_i \quad \textit{for} \quad i = 1, \ldots, \ell$$

and

$$\phi(\zeta_i) \neq \zeta_i \quad \textit{for} \quad i = \ell + 1, \ldots, m.$$

Then ϕ has exactly

$$2 + \sum_{i=1}^{\ell} \left(N(\phi, \boldsymbol{v}_i) - 2\right)$$

fixed points of Type I in U, counting multiplicity.

To prove Theorem 10.2, we need the following analogue of the Residue Theorem for open affinoids. It appears in [**114**, Section 2.1.1] and in a slightly different form as [**66**, Theorem 2.3.3(2)].

Proposition 10.3 (Residue Theorem for open affinoids). *Let $\psi \in \mathbb{C}_v(z)$ be a nonconstant rational function, and let $U \subseteq \mathbb{P}^1_{\mathrm{an}}$ be an open connected Berkovich affinoid with boundary points $\zeta_1, \ldots, \zeta_n$. For each $i = 1, \ldots, m$, let $\boldsymbol{v}_i$ be the direction at ζ_i containing U. Then*

$$\sum_{x \in U \cap \mathbb{P}^1(\mathbb{C}_v)} \operatorname{ord}_x(\psi) = \sum_{i=1}^{m} \operatorname{ord}_{\boldsymbol{v}_i}(\psi),$$

where $\operatorname{ord}_{\boldsymbol{v}_i}(\psi)$ *is the* order of vanishing of ψ in $\boldsymbol{v}_i$, *given by*

$$\operatorname{ord}_{\boldsymbol{v}_i}(\psi) := \begin{cases} -\operatorname{wdeg}_{W_i}(\psi) & \textit{if } \infty \in \boldsymbol{v}_i, \\ \operatorname{wdeg}_{W_i}(\psi) & \textit{if } \infty \notin \boldsymbol{v}_i, \end{cases}$$

where $\operatorname{wdeg}_{W_i}(\psi)$ *is the common inner and outer Weierstrass degree of ψ on an annulus W_i extending from ζ_i into $\boldsymbol{v}_i$, with no poles or zeros of ψ.*

Proof. This is left to the reader; see Exercise 10.1. □

Proof of Theorem 10.2.

Step 1. Choose a coordinate so that $\infty \in U$ and $\phi(\infty) \neq \infty$. Since $\phi : U \to U$ is bijective, it has exactly one pole in U. Thus, the rational function $\phi(z) - z$ has exactly two poles in U: one at ∞ and one at the unique pole of ϕ in U. On the other hand, the zeros of $\phi(z) - z$ are precisely the fixed points of ϕ. Writing F for the number of fixed points of ϕ in U, counted with multiplicity, and applying Proposition 10.3 to $\psi(z) := \phi(z) - z$ on U, we then have

$$F - 2 = \sum_{x \in U \cap \mathbb{P}^1(\mathbb{C}_v)} \operatorname{ord}_x(\psi) = \sum_{i=1}^{m} \operatorname{ord}_{\boldsymbol{v}_i}(\psi).$$

Thus, it suffices to show that $\operatorname{ord}_{\boldsymbol{v}_i}(\psi) = N(\phi, \boldsymbol{v}_i) - 2$ for $i = 1, \ldots, \ell$, and $\operatorname{ord}_{\boldsymbol{v}_i}(\psi) = 0$ for $i = \ell + 1, \ldots, m$.

Step 2. Given $i \geq \ell + 1$, write $\zeta_i = \zeta(a, r)$ and $\phi(\zeta_i) = \zeta(b, s) \neq \zeta_i$. Since ζ_i is not fixed, the disks $\overline{D}_{\mathrm{an}}(a, r)$ and $\overline{D}_{\mathrm{an}}(b, s)$ are disjoint; that is, $r, s < |b - a|$.

Let $W_i \subseteq \mathbb{C}_v$ be an annulus of the form $W_i := \{r < |z - a| < \lambda r\}$, where $\lambda > 1$ is small enough that the image $\phi(W_i)$ is the annulus given by

$$\phi(W_i) = \{s < |z - b| < \lambda s\}.$$

(Note that the outer radius of the image annulus is indeed λs, because U is an indifferent component, and therefore $\deg_{\zeta_i, \boldsymbol{v}_i}(\phi) = 1$.) Decrease $\lambda > 1$ further if necessary so that $\lambda r, \lambda s < |b - a|$.

For any $z \in W_i$, we have

$$\psi(z) = \phi(z) - z = \big(\phi(z) - b\big) - (z - a) + (b - a).$$

Thus, since

$$|\phi(z) - b| < \lambda s < |b - a| \quad \text{and} \quad |z - a| < \lambda r < |b - a|,$$

we have $|\psi(z)| = |b - a|$ for all $z \in W_i$, which is constant. Therefore, the common inner and outer Weierstrass degree of ψ on W_i is $\operatorname{wdeg}_{W_i}(\psi) = 0$, and hence $\operatorname{ord}_{\boldsymbol{v}_i}(\psi) = 0$.

Step 3. Given $i \leq \ell$, write $\zeta_i = \zeta(a, r)$. By hypothesis, we have $\phi(\zeta_i) = \zeta_i$. Moreover, $\boldsymbol{v}_i$ contains ∞, and because $\phi(U) = U$, we have $\phi_\#(\boldsymbol{v}_i) = \boldsymbol{v}_i$.

Since ζ_i is of Type II, we may write $r = |b|$ for some $b \in \mathbb{C}_v^\times$. Define

$$\theta(z) = \frac{b}{z - a} \in \mathrm{PGL}(2, \mathbb{C}_v),$$

which maps ζ_i to the Gauss point $\zeta(0, 1)$ with $\theta_\#(\boldsymbol{v}_i) = \vec{v}(0)$. Define

$$\Phi(z) := \theta \circ \phi \circ \theta^{-1}(z) = a + \frac{b}{\phi(a + b/z)},$$

which fixes the Gauss point, with $\Phi_\#(\vec{v}(0)) = \vec{v}(0)$.

Let $N := N(\phi, \boldsymbol{v}_i) \in \{1, \ldots, d+1\}$, so that by Definition 10.1 we have

$$\overline{\Phi}(z) - z = \frac{z^N P(z)}{Q(z)},$$

where $P, Q \in k[z]$ are polynomials defined over the residue field k, with $P(0), Q(0) \neq 0$. Expanding Φ as a Laurent series on an annulus W' of the form $W' := \{\lambda^{-1} < |z| < 1\}$ for $\lambda > 1$ small enough, we have

$$\Phi(z) = z + c_N z^N + h(z),$$

where $|c_N| = 1$ with $\overline{c}_N = P(0)/Q(0)$, and where

$$h(z) = \sum_{n \in \mathbb{Z}} c'_n z^n$$

satisfies $|h(z)| < |z|^N$ for all $z \in W'$. Note that if $N = 1$, then $|c_N + 1| = 1$, since otherwise we would have $\overline{\Phi} = 0$, contradicting the fact that Φ fixes the Gauss point.

Let $W_i := \theta^{-1}(W')$, which is an annulus $\{r < |z - a| < \lambda r\}$ extending from ζ_i into the direction $\boldsymbol{v}_i$. Then for $z \in W_i$, we have

$$\begin{aligned}\phi(z) - z = \theta^{-1} \circ \Phi \circ \theta(z) - z &= \frac{b - (z-a)\Phi\big(\theta(z)\big)}{\Phi\big(\theta(z)\big)} \\ &= \frac{-(z-a)\big[c_N\big(\theta(z)\big)^N + h\big(\theta(z)\big)\big]}{\theta(z) + c_N\big(\theta(z)\big)^N + h\big(\theta(z)\big)},\end{aligned}$$

and hence $\psi(z) = \phi(z) - z$ satisfies

$$\big|\psi(z)\big| = \frac{|z-a|\big|b(z-a)^{-1}\big|^N}{|b(z-a)^{-1}|} = |b|^{N-1}|z-a|^{2-N}.$$

Therefore, in the notation of Proposition 10.3, we have

$$\operatorname{ord}_{\boldsymbol{v}_i}(\psi) = -\operatorname{wdeg}_{W_i}(\psi) = -(2-N) = N-2,$$

as desired. □

Corollary 10.4. *Let $\phi \in \mathbb{C}_v(z)$ be a rational function of degree $d \geq 2$, and let $U \subseteq \mathcal{F}_{\phi,\mathrm{an}}$ be an indifferent component fixed by ϕ.*

(a) *If U is a disk, then it contains at least one Type I fixed point.*

(b) *If U is not a disk, and if none of the boundary points are fixed by ϕ, then U contains exactly two Type I fixed points, counting multiplicity.*

(c) *If the residue field k is algebraic over a finite field, then U contains infinitely many periodic points.*

Proof.

(a) The disk U has only one boundary point ζ_1, which is necessarily fixed. Theorem 10.2 shows that the number of Type I fixed points in U is $N(\phi, \boldsymbol{v}_1) \geq 1$.

(b) Since U has no fixed boundary points, Theorem 10.2 yields two Type I fixed points in U.

(c) This is left to the reader; see Exercise 10.2.

□

10.2. Indifferent components in mixed characteristic

When our field $\mathbb{C}_v$ is of mixed characteristic, like $\mathbb{C}_p$, Rivera-Letelier was able to say even more about the periodic points and dynamics in indifferent periodic components; see [**114**, Section 3.2].

Definition 10.5. Let $\phi(z) \in \mathbb{C}_v(z)$, where $\mathbb{C}_v$ has characteristic 0 but residue characteristic $p > 0$. We say that a point $x \in \mathbb{C}_v$ is *quasi-periodic* under ϕ if there is a radius $r > 0$ and an integer $m \geq 1$ such that the limit

$$L_\phi(z) := \lim_{n\to\infty} \frac{\phi^{mp^n}(z) - z}{mp^n} \tag{10.1}$$

converges uniformly on $\overline{D}(x, r)$. In that case, the function $L_\phi \in \overline{\mathcal{A}}(x, r)$ is called the *iterative logarithm* of ϕ.

We say that ∞ is *quasi-periodic* under ϕ if there is a linear fractional transformation $\eta \in \mathrm{PGL}(2, \mathbb{C}_v)$ such that $\eta(\infty)$ is quasi-periodic under the map $\eta \circ \phi \circ \eta^{-1}$.

The *quasi-periodicity domain* of ϕ is the set $\mathcal{Q}(\phi)$ of all $x \in \mathbb{P}^1(\mathbb{C}_v)$ that are quasi-periodic under ϕ.

Proposition 10.6. *Let* $\operatorname{char} \mathbb{C}_v = 0$ *and* $\operatorname{char} k = p > 0$. *Let* $\phi(z) \in \mathbb{C}_v(z)$.

(a) *For any* $x \in \mathbb{C}_v$, *integer* $m \geq 1$, *and radius* $r > 0$ *for which the limit of equation* (10.1) *converges uniformly on* $\overline{D}(x, r)$, *it also converges uniformly to the same function* $L_\phi(z)$ *with* ℓm *in place of* m, *for any integer* $\ell \geq 1$.

(b) *For* $\eta \in \mathrm{PGL}(2, \mathbb{C}_v)$, *let* $\psi = \eta \circ \phi \circ \eta^{-1}$. *Then*
$$\mathcal{Q}(\psi) = \eta\big(\mathcal{Q}(\phi)\big) \quad \textit{and} \quad L_\psi \circ \eta = \eta' \cdot L_\phi.$$

(c) $\mathcal{Q}(\phi)$ *is an open subset of* $\mathbb{P}^1(\mathbb{C}_v)$.

(d) ϕ *maps* $\mathcal{Q}(\phi)$ *bijectively onto itself.*

(e) *For any integer* $n \geq 1$, *we have*
$$\mathcal{Q}(\phi^n) = \mathcal{Q}(\phi), \quad L_{\phi^n} = nL_\phi, \quad \textit{and} \quad L_\phi \circ \phi^n = (\phi^n)' \cdot L_\phi.$$

Proof. This is left to the reader; see Exercises 10.6 and 10.7. □

The motivation behind Definition 10.5 is that if ϕ is analytically conjugate to $z \mapsto z+1$ on a neighborhood of x, then every p^n-th iterate is conjugate to $z \mapsto z+p^n$. Thus, the iterates of x land arbitrarily close to x. Moreover, if the conjugacy h such that $h(\phi(z)) = h(z)+1$ is unknown, we can recover it, because

$$\lim_{n\to\infty} \frac{\phi^{p^n}(z)-z}{p^n} = \lim_{n\to\infty} \frac{h^{-1}(h(z)+p^n) - h^{-1}(h(z))}{p^n} = \frac{1}{h'(z)}.$$

To use this idea, we first need the following technical lemmas.

Lemma 10.7. *Let $\phi(z) \in \mathbb{C}_v(z)$, where $\operatorname{char}\mathbb{C}_v = 0$ and $\operatorname{char} k = p > 0$. Let $x \in \mathcal{Q}(\phi) \smallsetminus \{\infty\}$. Then there exist a real number $r > 0$ and an integer $M \geq 1$ such that*

$$\left\| \phi^M(z) - z \right\|_{\zeta(x,r)} \leq |p|r.$$

Proof. Since $x \in \mathcal{Q}(\phi)$, there are an integer $m \geq 1$ and a radius $r > 1$ such that the limit of equation (10.1) converges uniformly to a power series $L_\phi(z)$ on $\overline{D}(x,r)$. For all $n \geq 1$ large enough, then, we have

$$\left\| \frac{\phi^{mp^n}(z) - z}{mp^n} \right\|_{\zeta(x,r)} \leq \max\left\{1, \|L_\phi\|_{\zeta(x,r)}\right\}.$$

Increasing n if necessary so that $|p|^n \|L_\phi\|_{\zeta(x,r)} \leq |p|r$, then, we have

$$\left\|\phi^{mp^n}(z) - z\right\|_{\zeta(x,r)} \leq |mp|r \leq |p|r.$$

Hence, choosing $M = mp^n$, we are done. □

Lemma 10.8. *Let $\phi(z) \in \mathbb{C}_v(z)$, where $\operatorname{char}\mathbb{C}_v = 0$ and $\operatorname{char} k = p > 0$. Let $x \in \mathbb{C}_v$ and $r > 0$. Suppose that there is a real number $0 < \gamma < 1$ such that*

$$\|\phi(z) - z\|_{\zeta(x,r)} \leq \gamma r.$$

Define a sequence of functions $\{T_n\}_{n\geq 0}$ on $\overline{D}(x,r)$ by

$$T_0(z) = z \quad \textit{and} \quad T_{n+1}(z) = T_n\big(\phi(z)\big) - T_n(z).$$

Then

(a) *For every $n \geq 0$,*

$$\phi^n = \sum_{i=0}^{n} \binom{n}{i} T_i \quad \textit{and} \quad T_n = \sum_{i=0}^{n} (-1)^{n-i} \binom{n}{i} \phi^i.$$

(b) *For every $n \geq 1$, $\|T_n\|_{\zeta(x,r)} \leq \gamma^n r$.*

(c) *The sequences*

$$\left\{\frac{\phi^{p^n}(z)-z}{p^n}\right\}_{n\geq 0} \quad \textit{and} \quad \left\{\sum_{i=1}^{n}(-1)^{i-1}\frac{T_i}{i}\right\}_{n\geq 1}$$

both converge uniformly on $\overline{D}(x,r)$ *to a limit* $L_\phi(z)\in\overline{\mathcal{A}}(x,r)$.

(d) $\overline{D}(x,r)\subseteq\mathcal{Q}(\phi)$.

Proof.

(a) We leave this simple induction to the reader; see Exercise 10.8.

(b) We proceed by induction on n. We have $\|T_1\|_{\zeta(x,r)}\leq\gamma r$ by hypothesis. Assuming the inequality holds for some $n\geq 1$, we have

$$\begin{aligned}\|T_{n+1}\|_{\zeta(x,r)} &= \|T_n(\phi(z))-T_n(z)\|_{\zeta(x,r)} \leq \frac{1}{r}\|T_n\|_{\zeta(x,r)}\|\phi(z)-z\|_{\zeta(x,r)}\\ &\leq \frac{1}{r}(\gamma^n r)(\gamma r)=\gamma^{n+1}r,\end{aligned}$$

where the first inequality is by Corollary 3.21, and the second is by the hypothesis and the inductive assumption.

(c) The uniform convergence of the second sequence to some limit $L_\phi(z)$ is immediate from part (b). In addition, for any $n\geq 0$,

$$\begin{aligned}&\left\|\frac{\phi^{p^n}(z)-z}{p^n}-\sum_{j=1}^{p^n}(-1)^{j-1}\frac{T_j(z)}{j}\right\|_{\zeta(x,r)}\\ &\qquad=\left\|\sum_{j=1}^{p^n}\left[\frac{1}{p^n}\binom{p^n}{j}-\frac{(-1)^{j-1}}{j}\right]T_j\right\|_{\zeta(x,r)}\\ &\qquad\leq |p|^n r\max\left\{\gamma^j|ij|^{-1}\,:\,1\leq i<j\leq p^n\right\},\end{aligned}\tag{10.2}$$

where the equality is by part (a), and the inequality is by Exercise 2.17(a) and part (b). In addition, the expression on the right-hand side of inequality (10.2) approaches 0 as $n\to\infty$, by Exercise 2.17(b). Hence, $(\phi^{p^n}(z)-z)/p^n$ also approaches the same limit $L_\phi(z)$ uniformly on $\overline{D}(x,r)$.

(d) This is immediate from part (c). □

Theorem 10.9. *Let* $\phi(z)\in\mathbb{C}_v(z)$ *be a rational function of degree* $d\geq 2$, *where* $\operatorname{char}\mathbb{C}_v=0$ *and* $\operatorname{char}k=p>0$. *Then* $\mathcal{Q}(\phi)$ *is contained in the indifference domain of* ϕ. *Moreover, if* k *is algebraic over a finite field, then* $\mathcal{Q}(\phi)$ *is equal to the set of Type I points in the indifference domain of* ϕ.

Proof. Given $x\in\mathcal{Q}(\phi)$, assume after a change of coordinates that $x=0$. By Lemma 10.7, there exist $r>0$ and $M\geq 1$ such that

$$|\phi^M(y)-y|\leq|p|r\quad\text{for all } y\in\overline{D}(0,r).$$

Thus, ϕ^M must map $D(0,r)$ onto itself with Weierstrass degree 1. Hence, $D(0,r)$ is contained in the indifference domain, proving the first statement.

For the second, given a Type I point x in the indifference domain, there is (by definition) an open connected affinoid W such that $x \in W$ and ϕ maps W bijectively onto itself. After a change of coordinates, we may assume that $x = 0$, that $\infty \notin W$, and that $\overline{D}(0,1) \subseteq W$. By the hypothesis on k and the generalization of Lemma 9.8 in Exercise 9.6, there is some $m \geq 1$ such that ϕ^m fixes $\zeta(0,1)$. Hence, ϕ^m maps $\overline{D}(0,1)$ onto itself with Weierstrass degree 1, and therefore we may write

$$\phi^m(z) = \sum_{j\geq 0} a_j z^j, \quad \text{with } |a_0| \leq 1,\ |a_1| = 1, \text{ and } |a_j| < 1 \text{ for } j \geq 2.$$

By hypothesis, the reductions $\overline{a}_0, \overline{a}_1 \in k$ lie in a finite field, and hence some iterate of $\overline{\phi^m}(z) = \overline{a}_0 + \overline{a}_1 z$ is $z \mapsto z$. Increasing m, then, we may assume that $|a_0|, |a_1 - 1| < 1$. We may therefore choose $\gamma \in (0,1)$ such that

$$\gamma \geq \max\left\{|a_0|^{1/2}, |a_1 - 1|\right\}.$$

Thus, $|a_j|\gamma^j \leq \gamma^2$ for $j = 0$ and $j \geq 2$, and $|a_1 - 1|\gamma \leq \gamma^2$. Hence,

$$\left\|\phi^m(z) - z\right\|_{\zeta(0,\gamma)} \leq \gamma^2.$$

By Lemma 10.8(d), we have $\overline{D}(0,\gamma) \subseteq \mathcal{Q}(\phi^m)$, and therefore $0 \in \mathcal{Q}(\phi)$. ☐

Theorem 10.10. *Let $\phi(z) \in \mathbb{C}_v(z)$ be a nonconstant rational function, where* $\operatorname{char}\mathbb{C}_v = 0$ *and* $\operatorname{char} k = p > 0$. *Let $x \in \mathcal{Q}(\phi)$.*

(a) *$L_\phi(x) = 0$ if and only if x is preperiodic. In that case, x is an indifferent periodic point.*

(b) *$L_\phi(x) = L'_\phi(x) = 0$ if and only if x is a rationally indifferent periodic point of ϕ.*

(c) *For every $x \in \mathcal{Q}(\phi)$ for which $L_\phi(x) \neq 0$, there exist $m \geq 1$ and a neighborhood U of x such that*

$$H_\phi\big(\phi^m(z)\big) = H_\phi(z) + m \quad \text{for all } z \in U,$$

where $H_\phi : U \to \overline{D}(0,|m|)$ is an antiderivative of $1/L_\phi$ that is injective on U.

Corollary 10.11. *Let $\phi(z) \in \mathbb{C}_v(z)$ be a rational function of degree $d \geq 2$, where* $\operatorname{char}\mathbb{C}_v = 0$ *and* $\operatorname{char} k = p > 0$. *The all nonrepelling periodic points of ϕ in $\mathbb{P}^1(\mathbb{C}_v)$ are isolated.*

Proof. Given an attracting periodic point x of ϕ, there are no other periodic points in some neighborhood of x, by Proposition 4.3(a). Given an indifferent periodic point x, we have $x \in \mathcal{Q}(\phi)$, by Theorem 10.9. Therefore, by Theorem 10.10(a), x is a zero of L_ϕ. Moreover, L_ϕ is a power series converging on a neighborhood U of x, with zeros precisely at the periodic points in U. Since ϕ has countably many periodic points (because $d \geq 2$), L_ϕ cannot be trivial, and hence its zeros are isolated. In particular, there are no other periodic points in some neighborhood of x. □

Proof of Theorem 10.10.

(a) By Proposition 10.6(d), $x \in \mathcal{Q}(\phi)$ is preperiodic if and only if it is periodic. Thus, the reverse implication is immediate: if x is periodic, then $L_\phi(x) = 0$ by definition of L_ϕ. Moreover, it follows from Lemma 10.7 that a periodic point in $\mathcal{Q}(\phi)$ must be indifferent.

Conversely, suppose $L_\phi(x) = 0$. After a change of coordinates, we may assume $x = 0$. By hypothesis, there are an integer $m \geq 1$ and a radius $r > 0$ such that the limit of equation (10.1) converges uniformly on $\overline{D}(0, r)$. Replacing ϕ by an iterate, we may assume that the integer m is $m = 1$.

For each $n \geq 1$, define $g_n(z) := p^{-n}(\phi^{p^n}(z) - z)$. By the uniform convergence of (10.1), there is some $C > 0$ such that $\|g_n\|_{\zeta(0,r)} \leq C$ for all $n \geq 1$. By Exercise 10.6(c) with $j = p$, we have

$$\left\| \left(\phi^{p \cdot p^n}(z) - z\right) - p \cdot p^n g_n(z) \right\|_{\zeta(0,r)} \leq \frac{C}{r} |p|^{2n} \|g_n\|_{\zeta(0,r)}.$$

for sufficiently large n. Hence, for all large enough n, we have

$$\|g_{n+1}\|_{\zeta(0,r)} = \|g_n\|_{\zeta(0,r)}.$$

If g_n is identically zero for some such n, then $\phi^{p^n}(z)$ is identically equal to z on $\overline{D}(0, r)$, and in particular 0 is periodic. (And $\deg(\phi) = 1$.) Otherwise, the norm $\|g_n\|_{\zeta(0,r)}$ is eventually constant and nonzero. The limit L_ϕ must have the same nonzero norm and hence is a nontrivial power series. Thus, its zeros are isolated. In particular, for some small enough $s > 0$, the only zero of L_ϕ in $\overline{D}(0, s)$ is 0 itself. Further increasing n if necessary, we have

$$\left\|\phi^{p^n}(z) - z\right\|_{\zeta(0,r)} \leq C|p|^n < s,$$

and hence $\phi^{p^n}(0) \in \overline{D}(0, s)$. Moreover, by the third equation of Proposition 10.6(e), we have

$$L_\phi\big(\phi^{p^n}(0)\big) = \big(\phi^{p^n}\big)'(0) \cdot L_\phi(0) = 0.$$

That is, $\phi^{p^n}(0)$ is a zero of L_ϕ in $\overline{D}(0, s)$, and hence $\phi^{p^n}(0) = 0$, as desired.

(b) This is left to the reader; see Exercise 10.9.

(c) Given $x \in \mathcal{Q}(\phi)$ with $L_\phi(x) \neq 0$, we may again assume that $x = 0$. Choose $m \geq 1$ and $r > 0$ so that the limit of equation (10.1) converges. There is some $s \in (0, r] \cap |\mathbb{C}_v^\times|$ such that L_ϕ has no zeros in $\overline{D}(0, s)$, and hence $1/L_\phi$ is analytic on $\overline{D}(0, s)$. Let $H_\phi(z) \in \mathbb{C}_v[[z]]$ be the (formal) antiderivative of $1/L_\phi$ with trivial constant term; then H_ϕ converges on

$D(0,s)$. Since $H'_\phi(0) = 1/L_\phi(0) \neq 0$, we may decrease s further if necessary so that H_ϕ maps $D(0,s)$ bijectively onto $D(0,|p|^n)$, for some $n \geq 1$.

By the convergence of the limit in equation (10.1), we may replace m by a multiple such that $\phi^m(0) \in D(0,s)$. For $z \in D(0,s)$, we have

$$\big(H_\phi \circ \phi^m\big)'(z) = \frac{(\phi^m)'(z)}{L_\phi(\phi^m(z))} = \frac{1}{L_\phi(z)} = H'_\phi(z),$$

where the second equality is by the third formula of Proposition 10.6(e). Since both H_ϕ and $H_\phi \circ \phi^m$ map $D(0,s)$ into $D(0,|p|^n)$, there must be some $c \in D(0,|p|^n)$ such that

$$H_\phi\big(\phi^m(z)\big) = H_\phi(z) + c \quad \text{for all } z \in D(0,s). \tag{10.3}$$

Furthermore,

$$\begin{aligned} L_\phi(0) &= \lim_{n\to\infty} \frac{\phi^{mp^n}(0) - 0}{mp^n} = \lim_{n\to\infty} \frac{H_\phi^{-1}(p^n c) - H_\phi^{-1}(0)}{mp^n} \\ &= \frac{c}{m}\big(H_\phi^{-1}\big)'(0) = \frac{c}{m}\cdot\frac{1}{H'_\phi(0)} = \frac{c}{m}\cdot L_\phi(0). \end{aligned}$$

Since $L_\phi(0) \neq 0$, we must have $c = m$. Choosing $U = D(0,s)$, equation (10.3) is then the desired result. □

Remark 10.12. In this section, we have made only modest changes (most notably Theorem 10.9) to generalize Rivera-Letelier's results of [**114**, Section 3.2] from $\mathbb{C}_p$ to arbitrary non-archimedean fields of mixed characteristic. However, similar iterative logarithms predate [**114**]. Écalle [**49**] developed a formal iterative logarithm in the 1970s over $\mathbb{C}$, in his study of the complex dynamics on parabolic components attached to a rationally indifferent periodic point. Moreover, the limit of Definition 10.5 nearly coincides with Lubin's 1994 definition of the "Lie logarithm" in [**99**, Section 4]. The difference is that Lubin defined his Lie logarithm only for p-adic power series with a fixed point at the origin of the multiplier congruent to 1 modulo the maximal ideal. In that context, Lubin proved special cases of several results in this section, including Theorem 10.10.

10.3. Counting cycles of Fatou components

In another contrast with the complex setting, a non-archimedean rational function can have infinitely many periodic Fatou components, and in fact it usually does. For example, any map of explicit good reduction has infinitely many periodic residue classes. Still, there are some restrictions on what is possible, as we describe in this section. We begin with the following result, which first appeared as [**114**, Proposition 4.8].

Proposition 10.13 (Rivera-Letelier). *Let $\phi \in \mathbb{C}_v(z)$ be a rational function of degree $d \geq 2$. Then ϕ has at most $d-1$ cycles of attracting periodic components of Cantor type.*

The proof of Proposition 10.13 uses the following lemma, which first appeared as [**12**, Lemma 3.1.4] and later as [**114**, Lemme 4.10].

Lemma 10.14. *Let $\phi \in \mathbb{C}_v(z)$ be a rational function of degree $d \geq 1$, and let $D_1, \ldots, D_m \subseteq \mathbb{P}^1_{\text{an}}$ be disks whose closures are pairwise disjoint. Suppose that for each $i = 1, \ldots, m$, at least one connected component of $\phi^{-1}(D_i)$ is not a disk. Then $m \leq d-1$.*

Proof. This is left to the reader; see Exercise 10.10. □

Proof of Proposition 10.13. Given an attracting periodic component W of Cantor type and of minimal period M, there is a unique attracting periodic point $x_0 \in W$. Choose an open disk V_0 with $x_0 \in V_0 \subseteq W$.

Define $x_i = \phi^i(x_0)$ for $i = 1, \ldots, M-1$. For every $n \geq 1$, let V_n be the connected component of $\phi^{-n}(V_0)$ containing some x_i. Because W is of Cantor type, there is some minimal $n \geq 1$ such that V_n is not a disk. Thus, some iterate of W contains an open disk $V := V_{n-1}$ for which $\phi^{-1}(V)$ is not a finite union of disks. In addition, the unique boundary point in ∂V must lie in the Fatou set and hence in W, as otherwise $W = V$ would be a disk.

Suppose there are at least m distinct cycles of attracting periodic components of Cantor type. By the previous paragraph, each such cycle of components contains an open disk D_i such that $\phi^{-1}(D_i)$ is not a finite union of disks, and whose boundary point ζ_i is contained in the same cycle of Fatou components. Thus, the closures $\overline{D}_i = D_i \cup \{\zeta_i\}$ are pairwise disjoint. Therefore, by Lemma 10.14, we have $m \leq d-1$. □

It is currently unknown whether there is a bound for the number of cycles of indifferent periodic components that are not disks. No such bound has been proven, but the only known construction of maps with many such components (in [**114**, Section 6.1]; see Exercise 9.13) requires large degree. By contrast, the following result shows that it is quite common to have infinitely many periodic Fatou components that are disks.

Theorem 10.15. *Let $\phi \in \mathbb{C}_v(z)$ be a rational function, and let $\zeta \in \mathbb{P}^1_{\text{an}}$ be a periodic Type II point in the Julia set of ϕ of period $m \geq 1$.*

(a) *If the reduction of ϕ^m at ζ is inseparable, then*
- *infinitely many residue classes at ζ are attracting periodic Fatou components, and*
- *no indifferent components have ζ as a boundary point.*

(b) *If ζ is repelling and the reduction of ϕ^m at ζ is separable, then*
- *infinitely many residue classes at ζ are indifferent periodic Fatou components, and*
- *there at most $2d-2$ cycles of attracting components among the residue classes at ζ, and each such cycle contains a critical point of ϕ.*

(c) *If ζ is indifferent, then at most two residue classes at ζ are periodic Fatou components, and any such components are indifferent.*

Note that case (a) of Theorem 10.15, which first appeared as [**114**, Corollaire 4.7], can only occur if $\mathbb{C}_v$ has positive residue characteristic $p > 0$, and if $\deg(\phi) \geq p$. Case (b) first appeared as [**114**, Corollaire 5.16].

Proof of Theorem 10.15. After a change of coordinates, we may assume $\zeta = \zeta(0,1)$. Case (c) is immediate from Exercise 8.10(b). Thus, we assume hereafter that ζ is repelling.

By Exercise 8.10(a), there are infinitely many periodic residue classes at ζ. Only finitely many are bad directions (in the sense of Theorem 7.34), and hence infinitely many are periodic Fatou components of ϕ.

(a) By hypothesis, there is some $\psi \in k(z)$ such that $\overline{\phi^m}(z) = \psi(z^p)$, with $p = \operatorname{char} k \geq 2$. Thus, the local degrees in every direction $\boldsymbol{v}$ satisfy $\deg_{\zeta,\boldsymbol{v}}(\phi^m) \geq p$. It follows that a periodic Fatou component U with boundary point ζ must map onto itself (under some iterate of ϕ^m) with multiplicity at least $p \geq 2$. By Theorems 9.7 and 9.14, U must be an attracting component, yielding both desired statements.

(b) Since ϕ has $2d-2$ critical points, it suffices to show that each cycle of attracting residue classes at ζ contains a critical point. Given a residue class U at $\zeta(0,1)$ which is an attracting periodic Fatou component, there is some $n \geq 1$ such that $\phi^{mn}(U) = U$. Since $\overline{\phi^m}$ is separable, so is $\overline{\phi^{mn}} = \left(\overline{\phi^m}\right)^n$. By Exercise 8.14 then, ϕ^{mn} must have a critical point in U. Equivalently, ϕ has a critical point in some iterate of U, as desired. □

Corollary 10.16. *Let $\phi \in \mathbb{C}_v(z)$ be a rational function of degree $d \geq 2$.*

(a) *If ϕ has more than $3d-3$ attracting periodic cycles, then it has infinitely many.*

(b) *If the residue field of $\mathbb{C}_v$ is algebraic over a finite field, and if the indifference domain of ϕ is nonempty (for example, if ϕ has at least one indifferent periodic point), then ϕ has infinitely many periodic cycles of indifferent components.*

Part (a) of Corollary 10.16 first appeared as [**114**, Corollaire 4.9], and part (b) first appeared as [**114**, Corollaire 5.17].

Proof of Corollary 10.16.

(a) By Proposition 10.13, ϕ has at most $d-1$ attracting periodic cycles in domains of Cantor type. By Theorem 9.5, all other attracting Fatou components are disks. If any cycle of such components lacks a critical point of ϕ, then by Theorem 10.15, ϕ must have an inseparable periodic Type II point, and there are infinitely many attracting periodic cycles. Otherwise, all cycles of attracting components that are disks contain critical points; hence, there are at most $2d-2$ such cycles. Adding the cycles of Cantor type, we have proven our desired bound of $3d-3$.

(b) Let U be an indifferent periodic component of ϕ. By Theorem 9.7, U has at least one boundary point ζ, which is a periodic Type II point in the Julia set, of some period $m \geq 1$. By Theorem 8.7 and Remark 8.9, because the residue field is algebraic over a finite field, ζ must be repelling. By Theorem 10.15, the reduction of ϕ^m at ζ must be separable, and infinitely many residue classes at ζ must be indifferent periodic components of ϕ. □

Non-archimedean attracting components need not contain critical points; Theorem 10.15 illustrates this phenomenon for disks, and Exercise 10.12 does so for domains of Cantor type. However, if the residue characteristic is either larger than $\deg(\phi)$ or equal to zero, then the parallel with complex dynamics returns, as the following result shows.

Theorem 10.17. *Let $p = \operatorname{char} k$, and let $\phi \in \mathbb{C}_v(z)$ be a rational function of degree $d \geq 2$. Suppose that $p = 0$ or $p > d$. Then every attracting periodic point attracts a critical point of ϕ. In particular, there can be at most $2d-2$ attracting periodic cycles.*

Proof. Given an attracting periodic point a of ϕ, it suffices to show that there is a critical point of ϕ in the attracting basin of a. Replacing ϕ by an iterate, we may assume for the remainder of the proof that a is fixed.

Because $\phi'(a) \neq 1$, the multiplicity of a as a fixed point is 1. Since ϕ has $d+1$ fixed points in $\mathbb{P}^1(\mathbb{C}_v)$, there must be at least one other fixed point. Hence, after a change of coordinates, we may assume that $a = 0$, and that $\phi(\infty) = \infty$. If $\phi'(0) = 0$, then a itself is critical, and we are done. Thus, we assume for the remainder of the proof that $\lambda := \phi'(0)$ satisfies $0 < |\lambda| < 1$.

Let U be the smallest open disk containing 0 and no other zeros or poles of ϕ. Since a is not critical, there must be at least one zero of ϕ in $\mathbb{C}_v \smallsetminus \{0\}$. Therefore, U is a rational open disk, and after a final change of coordinates of the form $z \mapsto cz$, we may assume that $U = D_{\text{an}}(0,1)$. Note that ϕ maps U bijectively onto $D_{\text{an}}(0,|\lambda|) \subsetneq U$.

Let V be the connected component of $\phi^{-1}(U)$ containing 0, and hence containing U. Then V is an open affinoid of the form

$$V = D_{\mathrm{an}}(0,R) \smallsetminus \big(W_1 \cup \cdots \cup W_\ell\big),$$

where $R > 1$, $\ell \geq 0$, and each W_i is a disk of the form $\overline{D}_{\mathrm{an}}(b_i, s_i)$. Moreover, the disks W_i are disjoint from each other and from U. Let $\zeta_0 := \zeta(0,R)$, and for each $i = 1, \ldots, \ell$, let $\zeta_i = \zeta(b_i, s_i)$. In addition, for each $i = 0, \ldots, \ell$, let $\boldsymbol{v}_i$ be the direction at ζ_i containing V. Then $\phi(\zeta_i) = \zeta(0,1)$, with $\phi_\#(\boldsymbol{v}) = \vec{v}(0)$.

Each directional degree $m_i := \deg_{\zeta_i, \boldsymbol{v}_i}(\phi)$ is at most d and hence not divisible by p. By Proposition 7.33, $\phi : V \to U$ is M-to-1 for some integer $M \geq 1$. By considering the inverse images of $\zeta(0,s)$ for $s < 1$ sufficiently close to 1, we see that $m_0 + \cdots + m_\ell = M$.

Note that ϕ has Weierstrass degree m_0 on a thin annulus extending from ζ_0 toward 0. On the other hand, for each $i = 1, \ldots, \ell$, because of the reversal of direction, ϕ has Weierstrass degree $-m_i$ on a thin annulus extending from ζ_i toward ∞. By Exercise 3.22 and the fact that $p \nmid m_i$, we see that ϕ' has Weierstrass degree $m_0 - 1$ just inside ζ_0, but Weierstrass degree $-m_i - 1$ just outside ζ_i for $i = 1, \ldots, \ell$. Applying Exercise 3.21 to each such annulus,

$$m_0 - 1 = \#\big(\text{zeros of } \phi' \text{ in } D(0,R)\big) - \#\big(\text{poles of } \phi' \text{ in } D(0,R)\big),$$

and for each $i = 1, \ldots, \ell$,

$$m_i + 1 = \#\big(\text{poles of } \phi' \text{ in } \overline{D}(b_i, s_i)\big) - \#\big(\text{zeros of } \phi' \text{ in } \overline{D}(b_i, s_i)\big).$$

Summing the above equalities and observing that ϕ' has no poles in V, we obtain

$$M + \ell - 1 = \#\big(\text{zeros of } \phi' \text{ in } V\big). \tag{10.4}$$

Recall that $M \geq 1$; and if $\ell = 0$, then V is a disk, and by construction we must have $M \geq 2$. Either way, then, the left-hand side of equation (10.4) is positive, and hence ϕ has a critical point in V. It follows that some critical point of ϕ has a forward iterate in U, which is then attracted to the original fixed point $a = 0$ under iteration. □

Theorem 10.17 first appeared as [**25**, Corollary 1.6], where it was deduced as a consequence of the following more powerful statement, whose proof we omit here.

Theorem 10.18. *Let $\phi \in \mathbb{C}_v(z)$ be a rational function of degree $d \geq 2$, and suppose that ϕ has a fixed point $a \in \mathbb{P}^1(\mathbb{C}_v)$ with multiplier λ satisfying*

$$0 < |\lambda| < |\deg_{\zeta,\boldsymbol{v}}(\phi)|^d$$

for all $\zeta \in \mathbb{P}^1_{\mathrm{an}}$ and all directions $\boldsymbol{v}$ at ζ. Then there is a critical point that is not preperiodic but is in the immediate attracting basin of a.

Proof. See [**25**, Theorem 4.1]. See also Exercises 10.14–10.15. □

10.4. Infinitely many periodic components

Theorem 10.15 shows that $\phi \in \mathbb{C}_v(z)$ can easily have infinitely many periodic Fatou components, but the components it produces all share the same boundary point $\zeta \in \mathbb{P}^1_{\text{an}}$. In this section, we describe maps with infinitely many periodic Fatou components with *different* boundary points. Equivalently, by Theorem 10.15 and the other results of Section 10.3, we describe maps with infinitely many repelling periodic Type II points.

Example 10.19. Let $c \in \mathbb{C}_v$ with $0 < |c| < 1$, and consider the Lattès map

$$\phi(z) = \frac{z^4 - 8c^4 z - c^4}{4z^3 + z^2 + 4c^4}$$

from several previous examples. We leave the following claims to the reader in Exercise 10.18.

For any odd integer $n \geq 1$, there is at least one relatively prime integer m with $0 \leq m \leq 2n$ such that $\zeta := \zeta(0, |c|^{m/n})$ is a repelling periodic point of ϕ. Moreover, the reduction of the appropriate iterate ϕ^ℓ at ζ is separable if and only if the residue characteristic $p := \operatorname{char} k$ is not 2.

Thus, there are infinitely many different Type II repelling periodic points, each with infinitely many periodic Fatou component disks attached. Moreover, if $p = 2$, then all of these periodic components are attracting. Otherwise, if $p \neq 2$, then they are all indifferent.

In an even bigger contrast with the complex setting, it is possible to have an infinite sequence of periodic Fatou disks accumulating at a Type I Julia point, at least for the case that the residue characteristic p satisfies $0 < p \leq \deg(\phi)$. The following example illustrates this phenomenon.

Example 10.20. Let $\mathbb{C}_v$ have mixed characteristic, with residue characteristic $p > 0$. Define

$$\phi(z) = \frac{1}{p} z^p \big(z + (p-1)\big) = \frac{1}{p} z^{p+1} + \frac{(p-1)}{p} z^p,$$

which has

$$\phi'(z) = \Big(\frac{p+1}{p}\Big) z^p + (p-1) z^{p-1} = \frac{z^{p-1}}{p}\big((p+1)z + p(p-1)\big).$$

Then $\zeta(0, |p|^{1/(p-1)})$ is a Type II repelling fixed point of multiplier p, and $z = 1$ is a Type I repelling fixed point, with multiplier $\lambda := p + p^{-1}$. Let

$$U := \big\{ |p|^{1/(p-1)} < |\zeta| \leq |p|^{1/p} \big\} \quad \text{and} \quad V := \overline{D}_{\text{an}}(1, |p|).$$

Note that $\phi(U) = \{|p|^{1/(p-1)} < |\zeta| \leq 1\}$ and $\phi(V) = \overline{D}_{\text{an}}(0,1)$, both of which contain $U \cup V$. In addition, we have

$$|\phi(z)| = |p|^{-1}|z|^p \quad \text{and} \quad |\phi'(z)| = |p|^{-1}|z|^p \quad \text{for all } z \in U \cap \mathbb{C}_v$$

and

$$|\phi(z) - 1| = |p|^{-1}|z-1| \quad \text{and} \quad |\phi'(z)| = |p|^{-1} \quad \text{for all } z \in V \cap \mathbb{C}_v.$$

This is exactly the setting of Exercise 10.16.

For any $n \geq 1$, let $m := (p-1)(n+1)$, and let A be the symbol sequence that consists of the string

$$\underbrace{2,\ldots,2}_{n},\underbrace{1,\ldots,1}_{m}$$

repeated forever. By Exercise 10.16, there is either a point x or (indifferent) disk $\overline{D}(y,t)$ that is periodic of minimal period $P := m+n$ and with forward orbit through U and V given by A. Moreover, the derivative of ϕ^P at x, or at any point $x \in \overline{D}(y,t)$, as appropriate, satisfies

$$\left|(\phi^P)'(x)\right| = |p|^{1-\ell} < 1, \quad \text{where } \ell = \frac{p^m - 1}{p^{m-1}(p-1)^2}.$$

This small derivative makes the periodic indifferent disk impossible, yielding an attracting periodic point, and hence an attracting periodic component V_n, of period $n+m$. In addition, the distance from any point of V_n to 1 is $|\lambda|^{-n}|1| = |p|^n$. Thus, there is an infinite sequence of attracting periodic Fatou components that accumulate at the repelling fixed point $z=1$. The reader may note that any periodic symbol sequence A with a large enough proportion of 1's to 2's also leads to an attracting component; and opening with enough 2's in a row places it very close to $z=1$.

We now show that there are also indifferent components accumulating at $z=1$. For any $n \geq 1$, let

$$m := p + (p-1)n \quad \text{and} \quad j := p^{m-1}(p-1),$$

and let

$$N := p^m - (1 + p + \cdots + p^{m-1}) \geq 1.$$

Let B be the symbol sequence that consists of $j-1$ copies of the segment

$$\underbrace{2,\ldots,2}_{n},\underbrace{1,\ldots,1}_{m}$$

followed by a single copy of the segment

$$\underbrace{2,\ldots,2}_{n+N},\underbrace{1,\ldots,1}_{m},$$

with that entire string of $P := j(n+m) + N$ symbols then repeated forever. Note that the minimal period of the periodic sequence B is indeed P, because of the different final segment of the string. By similar reasoning as for the attracting components just above, there is a periodic point or disk with minimal period P, and with $|(\phi^P)'| = |p|^\ell$, where

$$\ell = j\Big[\frac{pm}{p-1} - \frac{p(p^m-1)}{p^m(p-1)^2} - m - n\Big] - N = 0.$$

Thus, Exercise 10.16 yields either a disk that ϕ^P maps bijectively to itself or else an indifferent periodic point of minimal period P. Either way, we have an indifferent periodic component W_n of period P, and the distance from any point of W_n to 1 is $|\lambda|^{-1}|1| = |p|^n$. That is, there is a sequence of indifferent periodic Fatou components accumulating at the repelling fixed point $z = 1$.

Example 10.20 requires $\mathbb{C}_v$ to be of mixed characteristic, but the same phenomenon can occur when $\operatorname{char}\mathbb{C}_v > 0$; see Exercise 10.17. In both cases, it is possible to have both indifferent and attracting Fatou components accumulating at a Type I repelling fixed point Also in both cases, we require that $0 < p \le \deg(\phi)$, where p is the residue characteristic. It is currently unknown whether or not similar behavior can occur if $p = 0$ or if $p > \deg(\phi)$.

Exercises for Chapter 10

10.1. Use the result of Exercise 3.21 to prove Proposition 10.3.

10.2. Prove Corollary 10.4(c).

(*Suggestion*: Use Exercise 1.24(c).)

10.3. Let K be a locally compact non-archimedean field, and let $\mathbb{C}_v$ be the completion of an algebraic closure of K. Let $\phi \in K(z)$, and let $a \in \mathcal{F}_{\phi,\mathrm{I}}$ be a point in the Type I Fatou set of ϕ. Prove that there is an open set $U \subseteq \mathbb{P}^1(\mathbb{C}_v)$ containing a such that the family $\{\phi^n : n \ge 0\}$ is normal on U. That is, prove that every sequence $\{\phi^{n_i}\}_{i\ge 1}$ has a subsequence that converges uniformly on U. (This proves a claim made in Remark 5.9.)

(*Suggestion*: Use Rivera-Letelier's classification (Theorem 9.14), and apply results of this chapter to the cases that the Fatou component of a is preperiodic to an attracting or indifferent component, or wandering.)

10.4. Show that Exercise 10.3 can fail without the local compactness hypothesis. More precisely, assume the residue field $\overline{k}$ of $\mathbb{C}_v$ is not algebraic over a finite field. Give an example of a rational function $\phi \in \mathbb{C}_v(z)$ and a point $a \in \mathcal{F}_{\phi,\mathrm{I}}$ in a wandering domain such that for any open set $U \subseteq \mathbb{P}^1(\mathbb{C}_v)$ containing a, the sequence $\phi^n|_U$ has no pointwise-convergent subsequence.

(*Suggestion*: Use a map of explicit good reduction, and show there is a wandering residue class.)

10.5. Repeat Exercise 10.4, this time with the Type I Fatou point a lying in an indifferent fixed component, rather than a wandering domain.

10.6. Let $\phi(z) \in \mathbb{C}_v(z)$, where $\operatorname{char} \mathbb{C}_v = 0$ and $\operatorname{char} k = p > 0$. Let $x \in \mathbb{C}_v$, $m \geq 1$, and $r > 0$ such that the limit of equation (10.1) converges uniformly on $\overline{D}(x,r)$.

(a) For each $n \geq 1$, define $g_n \in \mathbb{C}_v(z)$ by

$$\phi^{mp^n}(z) = z + mp^n g(z).$$

Prove that $g_n \in \overline{\mathcal{A}}(x,r)$, and that $\lim_{n\to\infty} \|g_n\|_{\zeta(x,r)} = 0$.

(b) Prove that ϕ^m maps $\overline{D}(x,r)$ onto itself with Weierstrass degree 1.

(c) Prove that for any $j \geq 1$ and any sufficiently large n,

$$\left\| \phi^{jmp^n}(z) - \big(z + jmp^n g_n(z)\big) \right\|_{\zeta(x,r)} \leq \frac{1}{r} |p|^{2n} \|g_n\|^2_{\zeta(x,r)}.$$

10.7. Prove Proposition 10.6.

(*Suggestion*: Use Exercise 10.6.)

10.8. Prove Lemma 10.8(a).

10.9. Let $\phi(z) \in \mathbb{C}_v(z)$, where $\operatorname{char} \mathbb{C}_v = 0$ and $\operatorname{char} k = p > 0$. Let x be an indifferent periodic point of period $m \geq 1$ and multiplier λ.

(a) Prove that $x \in \mathcal{Q}(\phi)$ if and only if there is some $n \geq 1$ such that $|\lambda^n - 1| < 1$.

(b) In that case, prove that

$$L'_\phi(x) = \frac{1}{mn} \lim_{j\to\infty} \frac{\lambda^{p^j} - 1}{p^j} = \frac{1}{mn} \sum_{i=1}^{\infty} \frac{(-1)^{i-1}}{i} (\lambda^n - 1)^i.$$

(*Note*: This value can reasonably be called $\frac{1}{mn} \log(\lambda^n)$.)

(c) Prove that a point $x \in \mathbb{P}^1(\mathbb{C}_v)$ is a rationally indifferent periodic point if and only if

$$x \in \mathcal{Q}(\phi) \quad \text{and} \quad L_\phi(x) = L'_\phi(x) = 0.$$

10.10. Prove Lemma 10.14.

(*Suggestion*: Let $W = \mathbb{P}^1_{\text{an}} \smallsetminus (D_1 \cup \cdots \cup D_m)$. How many connected components does W have?)

10.11. Prove Lemma 10.14 under the weaker hypothesis that the disks $D_1, \ldots, D_m$ are pairwise disjoint (as opposed to having disjoint closures) and that their union is not all of $\mathbb{P}^1_{\text{an}}$.

10.12. Suppose $\mathbb{C}_v$ has residue characteristic $p > 0$, and let

$$\phi(z) = a\Big(z^p + \frac{z}{z-1}\Big), \quad \text{where } |p|^{(p-1)/(p+1)} < |a| < 1.$$

Prove that the fixed point $z = 0$ lies in an attracting domain of Cantor type but does not attract any critical points. (This example first appeared as [**114**, Exemple 6.2], for $\mathbb{C}_v = \mathbb{C}_p$.)

10.13. Let $p = \operatorname{char} k$, and let $\phi \in \mathbb{C}_v[z]$ be a polynomial of degree $d \geq 2$. Suppose that $p \nmid \deg_\zeta(\phi)$ for all points $\zeta \in \mathbb{P}^1_{\text{an}}$. (Such a map ϕ is said to be *tame*; see Definition 11.22.) Prove that either

(a) ϕ has potential good reduction, or

(b) there is a (finite, Type I) critical point $c \in \mathbb{C}_v$ in the immediate attracting basin of ∞.

Deduce that at most $d - 2$ critical points lie in the filled Julia set $\mathcal{K}_{\phi,\text{an}}$.

(*Suggestion*: If ϕ does not have potential good reduction, prove that the Fatou component U at ∞ is a domain of Cantor type. Then use the ideas of the proof of Theorem 10.17 to show that there is a critical point in U besides ∞.)

(*Note*: This is essentially [**137**, Proposition 2.11]; see also [**83**, Corollary 2.11]. Observe that if $p = 0$ or $p > d$, then ϕ is necessarily tame.)

10.14. Let $\phi \in \mathbb{C}_v(z)$ be a nonconstant rational function. Define a function $\delta(\phi, \cdot) : \mathbb{H} \to \mathbb{R}$ by

$$\delta(\phi, \zeta) := \log \operatorname{diam}(\zeta) + \log \|\phi'\|_\zeta - \log \|\phi\|_\zeta.$$

(a) Let $\zeta \in \mathbb{H}$, and let $\boldsymbol{v}$ be a direction at ζ. Suppose that $\phi_\#(\boldsymbol{v})$ contains either 0 or ∞, but not both. Prove that

$$\log \big| \deg_{\zeta,\boldsymbol{v}}(\phi) \big| \leq \delta(\phi, \zeta) \leq 0.$$

(b) Prove that $\delta(\phi, \cdot)$ is continuous with respect to the hyperbolic metric on $\mathbb{H}$ and is upper semicontinuous with respect to the Gel'fand topology.

(c) Prove that on any closed interval in $\mathbb{H}$, $\delta(\phi, \cdot)$ is piecewise linear with respect to the hyperbolic metric, with integer slopes.

The function $\delta(\phi, \cdot)$ is called the *distortion* of ϕ. It appeared in exponential form in [**20**], and it is used in the proof of [**25**, Theorem 4.1].

10.15. This exercise sketches the proof of Theorem 10.18 (i.e., [**25**, Theorem 4.1]). Suppose we are given ϕ, d, a, and λ as in Theorem 10.18. Change coordinates and construct the disk U and affinoid V (and disks $W_i = \overline{D}(b_i, s_i)$ and radius $R > 1$) as in the proof of Theorem 10.17. Let $M \geq 1$ be the degree with which ϕ maps V onto U. Define $G : \mathbb{H} \to \mathbb{R}$ by

$$G(\zeta) := M\delta(\phi, \zeta) + \log \|\phi\|_\zeta,$$

where $\delta(\phi, \cdot)$ is the distortion function of Exercise 10.14.

(a) Prove that $G(\zeta(0,1)) = \log|\lambda|$, that $G(\zeta(0,R)) > \log|\lambda|$, and that $G(\zeta(b_i, s_i)) > \log|\lambda|$ for each i.
(*Suggestion*: Use the hypotheses and Exercise 10.14(a).)

(b) Prove that along any closed interval $[\zeta_0, \zeta_1]$ in $\mathbb{H}$, G is continuous and piecewise linear with integer slopes, with respect to the hyperbolic metric (cf. Exercise 10.14(b)–(c)).

(c) Show that there is a closed affinoid $V' \subseteq V$ containing U such that at every boundary point ξ of V', G increases in the direction at ξ that does not intersect V'.
(*Suggestion*: Choose the boundary points of V' by moving along the intervals from $\zeta(0,1)$ to the boundary points of V until G starts to increase.)

(d) By counting critical points, poles, and distinct zeros of ϕ in various disks and affinoids in a manner similar to the proof of Theorem 10.17, prove that V' must contain a critical point α of ϕ that is not a zero of ϕ.

(e) Prove that $\phi^n(\alpha) \to 0$, and that α is not preperiodic.

10.16. This exercise applies the symbolic dynamics ideas of Exercise 1.26 to the existence of certain periodic orbits in non-archimedean dynamics.

Let $\phi \in \mathbb{C}_v(z)$, let $0 < r < R < |a|$ and $0 < S < |a|$, and let

$$U = \{r < |\zeta| \le R\} \quad \text{and} \quad V = \overline{D}_{\text{an}}(a, S)$$

be a half-open annulus and a closed disk. Suppose there are integers $e \ge 2$ and $f \ge 0$ and coefficients $b, c, \lambda \in \mathbb{C}_v^\times$ such that

$$|\phi(z)| = |b||z|^e \quad \text{and} \quad |\phi'(z)| = |c||z|^f \quad \text{for all } z \in U \cap \mathbb{C}_v,$$

$$|\phi(z) - a| = |\lambda||z - a| \quad \text{and} \quad |\phi'(z)| = |\lambda| \quad \text{for all } z \in V \cap \mathbb{C}_v,$$

and

$$|a| = |b|R^e = |\lambda|S \quad \text{and} \quad |b|r^e = r.$$

(a) Prove that $\phi(U) \supseteq U \cup V$, that $\phi(V) \supseteq U \cup V$, and that $\zeta(0,r)$ and a are repelling fixed points of ϕ.

(b) Let $Y_1 := \overline{U} = U \cup \{\zeta(0,r)\}$ and $Y_2 := V$, and let $\Sigma := \prod_{n \ge 0}\{1,2\}$. Let $A \in \Sigma$ be a symbol sequence that begins with n 2's (in positions 0 to $n-1$) followed by a 1. Prove that $|x - a| = |\lambda|^{-n}|a|$ for every x in the cylinder set $C_{A,n}$.

(c) Prove that for every *periodic* symbol sequence $A \in \Sigma$ of minimal period $P \ge 1$, the set Z_A of Exercise 1.26 contains either a periodic Type I point x or a disk $\overline{D}(x,t)$ that ϕ^P maps bijectively onto itself (or both), of minimal period P.
(*Note*: The point/disk need not be unique.)

(d) With Σ as in part (b), let $A \in \Sigma$ be a symbol sequence that begins

$$\underbrace{2,\ldots,2}_{n_1},\underbrace{1,\ldots,1}_{m_1},\underbrace{2,\ldots,2}_{n_2},\underbrace{1,\ldots,1}_{m_2},\ldots\underbrace{2,\ldots,2}_{n_j},\underbrace{1,\ldots,1}_{m_j},$$

where $m_1,\ldots,m_j$ and $n_1,\ldots,n_j$ are positive integers. Let

$$P := \sum_{i=1}^{j}(m_i+n_i),$$

and let x be an element of the cylinder set $C_{A,P-1}$. Prove that

$$\left|(\phi^P)'(x)\right| = |a|^L|b|^{-M}|c|^{m_1+\cdots+m_j}|\lambda|^{n_1+\cdots+n_j},$$

where

$$L := \sum_{i=1}^{j}\frac{f(e^{m_i}-1)}{e^{m_i}(e-1)} \quad\text{and}\quad M := \sum_{i=1}^{j}\left[\frac{fm_i}{e-1}-\frac{f(e^{m_i}-1)}{e^{m_i}(e-1)^2}\right].$$

10.17. Let $\mathbb{C}_v$ have residue characteristic $p>0$. Fix $a\in\mathbb{C}_v$ with $0<|a|<1$ and $|p|\le|a|$. Let

$$\phi(z) = (a^p-a)z^{p+1}+z^p = z^p\big((a^p-a)z+1\big),$$

and define

$$U := \{1<|\zeta|\le|a|^{-1/p}\} \quad\text{and}\quad V := \overline{D}_{\mathrm{an}}(1/a,|a|^{p-2}).$$

(a) Prove that $1/a$ is a Type I repelling fixed point of ϕ of some multiplier $\lambda\in\mathbb{C}_v$, with $|\lambda| = |a|^{1-p}$.

(b) Prove that $\zeta(0,1)$ is a Type II repelling fixed point of ϕ of multiplier p.

(c) Prove that ϕ maps U p-to-1 onto $\{1<|\zeta|\le|a|^{-1}\}\supseteq U\cup V$, with

$$|\phi(z)| = |z|^p \quad\text{and}\quad |\phi'(z)| = |a||z|^p \quad\text{for all } z\in U.$$

(d) Prove that ϕ maps V bijectively onto $\overline{D}_{\mathrm{an}}(0,|a|^{-1})\supseteq U\cup V$, with

$$|\phi(z)-1/a| = |\lambda|\,|z-1/a| \quad\text{and}\quad |\phi'(z)| = |\lambda| \quad\text{for all } z\in V.$$

(e) Use Exercise 10.16 and the method of Example 10.20 to show that there is an infinite sequence of attracting periodic Fatou components accumulating at $z=1/a$.
(*Note*: The point a of Exercise 10.16 is the point $1/a$ of this exercise.)

(f) Use Exercise 10.16 and the method of Example 10.20 to show that there is an infinite sequence of indifferent periodic Fatou components accumulating at $z=1/a$.

10.18. Let $c \in \mathbb{C}_v$ with $0 < |c| < 1$, and let ϕ be the Lattès map

$$\phi(z) = \frac{z^4 - 8c^4 z - c^4}{4z^3 + z^2 + 4c^4}.$$

In this exercise, you will prove several claims made in Example 10.19.

(a) Let $\zeta \in \mathbb{P}^1_{\text{an}}$. Prove that ζ is a Julia preperiodic point of ϕ if and only if $\zeta = \zeta(0, |c|^q)$, where $q \in [0, 2] \cap \mathbb{Q}$. If ζ is periodic, prove the denominator of q is odd. For each odd $n \geq 1$, prove there is at least one relatively prime $m \in \{1, \ldots, 2n-1\}$ such that $\zeta(0, |c|^{m/n})$ is periodic.

(b) Let $\zeta \neq \zeta(0, 1)$ be a periodic point as in part (a), of minimal period $\ell \geq 1$. Prove that the reduction $\psi \in k(z)$ of ϕ^ℓ at ζ is a map of the form $z \mapsto az^{\pm 2^\ell}$, for some $a \in k^\times$.
(At $\zeta(0, 1)$, the reduction is $z^2/(4z + 1)$.)

(c) If $\operatorname{char} k \neq 2$, prove that all six critical points of ϕ lie in strictly preperiodic Fatou components. Conclude that all periodic Fatou components are indifferent.

10.19. This exercise generalizes Example 10.19 and Exercise 10.18. It requires greater familiarity with elliptic curves.

Let E be an elliptic curve of multiplicative reduction over $\mathbb{C}_v$, such as $y^2 + xy = x^3 + a$, let $m \geq 2$, and let $\phi \in \mathbb{C}_v(z)$ be the flexible Lattès map for the multiplication-by-m map on E. (That is, if $\alpha : E \to \mathbb{P}^1$ is the quotient map by ± 1, then $\alpha \circ [m] = \phi \circ \alpha$.) Let $p := \operatorname{char} k$ be the residue characteristic.

(a) Prove that there are infinitely many periodic points in the Julia set of ϕ.

(b) Prove that at any such periodic point of minimal period $\ell \geq 1$, the reduction $\psi \in k(z)$ of ϕ^ℓ at ζ is separable if $p \nmid m$, and inseparable if $p | m$.

(c) If $p \nmid m$, prove that all $2m^2 - 2$ critical points of ϕ lie in strictly preperiodic Fatou components.
(*Suggestion*: Recall Exercise 4.20(b).)

(d) Conclude that all periodic Fatou components are indifferent if $p \nmid m$, or all are attracting if $p | m$.

Chapter 11

Wandering Domains

As noted in the introduction to Chapter 9, Sullivan proved in 1985 that in *complex* dynamics, the Fatou sets of rational functions do not have wandering domains. However, a complete understanding of the existence of wandering domains in *non-archimedean* dynamics remains an open problem. More precisely, there are some No Wandering Domains results assuming certain hypotheses, but there are also constructions of wandering domains in other circumstances. Much is known, but many questions remain.

11.1. Wandering domains are eventually disks

Definition 11.1. Let $\phi \in \mathbb{C}_v(z)$ be a rational function of degree $d \geq 2$ with Fatou set $\mathcal{F}_{\phi,\mathrm{an}}$. A *wandering domain* of ϕ is a connected component U of $\mathcal{F}_{\phi,\mathrm{an}}$ with the property that the iterates $U, \phi(U), \phi^2(U), \ldots$ are all distinct.

Theorem 11.2. *Let $\phi \in \mathbb{C}_v(z)$ be a rational function with Fatou set $\mathcal{F}_{\phi,\mathrm{an}}$, and let $U \subseteq \mathcal{F}_{\phi,\mathrm{an}}$ be a wandering domain of ϕ. Then there is some $N \geq 0$ such that $\phi^n(U)$ is a disk for all $n \geq N$.*

To prove Theorem 11.2, we need the following technical mapping property of connected affinoids.

Proposition 11.3. *Let $\phi \in \mathbb{C}_v(z)$ be a nonconstant rational function, and let $U \subseteq \mathbb{P}^1_{\mathrm{an}}$ be a connected Berkovich affinoid with at least two boundary points. Let $\delta > 0$ be the minimum hyperbolic distance between two distinct boundary points of U. Then for any two distinct boundary points $\xi_1, \xi_2 \in \partial(\phi(U))$, we have $d_{\mathbb{H}}(\xi_1, \xi_2) \geq \delta$.*

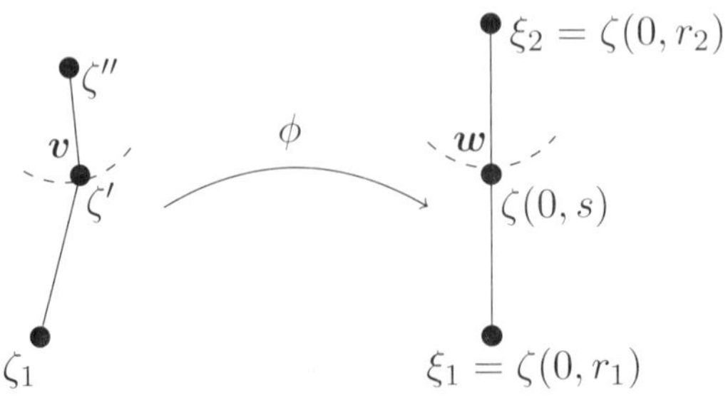

Figure 11.1. Proof of Proposition 11.3, Step 2.

Proof.

Step 1. Given distinct boundary points $\xi_1, \xi_2 \in \partial(\phi(U))$, we may assume after a change of coordinates that $\xi_i = \zeta(0, r_i)$, with $r_2 > r_1 > 0$. By Theorem 7.8(a), there exists $\zeta_1 \in \partial U$ with $\phi(\zeta_1) = \xi_1$. Define

$$\mathcal{S} := \left\{ r \in [r_1, r_2] : \begin{array}{l} \exists \zeta \in \overline{U} \text{ s.t. } \phi \text{ maps } [\zeta_1, \zeta] \\ \text{bijectively onto } [\xi_1, \zeta(0, r)] \end{array} \right\},$$

which is nonempty because $r_1 \in \mathcal{S}$. Define $s := \sup \mathcal{S}$. We claim that $s \in \mathcal{S}$.

Pick a sequence $\{s_n\}_{n \geq 1} \subseteq \mathcal{S}$ increasing to s, with a corresponding sequence of points $\{\zeta'_n\}_{n \geq 1} \subseteq \overline{U}$. Replacing the latter sequence by a convergent subsequence, we may assume there is some $\zeta' \in \overline{U}$ such that $\lim_{n \to \infty} \zeta'_n = \zeta'$. By continuity, we must have $\phi(\zeta') = \zeta(0, s)$.

In addition, by Theorem 7.34, any direction at ζ' containing any ζ'_n must also contain a preimage of 0; in particular, there are only finitely many such directions at ζ'. Again taking a subsequence, we may assume that every ζ'_n lies in the same direction $\boldsymbol{v}$ at ζ'. Since $\zeta'_n \to \zeta'$, there is some large enough n that ϕ is injective on $[\zeta'_n, \zeta']$, by Theorem 7.22. Since

$$\phi : [\zeta_1, \zeta'_n] \to [\xi_1, \zeta(0, s_n)] \quad \text{and} \quad \phi : [\zeta'_n, \zeta'] \to [\zeta(0, s_n), \zeta(0, s)]$$

are both bijective mappings, it follows that ϕ maps $[\zeta_1, \zeta']$ bijectively onto $[\xi_1, \zeta(0, s)]$. Hence, $s \in \mathcal{S}$, as claimed.

Step 2. Next, we claim that $s = r_2$. If not, then pick $\zeta' \in \overline{U}$ such that $\phi : [\zeta_1, \zeta'] \to [\xi_1, \zeta(0, s)]$ is bijective. Let $\boldsymbol{w}$ be the direction at $\zeta(0, s)$ containing ∞ (and hence containing ξ_2), and let $\boldsymbol{v}$ be a direction at ζ' such that $\phi_\#(\boldsymbol{v}) = \boldsymbol{w}$; see Figure 11.1. Again by Theorem 7.22, there is a point $\zeta'' \in \boldsymbol{v}$ such that ϕ maps $[\zeta', \zeta'']$ bijectively onto $[\zeta(0, s), \zeta(0, s+\varepsilon)]$ for some $\varepsilon > 0$. Hence, ϕ maps $[\zeta_1, \zeta'']$ bijectively onto $[\xi_1, \zeta(0, s + \varepsilon)]$. Therefore $s + \varepsilon \in \mathcal{S}$, contradicting the fact that $s = \sup \mathcal{S}$, and proving the claim.

Step 3. By the previous two steps, we have $r_2 \in \mathcal{S}$, and hence there is a point $\zeta_2 \in \overline{U}$ such that $\phi(\zeta_2) = \xi_2$, and ϕ maps $[\zeta_1, \zeta_2]$ bijectively onto

$[\xi_1, \xi_2]$. By Theorem 7.8(a), we have $\zeta_1, \zeta_2 \in \partial U$. Hence, by Exercise 7.14,

$$d_{\mathbb{H}}(\xi_1, \xi_2) \geq d_{\mathbb{H}}(\zeta_1, \zeta_2) \geq \delta. \qquad \square$$

Proof of Theorem 11.2. After a change of coordinates, we may assume that U contains $\mathbb{P}^1_{\mathrm{an}} \smallsetminus \overline{D}_{\mathrm{an}}(0,1)$. Considering $\phi(U)$ in place of U, we may assume for the remainder of the proof that the Julia set and all iterates of U (other than U itself) are contained in $\overline{D}_{\mathrm{an}}(0,1)$. Choose $t > 0$ small enough so that

$$\phi\big(D_{\mathrm{an}}(y_i, t)\big) \subseteq \mathbb{P}^1_{\mathrm{an}} \smallsetminus \overline{D}_{\mathrm{an}}(0,1)$$

for every pole $y_1, \ldots, y_\ell \in \mathbb{C}_v$ of ϕ. Thus,

$$\phi^n(U) \subseteq \overline{D}_{\mathrm{an}}(0,1) \smallsetminus \big(D_{\mathrm{an}}(y_1, t) \cup \cdots \cup D_{\mathrm{an}}(y_\ell, t)\big) \quad \text{for all } n \geq 0. \tag{11.1}$$

Step 1. For any integer $N \geq 0$ and any open connected affinoid V whose closure is contained in $\phi^N(U)$, we claim that there exist an integer $M \geq 0$ and an open disk D such that

$$\phi^M(V) \subseteq D \subseteq \overline{D} \subseteq \phi^{N+M}(U).$$

To prove the claim, which is trivial if V is already a disk, we assume that V has at least two boundary points. Let $\delta > 0$ be the smallest hyperbolic distance between two distinct boundary points of V.

For each $i = 1, \ldots, \ell$, consider the interval

$$I_i := \big[\zeta(y_i, t), \zeta(0,1)\big],$$

which has hyperbolic length $-\log t < \infty$. Any iterate $\phi^n(V)$ can only intersect the infinite interval $[y_i, \infty]$ along a subset of I_i, by equation (11.1). By Proposition 11.3, any nonempty intersection $\phi^n(V) \cap I_i$ must be an interval of length at least δ. Since I_i has finite length and the iterates $\phi^n(V)$ are disjoint, all but finitely many such intersections are empty. Thus, there is some $M \geq 0$ such that for all $n \geq M$, $\phi^n(V)$ does not intersect any of the intervals $[y_i, \infty]$.

Write

$$\phi^M(V) = D_{\mathrm{an}}(a, r_0) \smallsetminus \big(\overline{D}_{\mathrm{an}}(b_1, s_1) \cup \cdots \cup \overline{D}_{\mathrm{an}}(b_m, s_m)\big).$$

By the previous paragraph, ϕ does not have poles in any of the disks $\overline{D}_{\mathrm{an}}(b_i, s_i)$. Thus, $\phi(D_{\mathrm{an}}(a, r_0))$ is a disk of the form $D_{\mathrm{an}}(\phi(a), r_1)$, with

$$\phi^{M+1}(V) = D_{\mathrm{an}}(\phi(a), r_1) \smallsetminus \big(\overline{D}_{\mathrm{an}}(\phi(b_1), s'_1) \cup \cdots \cup \overline{D}_{\mathrm{an}}(\phi(b_q), s'_q)\big).$$

But there are again no poles in any of the disks $\overline{D}_{\mathrm{an}}(\phi(b_i), s'_i)$, and we can describe $\phi^{M+2}(V)$ similarly. Proceeding inductively, each $\phi^{M+n}(V)$ is an open disk $D_{\mathrm{an}}(\phi^n(a), r_n)$ with finitely many closed disks removed, and with $\phi^n(D_{\mathrm{an}}(a, r_0)) = D_{\mathrm{an}}(\phi^n(a), r_n)$.

Thus, defining $D := D_{\mathrm{an}}(a, r_0)$, we have $\phi^M(V) \subseteq D$. We also have $\phi^n(D) \subseteq \overline{D}_{\mathrm{an}}(0,1)$ for all $n \geq 0$, and hence $D \subseteq \mathcal{F}_{\phi,\mathrm{an}}$. Since D is connected and contains a, it must be a subset of the connected component $\phi^{N+M}(U)$ of $\mathcal{F}_{\phi,\mathrm{an}}$ containing a. Finally, the one boundary point of D is $\zeta(a, r_0)$, which is also a boundary point of $\phi^M(V)$ and hence belongs to $\phi^{M+N}(U)$, since $\overline{V} \subseteq \phi^N(U)$. Thus, $\overline{D} \subseteq \phi^{M+N}(U)$, proving the claim.

Step 2. If U is not a disk, there is some open connected affinoid V_1 with at least two boundary points and whose closure is contained in U, such that each of the components of $\mathbb{P}^1_{\mathrm{an}} \smallsetminus V_1$ intersects the Julia set. By the claim of Step 1, there exist an integer $M_1 \geq 0$ and a disk D_1 such that

$$\phi^{M_1}(V_1) \subseteq D_1 \subseteq \overline{D}_1 \subseteq \phi^{M_1}(U).$$

Let $V_1' \supseteq V_1$ be the connected component of $\phi^{-M_1}(D_1)$ containing V_1. Then $\overline{V'}_1 \subseteq U$, and V_1' cannot be a disk, by our assumption about V_1 and the Julia set. Thus, for some $j_1 \in \{1, \ldots, M_1\}$, the image $\phi^{j_1}(V_1')$ is a disk D_1', but the component $\phi^{j_1-1}(V_1')$ of the inverse image $\phi^{-1}(D_1')$ is not a disk.

If $\phi^{j_1}(U)$ is not a disk, then by similar reasoning there exist an integer $j_2 \geq 1$ and a connected affinoid V_2' with $\overline{V'}_2 \subseteq \phi^{j_1}(U)$ such that the connected affinoid $\phi^{j_2}(V_2') \subseteq \phi^{j_1+j_2}(U)$ is a disk D_2', but the component $\phi^{j_2-1}(V_2')$ of the inverse image $\phi^{-1}(D_2')$ is not.

Proceeding inductively in this fashion unless and until we hit an iterate of U which is a disk, we continue finding disks $D_1', D_2', \ldots$ whose closures are contained in distinct iterates of U, and whose preimages include components that are not disks. By Lemma 10.14, there can be at most $\deg(\phi) - 1$ such disks. Thus, there must be some $n \geq 0$ large enough that $\phi^n(U)$ is a disk. □

11.2. An expansion theorem

In this section we present some lemmas needed to prove our first No Wandering Domains result, Theorem 11.14. In particular, the upcoming Theorem 11.9 is a local expansion result at the heart of Theorem 11.14. Part of our analysis in the next two sections hinges on the distinction between "wild" and "tame" ramification, which we define now.

Definition 11.4. Let $p = \operatorname{char} k$ be the residue characteristic of $\mathbb{C}_v$, let $f(z)$ be a rational function or power series defined over $\mathbb{C}_v$, and let $a \in \mathbb{P}^1_{\mathrm{an}}$ be a ramification point of f. If p divides the multiplicity $\deg_a(f)$, we say that a is a *wild* critical point (or wild ramification point) of f; otherwise, we say a is a *tame* critical point (or tame ramification point) of f.

Recall from Section 1.4 that the multiplicity of f at $a \in \mathbb{C}_v$ is the order of vanishing of $f(z) - f(a)$ at $z = a$. Thus, expanding f as a power series

$$f(z) = c_0 + c_m(z-a)^m + c_{m+1}(z-a)^{m+1} + \cdots$$

with $m \geq 2$ and $c_d \neq 0$, the critical point a is wild if and only if $p|m$, or equivalently if $|m| < 1$.

The following lemma illustrates a key distinction in the behavior of functions near wild critical points in $\mathbb{C}_v$ versus tame ones.

Lemma 11.5. *Let $p = \operatorname{char} k$ be the residue characteristic of $\mathbb{C}_v$, let $a \in \mathbb{C}_v$, and let $r > 0$. Let $d \geq 1$ be an integer not divisible by $\operatorname{char} \mathbb{C}_v$, so that $|d| > 0$. Let*

$$f(z) = c_0 + \sum_{n \geq d} c_n (z-a)^n \in \mathbb{C}_v[[z-a]]$$

be a power series converging on $\overline{D}(a,r)$, with $|c_n| r^{n-d} < |dc_d|$ for all $n > d$. Then for any $b \in \overline{D}(a,r) \smallsetminus \{a\}$ and any $0 < \rho < |b-a|$, the disk

$$f\big(D_{\mathrm{an}}(b,\rho)\big) \quad \text{has diameter at least} \quad \big|dc_d(b-a)^{d-1}\big|\rho.$$

Moreover, if

$$\rho \leq \begin{cases} |p|^{1/(p-1)}|b-a| & \text{if } p|d, \\ |b-a| & \text{if } p \nmid d, \end{cases}$$

then f maps $D_{\mathrm{an}}(b,\rho)$ onto $D_{\mathrm{an}}\big(f(b), |dc_d(b-a)^{d-1}|\rho\big)$ with Weierstrass degree 1.

In Lemma 11.5, there are really two disks of interest: the closed disk $\overline{D}_{\mathrm{an}}(a,|b-a|)$ containing both the critical point a and the second point b, and the smaller disk $D_{\mathrm{an}}(b,\rho)$. By Theorem 3.15, the image $f(\overline{D}_{\mathrm{an}}(a,|b-a|))$ of the larger disk has diameter $|c_d(b-a)^d|$. Thus, while the ratio of the diameters of the original disks is $\rho/|b-a|$, the ratio of the diameters of the image disks is $|d|\rho/|b-a|$. In particular, if a is a tame critical point, i.e., if $|d| = 1$, then the two ratios are the same! The upshot, therefore, is that although ratios of diameters can be distorted in the vicinity of a wild critical point a, there is no distortion when a is tame (assuming the smaller disk does not contain a). Moreover, even when a is wild, the distortion factor is bounded below by $|d|$.

Proof of Lemma 11.5. Replacing $f(z)$ by $f(z+a)$, we may assume that $a = 0$. Expand f as a power series centered at b. Writing $z \in D(b,\rho)$ as $z = b + x$, we have

$$f(b+x) - f(b) = \sum_{n \geq d} c_n\big((b+x)^n - b^n\big) = \sum_{n \geq d} \sum_{i=1}^{n} \binom{n}{i} c_n b^{n-i} x^i = \sum_{i \geq 1} C_i x^i,$$

where

$$C_i := \sum_{n \geq \max\{i,d\}} \binom{n}{i} c_n b^{n-i}.$$

For any $i \geq 1$ and $n \geq \max\{i, d+1\}$, we have

$$\left|\binom{n}{i} c_n b^{n-i}\right| \rho^{i-1} \leq \left|c_n b^{d-1}\right| r^{n-d} < \left|d c_d b^{d-1}\right|. \tag{11.2}$$

In particular, for $i = 1$, it follows that $|C_1| = |dc_d b^{d-1}|$. The first statement of the lemma, that the image disk has diameter at least $|C_1|\rho$, is now immediate from Theorem 3.15.

For $2 \leq i \leq d$ with $p|d$, write $i = p^e \ell$ with $e \geq 0$ and $p \nmid \ell$. Then

$$\frac{i-1}{p-1} \geq \frac{p^e - 1}{p-1} = 1 + p + \cdots + p^{e-1} \geq e,$$

and hence

$$\left|\binom{d}{i} c_d b^{d-i}\right| \rho^{i-1} \leq \left|\frac{d}{p^e} \cdot c_d b^{d-1}\right| \cdot |p|^{(i-1)/(p-1)} \leq \left|d c_d b^{d-1}\right|, \tag{11.3}$$

by Exercise 2.15(c) and the fact that $\rho \leq |p|^{1/(p-1)}|b|$.

On the other hand, for $2 \leq i \leq d$ with $p \nmid d$,

$$\left|\binom{d}{i} c_d b^{d-i}\right| \rho^{i-1} \leq |c_d b^{d-1}| = |d c_d b^{d-1}| \tag{11.4}$$

since $\rho \leq |b|$ and $|d| = 1$.

Combining the bounds of (11.2), (11.3), and (11.4) with the fact that $|C_1| = |dc_d b^{d-1}|$, we have $|C_i|\rho^i \leq |C_1|\rho$ for all $i \geq 1$. By Theorem 3.15, then, f is one-to-one on $D(b, \rho)$, and its image is $D(f(b), |C_1|\rho)$. The analogous statement for the corresponding Berkovich disks is immediate from Theorem 7.8. $\square$

We need some more notation specific to the results and proofs of this section and Section 11.3. First, given a point $x \in \mathbb{C}_v$ and a set $S \subseteq \mathbb{P}^1_{\mathrm{an}}$ for which $S \cap \mathbb{C}_v \neq \varnothing$, we denote by $\mathrm{dist}_{\mathrm{I}}(x, S)$ the distance

$$\mathrm{dist}_{\mathrm{I}}(x, S) := \inf\left\{|x - y| : y \in S \cap \mathbb{C}_v\right\}.$$

(We will only use this notation when S is either a disk or a subset of $\mathbb{C}_v$.) Second, we make the following definition.

Definition 11.6. Let K be a complete subfield of $\mathbb{C}_v$, and let $\phi \in K(z)$ be a rational function defined over K. Suppose the Julia set $\mathcal{J}_{\phi,\mathrm{an}}$ of ϕ satisfies $\mathcal{J}_{\phi,\mathrm{an}} \subseteq \overline{D}_{\mathrm{an}}(0,1)$. We say a Type I Julia point $x \in \mathcal{J}_{\phi,\mathrm{an}} \cap \mathbb{C}_v$ is *locally expanding* for (ϕ, K) if for every $\varepsilon > 0$ there exist $M, r > 0$ with the following property. For any disk $U \subseteq \mathcal{F}_{\phi,\mathrm{an}}$ for which

(a) $U \subseteq \overline{D}_{\mathrm{an}}(x, r)$,

(b) $U \cap K \neq \varnothing$, and

(c) $\mathrm{diam}(U) \geq \varepsilon\, \mathrm{dist}_{\mathrm{I}}(x, U)$, there is some integer $n \geq 0$ such that $\operatorname{diam}(\phi^n(U)) \geq M$.

Loosely speaking, a Fatou disk U intersecting K and lying near a locally expanding point x must eventually grow to diameter at least M under iteration. Note that this property holds for arbitrarily small U, provided the ratio $\operatorname{diam}(U)/\operatorname{dist}_{\mathrm{I}}(x,U)$ is bounded below by ε. Thus, although M depends on ε, the essential property of Definition 11.6 is that an *arbitrarily small* disk U intersecting K must eventually grow to diameter M, provided it satisfies the ratio restriction.

Lemma 11.7. *Let K be a complete subfield of $\mathbb{C}_v$, and let $\phi \in K(z)$ be a rational function defined over K with Julia set $\mathcal{J}_{\phi,\mathrm{an}} \subseteq \overline{D}_{\mathrm{an}}(0,1)$. Let $x \in \mathcal{J}_{\phi,\mathrm{an}} \cap K$. If* $\operatorname{char}\mathbb{C}_v > 0$, *suppose further that x is not a wild critical point of ϕ. If $\phi(x)$ is locally expanding for (ϕ, K), then x is also locally expanding for (ϕ, K).*

Proof. Write ϕ as a power series

$$\phi(z) = c_0 + \sum_{n \geq d} c_n (z-x)^n \tag{11.5}$$

centered at x, with $c_d \neq 0$. By hypothesis, d is not divisible by $\operatorname{char}\mathbb{C}_v$, and therefore $|d| > 0$. Hence, there is some $s > 0$ so that the series (11.5) converges on $\overline{D}(x,s)$, and with $|c_n|s^{n-d} < |dc_d|$ for all $n > d$.

Given $\varepsilon > 0$, then because $\phi(x)$ is locally expanding for (ϕ, K), there exist $M, r > 0$ that satisfy Definition 11.6 for the Julia point $\phi(x)$ and the ratio $|d|\varepsilon > 0$. Decrease s if necessary so that $|c_d|s^d \leq r$, and hence $\phi(\overline{D}(x,s)) \subseteq \overline{D}(\phi(x), r)$. We now show that M and s satisfy Definition 11.6 for the Julia point x and ratio ε.

Given a disk $U \subseteq \mathcal{F}_{\phi,\mathrm{an}}$ contained in $\overline{D}_{\mathrm{an}}(x,s)$ that contains a K-rational point and that satisfies $\operatorname{diam}(U) \geq \varepsilon \operatorname{dist}_{\mathrm{I}}(x,U)$, then

$$\operatorname{dist}_{\mathrm{I}}\big(\phi(x), \phi(U)\big) = |c_d| \operatorname{dist}_{\mathrm{I}}(x,U)^d,$$

while by Lemma 11.5, we have

$$\operatorname{diam}\big(\phi(U)\big) \geq |dc_d| \operatorname{dist}_{\mathrm{I}}(x,U)^{d-1} \operatorname{diam}(U).$$

Thus,

$$\frac{\operatorname{diam}\big(\phi(U)\big)}{\operatorname{dist}_{\mathrm{I}}\big(\phi(x),\phi(U)\big)} \geq \frac{|d| \operatorname{diam}(U)}{\operatorname{dist}_{\mathrm{I}}(x,U)} \geq |d|\varepsilon,$$

confirming that the disk $\phi(U) \subseteq \mathcal{F}_{\phi,\mathrm{an}}$ satisfies conditions (a)–(c) of Definition 11.6 for the point $\phi(x)$ and the ratio $|d|\varepsilon$. Hence, there is some integer $n \geq 0$ such that $\operatorname{diam}(\phi^{n+1}(U)) \geq M$. □

To state the main theorem of this section, we need one last bit of specific notation, to describe certain sets of critical points. Recall from Section 1.1 that $\mathrm{Orb}_\phi^+(y)$ denotes the forward orbit of y under ϕ.

Definition 11.8. Let $\phi \in \mathbb{C}_v(z)$ be a rational function with Julia set $\mathcal{J}_{\phi,\mathrm{an}}$. Let C_t be the set of tame critical points of ϕ in $\mathcal{J}_{\phi,\mathrm{an}} \cap \mathbb{P}^1(\mathbb{C}_v)$, and let C_w be the set of wild critical points of ϕ in $\mathcal{J}_{\phi,\mathrm{an}} \cap \mathbb{P}^1(\mathbb{C}_v)$.

Define the sets $S_0, S_1, S_2, \ldots$ and $T_0, T_1, T_2, \ldots$ inductively as follows. Let $S_0 := \varnothing$ and $T_0 := C_\mathrm{w}$, and for each $i \geq 1$, let

$$S_i := \left\{ x \in C_\mathrm{t} \cup C_\mathrm{w} : T_{i-1} \cap \overline{\mathrm{Orb}_\phi^+(\phi(x))} = \varnothing \right\}$$

and $T_i := T_{i-1} \smallsetminus S_i$.

Thus, any $y \in T_i$ belongs to T_{i-1} and either eventually maps to or accumulates at a point of T_{i-1}, possibly itself. Clearly,

$$S_0 \subseteq S_1 \subseteq S_2 \subseteq \cdots \subseteq C_\mathrm{t} \cup C_\mathrm{w} \quad \text{and} \quad C_\mathrm{w} = T_0 \supseteq T_1 \supseteq T_2 \supseteq \cdots.$$

In particular, we have $T_i = C_\mathrm{w} \smallsetminus S_i$ for all $i \geq 0$.

Theorem 11.9. *Let K be a locally compact subfield of $\mathbb{C}_v$ with residue characteristic $p \geq 2$. Let $\phi \in K(z)$ be a rational function defined over K with Julia set $\mathcal{J}_{\phi,\mathrm{an}}$ such that $\mathcal{J}_{\phi,\mathrm{an}} \subseteq \overline{D}_\mathrm{an}(0,1)$. Using the notation from Definition* 11.8*:*

(a) *If $\mathrm{char}\,\mathbb{C}_v = 0$, then for any $i \geq 0$, every point $x \in S_i$ is locally expanding for (ϕ, K).*

(b) *If $\mathrm{char}\,\mathbb{C}_v = p$, assume further that $C_\mathrm{w} = \varnothing$. Then every point $x \in S_1 = C_\mathrm{t}$ is locally expanding for (ϕ, K).*

The proof of Theorem 11.9 is long and technical, but the underlying idea is relatively simple, as follows. Given a (Julia) point $x \in S_i$ and a disk U contained in $\mathcal{F}_{\phi,\mathrm{an}}$, containing a K-rational point b, and satisyfing $\mathrm{diam}(U) \geq \varepsilon\, \mathrm{dist}_\mathrm{I}(x, U)$, consider the disks

$$V_\rho := D_\mathrm{an}(b, \rho) \quad \text{and} \quad \overline{V}_\sigma := \overline{D}_\mathrm{an}(b, \sigma),$$

where $\rho, \sigma \in |K^\times|$ with $\rho \leq \mathrm{diam}(U)$ and $\sigma \geq |x-b|$. Consider the diameters

$$\rho_j := \mathrm{diam}\left(\phi^j(V_\rho)\right) \quad \text{and} \quad \sigma_j := \mathrm{diam}\left(\phi^j(\overline{V}_\sigma)\right)$$

of the iterates of these two disks. Because V_σ contains the Julia point x, the diameters σ_j must eventually become arbitrarily large. However, whenever ϕ is one-to-one on the disk $\phi^j(\overline{V}_\sigma)$, the ratios

$$e_j := \frac{\rho_j}{\sigma_j} \quad \text{and} \quad e_{j+1} := \frac{\rho_{j+1}}{\sigma_{j+1}}$$

must coincide. Thus, if we suppose for simplicity that there is some radius $R > 0$ for which ϕ is one-to-one on $D_{\mathrm{an}}(a, R)$ for every Type I Julia point $a \in \mathcal{J}_{\phi,\mathrm{I}}$, then ρ_j must eventually grow as large as $M := \varepsilon R$, as desired.

Of course, things get more complicated when we allow critical points in the Julia set. However, Lemma 11.5 shows that we still have $e_{j+1} = e_j$ even when there is a tame critical point y in $\phi^j(\overline{V}_\sigma)$, as long as y is at distance σ_j from the Fatou disk $\phi^j(V_\rho)$. The same is true if there is a wild critical point outside $\phi^j(\overline{V}_\sigma)$, but not too close. On the other hand, if $\phi^j(V_\rho)$ lands too close to a (Julia) critical point y, then the ratio e_{j+1} may decrease; but in that case we can replace $\phi^j(\overline{V}_\sigma)$ by the smaller disk $\overline{D}_{\mathrm{an}}(\phi^j(b), |\phi^j(b) - y|)$, which contains the Julia point y. Thus, the technical details of the proof are ultimately devoted to showing that this switch to a smaller disk can happen only a finite number of times. Then, sometime after the last switch, some ρ_{j+n} must be as large as M.

Proof of Theorem 11.9. Define

$$\alpha := \begin{cases} |p|^{1/(p-1)} & \text{if } \operatorname{char} \mathbb{C}_v = 0, \\ 1 & \text{if } \operatorname{char} \mathbb{C}_v = p. \end{cases} \tag{11.6}$$

Replacing K by a finite extension, we may assume that all critical points of ϕ in $\mathbb{C}_v$ are defined over K, and that $\alpha \in |K^\times|$. Since K is locally compact, it is discretely valued, and hence we may choose a uniformizer π for K.

We proceed by induction on i. The statement is vacuously true for $i = 0$. Assuming the statement is true for some $i - 1 \geq 0$, we now prove it for i. (Of course, if $\operatorname{char} \mathbb{C}_v = p > 0$, we assume $i = 1$.)

Given $x \in S_i$, the orbit $\mathrm{Orb}_\phi^+(x)$ is contained in the Julia set and hence cannot contain any periodic critical points, which would be superattracting and, hence, Fatou. Because there are only finitely many critical points, there must be some $N \geq 0$ such that $\mathrm{Orb}_\phi^+(\phi^N(x))$ contains no critical points of ϕ. Moreover, if $\operatorname{char} \mathbb{C}_v > 0$, then by hypothesis, none of the iterates of x is a wild critical point. By repeated application of Lemma 11.7 then, it suffices to show that $\phi^N(x)$ is locally expanding for (ϕ, K). Thus, we may assume for the remainder of the proof that x is a forward iterate of a point of S_i and that $\mathrm{Orb}_\phi^+(x)$ contains no critical points.

Step 1. To choose a suitably small radius to use in the proof, we now cover the compact set $\mathcal{J}_{\phi,\mathrm{an}} \cap K$ with certain small disks.

For each $y \in \mathcal{J}_{\phi,\mathrm{an}} \cap K$, pick $s_y > 0$ so that the power series expansion $\phi(z) = c_0 + \sum_{n \geq d} c_n (z - y)^n$ converges for $z \in D(y, s_y)$, with $d \geq 1$, $c_d \neq 0$, and

$$|c_n| s_y^{n-d} \leq |d c_d| \quad \text{for all } n > d.$$

The family of open disks $\{D(y, s_y) : y \in \mathcal{J}_{\phi,\mathrm{an}} \cap K\}$ covers $\mathcal{J}_{\phi,\mathrm{an}} \cap K$, and hence there is a finite subcover. Let $R > 0$ be the minimum radius of all the disks in this subcover.

Claim 11.10. For any point $c \in \mathcal{J}_{\phi,\mathrm{an}} \cap K$ and any radius $0 < s < \alpha R$, either ϕ is one-to-one on $\overline{D}_{\mathrm{an}}(c, s)$ or else $\mathrm{dist}_{\mathrm{I}}(c, C_{\mathrm{t}} \cup C_{\mathrm{w}}) < R$.

We devote the rest of Step 1 to proving Claim 11.10. Given c and s as in the claim, we have $c \in D(y, s_y)$ for some disk $D(y, s_y)$ in the finite subcover described just above.

If y is not a critical point of ϕ, then by our choice of $s_y \geq R > s$, ϕ has Weierstrass degree 1 on $D(y, s_y) \supseteq \overline{D}(c, s)$, as desired.

If $y \in C_{\mathrm{t}}$ with $|y - c| \geq R$, then ϕ is one-to-one on $\overline{D}_{\mathrm{an}}(c, s)$ by Lemma 11.5, since $s < R \leq |y - c|$.

If $y \in C_{\mathrm{w}}$ with $|y - c| \geq R$, then since $C_{\mathrm{w}} \neq \varnothing$, we have $\mathrm{char}\, \mathbb{C}_v = 0$ by hypothesis. Since $s < \alpha R \leq \alpha|y - c|$, Lemma 11.5 shows that ϕ is one-to-one on $\overline{D}_{\mathrm{an}}(c, s)$.

If none of the above possibilities occurs, then

$$\mathrm{dist}_{\mathrm{I}}(c, C_{\mathrm{t}} \cup C_{\mathrm{w}}) \leq |y - c| < R,$$

proving Claim 11.10.

Step 2. Given $\varepsilon > 0$, we wish to choose M and r for Definition 11.6. By decreasing ε if necessary, we may assume that $\varepsilon \in |K^\times|$ with $\varepsilon < 1$.

Invoking our inductive hypothesis that every $y \in S_{i-1}$ is locally expanding for (ϕ, K), there are real numbers $M, r' > 0$ that satisfy Definition 11.6 for every $y \in S_{i-1}$ and for $\alpha\varepsilon > 0$. That is, for each $y \in S_{i-1}$ and every disk $U \subseteq \mathcal{F}_{\phi,\mathrm{an}} \cap \overline{D}_{\mathrm{an}}(y, r')$ with $U \cap K \neq \varnothing$ and $\mathrm{diam}(U) \geq \alpha\varepsilon\, \mathrm{dist}_{\mathrm{I}}(y, U)$, some iterate of U has diameter at least M.

Pick $r > 0$ such that for each critical point $a \in \mathbb{C}_v$ of ϕ at which $\mathrm{Orb}_\phi^+(x)$ does not accumulate, we have

$$|\phi^n(x) - a| > r/\alpha \quad \text{for all } n \geq 0.$$

Decreasing r if necessary, we may also ensure that $r \leq \alpha \min\{R, r'\}$. Similarly, decrease M if necessary so that $M < |\pi| r\varepsilon$. We wish to show that this choice of M and r satisfies Definition 11.6 for x and ε.

Step 3. Given an arbitrary disk $U \subseteq \mathcal{F}_{\phi,\mathrm{an}} \cap \overline{D}_{\mathrm{an}}(x, r)$ that contains a K-rational point and satisfies $\mathrm{diam}(U) \geq \varepsilon\, \mathrm{dist}_{\mathrm{I}}(x, U)$, pick $b \in U \cap K$. We now define two sequences of radii, $\{\rho_j\}_{j \geq 0}$ and $\{\sigma_j\}_{j \geq 0}$.

Let ρ_0 be the largest value in $|K^\times|$ such that $D_{\mathrm{an}}(b, \rho_0) \subseteq U$. For each $j \geq 1$, define

$$\rho_j := \mathrm{diam}\left(\phi^j\left(D_{\mathrm{an}}(b, \rho_0)\right)\right) \leq \mathrm{diam}\left(\phi^j(U)\right).$$

In addition, let

$$\sigma_0 := \operatorname{dist}_{\mathrm{I}}(x, U) = |x - b| \in |K^\times|.$$

Given σ_j, define σ_{j+1} inductively as follows. If $\operatorname{dist}_{\mathrm{I}}(\phi^j(b), C_{\mathrm{t}}) \geq \sigma_j$, i.e., if there are no tame Julia critical points in $D(\phi^j(b), \sigma_j)$, then define

$$\sigma_{j+1} := \operatorname{diam}\Big(\phi\big(D_{\mathrm{an}}(\phi^j(b), \sigma_j)\big)\Big). \tag{11.7}$$

Otherwise, i.e., if there are tame Julia critical points in $D(\phi^j(b), \sigma_j)$, then define

$$\sigma'_j := \operatorname{dist}_{\mathrm{I}}\big(\phi^j(b), C_{\mathrm{t}}\big) = \min\big\{\big|a - \phi^j(b)\big| : a \in C_t\big\} < \sigma_j$$

and

$$\sigma_{j+1} := \operatorname{diam}\Big(\phi\big(D_{\mathrm{an}}(\phi^j(b), \sigma'_j)\big)\Big). \tag{11.8}$$

For any $j \geq 0$, we claim that the closed disk $\overline{D}(\phi^j(b), \sigma_j)$ contains a point of $\mathcal{J}_{\phi,\mathrm{an}} \cap K$. Indeed, for $j = 0$, this disk contains the original point x. Proceeding inductively, assuming the jth disk contains a Type I Julia point w, there are two possibilities. If C_{t} lies outside the open disk $D(\phi^j(b), \sigma_j)$, then the closed disk $\overline{D}(\phi^{j+1}(b), \sigma_{j+1})$ contains $\phi(w) \in \mathcal{J}_{\phi,\mathrm{an}} \cap K$. Otherwise, the smaller disk $\overline{D}(\phi^j(b), \sigma'_j)$ contains some $a \in C_{\mathrm{t}} \subseteq \mathcal{J}_{\phi,\mathrm{an}} \cap K$. Thus, $\overline{D}(\phi^{j+1}(b), \sigma_{j+1})$ contains $\phi(a) \in \mathcal{J}_{\phi,\mathrm{an}} \cap K$, proving our claim.

Another simple induction shows that $\rho_j, \sigma_j \in |K^\times|$ for all $j \geq 0$. Thus, defining the sequence of ratios $\{e_j\}_{j \geq 0}$ by

$$e_j := \rho_j / \sigma_j,$$

we have $e_j \in |K^\times|$ for all $j \geq 0$. Note that $e_j \leq 1$, because

$$D_{\mathrm{an}}\big(\phi^j(b), \rho_j\big) \subseteq \phi^j(U) \subseteq \mathcal{F}_{\phi,\mathrm{an}},$$

but as we saw in the previous paragraph, the disk $\overline{D}_{\mathrm{an}}(\phi^j(b), \sigma_j)$ contains a point of $\mathcal{J}_{\phi,\mathrm{an}}$ and hence must have radius at least as large. On the other hand, by our choice of U, we have $\operatorname{diam}(U) \geq \varepsilon\sigma_0$. Since $\varepsilon, \sigma_0 \in |K^\times|$ and since ρ_0 was the *largest* value in $|K^\times|$ such that $D_{\mathrm{an}}(b, \rho_0) \subseteq U$, then we must have $\rho_0 \geq \varepsilon\sigma_0$. Hence, $\varepsilon \leq e_0 \leq 1$.

The remainder of the proof hinges on the following essential claim.

Claim 11.11. For any $j \geq 0$, if

- $e_j \geq \varepsilon$,
- $\sigma_0, \sigma_1, \ldots, \sigma_j \leq |\pi| r$, and
- $\operatorname{dist}_{\mathrm{I}}(\phi^j(b), S_{i-1}) > \sigma_j / \alpha$;

then

(a) if $\operatorname{dist}_{\mathrm{I}}(\phi^j(b), C_t) \geq \sigma_j$, then $e_{j+1} = e_j$;

(b) otherwise, $e_{j+1} \geq |\pi|^{-1} e_j$.

Step 4. In this step, assume the truth of Claim 11.11, which we will prove in Step 6; we now deduce Theorem 11.9. Note in particular that unless and until we reach some j for which the hypotheses of Claim 11.11 fail, we have $\varepsilon \leq e_0 \leq e_1 \leq \cdots \leq e_j$.

For each $j \geq 0$, the disk

$$D_{\text{an}}\big(\phi^j(b), |\pi|^{-1}\sigma_j\big) \supseteq \overline{D}\big(\phi^j(b), \sigma_j\big)$$

intersects $\mathcal{J}_{\phi,\text{an}}$, and hence some iterate $\phi^n(D_{\text{an}}(\phi^j(b), |\pi|^{-1}\sigma_j))$ has diameter greater than r. Applying Claim 11.11 inductively, there is some minimal $j \geq 0$ for which at least one of the following three possibilities occurs:

(A) $\sigma_j > |\pi| r$; or

(B) there is some $x' \in S_{i-1}$ such that $|\phi^j(b) - x'| \leq \sigma_j/\alpha$; or

(C) there is some $a \in C_{\text{t}}$ such that $|\phi^j(b) - a| < \sigma_j$.

If (A) occurs, we have

$$\rho_j = \sigma_j e_j > |\pi| r\varepsilon > M,$$

since $e_j \geq e_0 \geq \varepsilon$. Thus, $\text{diam}(\phi^j(U)) \geq \rho_j > M$, as desired.

If (B) occurs but not (A), then $\phi^j(U) \subseteq \mathcal{F}_{\phi,\text{an}}$ is a disk with

$$\text{(11.9)} \qquad \text{dist}_{\text{I}}\big(x', \phi^j(U)\big) = \big|\phi^j(b) - x'\big| \leq \sigma_j/\alpha \leq r/\alpha \leq r',$$

where r' is the radius we chose early in Step 2. Note that

- $\phi^j(U) \subseteq \overline{D}_{\text{an}}(x', r')$ by inequality (11.9),
- $\phi^j(U) \cap K$ is nonempty since it contains $\phi^j(b)$, and
- $\text{diam}(\phi^j(U)) \geq \rho_j = e_j\sigma_j \geq \varepsilon\sigma_j \geq \alpha\varepsilon \, \text{dist}_{\text{I}}(x', \phi^j(U))$,

where the last inequality is again by (11.9). Since $x' \in S_{i-1}$ is locally expanding, and by our choice of r' and M from Step 2, then, there is some $n \geq 0$ such that $\text{diam}(\phi^{j+n}(U)) \geq M$, as desired.

Finally, if (C) occurs, then by the conclusions of Claim 11.11, we see that $e_{j+1} \geq |\pi|^{-1} e_j > e_j$. However, recalling the definitions of σ_j', σ_{j+1}, and e_{j+1} from Step 3, the disk $\overline{D}(\phi^j(b), \sigma_j')$ intersects $\mathcal{J}_{\phi,\text{an}}$ at $a \in C_{\text{t}}$, and hence

$$\overline{D}\big(\phi^{j+1}(b), \sigma_{j+1}\big) \quad \text{intersects} \quad \mathcal{J}_{\phi,\text{an}} \quad \text{at} \quad \phi(a).$$

Therefore, some iterate of the disk $D_{\text{an}}(\phi^{j+1}(b), |\pi|^{-1}\sigma_{j+1})$ must have diameter greater than r. Thus, we may continue to apply Claim 11.11 to the iterates $j+1, j+2, \ldots$ with the assurance that there must be some minimal $j' \geq j+1$ at which one of cases (A)–(C) arises again.

Continuing this process, at each stage either we arrive at case (A) or (B), giving an iterate of U of diameter at least M, or else we arrive at case (C), causing e_j to increase. Recall, however, that $e_j \le 1$ for all $j \ge 0$, and that each increase of e_j is by a factor of at least $|\pi|^{-1} > 1$. Thus, we can only arrive at case (C) a finite number of times. Hence, the process must eventually arrive at case (A) or (B), giving us the desired iterate of U of diameter at least M.

Step 5. In order to prove Claim 11.11, we first need to prove the following technical statement:

$$\text{For any } j \ge 0 \text{ such that } \sigma_0, \sigma_1, \ldots, \sigma_j < r, \tag{11.10}$$
$$\text{there is some } n \ge 0 \text{ such that } \left|\phi^n(x) - \phi^j(b)\right| \le \sigma_j.$$

To see this, we proceed by induction on j. If $j = 0$, then we may use $n = 0$ in (11.10), since $|x - b| = \sigma_0$ by definition of σ_0.

Assume (11.10) is true for some $j \ge 0$; we now prove it for $j+1$. Suppose first that there are no tame Julia critical points in $D(\phi^j(b), \sigma_j)$, so that

$$\sigma_{j+1} = \operatorname{diam}\Big(\phi\big(D_{\mathrm{an}}(\phi^j(b), \sigma_j)\big)\Big)$$

as in equation (11.7). Since ϕ has no poles in the closed disk

$$\overline{D}\big(\phi^j(b), \sigma_j\big) \subseteq D\big(\phi^j(b), R\big),$$

(which in turn is because $\sigma_j < r \le R$), Corollary 3.18 shows that

$$\sigma_{j+1} = \operatorname{diam}\Big(\phi\big(\overline{D}_{\mathrm{an}}(\phi^j(b), \sigma_j)\big)\Big).$$

Since our inductive hypothesis gives some $n \ge 0$ such that

$$\left|\phi^n(x) - \phi^j(b)\right| \le \sigma_j,$$

we therefore have $|\phi^{n+1}(x) - \phi^{j+1}(b)| \le \sigma_{j+1}$, as desired.

Otherwise, there is a tame critical point $y \in C_{\mathrm{t}}$ with $|\phi^j(b) - y| < \sigma_j$. Since $\sigma_j < r \le R$, there is only one such critical point y. Moreover, by our definition of r, y must be an accumulation point of the orbit $\mathrm{Orb}_\phi^+(x)$. Hence, its image $\phi(y)$ must also be an accumulation point of the same orbit. In particular, there must be some $n \ge 0$ such that $|\phi^n(x) - \phi(y)| \le \sigma_{j+1}$. Moreover, the definition of σ_j' in Step 3 gives $\sigma_j' = |y - \phi^j(b)|$, and therefore $|\phi(y) - \phi^{j+1}(b)| \le \sigma_{j+1}$, by the definition of σ_{j+1} from equation (11.8) and by Corollary 3.18 again. Thus,

$$\left|\phi^n(x) - \phi^{j+1}(b)\right| \le \max\left\{\left|\phi^n(x) - \phi(y)\right|, \left|\phi(y) - \phi^{j+1}(b)\right|\right\} \le \sigma_{j+1},$$

completing our proof of (11.10).

Step 6. Finally, we now prove Claim 11.11. Given $j \ge 0$ satisfying the hypotheses of Claim 11.11, define

$$V_\rho := D_{\mathrm{an}}\big(\phi^j(b), \rho_j\big) \quad \text{and} \quad V_\sigma := D_{\mathrm{an}}\big(\phi^j(b), \sigma_j\big).$$

As we noted in the middle of Step 3, the disk $\overline{D}(\phi^j(b), \sigma_j)$ must contain a Type I Julia point $c \in \mathcal{J}_{\phi,\mathrm{an}} \cap K$. By the hypotheses, we have

$$|\pi|^{-1}\sigma_j \leq r \leq \alpha R.$$

We consider four cases.

Case 1. Suppose ϕ is one-to-one on V_σ. Then

$$\rho_{j+1} = \operatorname{diam}\big(\phi(V_\rho)\big) = e_j \operatorname{diam}\big(\phi(V_\sigma)\big) = e_j\sigma_{j+1},$$

where the second equality is by Proposition 3.20, and the third is by our definition of σ_{j+1}, given the lack of tame critical points in V_σ. Therefore, $e_{j+1} = e_j$, and clearly $\operatorname{dist}_{\mathrm{I}}(\phi^j(b), C_{\mathrm{t}}) \geq \sigma_j$.

Case 2. Suppose there is a wild critical point $a \in D(c, R) \cap C_{\mathrm{w}}$. Then we now wish to show that

$$\big|\phi^j(b) - a\big| > \sigma_j/\alpha. \tag{11.11}$$

If $a \in S_{i-1}$, we already know inequality (11.11) holds by the hypotheses of Claim 11.11. Thus, we may assume that $a \in C_{\mathrm{w}} \smallsetminus S_{i-1} = T_{i-1}$.

By (11.10), there is some $n \geq 0$ such that $|\phi^n(x) - \phi^j(b)| \leq \sigma_j$. On the other hand, since x is a forward iterate of a point of S_i, we know by definition of S_i that its orbit $\operatorname{Orb}^+_\phi(x)$ does not accumulate at a. By our choice of r in Step 2,

$$\big|\phi^n(x) - a\big| > r/\alpha \geq r > \sigma_j \geq \big|\phi^n(x) - \phi^j(b)\big|,$$

and hence

$$\big|\phi^j(b) - a\big| = \big|\phi^n(x) - a\big| > r/\alpha > \sigma_j/\alpha,$$

proving the desired inequality (11.11).

Expanding ϕ as a power series centered at a, inequality (11.11) and Lemma 11.5 together imply that ϕ is one-to-one on V_σ. Thus, we are in fact in Case 1, giving us $e_{j+1} = e_j$ and $\operatorname{dist}_{\mathrm{I}}(\phi^j(b), C_{\mathrm{t}}) \geq \sigma_j$.

Case 3. Suppose there is a tame critical point $a \in D(c, R) \smallsetminus V_\sigma$. By Lemma 11.5 again, ϕ is one-to-one on V_σ. Again we are actually in Case 1, and hence $e_{j+1} = e_j$ and $\operatorname{dist}_{\mathrm{I}}(\phi^j(b), C_{\mathrm{t}}) \geq \sigma_j$.

Case 4. Since $\sigma_j < r \leq \alpha R$, Claim 11.10 for c and σ_j shows that the only remaining possibility is that there is a tame critical point $a \in V_\sigma \cap C_{\mathrm{t}}$. As in Step 3, define $\sigma'_j := |\phi^j(b) - a|$, so that

$$\operatorname{dist}_{\mathrm{I}}\big(\phi^j(b), C_{\mathrm{t}}\big) = |\phi^j(b) - a| = \sigma'_j < \sigma_j.$$

By Lemma 11.5, ϕ is one-to-one on the open disk $D_{\mathrm{an}}(\phi^j(b), \sigma'_j)$. Proposition 3.20 therefore yields

$$e_{j+1} = \frac{\rho_{j+1}}{\sigma_{j+1}} = \frac{\operatorname{diam}(\phi(V_\rho))}{\operatorname{diam}(\phi(D_{\mathrm{an}}(\phi^j(b), \sigma'_j)))} = \frac{\rho_j}{\sigma'_j} > \frac{\rho_j}{\sigma_j} = e_j.$$

However, because $e_{j+1}, e_j \in |K^\times|$, we must have $e_{j+1} \geq |\pi|^{-1} e_j$, proving Claim 11.11 and, hence, Theorem 11.9. □

Remark 11.12. In Theorem 11.9, the hypothesis that K is locally compact (i.e., that K is complete), discretely valued, and has finite residue field, is not entirely needed. The discrete valuation is essential (in Step 4, and again in Case 4 of Step 6), while the completeness is needed as an artifact of our phrasing of Definition 11.6. Otherwise, local compactness is used only in Step 1 of the proof, to choose the radius R. Exercise 11.6 shows that R can be chosen with the desired properties even if K is not locally compact. Similarly, the restriction of part (b) of Theorem 11.9, that $C_{\mathrm{w}} = \varnothing$ if $\operatorname{char} \mathbb{C}_v > 0$, can also be relaxed; see Exercises 11.4 and 11.5.

However, we need the full strength of both of these hypotheses—that K is locally compact and that $C_{\mathrm{w}} = \varnothing$ if $\operatorname{char} \mathbb{C}_v > 0$—to prove Theorems 11.14 and 11.15 in the next section. (See, for example, Exercise 11.7.) Thus, in the interests of simplifying the already complicated proof of what is really just a technical lemma, we have added these same hypotheses to Theorem 11.9 even though they are not strictly needed.

11.3. No wandering domains for p-adic fields: A theorem

Definition 11.13. Let X be a topological space, let $f : X \to X$ be a continuous map, and let $a \in X$. We say a is *recurrent* (under f) if

$$a \in \overline{\operatorname{Orb}_f^+(f(a))},$$

i.e., if a lies in the closure of the forward orbit of $f(a)$.

All periodic points are clearly recurrent, but it is also possible for non-periodic points to be recurrent.

Theorem 11.14 (Benedetto). *Let K be a locally compact subfield of $\mathbb{C}_v$ with residue characteristic $p \geq 2$. Let $\phi \in K(z)$ be a rational function defined over K with Type I Julia set $\mathcal{J}_{\phi,\mathrm{I}}$ and Berkovich Fatou set $\mathcal{F}_{\phi,\mathrm{an}}$. Suppose that*

- *$\mathcal{J}_{\phi,\mathrm{I}}$ contains no wild critical points, if* $\operatorname{char} K = p$, *or*
- *$\mathcal{J}_{\phi,\mathrm{I}}$ contains no recurrent wild critical points, if* $\operatorname{char} K = 0$.

Then

(a) *If the algebraic closure $\overline{K}$ of K is dense in $\mathbb{C}_v$, then $\mathcal{F}_{\phi,\mathrm{an}}$ has no wandering domains.*

(b) *Only finitely many periodic components of $\mathcal{F}_{\phi,\mathrm{an}}$ contain Type I points in $\mathbb{P}^1(K)$.*

Theorem 11.14 is a corollary of the following expansion result.

Theorem 11.15. *Let K and ϕ be as in Theorem 11.14, and let $\mathcal{J}_{\phi,\mathrm{an}}$ be the Julia set of ϕ. Assume that $\mathcal{J}_{\phi,\mathrm{an}} \subseteq \overline{D}_{\mathrm{an}}(0,1)$. Then there is a positive constant $M > 0$ (depending only on ϕ and K) with the following property:*

Let U be a Fatou component contained in $\overline{D}_{\mathrm{an}}(0,1)$ and containing a K-rational point. Then there is some $n \geq 0$ and some $c \in K$ such that $\phi^n(U)$ contains $D_{\mathrm{an}}(c, M)$.

Proof. Let $p \geq 2$ be the residue characteristic of $\mathbb{C}_v$. As in equation (11.6), define $\alpha := |p|^{1/(p-1)}$ if $\operatorname{char} \mathbb{C}_v = 0$, or $\alpha := 1$ if $\operatorname{char} \mathbb{C}_v = p$. Let C_{t} and C_{w} be the sets of Type I Julia critical points of ϕ that are tame and wild, respectively. Replacing K by a finite extension, we may assume that $C_{\mathrm{t}} \cup C_{\mathrm{w}} \subseteq K$, and that $\alpha \in |K^\times|$. Let π be a uniformizer for K.

Step 1. We claim that all points of $C_{\mathrm{t}} \cup C_{\mathrm{w}}$ are locally expanding for (ϕ, K). If $\operatorname{char} \mathbb{C}_v > 0$, this is immediate from Theorem 11.9(b), since our hypotheses in that case state that $C_{\mathrm{w}} = \varnothing$.

To prove the claim if $\operatorname{char} \mathbb{C}_v = 0$, recall the sets T_i from Definition 11.8: $T_0 := C_{\mathrm{w}}$, and any $x \in T_i$ is an element of C_{w} for which $\overline{\operatorname{Orb}_\phi^+(\phi(x))}$ contains an element of T_{i-1}. Let $m := \#C_{\mathrm{w}}$.

If T_m were nonempty, then there would be a point $a_m \in T_m$, and hence a finite sequence $a_0, a_1, \ldots, a_m$ with each $a_i \in T_i$, and with

$$a_i \in \overline{\operatorname{Orb}_\phi^+(\phi(a_{i+1}))} \quad \text{for each } 0 \leq i \leq m-1. \tag{11.12}$$

Since $a_0, \ldots, a_m \in C_{\mathrm{w}}$ and $m = \#C_{\mathrm{w}}$, there must be some $0 \leq i < j \leq m$ for which $a_i = a_j$. It is immediate from (11.12) that $a_i \in \overline{\operatorname{Orb}_\phi^+(\phi(a_j))}$. Thus, a_i is a recurrent wild Type I Julia critical point, contradicting the hypotheses. Therefore, $T_m = \varnothing$, and hence $C_{\mathrm{t}} \cup C_{\mathrm{w}} = S_{m+1}$. Our claim now follows from Theorem 11.9(a).

Step 2. By the claim of Step 1, there exist $M, r > 0$ satisfying Definition 11.6 for (ϕ, K) for each Type I Julia critical point, with $\varepsilon := |\pi|\alpha > 0$. Choose $R > 0$ as in Step 1 of the proof of Theorem 11.9. Decrease R if necessary so that $R \leq \min\{1, r\}$ and $R \in |K^\times|$. Define

$$W := \bigcup_{x \in \mathcal{J}_{\phi,\mathrm{an}} \cap K} D_{\mathrm{an}}(x, R).$$

We claim that there are only finitely many Fatou components contained in $\overline{D}_{\mathrm{an}}(0,1)$ and that contain K-rational points, but which are *not* contained in W. To see this, suppose not. Choose a sequence $\{U_i\}_{i \geq 1}$ of distinct such components, and choose $a_i \in U_i \cap K$ for each i. Then the sequence $\{a_i\}_{i \geq 1}$ has an accumulation point a in the compact set $K \cap \overline{D}(0,1)$. If the point a were Fatou, its Fatou component would intersect and hence coincide with infinitely many of the U_i, a contradiction. Thus, $a \in \mathcal{J}_{\phi,\mathrm{an}} \cap K$.

Infinitely many of the points a_i lie in $D_{\mathrm{an}}(a, R)$, but since $U_i \not\subseteq W$, we have $\zeta(a, R) \in U_i$ for each such i, because U_i is connected. Thus, these infinitely many components U_i intersect and hence coincide. Once again we have a contradiction, proving our claim.

For each of the finitely many Fatou components U_i described in the previous paragraph (i.e., contained in $\overline{D}_{\mathrm{an}}(0,1)$, containing K-rational points, and not contained in W), choose a point $a_i \in U_i \cap K$ and a radius $M_i > 0$ such that $D_{\mathrm{an}}(a_i, M_i) \subseteq U_i$. Let $M' > 0$ be the minimum of these finitely many radii M_i. Decrease M if necessary so that $0 < M \leq \max\{R, M'\}$.

Step 3. The rest of the proof centers on the following statement.

Claim 11.16. Let $U \subseteq \overline{D}_{\mathrm{an}}(0,1)$ be a Fatou component that contains a K-rational point b. Then there is some $n \geq 0$ such that either

(a) $\phi^n(U) \not\subseteq W$, or

(b) $\phi^n(U)$ contains the disk $D_{\mathrm{an}}(\phi^n(b), R)$, or

(c) there exist a disk V with $b \in V \subseteq U$ and some $y \in C_{\mathrm{t}} \cup C_{\mathrm{w}}$ such that
$$\operatorname{diam}\big(\phi^n(V)\big) \geq |\pi|\alpha \operatorname{dist}_{\mathrm{I}}\big(y, \phi^n(V)\big).$$

We prove Claim 11.16 by contradiction. Given a Fatou component U contained in $\overline{D}_{\mathrm{an}}(0,1)$ and containing a K-rational point b, suppose that conclusions (a), (b), and (c) of Claim 11.16 fail for every $n \geq 0$. Let U_0 be the largest open disk contained in U and containing b; write
$$U_0 := D_{\mathrm{an}}\big(b, \rho_0\big).$$

Define two sequences of radii, $\{\rho_n\}_{n\geq 0}$ and $\{\sigma_n\}_{n\geq 0}$, as follows. First, let
$$\rho_n := \operatorname{diam}\big(\phi^n(U_0)\big) \qquad \text{for each } n \geq 0,$$
so that $\phi^n(U_0) = D_{\mathrm{an}}(\phi^n(b), \rho_n)$. Second, let σ_0 be the smallest value in $|K^\times|$ which is strictly larger than ρ_0, and define σ_n by

$$\sigma_n := \operatorname{diam}\Big(\phi^n\big(D_{\mathrm{an}}(b, \sigma_0)\big)\Big) \qquad \text{for each } n \geq 1, \tag{11.13}$$

so that $\phi(D_{\mathrm{an}}(\phi^n(b), \sigma_n)) = D_{\mathrm{an}}(\phi^{n+1}(b), \sigma_{n+1})$. Our immediate goal is to show that $\sigma_n \in |K^\times|$ and $\rho_n/\sigma_n = \rho_0/\sigma_0$ for every $n \geq 0$.

We proceed by induction on n. The case $n = 0$ is trivial. Given that $\rho_n/\sigma_n = \rho_0/\sigma_0 \in [|\pi|, 1)$ with $\sigma_n \in |K^\times|$ for some particular n, it must be that σ_n is the smallest value in $|K^\times|$ that is strictly larger than ρ_n. Therefore, since $\rho_n < R \in |K^\times|$, we have $\sigma_n \leq R$, and hence $\sigma_{n+1} \in |K^\times|$ by equation (11.13) and Theorem 3.15.

Moreover, because $\phi^n(U)$ fails Claim 11.16(a)–(c), we have
$$\phi^n(U_0) = D_{\mathrm{an}}\big(\phi^n(b), \rho_n\big) \subseteq W,$$

with

$$\rho_n < \max\left\{R, |\pi|\alpha \operatorname{dist}_{\mathrm{I}}\left(\phi^n(b), C_{\mathrm{t}} \cup C_{\mathrm{w}}\right)\right\}. \tag{11.14}$$

Since $\sigma_n \leq R$ as we noted above, the disk $D_{\mathrm{an}}(\phi^n(b), \sigma_n)$ lies completely inside one of the disks $D_{\mathrm{an}}(y, s_y)$ in the finite subcover used to define R in Step 1 of the proof of Theorem 11.9. If y is not a critical point, then ϕ is one-to-one on $D_{\mathrm{an}}(y, s_y)$ and hence on $D_{\mathrm{an}}(\phi^n(b), \sigma_n)$. The only other possibility is that $y \in C_{\mathrm{t}} \cup C_{\mathrm{w}}$, in which case

$$\sigma_n \leq \frac{\rho_n}{|\pi|} < \alpha\left|\phi^n(b) - y\right|,$$

where the second inequality is by (11.14). By Lemma 11.5, then, ϕ is again one-to-one on $D(\phi^n(b), \sigma_n)$. Thus, $\rho_{n+1}/\sigma_{n+1} = \rho_n/\sigma_n$ by Proposition 3.20, achieving the goal stated after equation (11.13).

As a consequence, for each $n \geq 0$, we have

$$\sigma_n = \frac{\sigma_0 \rho_n}{\rho_0} \leq |\pi|^{-1}\rho_n < |\pi|^{-1}R \leq |\pi|^{-1}.$$

Thus,

$$\bigcup_{n\geq 0} \phi^n\left(D_{\mathrm{an}}(b, \sigma_0)\right) = \bigcup_{n\geq 0} D_{\mathrm{an}}\left(\phi^n(b), \sigma_n\right) \subseteq D_{\mathrm{an}}(0, |\pi|^{-1}),$$

and therefore $D_{\mathrm{an}}(b, \sigma_0) \supsetneq U_0$ is contained in the Fatou set of ϕ and hence in U. However, U_0 was the largest open disk containing b and contained in U. Claim 11.16 follows from this contradiction.

Step 4. We are now ready to prove the theorem. Given a Fatou component U contained in $\overline{D}_{\mathrm{an}}(0,1)$ and containing a K-rational point, there must be some $n \geq 0$ such that one of conclusions (a)–(c) of Claim 11.16 holds.

If (a) holds, then $\phi^n(U) \not\subseteq W$, and hence $\phi^n(U)$ is either one of the finitely many Fatou components described in Step 2 or the Fatou component containing $\mathbb{P}^1_{\mathrm{an}} \smallsetminus \overline{D}_{\mathrm{an}}(0,1)$. Either way, $\phi^n(U)$ contains a disk

$$D_{\mathrm{an}}(c, M') \supseteq D_{\mathrm{an}}(c, M),$$

for some $c \in K$.

If (b) holds, then $\phi^n(U)$ contains the disk $D_{\mathrm{an}}(c, R) \supseteq D_{\mathrm{an}}(c, M)$, where $c := \phi^n(b) \in K$.

Finally, if (c) holds, then there is a disk V with $b \in V \subseteq U$ such that

$$\operatorname{diam}\left(\phi^n(V)\right) \geq |\pi|\alpha \operatorname{dist}_{\mathrm{I}}\left(y, \phi^n(U)\right) \quad \text{for some } y \in C_{\mathrm{t}} \cup C_{\mathrm{w}}.$$

By our choice of M in Step 2, which came from the local expansion of y for (ϕ, K) from Step 1, there is some $m \geq 0$ such that

$$\operatorname{diam}\left(\phi^{n+m}(V)\right) \geq M. \qquad \square$$

Proof of Theorem 11.14.

(a) Given a component U of $\mathcal{F}_{\phi,\mathrm{an}}$, suppose that U is a wandering domain. Since $\mathbb{P}^1(\overline{K})$ is dense in $\mathbb{P}^1(\mathbb{C}_v)$, there is some $x \in U \cap \mathbb{P}^1(\overline{K})$. Replacing K by $K(x)$ if necessary, we may assume that $U \cap \mathbb{P}^1(K)$ is nonempty. By a coordinate change in $\mathrm{PGL}(2,K)$, we may assume that $\infty \in U$, and, by a further coordinate change of the form $z \mapsto cz$, that

$$\mathbb{P}^1_{\mathrm{an}} \smallsetminus U \subseteq \overline{D}_{\mathrm{an}}(0,1).$$

In particular, the Julia set $\mathcal{J}_{\phi,\mathrm{an}}$ is contained in $\overline{D}_{\mathrm{an}}(0,1)$.

Let $M > 0$ be the constant of Theorem 11.15 for ϕ and K. Applying that theorem to $\phi(U)$, some iterate $\phi^{n_1}(U)$ contains a disk $D_{\mathrm{an}}(c_1, M)$, with $c_1 \in K$ and $n_1 \geq 1$. The same theorem applied to $\phi^{1+n_1}(U)$ shows that some iterate $\phi^{n_2}(U)$ contains a disk $D_{\mathrm{an}}(c_2, M)$, with $c_2 \in K$ and $n_2 > n_1 \geq 1$. Continuing in this fashion, there is a strictly increasing sequence of positive integers

$$n_1 < n_2 < n_3 < \cdots$$

such that each $\phi^{n_i}(U)$ contains a disk $D_{\mathrm{an}}(c_i, M)$ with $c_i \in K$. Thus, the sequence

$$\{c_i\}_{i\geq 1} \subseteq K \cap \overline{D}(0,1)$$

has the property that any two elements of the sequence are distance at least M apart. However, $K \cap \overline{D}(0,1)$ is compact, and hence the sequence must have a limit point. This contradiction shows no such U can exist.

(b) This is similar to part (a); see Exercise 11.8. □

Theorem 11.14 relies heavily on the fact that the field K over which ϕ is defined is locally compact. This hypothesis is essential; without it, there are a number of situations in which wandering domains can occur. In particular, over a field of residue characteristic $p > 0$ that is *not* discretely valued, the following result shows that a polynomial of degree larger than p can have wandering domains. The construction in the proof (which we only sketch here) uses wild ramification in an essential way, but it works both for mixed characteristic fields like $\mathbb{C}_p$ and for fields of positive characteristic.

Theorem 11.17 (Benedetto). *Assume that $\mathbb{C}_v$ has residue characteristic $p > 0$. Then there is a polynomial $\phi(z) \in \mathbb{C}_v[z]$ of degree $p+1$ such that ϕ has wandering domains.*

Sketch and references for the proof. See [**17**] for $\mathbb{C}_v = \mathbb{C}_p$ and [**19**] for general such $\mathbb{C}_v$. The idea is to consider a one-parameter family such as

$$\phi_a(z) = (1-a)z^{p+1} + az^p$$

for $a \in \mathbb{C}_v$ with $|a| > 1$. Then ϕ_a has a Type I repelling fixed point at $z = 1$ and a Type II repelling periodic point of multiplier p at $\zeta(0,r)$ for a certain radius $0 < r < 1$. Thus, points in the annulus $U = \{r < |\zeta| < 1\}$ are pushed outward, but due to the wild ramification, $|\phi'|$ is small in U. On the other hand, ϕ_a is bijective on $V = D_{\mathrm{an}}(1,1)$, stretching distances by $|a| > 1$. Hence, we are in essentially the symbolic dynamics situation of Exercises 10.16 and 10.17.

We then choose a disk D in U and appropriate sequences of integers $m_1, m_2, \ldots$ and $n_1, n_2, \ldots$, along with parameters $a_1, a_2, \ldots \in \mathbb{C}_v$ so that the orbit of D under ϕ_{a_j} spends m_1 iterates in U, then n_1 in V, then m_2 in U, etc., through n_j in V, finally landing on the repelling fixed point at $z = 1$ itself. By choosing the sequences carefully, we can guarantee that the iterates of D are sufficiently shrunk by their time spent in U to emerge still small from their time spent in V, while also guaranteeing that $\{a_i\}_{i\geq 1}$ is a Cauchy sequence in $\mathbb{C}_v$. The map ϕ_a for the limiting parameter a then has a wandering domain that contains the original disk D. □

Remark 11.18. Rivera-Letelier showed in [**119**] that it is possible to choose the parameter $a \in \mathbb{C}_p$ in the proof of Theorem 11.17 so that the wandering domain of ϕ_a is a disk of essentially whatever radius we like. In particular, we can force the wandering domain U to be either a rational disk or an irrational disk. In the latter case, in light of Theorem 8.15(f), the attracting basin of ∞ is a domain of Cantor type whose boundary includes some Type III points, namely the unique boundary point ζ of U, as well as all other points in the grand orbit of ζ. Currently, this construction—of a wandering domain that is an irrational disk—is the only known way to produce an attracting domain of Cantor type with a point of Type III in its boundary.

The other hypotheses of Theorem 11.14 place restrictions on wild critical points in the Julia set. It is therefore natural to ask about maps for which these conditions are violated. Note that in general, one would expect critical points to lie in the Fatou set; after all, the map contracts distances substantially in the neighborhood of a critical point. For example, a periodic critical point is superattracting and hence Fatou.

Of course, there are ways to force a critical point to lie in the Julia set. For example, $\phi(z) = \pi^{-1}(z^{p+1} - z^p) + 1$, where $0 < |\pi| < 1$, sends the wild critical point $z = 0$ to the repelling fixed point $z = 1$. In particular, if $\mathrm{char}(\mathbb{C}_v) = p > 0$, then this map violates hypothesis (i) of Theorem 11.14.

It is far less obvious that in mixed characteristic, there exist maps with recurrent wild Julia critical points. In [**118**], Rivera-Letelier constructed maps with this property; in fact, he did it for polynomials in $\mathbb{Q}_p[z]$. Specifically, for any degree $d \geq 3$, he worked with the family

$$\psi_t(z) = -p^{1-d} z(z-1)^{d-1} + t.$$

When $t = 0$, the critical point at $z = 1$ maps under ψ_0 to the repelling fixed point at $z = 0$; but for most small perturbations of t, the critical point at 1 lies in the Fatou set of ψ_t. This situation, when a preperiodic Julia critical point becomes Fatou under a small perturbation, is called a *Misiurewicz bifurcation*.

Following McMullen's work [**103**] for $\mathbb{C}$, Rivera-Letelier showed in [**118**] over $\mathbb{C}_v$ that the Misiurewicz bifurcation generates a *Misiurewicz cascade*. That is, for some parameter t_1 close to 0, the critical point at $z = 1$ eventually lands on a repelling cycle of longer period that includes at least one point close to $z = 1$. Then, for some parameter t_2 close to t_1, the critical point eventually lands on an even longer repelling cycle, including an iterate even closer to $z = 1$. By careful selection of these inductively chosen parameters $t_1, t_2, \ldots$, one can guarantee that the sequence $\{t_n\}_{n\geq 1}$ is Cauchy, and even that it is contained in the base field $\mathbb{Q}_p$. For the resulting limit parameter $t := \lim_{n\to\infty} t_n \in \mathbb{Q}_p$, the map ψ_t has a recurrent Julia critical point at $z = 1$; and if $p|d$, then it is wild.

Thus, even over locally compact fields, there are maps in equal characteristic with wild Julia critical points, and there are maps in mixed characteristic with recurrent wild Julia critical points. Very little is known about the existence or nonexistence of wandering domains for such maps.

11.4. Wandering domains for tame maps

If the residue field of $\mathbb{C}_v$ is *not* algebraic over a finite field, then any map $\phi \in \mathbb{C}_v(z)$ of good reduction has wandering domains, corresponding to the wandering points of $\mathbb{P}^1(\overline{k})$ under the reduced map $\overline{\phi}$. More generally, as we will see in Theorem 11.20, this phenomenon occurs whenever there is a periodic point of Type II in the Julia set. For example, over such a field, the map $z \mapsto z^2$ has infinitely many wandering points over the residue field $\overline{k}$, and hence both the cubic polynomial of Example 8.18 and the rational function of Example 8.19 have corresponding wandering domains. These observations motivate the following definition.

Definition 11.19. Let $\phi \in \mathbb{C}_v(z)$ be a rational function of degree at least 2, let $\mathcal{F}_{\phi,\mathrm{an}}$ be the Berkovich Fatou set of ϕ, and let $\zeta \in \mathbb{P}^1_{\mathrm{an}}$ be a Type II periodic point of ϕ of minimal period m. We say that a wandering component U of $\mathcal{F}_{\phi,\mathrm{an}}$ is in the *attracting basin of* ζ if there is some integer $N \geq 0$ such that for all $n \geq 0$,

$$\phi^{N+nm}(U) \text{ is a residue class at } \zeta.$$

The use of the term "attracting" is justified in Definition 11.19, because for any point $\xi \in U$, it is easy to check that in the Gel'fand topology, $\lim_{n\to\infty} \phi^{N+nm}(\xi) = \zeta$. The Type II point ζ must of course be in the Julia set, or else the component U would contain ζ and hence be periodic. In that case, however, the existence of a wandering domain in the attracting basin of ζ is entirely dependent on the residue field, as shown in our next result, which is essentially Theorem A of [**18**].

Theorem 11.20 (Benedetto). *Let $\phi \in \mathbb{C}_v(z)$ be a rational function of degree at least 2, and let $\zeta \in \mathbb{P}^1_{\mathrm{an}}$ be a Type II periodic point of ϕ lying in the Julia set of ϕ. Then ϕ has wandering domains in the attracting basin of ζ if and only if the residue field k of $\mathbb{C}_v$ is* not *algebraic over a finite field.*

The proof of Theorem 11.20 requires the following lemma.

Lemma 11.21. *Let k be a field, and let $f(z) \in k(z)$ be a nonconstant rational function. Suppose that $f^n(z)$ is not the identity function for any $n \geq 1$. Then all points of $\mathbb{P}^1(k)$ are preperiodic under ϕ if and only if k is algebraic over a finite field.*

Proof. This is left to the reader. The reverse implication is straightforward, but the forward implication requires some facts from the theory of Weil heights; see [**18**, Lemma 4.1] or Exercises 11.9–11.12. □

Proof of Theorem 11.20. Changing coordinates and replacing ϕ by an iterate if necessary, we may assume that $\zeta = \zeta(0,1)$ is the Gauss point, and that it is fixed by ϕ.

To prove the forward implication, suppose that k is algebraic over a finite field. Then every point of $\mathbb{P}^1(k)$ is preperiodic under $\overline{\phi}$, by Lemma 11.21. Hence, ϕ has no wandering residue classes at ζ.

Conversely, suppose k is not algebraic over a finite field. By Theorem 8.7, since ζ is in the Julia set, the reduction $\overline{\phi}$ is not of finite order. Therefore, by Lemma 11.21, $\overline{\phi}$ has a wandering point $\overline{x} \in \mathbb{P}^1(\overline{k})$. Choose $M \geq 0$ large enough that $\overline{\phi}^n(\overline{x})$ does not lie in the finite set T of bad directions of Theorem 7.34 for any $n \geq M$. By Theorem 7.34, the residue class corresponding to $\overline{\phi}^M(\overline{x})$ is a wandering domain in the attracting basin of ζ. □

Definition 11.22. Let $\phi \in \mathbb{C}_v(z)$, and let p be the residue characteristic of $\mathbb{C}_v$. Suppose that for every point $\zeta \in \mathbb{P}^1_{\mathrm{an}}$, the local degree $\deg_\zeta(\phi)$ is not divisible by p. Then we say that ϕ is a *tame* map.

Observe that:

- any composition of tame maps is tame.
- if $\deg(\phi) < p$, then ϕ is a tame map.
- if $\mathbb{C}_v$ has residue characteristic 0, then *all* nonconstant maps $\phi \in \mathbb{C}_v(z)$ are tame.

Theorem 11.20 shows that wandering domains are possible, and indeed common, over certain fields. Conversely, the following result (from [**18**, Theorem B]) shows that in many circumstances, attracting basins of periodic Type II points are the only source of wandering domains.

Theorem 11.23 (Benedetto). *Let K be a discretely valued subfield of $\mathbb{C}_v$ with algebraic closure $\overline{K} \subseteq \mathbb{C}_v$, and let $\phi \in K(z)$ be a tame map. Then any wandering domains of ϕ containing $\overline{K}$-rational points lie in attracting basins of Type II Julia periodic points.*

Proof. Let U be a wandering domain of ϕ containing a $\overline{K}$-rational point a. Extending K to $K(a)$, we may assume that $a \in \mathbb{P}^1(K)$. After a K-rational change of coordinates, we may then assume that U contains $\mathbb{P}^1_{\text{an}} \smallsetminus \overline{D}_{\text{an}}(0,1)$. Write $U_n := \phi^n(U)$. Then $U_n \subseteq \overline{D}(0,1)$ for all $n \geq 1$, and U_n is an open disk for all sufficiently large n, by Theorem 11.2. Considering U_j in place of U for some large enough j, we may assume that U and all of its iterates U_n are open disks contained in $\overline{D}(0,1)$. We may also assume that U and its iterates do not contain any of the (finitely many Type I) critical points of ϕ.

Write $U_n = D_{\text{an}}(\phi^n(a), r_n)$ for each $n \geq 0$. Define $\zeta := \zeta(a, r_0)$ to be the unique boundary point of U.

Replacing K by a finite extension, we may assume that all poles and critical points of ϕ are K-rational. Let π be a uniformizer for K, so that $|K^\times| = |\pi|^{\mathbb{Z}}$. Let p be the residue characteristic of $\mathbb{C}_v$.

Step 1. Our first goal is to show that $r_n \in |K^\times|$ for every $n \geq 0$. Let s_0 be the smallest element of $|K^\times|$ such that $s_0 \geq r_0$, and define $s_n := r_n \cdot s_0 / r_0$. We claim that $\phi^n(D_{\text{an}}(a, s_0)) = D_{\text{an}}(\phi^n(a), s_n)$ for all $n \geq 0$.

The claim is clear for $n = 0$. Assuming it is true for some $n \geq 0$, we have $s_n \in |K^\times|$. Our assumptions about U_n imply that ϕ has no poles or critical points in $D(\phi^n(a), r_n)$. Moreover, our assumptions about K imply there are no poles or critical points in the annulus $\{r_n \leq |z - \phi^n(a)| < s_n\}$ either. Hence, ϕ has a power series expansion on $D := D(\phi^n(a), s_n)$ with no critical points. Since ϕ is a tame map, we have $p \nmid \text{wdeg}_D(\phi)$. Thus, Proposition 3.9 yields

$$\text{wdeg}_D(\phi) = 1 + \text{wdeg}_D(\phi') = 1,$$

where the second equality is because there are no critical points in D. By Proposition 3.20,

$$r_{n+1} = \text{diam}\Big(\phi\big(D(\phi^n(a), r_n)\big)\Big) = \frac{\text{diam}\left(\phi(D(\phi^n(a), s_n))\right)}{s_n} \cdot r_n,$$

and hence the image disk $\phi(D(\phi^n(a), s_n))$ has radius $s_n \cdot r_{n+1}/r_n = s_{n+1}$, proving the claim.

It is immediate from the claim that all the iterates $\phi^n(D_{\mathrm{an}}(a,s_0))$ are contained in $D_{\mathrm{an}}(0,s_0/r_0)$, and hence the disk $D_{\mathrm{an}}(a,s_0)$ is contained in the Fatou set. However, because the original disk $U = D_{\mathrm{an}}(a,r_0)$ is the Fatou component containing a, we must have $r_0 = s_0 \in |K^\times|$. It follows immediately that $r_n \in |K^\times|$ for all $n \geq 0$.

Step 2. For each $n \geq 0$, define $V_n := \overline{D}_{\mathrm{an}}(\phi^n(a), r_n)$. We claim that there are infinitely many $n \geq 0$ for which V_n contains a pole or critical point of ϕ.

If not, then after replacing U by an iterate $\phi^m(U)$, we may assume that none of the disks V_n contains a pole or critical point. By our assumptions about K, there are no poles or critical points in any of the annuli

$$\{r_n < |z - \phi^n(a)| < |\pi|^{-1} r_n\}$$

either. Thus, none of the open disks $V_n' := D_{\mathrm{an}}(\phi^n(a), |\pi|^{-1}r_n)$ contain poles or critical points of ϕ. By the same inductive argument as in Step 1, it follows that $V_0' \supsetneq U$ is contained in the Fatou set, contradicting the fact that U is a Fatou component, and proving the claim.

Step 3. Next, we claim that if only finitely many of the closed disks V_n contain poles, then the boundary point ζ of U must be preperiodic.

By Step 2, the assumption that only finitely many of the V_n contain poles implies that infinitely many of them contain critical points. Since ϕ has only finitely many critical points, there must be integers $N > M \geq 0$ such that $V_M \cap V_N$ is nonempty, and such that V_n contains no poles for all $n \geq M$. Using $\phi^M(U)$ in place of U, we may assume that $M = 0$.

By Theorem 3.15 and the facts that $\phi(U_n) = U_{n+1}$ and that V_n contains no poles, we have $\phi(V_n) = V_{n+1}$ for all $n \geq 0$. Furthermore, because V_0 and V_N are intersecting disks in $\mathbb{C}_v$, we have either $V_0 \supsetneq V_N$ or $V_0 \subseteq V_N$.

Suppose $V_0 \supsetneq V_N$. Then $\phi^N(a) \in V_0$ and $r_N < r_0$. Define

$$U' := D_{\mathrm{an}}\big(\phi^N(a), r_0\big).$$

Then $V_N \subseteq U' \subseteq V_0$, and hence

$$\phi^N(U') \subseteq \phi^N(V_0) = V_N \subseteq U'.$$

Therefore, U' is contained in a periodic component of the Fatou set. However, $U_N \subseteq U'$, contradicting the hypothesis that U is wandering.

Thus, we must have $V_0 \subseteq V_N$. It follows immediately that

$$V_0 \subseteq V_N \subseteq V_{2N} \subseteq V_{3N} \subseteq \cdots .$$

If all the inclusions are proper, then the corresponding sequence of radii $\{r_{iN}\}_{i\geq 0}$ is strictly increasing. Because each radius r_n belongs to the discrete group $|K^\times|$, there must be some $i \geq 0$ such that $r_{iN} > 1$, contradicting the assumption that $U_{iN} \subseteq \overline{D}_{\text{an}}(0,1)$. Therefore, there must be some $j > i \geq 0$ such that $V_{jN} = V_{iN}$. Hence, $\phi^{jN}(\zeta) = \phi^{iN}(\zeta)$, proving our claim.

Step 4. Our final claim is that even if infinitely many of the closed disks V_n contain poles, the boundary point ζ is still preperiodic.

Since there are only finitely many poles, there must be a pole $y \in K$ and an infinite set I of nonnegative integers such that $y \in V_n$ for all $n \in I$. Since y is a pole, there is some $s > 0$ such that $\phi(D_{\text{an}}(y,s)) \subseteq \mathbb{P}^1_{\text{an}} \smallsetminus \overline{D}_{\text{an}}(0,1)$. Therefore, no U_n can intersect $D(y,s)$, or else U_{n+1} would not be contained in $\overline{D}_{\text{an}}(0,1)$. In particular, $y \in V_n \smallsetminus U_n$ for every $n \in I$, and hence

$$s < |y - \phi^n(a)| = r_n \leq 1 \quad \text{for every } n \in I.$$

However, every radius r_n belongs to the discrete group $|K^\times|$, which has finite intersection with the interval $(s,1]$. Thus, there must be some $M, N \in I$ with $N > M$ and $r_M = r_N$. Since y lies in both V_M and V_N, we have $V_M = V_N$, and hence $\phi^M(\zeta) = \phi^N(\zeta)$, proving the claim.

Step 5. By the results of Steps 3 and 4, we know that the boundary point ζ of U is preperiodic. In addition, since $r_n \in |K^\times|$ by Step 1, the point ζ is of Type II. Replacing U by $\phi^M(U)$ and ϕ by an appropriate iterate ϕ^n, we may assume that ϕ fixes ζ. Thus, $\phi^n(U) = D_{\text{an}}(\phi^n(a), r_0)$ is a residue class at ζ for every $n \geq 0$. That is, U is in the attracting basin of ζ. □

11.5. Wandering domains for tame polynomials

Theorem 11.23 requires that the tame map be defined over a discretely valued subfield of $\mathbb{C}_v$. Trucco proved the same conclusion for tame *polynomials* under weaker restrictions on the coefficient field. The following result first appeared as [**137**, Theorem A], and its corollary as [**137**, Corollary B].

Theorem 11.24 (Trucco). *Suppose that $\mathbb{C}_v$ contains a discretely valued field K whose algebraic closure is dense in $\mathbb{C}_v$. Let $\phi \in \mathbb{C}_v[z]$ be a tame polynomial with Julia set $\mathcal{J}_{\phi,\text{an}}$. Then ϕ has no Julia points of Types III or IV, and the set of Julia points not of Type I can be written as a disjoint union of finitely many grand orbits*

$$\mathcal{J}_{\phi,\text{an}} \cap \mathbb{H} = \text{GO}(\zeta_1) \cup \cdots \cup \text{GO}(\zeta_m),$$

where $0 \leq m \leq \max\{1, d-2\}$, and each ζ_i is a repelling periodic point of Type II.

Corollary 11.25 (Trucco). *Let $\mathbb{C}_v$, K, and ϕ be as in Theorem 11.24. Then any wandering domain of ϕ lies in the attracting basin of a Type II Julia periodic point.*

The hypothesis that there is a discretely valued field K whose algebraic closure is dense in $\mathbb{C}_v$ applies both to the p-adic field $\mathbb{C}_p$ and to the completion of an algebraic closure of any Laurent series field $\mathbb{F}((t))$. Indeed, Trucco's main application in [**137**] was to the field of Puiseux series over $\mathbb{C}$. There are, however, larger fields $\mathbb{C}_v$ which do not have this property; for such fields, very little is known about wandering domains, even for polynomials.

Before proving Theorem 11.24 and Corollary 11.25, we present two simpler propositions about non-archimedean polynomial dynamics. They hold in much greater generality, and they are of interest in their own right.

Proposition 11.26. *Let $\phi \in \mathbb{C}_v[z]$ be a polynomial of degree at least 2.*

(a) *There is a smallest disk $U \subseteq \mathbb{A}^1_{\mathrm{an}}$ that contains the Julia set $\mathcal{J}_{\phi,\mathrm{an}}$.*

(b) *U is a rational closed disk.*

(c) *$\phi(U) \supseteq U$, with equality if and only if ϕ has potential good reduction.*

Proof. This is left to the reader; see Exercise 11.13. □

Proposition 11.27. *Let $\phi \in \mathbb{C}_v[z]$ be a tame polynomial of degree $d \geq 2$. Suppose that ϕ does not have potential good reduction. Then*

(a) *there are at most $d-2$ points of $\mathcal{J}_{\phi,\mathrm{an}} \cap \mathrm{Ram}(\phi)$, and*

(b) *there are at most $d-2$ orbits of periodic Julia points that are* not *of Type I. Moreover, all such points are repelling and of Type II.*

Proof.

(a) This is left to the reader; see Exercise 11.14.

(b) By Theorem 8.7(a), a periodic Julia point $\zeta \in \mathcal{J}_{\phi,\mathrm{an}}$ that is not of Type I must be of Type II. We claim that ζ is a repelling periodic point.

To prove this claim, change coordinates and replace ϕ by an iterate so that that $\zeta = \zeta(0,1)$ is the Gauss point and is fixed by ϕ. Thus, the reduction $\overline{\phi} \in k[z]$ at ζ is a nonconstant polynomial. Because ϕ has no poles in $\mathbb{C}_v$, the only possible bad direction $\boldsymbol{v}$ at ζ is the direction $\vec{v}(\infty)$ containing ∞. If $\vec{v}(\infty)$ is actually a good direction, then by Theorem 7.34, ϕ has good reduction, contradicting our hypotheses. Therefore, $\vec{v}(\infty)$ is the unique bad direction at $\zeta(0,1)$. Since $\overline{\phi}$ is a polynomial, we have $\phi_{\#}(\vec{v}(\infty)) = \vec{v}(\infty)$. By Theorem 8.7(b), it follows that $\zeta(0,1)$ is not indifferent. Hence, $\zeta = \zeta(0,1)$ is a repelling fixed point of ϕ, proving the claim.

Returning to the main proof (i.e., before we changed coordinates and replaced ϕ by an iterate), note that each Type II repelling periodic orbit contains at least one point ζ with $\deg_\zeta(\phi) \geq 2$. Such a point necessarily lies in $\mathcal{J}_{\phi,\mathrm{an}} \cap \mathrm{Ram}(\phi)$. By part (a), there are at most $d-2$ such points, and therefore at most $d-2$ such orbits. □

Proof of Theorem 11.24.

Step 1. By Proposition 11.26, there is a smallest disk U containing the Julia set $\mathcal{J}_{\phi,\mathrm{an}}$ of ϕ. After a change of coordinates, we may assume this (rational closed) disk is $\overline{D}_{\mathrm{an}}(0,R)$ for some $R \in |\mathbb{C}_v^\times|$, and we have $\phi(\overline{D}_{\mathrm{an}}(0,R)) = \overline{D}_{\mathrm{an}}(0,S)$ for some $S \geq R$, with equality if and only if ϕ has potential good reduction.

Our desired result follows easily if ϕ has potential good reduction. Thus, we assume for the rest of the proof that $S > R$. For each Type I critical point $c \in \overline{D}(0,R) \smallsetminus \mathcal{K}_{\phi,\mathrm{an}}$ in U but not in the filled Julia set, we have $R < |\phi^n(c)| \leq S$ for some unique integer $n \geq 1$. Let $r_1 > r_2 > \cdots > r_{q-1}$ be the (finitely many) such values of $|\phi^n(c)|$ in (R,S), and let $r_0 := S$. (In fact, $r_0 = S$ is also necessarily the value of $|\phi^n(c)|$ for some such c and n, as can be shown using Exercise 10.13; but we do not need that fact here.)

Let $d := \deg(\phi)$, and let W be the Fatou component containing ∞. Then W is totally invariant, since $\infty \in W$ maps to itself d-to-1. It is also a domain of Cantor type, since ϕ does not have potential good reduction.

Claim 11.28. Let $\zeta \in W$. If the set $\mathcal{J}_{\phi,\mathrm{an}} \cup \{\infty\}$ intersects at least three different directions at ζ, then there is some $n \geq 0$ and some $j \in \{0,\ldots,q-1\}$ such that $\phi^n(\zeta) = \zeta(0,r_j)$.

Claim 11.28 is essentially [**137**, Proposition 4.3]. We leave the proof to the reader; see Exercise 11.15.

Step 2. We set some notation. For each for $\xi \in \mathcal{J}_{\phi,\mathrm{an}}$, define the sequence of all the distinct points $\{G_n(\xi)\}_{n\geq 0}$ in $\mathbb{H}$

$$G_0(\xi) := \zeta(0,R) \succ G_1(\xi) \succ G_2(\xi) \succ \cdots$$

such that for each $n \geq 1$, $G_n(\xi)$ is the maximal point in the interval $(\xi, G_{n-1}(\xi)) \subseteq \mathbb{H}$ for which $\phi^i(G_n(\xi)) = \zeta(0,r_i)$ for some $i \geq 1$ and some $j \in \{0,\ldots,q-1\}$. (Here, "maximal" is with respect to the relation $\preceq$ of lying above in $\mathbb{H}$.) That is, the points $G_n(\xi)$ are the unique points in the interval $(\xi,\zeta(0,R)] \subseteq \mathbb{H}$ in the backward orbits of the points $\zeta(0,r_j)$. Hence, by Claim 11.28, any branch point of the convex hull of $\mathcal{J}_{\phi,\mathrm{an}} \cup \{\infty\}$ in the interval (ξ,∞) must be one of the points $G_n(\xi)$. See Figure 11.2, which shows a portion of the convex hull of $\mathcal{J}_{\phi,\mathrm{an}} \cup \{\infty\}$ for an example with $q = 3$. Trucco calls $\{G_n(\xi)\}_{n\geq 0}$ the *geometric sequence* of ξ.

Claim 11.29. Let $\xi \in \mathcal{J}_{\phi,\mathrm{an}}$, and let $n \geq 0$. Then

(a) $\phi\big(G_n(\xi)\big) = \begin{cases} G_{n-q}\big(\phi(\xi)\big) \text{ if } n \geq q, \\ \zeta(0,r_n) \text{ if } n \leq q-1; \end{cases}$

(b) $\deg_\zeta(\phi)$ is constant on the half-open interval $[G_{n+1}(\zeta_0), G_n(\zeta_0))$.

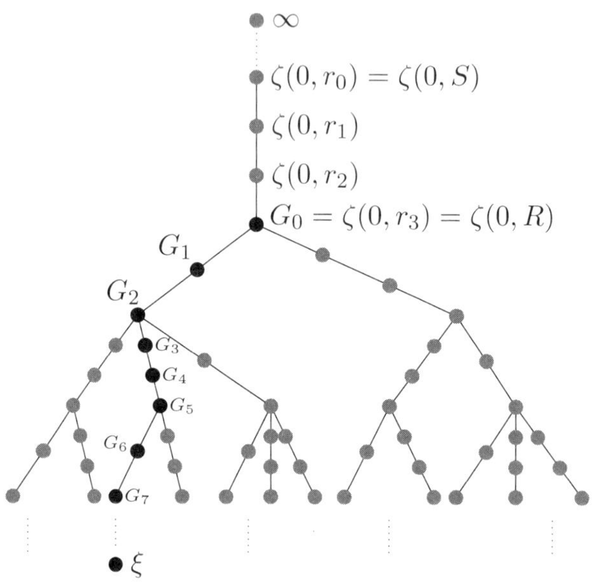

Figure 11.2. The sequence $\{G_n(\xi)\}_{n\geq 0}$ in the proof of Theorem 11.24.

Part (a) of Claim 11.29 is essentially [**137**, Proposition 4.7], and part (b) is [**137**, Proposition 4.8]. We leave the proof to the reader; see Exercise 11.16.

For any two distinct Julia points $\zeta \neq \zeta' \in \mathcal{J}_{\phi,\text{an}}$, define

$$\operatorname{lcg}(\zeta,\zeta') := \text{ the maximal } j \geq 0 \text{ such that } G_j(\zeta) = G_j(\zeta').$$

Trucco calls $\operatorname{lcg}(\zeta,\zeta')$ the *largest common geometric level* of ζ and ζ'.

Given an arbitrary $\xi \in \mathcal{J}_{\phi,\text{an}}$, let $\eta \in \mathbb{P}^1_{\text{an}}$ be an ω-limit point of ξ. That is, η is the limit of some subsequence of $\{\phi^n(\xi)\}_{n\geq 0}$. Then $\eta \in \mathcal{J}_{\phi,\text{an}}$, since $\mathcal{J}_{\phi,\text{an}}$ is closed. Moreover, because some iterates $\phi^n(\xi)$ approach η, and because each Type II point $G_j(\eta)$ bounds a disk containing η, we have

$$\limsup_{j\to\infty} \operatorname{lcg}\left(\phi^j(\xi),\eta\right) = +\infty. \tag{11.15}$$

Define two sequences $\{k_n\}_{n\geq 0}$ and $\{\ell_n\}_{n\geq 0}$ of nonnegative integers inductively by

$$k_0 := 0 \quad \text{and} \quad \ell_0 := \operatorname{lcg}(\xi,\eta) \geq 0,$$

and for $n \geq 1$,

$$k_n := \min\left\{j \geq 0 : \operatorname{lcg}\left(\phi^j(\xi),\eta\right) > \ell_{n-1}\right\} \quad \text{and} \quad \ell_n := \operatorname{lcg}\left(\phi^{k_n}(\xi),\eta\right).$$

If $\phi^{k_n}(\xi) = \eta$ for some $n \geq 0$, then $\ell_n = \infty$, and we stop both sequences at this finite stage. Otherwise, by equation (11.15) and the fact that $\phi^j(\xi) \neq \eta$

for all $j \geq 0$, each successive k_n is indeed defined. In either case, both sequences are strictly increasing.

Step 3. We can now use the geometric sequences $\{G_n(\xi)\}$ and $\{G_n(\eta)\}$ to prove three claims which are the heart of Trucco's proof.

Claim 11.30. *For all $n \geq 1$ for which k_n is defined, and for each level $N > \ell_{n-1} + qk_n$, none of the Type II points*

$$\{\phi^i(G_N(\xi)) : 0 \leq i < k_n\}$$

lie above or below any of the others. (That is, none of these k_n points are related by $\preceq$.)

If Claim 11.30 were false, there would be integers $0 \leq i < m < k_n$ such that $\phi^i(G_N(\xi)) \preceq \phi^m(G_N(\xi))$. Rewriting both points according to Claim 11.29(a), we have $G_{N-qi}(\phi^i(\xi)) \preceq G_{N-qm}(\phi^m(\xi))$. By definition of the sequence $\{G_j(\phi^i(\xi))\}_{j\geq 0}$, it follows that

$$G_{N-qm}(\phi^m(\xi)) = G_{N-qm}(\phi^i(\xi)).$$

Thus,

$$\begin{aligned} \phi^{k_n}(G_N(\xi)) &= \phi^{k_n-m}\Big(G_{N-qm}(\phi^m(\xi))\Big) = \phi^{k_n-m}\Big(G_{N-qm}(\phi^i(\xi))\Big) \\ &= \phi^{k_n-(m-i)}(G_{N-q(m-i)}(\xi)), \end{aligned} \tag{11.16}$$

where the first and third equalities are by Claim 11.29(a). However, by the definition of k_n and by Claim 11.29(a) again, we have

$$\phi^{k_n}(G_N(\xi)) = G_{N-qk_n}(\phi^{k_n}(\xi)) \preceq G_{1+\ell_{n-1}}(\phi^{k_n}(\xi)) = G_{1+\ell_{n-1}}(\eta).$$

Therefore, by equation (11.16), we have

$$\phi^{k_n-(m-i)}(\xi) \preceq \phi^{k_n-(m-i)}(G_{N-q(m-i)}(\xi)) = \phi^{k_n}(G_N(\xi)) \preceq G_{1+\ell_{n-1}}(\eta),$$

contradicting the minimality of k_n, and proving Claim 11.30.

Claim 11.31. *For n and N as in Claim 11.30, we have*

$$\deg_{G_N(\xi)}(\phi^{k_n}) \leq d^{d-1}.$$

Claim 11.31 is essentially [**137**, Lemma 7.4]. Its proof, which we leave to the reader, is a direct application of Claim 11.30; see Exercise 11.17.

Claim 11.32. *For any $\xi \in \mathcal{J}_{\phi,\mathrm{an}}$ and any ω-limit point η of ξ, we have*

$$d_{\mathbb{H}}(\eta, \zeta(0,R)) \leq d^{d-1} \cdot d_{\mathbb{H}}(\xi, \zeta(0,R)).$$

Claim 11.32 is [**137**, Proposition 7.1]. To prove it, write $G_0 := \zeta(0,R)$, and write the Berkovich interval (η, G_0) as the disjoint union

$$(\eta, G_0) = [G_{\ell_0}(\eta), G_0) \cup \bigcup_{n\geq 1} [G_{\ell_n}(\eta), G_{\ell_{n-1}}(\eta)). \tag{11.17}$$

The reader may check, using various earlier claims, that

$$(11.18)\qquad d_{\mathbb{H}}\big(G_{\ell_n}(\eta), G_{\ell_{n-1}}(\eta)\big) \le d^{d-1}\cdot d_{\mathbb{H}}\big(G_{\ell_n+qk_n}(\xi), G_{\ell_{n-1}+qk_n}(\xi)\big).$$

(See Exercise 11.18.) Summing over the intervals in equation (11.17), we have

$$\begin{aligned} d_{\mathbb{H}}(\eta, G_0) &\le d_{\mathbb{H}}\big(G_{\ell_0}(\eta), G_0\big) + d^{d-1}\sum_{n\ge 1} d_{\mathbb{H}}\big(G_{\ell_n+qk_n}(\xi), G_{\ell_{n-1}+qk_n}(\xi)\big)\\ &\le d_{\mathbb{H}}\big(G_{\ell_0}(\xi), G_0\big) + d^{d-1}\sum_{n\ge \ell_0+qk_1} d_{\mathbb{H}}\big(G_{n+1}(\xi), G_n(\xi)\big)\\ &\le d^{d-1}\cdot d_{\mathbb{H}}\big(\xi, G_0\big), \end{aligned}$$

completing the proof of Claim 11.32. Thus, each ω-limit point η satisfies

$$(11.19)\qquad \log\operatorname{diam}(\eta) \ge C_{R,d} + d^{d-1}\log\operatorname{diam}(\xi),$$

where $C_{R,d}$ is an explicit constant that depends only on R and d.

Step 4. We assume for the rest of the proof that $\xi \in \mathcal{J}_{\phi,\mathrm{an}}$ is not of Type I. Our goal in this step is to show that the ω-limit set of ξ includes a periodic point of Type II. To do so, we need two more claims.

Claim 11.33. There is an ω-limit point of ξ in $\operatorname{Ram}(\phi)\cap\mathcal{J}_{\phi,\mathrm{an}}\cap\mathbb{H}$.

To prove Claim 11.33 (which is [**137**, Proposition 3.14]), suppose no ω-limit point of ξ is a ramification point of ϕ. In particular, by Proposition 11.27(a), there would be at most finitely many $n\ge 0$ such that $\phi^n(\xi)\in\operatorname{Ram}(\phi)$. Thus, replacing ξ by an iterate, we may assume that $\phi^n(\xi)\notin\operatorname{Ram}(\phi)$ for all $n\ge 0$.

For each Type I critical point c in the Fatou component W at ∞, choose an integer N_c big enough that $|\phi^n(c)| > R$ for all $n\ge N_c$. For each ramification point η in $\mathcal{J}_{\phi,\mathrm{an}}$ (of which there are only finitely many, again by Proposition 11.27(a)), our supposition further implies that there is some integer M_η such that

$$G_n\big(\phi^j(\xi)\big) \ne G_n(\eta) \quad \text{for all} \quad j\ge 0 \quad \text{and} \quad n\ge M_\eta.$$

Let M be the maximum of all the M_η and all the qN_c. Then $G_n(\phi^j(\xi))$ does not lie above any critical points of ϕ, for all $j\ge 0$ and all $n\ge M$. As before, by Proposition 3.9 and the fact that ϕ is tame, it follows that $G_n(\phi^j(\xi))$ is not a ramification point itself.

Increase M if necessary so that M is a multiple of q, and write $M = qN$. Then for all $n\ge 0$, Claim 11.29 yields

$$\begin{aligned} d_{\mathbb{H}}\big(G_{M+qn+1}(\xi), G_{M+qn}(\xi)\big) &= d_{\mathbb{H}}\Big(G_{M+1}\big(\phi^n(\xi)\big), G_M\big(\phi^n(\xi)\big)\Big)\\ &\ge d^{-(N+1)}\cdot d_{\mathbb{H}}\big(\zeta(0,r_1), \zeta(0,r_0)\big). \end{aligned}$$

Therefore,

$$\begin{aligned} d_{\mathbb{H}}\big(\xi, G_M(\xi)\big) &\geq \sum_{n\geq 0} d_{\mathbb{H}}\big(G_{M+qn+1}(\xi), G_{M+qn}(\xi)\big) \\ &\geq \sum_{n\geq 0} d^{-(N+1)} \cdot d_{\mathbb{H}}\big(\zeta(0,r_1), \zeta(0,r_0)\big) = \infty. \end{aligned}$$

However, since $\xi, G_M(\xi) \in \mathbb{H}$, we have $d_{\mathbb{H}}(\xi, G_M(\xi)) < \infty$, a contradiction. Thus, our original supposition must be false, and ξ must have an ω-limit point $\eta \in \operatorname{Ram}(\phi)$. Then $\eta \in \mathcal{J}_{\phi,\mathrm{an}}$ since the Julia set is closed, and $\eta \in \mathbb{H}$ by Claim 11.32, completing the proof of Claim 11.33.

Claim 11.34. There is an ω-limit point of ξ that is a Type II repelling periodic point of ϕ.

We begin our proof of Claim 11.34, which is essentially [**137**, Proposition 7.7], by considering an ω-limit point $\eta \in \operatorname{Ram}(\phi) \cap \mathcal{J}_{\phi,\mathrm{an}} \cap \mathbb{H}$ of ξ, according to Claim 11.33. Clearly, every iterate $\phi^n(\eta)$ is also an ω-limit point of ξ, and hence by inequality (11.19), there is some $t > 0$ such that $\operatorname{diam}(\phi^n(\eta)) > t$ for all $n \geq 0$.

By Exercise 11.19, since the algebraic closure $\overline{K}$ of the discretely valued field K is dense in $\mathbb{C}_v$, we may perturb ϕ to a tame polynomial $\psi \in \overline{K}[z]$ with $\deg(\psi) = d$ for which $\mathcal{J}_{\psi,\mathrm{an}} \subseteq \overline{D}_{\mathrm{an}}(0,R)$, and such that

$$\big|\psi(x) - \phi(x)\big| < t \quad \text{for all } x \in \overline{D}(0,R).$$

Thus, a simple induction shows that $\phi^n(\eta) = \psi^n(\eta)$ for every $n \geq 0$. In addition, any open disk V containing η must have $\phi^n(V) = \psi^n(V)$ for every $n \geq 0$; therefore $\eta \in \mathcal{J}_{\psi,\mathrm{an}}$.

Observe that η must be of Type II or III. Indeed, if η were of Type IV, there would be a small disk D containing η but no Type I critical points of ϕ. Thus, by Proposition 3.9, ϕ would be one-to-one on D, and hence $\deg_\eta(\phi) = 1$. But $\deg_\eta(\phi) \geq 2$, since $\eta \in \operatorname{Ram}(\phi)$, confirming that η must be of Type II or III. Write $\eta = \zeta(a,r)$ with $a \in \overline{K}$ and $r > 0$.

Replacing K by a finite extension if necessary, we may assume that $\psi \in K[z]$. If η were a wandering point of ϕ, and hence also of ψ, then the open disk $D_{\mathrm{an}}(a,r)$ would be a wandering domain of ψ that is not in the attracting basin of a Type II point, contradicting Theorem 11.23. Thus, η must be preperiodic under ψ and hence under ϕ.

Replacing η by an iterate (which is still an ω-limit point of ξ), we may assume that η is a Julia periodic point of ϕ in $\mathbb{H}$. As previously noted, η must therefore be a Type II repelling periodic point of ϕ, completing the proof of Claim 11.34.

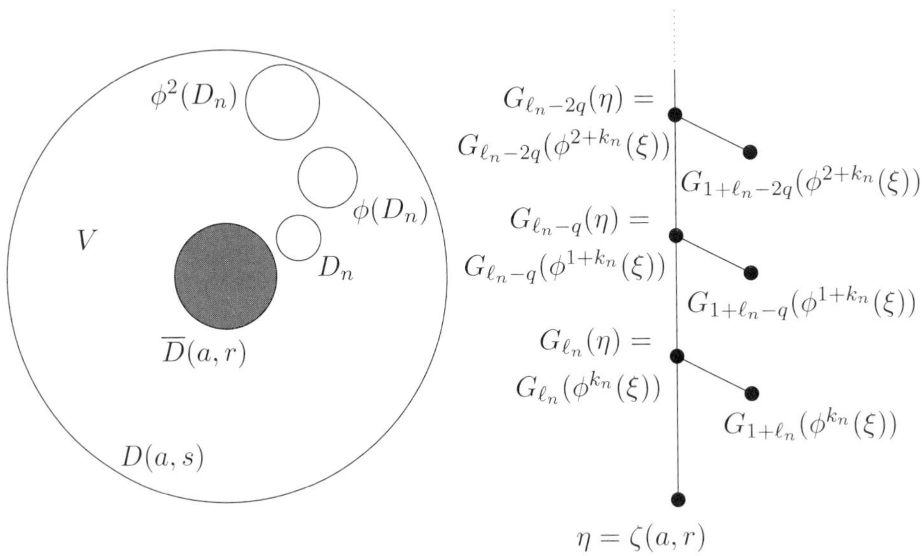

Figure 11.3. Step 5 of the proof of Theorem 11.24. The first e_n iterates of D_n, the disk bounded by $G_{1+\ell_n}(\phi^{k_n}(\xi))$, lie in $V = \{r < |\zeta - a| < s\}$.

Step 5. In this final step, we show that ξ itself must be of Type II and preperiodic to a repelling periodic point. To do so, we follow the proof of [**137**, Proposition 7.6].

Replacing ϕ by an iterate, we may assume that the Type II repelling periodic ω-limit point η of Claim 11.34 is fixed. It suffices to show that some iterate of ξ is η. To prove this, suppose not; then the sequences $\{k_n\}_{n\geq 0}$ and $\{\ell_n\}_{n\geq 0}$ are defined for all $n \geq 0$.

Write $\eta = \zeta(a,r)$, and choose $s > r$ small enough that ϕ has no zeros or critical points in the annulus

$$V := \{r < |\zeta - a| < s\}.$$

Let $m \geq 2$ be the common inner and outer Weierstrass degree of ϕ on V (see Proposition 3.32(b)). Then

$$\big|\phi(x) - a\big| = r^{1-m}|x-a|^m \quad \text{for all} \quad x \in V$$

by Theorem 3.33.

Choose $M \geq 0$ large enough that $G_{qM}(\eta) \in D_{\mathrm{an}}(a,s)$. For any $n \geq 0$ for which $\ell_n \geq qM$, we have

$$G_{\ell_n}(\eta) = G_{\ell_n}\big(\phi^{k_n}(\xi)\big) \in V,$$

but, by definition of ℓ_n, we have

$$G_{1+\ell_n}(\eta) \neq G_{1+\ell_n}\big(\phi^{k_n}(\xi)\big),$$

although both points still lie in V. Thus, the closed disk D_n bounded by $G_{1+\ell_n}(\phi^{k_n}(\xi))$ lies entirely inside the annulus V. Since ϕ has no critical points in V, Proposition 3.9 shows that ϕ is injective on D_n. Repeatedly applying Theorem 3.33 and Claim 11.29(a), it follows that for each $i = 0, \ldots, e_n := \lfloor \ell_n/q \rfloor - M$, the disk $\phi^i(D_n)$ bounded by the point

$$\phi^i\Big(G_{1+\ell_n}\big(\phi^{k_n}(\xi)\big)\Big) = G_{1+\ell_n-qi}\big(\phi^{i+k_n}(\xi)\big)$$

lies inside V, and hence ϕ is injective on $\phi^i(D_n)$. See Figure 11.3, which shows the disks $\phi^i(D_n)$ inside V on the left, and which shows the corresponding portion of the tree of Figure 11.2 on the right.

Given $n \geq 1$ such that $\ell_n \geq qM$, let

$$N_n := 1 + \ell_n + qk_n > \ell_{n-1} + qk_n.$$

By Claims 11.29(a) and 11.31, we have

$$\phi^{k_n} : G_{N_n}(\xi) \mapsto G_{1+\ell_n}\big(\phi^{k_n}(\xi)\big) \quad \text{with local degree } \leq d^{d-1}.$$

In addition, by the discussion above, we have

$$\phi^{e_n} : G_{1+\ell_n}\big(\phi^{k_n}(\xi)\big) \mapsto G_{qM+j}\big(\phi^{k_n+e_n}(\xi)\big) \quad \text{with local degree } = 1,$$

where $j := 1 + \ell_n - qe_n - qM \in \{1, \ldots, q\}$. Moreover,

$$\phi^{M+1} : G_{qM+j}\big(\phi^{k_n+e_n}(\xi)\big) \mapsto \zeta(0, r_j) \quad \text{with local degree } \leq d^{M+1},$$

where we define $r_q := R$ if $j = q$. Composing the above three functions, then

$$\phi^{k_n+e_n+M+1} : G_{N_n}(\xi) \mapsto \zeta(0, r_j) \quad \text{with local degree } \leq d^{d+M}.$$

By Claim 11.29(b), it follows that for every $n \geq 1$ such that $\ell_n \geq qM$, we have

$$d_{\mathbb{H}}\big(G_{N_n}(\xi), G_{N_n-1}(\xi)\big) \geq d^{-(d+M)} \cdot \min\big\{ \log(r_{j-1}/r_j) : 1 \leq j \leq q \big\}.$$

Therefore,

$$\begin{aligned} d_{\mathbb{H}}(\xi, \zeta(0,R)) &= \sum_{n\geq 1} d_{\mathbb{H}}\big(G_n(\xi), G_{n-1}(\xi)\big) \geq \sum_{n\geq qM} d_{\mathbb{H}}\big(G_{N_n}(\xi), G_{N_n-1}(\xi)\big) \\ &\geq \sum_{n\geq qM} d^{-(d+M)} \cdot \min\big\{ \log(r_{j-1}/r_j) : 1 \leq j \leq q \big\} = \infty, \end{aligned}$$

and hence ξ must be of Type I, contradicting our assumption from the start of Step 4. Thus, some iterate of ξ is the Type II repelling periodic point η, completing the proof of Theorem 11.24. □

Proof of Corollary 11.25. Suppose $U \subseteq \mathcal{F}_{\phi,\mathrm{an}}$ is a wandering domain of ϕ. By Theorem 11.2, some iterate $\phi^n(U)$ is a disk. (In fact, by Exercise 8.25, U is already a disk, but we do not need that fact here.) Let $\zeta \in \mathbb{H}$ be the unique boundary point of U. Then ζ is necessarily in the Julia set, as otherwise U would contain an open neighborhood of ζ.

By Theorem 11.24, ζ is preperiodic to a repelling periodic point of Type II. Thus, there are integers $N \geq 0$ and $m \geq 1$ such that $\phi^N(U)$ is a rational open disk $D_{\text{an}}(b, s)$, and the boundary point $\zeta(b, s)$ is fixed by ϕ^m. Thus, every $\phi^{N+nm}(U)$ is a residue class at $\zeta(b, s)$. □

The proof of Theorem 11.24 relies heavily on the fact that ϕ is a polynomial. Nevertheless, the combination of Theorems 11.23 and 11.24 suggests that something similar should be true for tame rational functions. On the other hand, Lattès maps like those of Example 8.20 have infinitely many periodic points of Type II in their Julia sets. In addition, if $|\mathbb{C}_v^\times| \neq (0, \infty)$, such maps also have Type III Julia points. Thus, it is too much to ask that a conclusion quite as strong as that of Theorem 11.24 holds for rational functions. Instead, we are motivated to state the following conjecture.

Conjecture 11.35. *Suppose that $\mathbb{C}_v$ contains a discretely valued field K whose algebraic closure is dense in $\mathbb{C}_v$. Let $\phi \in \mathbb{C}_v(z)$ be a tame map. Then*

(a) *ϕ has no Julia points of Type IV, and*

(b) *any wandering domain of ϕ lies in the attracting basin of a Type II Julia periodic point.*

Exercises for Chapter 11

11.1. Assume that $\operatorname{char} \mathbb{C}_v = 0$ but $\mathbb{C}_v$ has positive residue characteristic $p > 0$. Prove the following generalization of Lemma 11.5.

Let $r > 0$, $d \geq 1$, $b \in \overline{D}(0, r) \smallsetminus \{0\}$, and $0 < \rho < |b|$. Write $d = p^j m$ with $j \geq 0$ and $p \nmid m$. Let

$$f(z) = \sum_{n \geq d} c_n z^n \in \mathbb{C}_v[[z]]$$

be a power series convergent on $\overline{D}(0, r)$, with $|c_n| r^{n-d} < |d c_d|$ for all $n > d$.

(a) Prove that $f(D_{\text{an}}(b, \rho)) = D_{\text{an}}(f(b), s)$, where

$$s := |d c_d b^d| \max \left\{ |p|^{-i} \left(\frac{\rho}{|b|} \right)^{p^i} : 0 \leq i \leq j \right\}.$$

(b) Prove that the mapping $f : D_{\text{an}}(b, \rho) \to D_{\text{an}}(f(b), s)$ is p^i-to-1, where $i \in \{0, 1, \ldots, j\}$ is the smallest value for which the maximum defining s is attained.

(*Suggestion*: Use Exercise 2.15(d).)

11.2. Suppose that $\operatorname{char}\mathbb{C}_v = p > 0$, and define $f(z) := z^p + z^{p+1}$. Let $b \in D(0,1) \smallsetminus \{0\}$ and $0 < \rho < |b|$. Prove that

$$f\big(D_{\text{an}}(b,\rho)\big) = D_{\text{an}}\big(f(b), s\big), \quad \text{where} \quad s := \max\big\{|b|^p\rho, \rho^p\big\}.$$

11.3. Prove the following analogue of Lemma 11.5 for the case that $|d| = 0$.

Suppose that $\operatorname{char}\mathbb{C}_v = p > 0$, let $a \in \mathbb{C}_v$, and let $r > 0$. Let $d \geq p$ be an integer divisible by p, and let $e > d$ be an integer prime to p. Let

$$f(z) = c_0 + \sum_{n \geq d} c_n(z-a)^n \in \mathbb{C}_v[[z-a]]$$

be a power series converging on $\overline{D}(a,r)$, with

- $|c_n|r^{n-d} < |c_d|$ for all $n > d$,
- $|c_n|r^{n-e} < |c_e|$ for all $n > e$, and
- $|nc_n| = 0$ for all $0 < n < e$.

Then for any $b \in \overline{D}(a,r) \smallsetminus \{a\}$ and any $0 < \rho < |b-a|$, we have

$$\operatorname{diam}\Big(f\big(D_{\text{an}}(b,\rho)\big)\Big) \geq \big|c_e(b-a)^{e-1}\big|\rho.$$

Moreover, if $\rho \leq \beta|b-a|$, where

$$\beta := \left(\left|\frac{c_e}{c_d}\right| \cdot |b-a|^{e-d}\right)^{1/(p-1)},$$

then the mapping $f : D_{\text{an}}(b,\rho) \to D_{\text{an}}\big(f(b), |c_e(b-a)^{e-1}|\rho\big)$ is one-to-one and onto.

11.4. Prove Lemma 11.7 without the assumption that x is not wild when $\operatorname{char}\mathbb{C}_v > 0$.

That is, assume $\operatorname{char}\mathbb{C}_v > 0$, let K be a complete subfield of $\mathbb{C}_v$, let $\phi \in K(z)$ have Julia set $\mathcal{J}_{\phi,\text{an}} \subseteq \overline{D}_{\text{an}}(0,1)$, and let $x \in \mathcal{J}_{\phi,\text{an}} \cap K$ be a wild critical point of ϕ. If $\phi(x)$ is locally expanding for ϕ, prove that x is also locally expanding for ϕ.

(*Suggestion*: With notation as in Exercise 11.3, write $d = p^j m$ with $p \nmid m$. If we first specify that $\rho/|b-a| \geq \varepsilon$, use Exercise 2.15(b) to show that we can then choose $|b-a|$ small enough so that f is p^j-to-1 on $D_{\text{an}}(b,\rho)$. That should lead to showing that applying f raises the ratio of diameter to distance (cf. Definition 11.6) to the power p^j.)

11.5. Generalize Theorem 11.9(b) as follows: Let K, p, ϕ, and $\mathcal{J}_{\phi,\text{an}}$ be as in Theorem 11.9. Assume that $\operatorname{char}\mathbb{C}_v = p$. Prove that any $x \in C_t$ such that $C_{\text{w}} \cap \overline{\operatorname{Orb}_\phi^+\big(\phi(x)\big)} = \varnothing$ is locally expanding for (ϕ, K).

(Do not rewrite the whole proof; only rewrite the few steps where changes are necessary. Exercises 11.3 and 11.4 may be useful.)

11.6. Let $\phi \in \mathbb{C}_v(z)$ be a separable rational function (i.e., $\phi(z)$ is not of the form $\psi(z^p)$ for $\psi \in \mathbb{C}_v(z)$ and $p = \operatorname{char} \mathbb{C}_v$) with finite critical points $a_1, \ldots, a_m \in \mathbb{C}_v$ of multiplicities $d_1, \ldots, d_m$, respectively. Assume that none of $a_1, \ldots, a_m$ is a pole of ϕ. Let $B_1, \ldots, B_m > 0$ be any positive constants. Prove that there are real numbers $R, \alpha > 0$ with the following property: for every point $x \in \overline{D}(0,1)$, either

(a) ϕ is one-to-one on $D(x, \alpha R)$, or

(b) $D(x, R)$ contains a unique critical point a_i, the power series expansion

$$\phi(z) = c_0 + \sum_{n \geq d_i} c_n (z - a_i)^n$$

converges on $D(x, R)$, and $|c_n| R^{n-d_i} \leq B_i$ for all $n > d_i$.

11.7. Suppose that $\operatorname{char} \mathbb{C}_v = p > 0$. Let $c \in \mathbb{C}_v$ with $|c| > 1$, and define

$$\phi(z) = cz^{p+1} - cz^p + 1.$$

(a) Prove that 1 is a repelling fixed point, that 0 is a wild Julia critical point mapping to 1, and that the Julia set $\mathcal{J}_{\phi,\mathrm{an}}$ of ϕ is contained in $\overline{D}_{\mathrm{an}}(0,1)$.

(b) For any complete subfield K of $\mathbb{C}_v$, prove that the point 1 is locally expanding for (ϕ, K).

(c) Assume the complete subfield K is locally compact. Explain how Step 3 of the proof of Theorem 11.15 fails for ϕ, even taking into account the results of Exercises 11.4 and 11.6.

11.8. Prove Theorem 11.14(b).

11.9. The next few exercises are devoted to proving Lemma 11.21. In this exercise, prove the reverse implication of the lemma. That is, let k be a field, and let $f(z) \in k(z)$ be a nonconstant rational function. If k is algebraic over a finite field, prove that all points of $\mathbb{P}^1(k)$ are preperiodic under ϕ.

11.10. Let k be a field that is not algebraic over a finite field. Prove that k is a finite extension either of $\mathbb{Q}$ or of a function field $L(T)$, for some field L.

(*Suggestion*: Use the fact that if k is transcendental over some subfield L_0, then there is a transcendence basis $B \subseteq k$ such that k is algebraic over $L_0(B)$. See, for example, [**80**, Theorem 8.35].)

11.11. Let k be a field that is not algebraic over a finite field, and let $f(z) \in k(z)$ be a rational function of degree 1. Suppose that $f^n(z)$ is not the identity function for any $n \geq 1$. Prove that there are points of $\mathbb{P}^1(k)$ that are not preperiodic under f.

(*Suggestion*: Use Exercise 11.10 and the fact that f is conjugate over $\overline{k}$ either to $z \mapsto z + 1$ or to a map of the form $z \mapsto cz$.)

11.12 (This exercise requires familiarity with Weil heights.). Finish the proof of Lemma 11.21. That is, let k be a field that is not algebraic over a finite field, and let $f(z) \in k(z)$ be a rational function of degree $\deg(f) \geq 2$. Prove that there are points of $\mathbb{P}^1(k)$ that are not preperiodic under f.

(*Suggestion*: Use Exercise 11.10 to construct a Weil height function $h : \mathbb{P}^1(k) \to \mathbb{R}$ which takes on arbitrarily large values. Then show that no $x \in \mathbb{P}^1(k)$ of sufficiently large height $h(x)$ can be preperiodic under ϕ.)

11.13. Prove Proposition 11.26.

(*Suggestion*: Use the filled Julia set $\mathcal{K}_{\phi,\mathrm{an}}$ to define U.)

11.14. Let $\phi \in \mathbb{C}_v[z]$ be a polynomial of degree $d \geq 2$ with Julia set $\mathcal{J}_{\phi,\mathrm{an}}$.

(a) For any Type I point $a \in \mathbb{C}_v$, prove that at most one point of $\mathcal{J}_{\phi,\mathrm{an}}$ lies above a.

(b) If ϕ is tame and $\zeta \in \mathcal{J}_{\phi,\mathrm{an}} \cap \mathrm{Ram}(\phi)$, prove that there is a Type I critical point of ϕ lying below ζ.
(*Suggestion*: Use Proposition 3.9.)

(c) Use Exercise 10.13 to prove Proposition 11.27(a).

11.15. Prove Claim 11.28.

(*Suggestion*: See Exercise 10.13; use ideas from the proof of Theorem 10.17.)

11.16. Prove Claim 11.29.

11.17. Prove Claim 11.31.

(*Suggestion*: Use Proposition 3.9 and Claim 11.30 to show that at most $d-1$ of the points $\{\phi^i(G_N(\xi)) : 0 \leq i < k_n\}$ are ramification points of ϕ.)

11.18. Prove inequality (11.18) in the proof of Claim 11.32.

(*Suggestion*: Use Exercise 7.14, Claim 11.29(b), and Claim 11.31.)

11.19. Let $\phi(z) = a_d z^d + \cdots + a_0 \in \mathbb{C}_v[z]$ be a polynomial of degree $d \geq 2$, and let $R > r > 0$. Suppose that $\mathcal{J}_{\phi,\mathrm{an}} \subseteq \overline{D}_{\mathrm{an}}(0, R)$. Prove that there exists $\varepsilon > 0$ with the following property. For all

$$\psi = b_d z^d + \cdots + b_0 \in \mathbb{C}_v[z]$$

with $\deg(\psi) = d$ and with $|b_i - a_i| \leq \varepsilon$ for all $i = 0, \ldots, d$, we have

(a) $\mathcal{J}_{\psi,\mathrm{an}} \subseteq \overline{D}_{\mathrm{an}}(0, R)$, and

(b) for all $x \in \overline{D}(0, R)$, we have $|\psi(x) - \phi(x)| < r$.

Moreover, if ϕ is tame and $K \subseteq \mathbb{C}_v$ is a dense subfield, prove that there exists $\psi \in K[z]$ satisfying conditions (a) and (b) above and that is also tame.

Chapter 12

Repelling Points in Berkovich Space

In [**115**, Théorème B], Rivera-Letelier proved that any $\phi(z) \in \mathbb{C}_v(z)$ of degree at least 2 has a repelling fixed point in $\mathbb{P}^1_{\mathrm{an}}$. His proof, which uses Theorem 9.7, was stated for $\mathbb{C}_p$ but works for general $\mathbb{C}_v$. Separately, in [**117**, Lemme 9.3], he also proved the existence of fixed points in certain connected affinoids, although he did not show they were repelling. Baker and Rumely later generalized both results in [**8**, Theorem 10.83], reproduced here as Theorem 12.7, to show the existence of a *repelling* fixed point in such connected affinoids. In this chapter, we discuss these repelling fixed point theorems, as well as some generalizations and consequences.

12.1. Preliminary repelling fixed point lemmas

To prove results like those described above, we first need two technical lemmas that we present in this section. Both are weaker versions of the upcoming Theorem 12.5—they have the same conclusion, but stronger hypotheses.

Lemma 12.1. *Let $\phi \in \mathbb{C}_v(z)$ be a rational function of degree at least 2, let U be an open connected affinoid, let $\zeta_0 \in U$, and let $\boldsymbol{v}$ be a direction at ζ_0. Suppose that*

(a) $\phi(U) \supseteq \overline{U}$,

(b) $\phi(\partial U) \cap \phi(U) = \varnothing$,

(c) $\phi(\zeta_0) \neq \zeta_0$, *and*

(d) $\phi_{\#}(\boldsymbol{v}) \supseteq \boldsymbol{v}$.

Then ϕ has a repelling fixed point in $\boldsymbol{v} \cap U$.

Proof. Hypotheses (c) and (d) force ζ_0 to be of Type II or III; otherwise, we would have $\boldsymbol{v} = \mathbb{P}^1_{\text{an}} \smallsetminus \{\zeta_0\}$ and $\phi_\#(\boldsymbol{v}) = \mathbb{P}^1_{\text{an}} \smallsetminus \{\phi(\zeta_0)\}$, and the latter does not contain the former.

Step 1. Define the following subset of $\overline{\boldsymbol{v}} \cap U$:

$$X := \left\{ \zeta \in \overline{\boldsymbol{v}} \cap U : \begin{array}{l} \phi \text{ is injective on } [\zeta_0, \zeta], \text{ and} \\ \phi(\xi) \in [\phi(\zeta_0), \xi) \text{ for all } \xi \in [\zeta_0, \zeta] \end{array} \right\}. \tag{12.1}$$

Clearly, $\zeta_0 \in X$. Pick $\zeta_\infty \in \mathbb{P}^1_{\text{an}}$ such that the interval $[\zeta_0, \zeta_\infty)$ is contained in X and is maximal (under $\subseteq$) among intervals of that form contained in X. (Such a maximal interval must exist, by a simple argument using Zorn's lemma; we leave the details to the reader in Exercise 12.1.)

See Figure 12.1 for a diagram of the Berkovich points ζ_0 and ζ_∞, as well as a more general point $\zeta \in [\zeta_0, \zeta_\infty) \subseteq X$ and also the directions $\boldsymbol{v}$ and $\phi_\#(\boldsymbol{v})$. Note that any point $\zeta \in X$ necessarily lies in the closure $\overline{\boldsymbol{v}}$ of the direction $\boldsymbol{v}$; by hypotheses (c) and (d) then, ζ_0 lies between ζ and $\phi(\zeta_0)$. Equivalently,

$$\big[\phi(\zeta_0), \zeta\big) = \big[\phi(\zeta_0), \zeta_0\big] \cup (\zeta_0, \zeta) \quad \text{for all } \zeta \in X. \tag{12.2}$$

On the other hand, the particular point $\zeta \in X$ shown in Figure 12.1 also satisfies $\phi(\zeta) \in (\zeta_0, \zeta)$, which is *not* always the case. Instead, we can have either $\phi(\zeta) \in (\zeta_0, \zeta)$ or $\phi(\zeta) \in [\phi(\zeta_0), \zeta_0]$.

By equations (12.1) and (12.2), for any point $\zeta \in [\zeta_0, \zeta_\infty) \subseteq X$, we have

$$\begin{aligned} \phi(\zeta) \in \big[\phi(\zeta_0), \zeta\big) &= \big[\phi(\zeta_0), \zeta_0\big] \cup (\zeta_0, \zeta) \\ &\subseteq \big[\phi(\zeta_0), \zeta_0\big] \cup (\zeta_0, \zeta_\infty) = \big[\phi(\zeta_0), \zeta_\infty\big). \end{aligned}$$

That is, $\phi([\zeta_0, \zeta_\infty)) \subseteq [\phi(\zeta_0), \zeta_\infty)$. By the continuity of ϕ, taking closures yields

$$\phi(\zeta_\infty) \in \big[\phi(\zeta_0), \zeta_\infty\big]. \tag{12.3}$$

Step 2. We claim that $\zeta_\infty \in U$. By definition, $[\zeta_0, \zeta_\infty) \subseteq X \subseteq U$. Thus,

$$\zeta_\infty \in [\zeta_0, \zeta_\infty] \subseteq \overline{U} \subseteq \phi(U),$$

where the last inclusion is by hypothesis (a). In addition, $\phi(U)$ is an open connected affinoid (by Theorem 7.8), and we have $\phi(\zeta_0) \in \phi(U)$. It follows that the whole interval $[\phi(\zeta_0), \zeta_\infty]$ is contained in $\phi(U)$, since connected affinoids are uniquely path-connected, by Theorem 6.32. Therefore, by inclusion (12.3), we have $\phi(\zeta_\infty) \in \phi(U)$. By hypothesis (b), we must have $\zeta_\infty \notin \partial U$. Hence, $\zeta_\infty \in \overline{U} \smallsetminus \partial U \subseteq U$, as claimed.

Step 3. We claim that $\phi(\zeta_\infty) = \zeta_\infty$. If $\phi(\zeta_\infty) \neq \zeta_\infty$, then by inclusion (12.3), we must have $\phi(\zeta_\infty) \in [\phi(\zeta_0), \zeta_\infty)$. In that case, let $\boldsymbol{w}$ be the direction at $\phi(\zeta_\infty)$ containing ζ_∞. By Theorem 7.30, there is a direction $\boldsymbol{u}$ at ζ_∞ such that $\phi_\#(\boldsymbol{u}) = \boldsymbol{w}$.

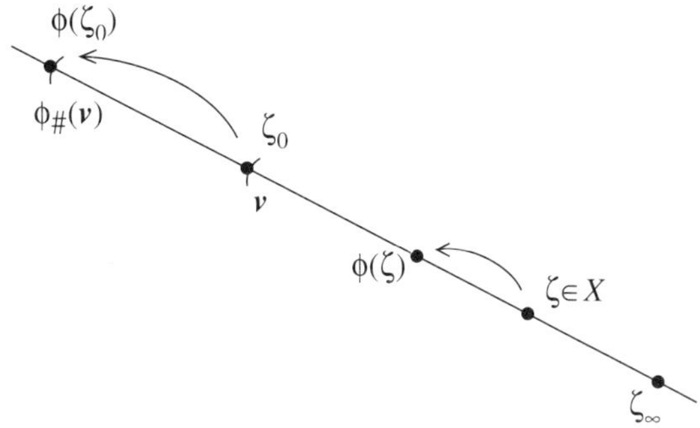

Figure 12.1. Step 1 of the proof of Lemma 12.1: $[\zeta_0, \zeta_\infty)$ is a maximal interval $[\zeta_0, \zeta)$ contained in X.

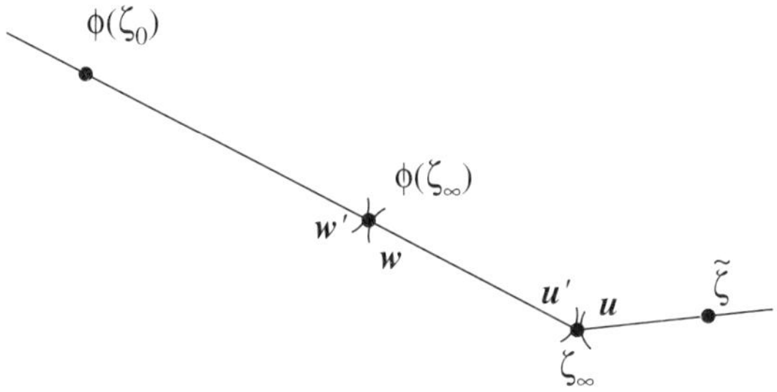

Figure 12.2. Step 3 of the proof of Lemma 12.1: $\phi_\#(\boldsymbol{u}) = \boldsymbol{w}$ and $\phi_\#(\boldsymbol{u}') = \boldsymbol{w}'$, so $\boldsymbol{u} \neq \boldsymbol{u}'$.

Let $\boldsymbol{u}'$ be the direction at ζ_∞ containing $\phi(\zeta_0)$, and let $\boldsymbol{w}'$ be the direction at $\phi(\zeta_\infty)$ containing $\phi(\zeta_0)$; see Figure 12.2. By the definitions of X and ζ_∞, ϕ must be injective on $[\zeta_0, \zeta_\infty)$, and hence by unique path-connectedness

$$\phi\big([\zeta_0, \zeta_\infty)\big) = \big[\phi(\zeta_0), \phi(\zeta_\infty)\big).$$

In particular, for every point $\zeta \in [\zeta_0, \zeta_\infty)$ sufficiently close to ζ_∞, the image $\phi(\zeta)$ lies in the direction $\boldsymbol{w}'$. Thus,

$$\phi_\#(\boldsymbol{u}') = \boldsymbol{w}' \neq \boldsymbol{w}.$$

In particular, the direction $\boldsymbol{u}$ at ζ_∞ is *not* the direction $\boldsymbol{u}'$ containing ζ_0. Hence, $\boldsymbol{u} \cap [\zeta_0, \zeta_\infty) = \varnothing$.

By Theorem 7.22, there is a point $\tilde{\zeta} \in \boldsymbol{u}$ such that ϕ maps $[\zeta_\infty, \tilde{\zeta}]$ homeomorphically onto $[\phi(\zeta_\infty), \phi(\tilde{\zeta})]$. Moreover, the interval $(\phi(\zeta_\infty), \phi(\tilde{\zeta})]$ lies entirely in the direction $\boldsymbol{w}$ at $\phi(\zeta_\infty)$. By moving $\tilde{\zeta}$ closer to ζ_∞ if necessary, we can guarantee that $\tilde{\zeta} \in U$ and that $\phi(\tilde{\zeta})$ lies between $\phi(\zeta_\infty)$ and ζ_∞.

We now show that $\tilde{\zeta} \in X$. First, recall that ϕ is injective on $[\zeta_0, \zeta_\infty)$; by the previous paragraph, ϕ is also injective on $[\zeta_\infty, \tilde{\zeta})$. Furthermore, because

$$\phi\big((\zeta_\infty, \tilde{\zeta})\big) \subseteq \big(\phi(\zeta_\infty), \phi(\tilde{\zeta})\big) \subseteq \boldsymbol{w}$$

and

$$\phi\big([\zeta_0, \zeta_\infty)\big) \subseteq \big[\phi(\zeta_0), \phi(\zeta_\infty)\big) \subseteq \boldsymbol{w}',$$

ϕ is in fact injective on the full interval $[\zeta_0, \tilde{\zeta})$, since $\boldsymbol{w} \cap \boldsymbol{w}' = \varnothing$.

Second, we already know that

$$\phi(\xi) \in \big[\phi(\zeta_0), \xi\big) \quad \text{for all } \xi \in [\zeta_0, \zeta_\infty),$$

because $[\zeta_0, \zeta_\infty) \subseteq X$. In addition, we noted at the start of Step 3 that $\phi(\zeta_\infty) \in [\phi(\zeta_0), \zeta_\infty)$. Finally, for all $\xi \in (\zeta_\infty, \tilde{\zeta})$, we have

$$\phi(\xi) \in \big[\phi(\zeta_\infty), \zeta_\infty\big] \subseteq \big[\phi(\zeta_0), \tilde{\zeta}\big),$$

proving that $\tilde{\zeta} \in X$.

As a result, the interval $[\zeta_0, \tilde{\zeta})$ is contained in X. However, the same interval properly contains $[\zeta_0, \zeta_\infty)$, contradicting the maximality of $[\zeta_0, \zeta_\infty)$ among such intervals. Thus, we have proven our claim, that $\phi(\zeta_\infty) = \zeta_\infty$.

Step 4. Finally, we claim that ζ_∞ is repelling. We consider two cases.

First, if ζ_∞ is of Type I, then after a change of coordinates, we may assume that $\zeta_\infty = 0$. Thus, all points in the interval $[\zeta_0, 0)$ sufficiently close to 0 are of the form $\zeta(0, r)$. If 0 is nonrepelling, then for r sufficiently close to 0, we have

$$\phi\big(\zeta(0, r)\big) = \zeta(0, s) \quad \text{for some} \quad s \leq r.$$

However, since $\zeta(0, r) \in [\zeta_0, 0) \subseteq X$, we must have

$$\phi\big(\zeta(0, r)\big) \in \big[\phi(\zeta_0), \zeta(0, r)\big),$$

contradicting the previous sentence. Hence, 0 must have been repelling.

Second, if ζ_∞ is of Type II, III, or IV, we again suppose that it is nonrepelling, so that ϕ has local degree 1 at ζ_∞. As in Step 3, let $\boldsymbol{u}'$ be the direction at ζ_∞ containing ζ_0, and hence containing the interval $[\zeta_0, \zeta_\infty)$. By what we have shown so far, we know that

$$\phi\big([\zeta_0, \zeta_\infty)\big) = \big[\phi(\zeta_0), \zeta_\infty\big),$$

and hence $\phi_\#(\boldsymbol{u}') = \boldsymbol{u}'$. By Theorem 7.30, the local degree of ϕ in the direction of $\boldsymbol{u}'$ is also 1. Therefore, by Theorem 7.22, we have

$$d_{\mathbb{H}}\big(\phi(\zeta), \zeta_\infty\big) = d_{\mathbb{H}}(\zeta, \zeta_\infty)$$

for all $\zeta \in [\zeta_0, \zeta_\infty)$ sufficiently close to ζ_∞. Since ζ and $\phi(\zeta)$ both lie in the interval $[\phi(\zeta_0), \zeta_\infty)$, it follows that $\phi(\zeta) = \zeta$ for all such ζ. However, each $\zeta \in [\zeta_0, \zeta_\infty)$ belongs to X and hence cannot be fixed by ϕ. By this contradiction, it follows that ζ_∞ is repelling. $\square$

Remark 12.2. We used Zorn's lemma to construct the point ζ_∞, but the construction can also be done a little more explicitly, by gradually extending a path $[\zeta_0, \zeta]$ in X until one reaches a fixed point. Rivera-Letelier uses essentially this strategy in his proof of [**117**, Lemme 9.3], as do Baker and Rumely in the first case (pages 344–345) of their proof of [**8**, Theorem 10.83]. Nevertheless, both constructions still use the Axiom of Choice implicitly.

Lemma 12.3. *Let $\phi \in \mathbb{C}_v(z)$ be a rational function of degree at least 2, and let U be an open connected affinoid. Suppose that*

(a) $\phi(U) \supseteq \overline{U}$,

(b) $\phi(\partial U) \cap \phi(U) = \varnothing$, *and*

(c) $\deg_{\zeta_0}(\phi) \geq 2$ *for some* $\zeta_0 \in U$.

Then ϕ has a repelling fixed point in U.

Proof. If ζ_0 is of Type I or IV, then $\deg_{\zeta_0}(\phi)$ is the local degree of ϕ at ζ_0 in the unique direction at ζ_0, by Theorem 7.29(c). It then follows from Definition 7.18 that there must be a point of Type II or III in a small enough disk containing ζ_0 (and, in particular, lying in U) with the same local degree. Hence, we may assume without loss that ζ_0 is of Type II or III.

If $\phi(\zeta_0) = \zeta_0$, then ζ_0 is the desired fixed point, and we are done. Thus, we assume for the remainder of the proof that $\phi(\zeta_0) \neq \zeta_0$. Much of the proof now parallels that of Lemma 12.1.

Step 1. Define

$$Y := \left\{ \zeta \in [\zeta_0, \phi(\zeta_0)] : \begin{array}{l} \deg_\zeta(\phi) \geq 2, \text{ and} \\ \phi(\xi) \in [\phi(\zeta_0), \xi] \text{ for all } \xi \in [\zeta_0, \zeta] \end{array} \right\}.$$

Clearly, $\zeta_0 \in Y$. Pick $\zeta_\infty \in \mathbb{P}^1_{\text{an}}$ such that the interval $[\zeta_0, \zeta_\infty)$ is contained in Y and is maximal (under $\subseteq$) among intervals of that form contained in Y. As in the proof of Lemma 12.1, such a maximal interval must exist. This time, however, Zorn's lemma is not required, because Y itself is contained in the interval $[\zeta_0, \phi(\zeta_0)]$. Hence, the desired maximal interval is simply the the union of all intervals $[\zeta_0, \zeta)$ for $\zeta \in Y$.

Step 2. We claim that $\zeta_\infty \in Y$. We have $[\zeta_0, \zeta_\infty) \subseteq [\zeta_0, \phi(\zeta_0)]$ by the definitions of Y and ζ_∞. Taking closures, we have $\zeta_\infty \in [\zeta_0, \phi(\zeta_0)]$ as well. In addition, $\deg_\xi(\phi) \geq 2$ for all $\xi \in [\zeta_0, \zeta_\infty)$. Therefore, $\deg_{\zeta_\infty}(\phi) \geq 2$ by the upper semicontinuity of deg from Theorem 7.29(a). Finally, by the continuity of ϕ, we have $\phi(\zeta_\infty) \in [\phi(\zeta_0), \zeta_\infty]$, as desired.

Step 3. We claim that $\zeta_\infty \in U$. Since $\phi(U)$ is connected and contains U, the whole interval $[\zeta_0, \phi(\zeta_0)]$ is contained in $\phi(U)$. In addition, because $[\zeta_0, \zeta_\infty] \subseteq Y$, we have

$$\phi(\xi) \in \big[\phi(\zeta_0), \xi\big] \subseteq \big[\zeta_0, \phi(\zeta_0)\big] \subseteq \phi(U) \quad \text{for all } \xi \in [\zeta_0, \zeta_\infty].$$

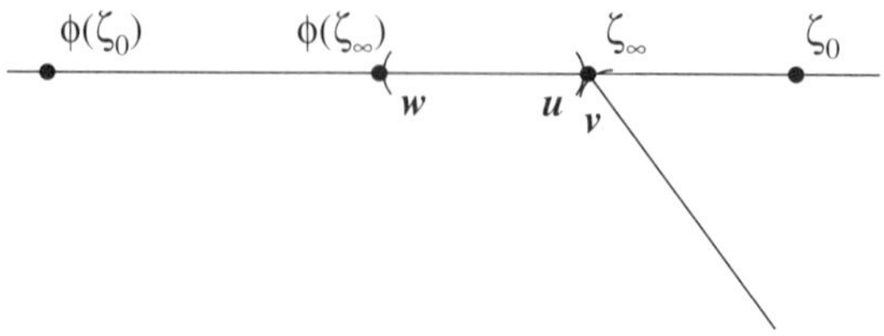

Figure 12.3. Proving Lemma 12.3, Step 4: There must be a direction $\boldsymbol{v} \neq \boldsymbol{u}$ at ζ_∞ such that $\phi_\#(\boldsymbol{v}) = \boldsymbol{w}$.

By hypotheses (b), $\xi \notin \partial U$ for all such ξ. Since U is connected and $\zeta_0 \in U$, it follows that $[\zeta_0, \zeta_\infty] \subseteq U$. In particular, $\zeta_\infty \in U$.

Step 4. If $\phi(\zeta_\infty) = \zeta_\infty$, then because $\deg_{\zeta_\infty}(\phi) \geq 2$, as we showed in Step 2, ζ_∞ itself is our desired repelling fixed point. Thus, we may assume without loss that ζ_∞ is not fixed. In other words, since $\zeta_\infty \in Y$, we must have $\phi(\zeta_\infty) \in (\zeta_\infty, \phi(\zeta_0)]$.

Let $\boldsymbol{w}$ be the direction at $\phi(\zeta_\infty)$ containing ζ_∞, let $\boldsymbol{u}$ be the direction at ζ_∞ containing $\phi(\zeta_\infty)$, and let $m := \deg_{\zeta_\infty}(\phi) \geq 2$; see Figure 12.3.

We claim that there is a direction $\boldsymbol{v}$ at ζ_∞ *other than* $\boldsymbol{u}$ such that $\phi_\#(\boldsymbol{v}) = \boldsymbol{w}$. Indeed, if no such direction $\boldsymbol{v}$ existed, then we would have

$$\phi_\#(\boldsymbol{u}) = \boldsymbol{w} \quad \text{and} \quad \deg_{\zeta_\infty, \boldsymbol{u}}(\phi) = m,$$

by Theorem 7.30. It would then be immediate from Definition 7.18(b) that ϕ maps a thin annulus extending from ζ_∞ in the direction $\boldsymbol{u}$ onto a thin annulus extending from $\phi(\zeta_\infty)$ in the direction $\boldsymbol{w}$ with Weierstrass degree m. In particular, all points ξ in the nonempty interval

$$\big(\zeta_\infty, \phi(\zeta_\infty)\big) \subseteq \big[\zeta_0, \phi(\zeta_0)\big]$$

close enough to ζ_∞ would map to points $\phi(\xi)$ in the same interval, but close to $\phi(\zeta_\infty)$; moreover, we would have

$$\deg_\xi(\phi) = m \geq 2 \quad \text{for all such } \xi.$$

In other words, the original set Y would include the whole interval $[\zeta_0, \xi]$, contradicting the maximality of ζ_∞ and proving our claim.

Since $\boldsymbol{v} \neq \boldsymbol{u}$, we have $\boldsymbol{v} \subseteq \boldsymbol{w} = \phi_\#(\boldsymbol{v})$. The existence of a repelling fixed point in U is now immediate from Lemma 12.1. □

Remark 12.4. Figure 12.3 shows the case where $\zeta_\infty \neq \zeta_0$, which can arise only when $\phi_\#(\boldsymbol{u}') = \boldsymbol{w}'$, where $\boldsymbol{u}'$ is the direction at ζ_0 containing $\phi(\zeta_0)$, and where $\boldsymbol{w}'$ is the direction at $\phi(\zeta_0)$ containing ζ_0. That is, if $\phi_\#(\boldsymbol{u}')$ is any direction at $\phi(\zeta_0)$ other than $\boldsymbol{w}$, then the set Y is simply $\{\zeta_0\}$, and hence $\zeta_\infty = \zeta_0$. In that case, the direction $\boldsymbol{v}$ found in Step 4 of the above proof is simply any direction at ζ_0 mapping to $\boldsymbol{w}'$.

12.2. Repelling fixed points exist

The main result of this section is the following.

Theorem 12.5. *Let $\phi \in \mathbb{C}_v(z)$ be a rational function of degree at least 2, and let U be an open connected affinoid. Suppose that*

$$\phi(U) \supseteq \overline{U} \quad \textit{and} \quad \phi(\partial U) \cap \overline{U} = \varnothing.$$

Then ϕ has a repelling fixed point of Type I or II in U.

Recall that by our definitions, the full space $U = \mathbb{P}^1_{\text{an}}$ is an open connected affinoid. In that case, of course, both conditions on $\phi(U)$ in Theorem 12.5 are vacuously true. Thus, we have the following corollary.

Corollary 12.6. *Let $\phi \in \mathbb{C}_v(z)$ be a rational function of degree at least 2. Then the Berkovich Julia set $\mathcal{J}_{\phi,\text{an}}$ of ϕ is nonempty.*

Proof. By Theorem 12.5 applied to $\mathbb{P}^1_{\text{an}}$, the map ϕ has a repelling fixed point $\zeta \in \mathbb{P}^1_{\text{an}}$. By Theorem 8.7(c), ζ must lie in the Julia set of ϕ. □

Note that Theorem 12.5 says nothing about whether there are repelling fixed points of Type I, of Type II, or some of each. In fact, almost anything can happen. For example, a map of explicit good reduction has a single repelling fixed point at the (Type II) Gauss point and no other repelling periodic points anywhere in $\mathbb{P}^1_{\text{an}}$. On the other hand, for $|\lambda| > 1$, the quadratic polynomial $\phi(z) = z^2 + \lambda z$ of Example 5.29 has two Type I repelling fixed points, as well as infinitely many Type I repelling periodic points, but no repelling Type II points. The cubic polynomial $\phi(z) = az^3 + z^2 + bz + c$ of Example 5.31, in residue characteristic not equal to 2, and with $0 < |a| < 1$ and $|b|, |c| \leq 1$, has both a repelling fixed point of Type II (the Gauss point) and one of Type I (near $-1/a$). It also has infinitely many Type I repelling periodic points but no other Type II repelling periodic points.

Other possible combinations of Type I and Type II repelling periodic points appear in Exercises 12.7–12.8. The only apparent restrictions on what combinations of Type I and Type II repelling periodic points are possible arise from two upcoming theorems. First, Theorem 12.10 implies that unless ϕ has potential good reduction, there must be infinitely many repelling periodic points. Second, Corollary 12.19 implies that if there is at least one repelling periodic point of Type I, then (except in certain rare circumstances) there must be infinitely many.

To prove Theorem 12.5, we first present a slightly weaker version in Theorem 12.7, which is Baker and Rumely's extension [**8**, Theorem 10.83] of Rivera-Letelier's earlier result [**117**, Proposition 9.3]. Theorems 12.5 and 12.7 differ only in their last hypothesis. Specifically, Theorem 12.7

requires that $\phi(\partial U) \cap \phi(U) = \varnothing$. By Corollary 7.10, this condition is equivalent to requiring that U is a connected component of $\phi^{-1}(\phi(U))$. The terminology for this condition varies; Baker and Rumely say that such U is "ϕ-saturated," whereas Rivera-Letelier says that ϕ is "proper" on U.

Theorem 12.7. *Let $\phi \in \mathbb{C}_v(z)$ be a rational function of degree at least 2, and let U be an open connected affinoid. Suppose that*

$$\phi(U) \supseteq \overline{U} \quad \text{and} \quad \phi(\partial U) \cap \phi(U) = \varnothing.$$

Then ϕ has a repelling fixed point of Type I or II in U.

Proof. In light of Theorem 8.7(a), it suffices to find a repelling fixed point in U, with no regard to its type.

If $U = \mathbb{P}^1_{\text{an}}$, then by the Riemann–Hurwitz formula (Theorem 1.9), ϕ must have a Type I ramification point (since $\deg(\phi) \geq 2$). We are then done by Lemma 12.3. Thus, we assume for the remainder of the proof that U is a connected affinoid of the form

$$U = \mathbb{P}^1_{\text{an}} \smallsetminus \big(D_1 \cup \cdots \cup D_n\big),$$

where $n \geq 1$, and each D_j is a closed (Berkovich) disk. Let ζ_j be the unique boundary point of D_j. Note that ζ_j is necessarily of Type II or III. Our hypotheses imply that $\phi(\zeta_j) \notin \phi(U)$, and hence that $\phi(\zeta_j) \notin \overline{U}$.

Let $S := \{1, \ldots, n\}$. Define $T : S \to S$ by setting $T(j)$ to be the unique $\ell \in S$ for which $\phi(\zeta_j) \in D_\ell$. (Such ℓ exists because $\phi(\zeta_j) \notin U$.)

Case 1. Suppose T has a fixed point. Without loss, we have $\phi(\zeta_1) \in D_1$. We would like to apply Lemma 12.1 to ζ_1 and the direction at ζ_1 containing U. However, because $\zeta_1 \notin U$, that lemma does not apply directly. Instead, we use a point close to ζ_1 but just inside U, as follows.

Because $\phi(\zeta_1) \notin \phi(U)$ and $\phi(U) \supseteq U$, the image $\phi(U)$ is contained in the direction at $\phi(\zeta_1)$ containing U. In particular, any thin enough annulus emanating from ζ_1 in the direction of U must map to a thin annulus V emanating from $\phi(\zeta_1)$, also in the direction of U, bearing in mind Theorem 7.12. In addition, $\phi(\zeta_1)$ must lie in the interior of D_1 because $\phi(\zeta_1) \notin \overline{U}$.

After a change of coordinates moving some Type I point of U to 0, and some appropriate Type I point in D_1 to ∞, we may assume that

- $\zeta_1 = \zeta(0, r)$,
- $\phi(\zeta_1) = \zeta(0, r')$,
- the annulus V is $V = \{s < |\zeta| < r\} = D_{\text{an}}(0, r) \smallsetminus \overline{D}_{\text{an}}(0, s)$, and
- $\phi(V) = \{s' < |\zeta| < r'\} = D_{\text{an}}(0, r') \smallsetminus \overline{D}_{\text{an}}(0, s')$,

where $0 < s < r < s' < r'$. Choose a radius $t \in (s, r)$, let $\zeta_0 := \zeta(0, t)$, and let $\boldsymbol{v}$ be the direction at ζ_0 containing 0. Then by Definition 7.18(b), we have

$$\phi(\zeta_0) = \zeta(0, t') \quad \text{for some } t' \in (s', r'),$$

and $\phi_\#(\boldsymbol{v})$ is the direction at $\phi(\zeta_0)$ containing 0. Thus, $\phi(\zeta_0) \neq \zeta_0$, and $\phi_\#(\boldsymbol{v}) \supseteq \boldsymbol{v}$. Because $\zeta_0 \in V \subseteq U$, we are done by Lemma 12.1.

Case 2. Otherwise, T has no fixed points. Because S is finite and nonempty, the function $T : S \to S$ must have a periodic cycle of some period $m \geq 2$. Without loss, T maps

$$1 \mapsto 2, \quad 2 \mapsto 3, \quad \ldots, \quad m-1 \mapsto m, \quad m \mapsto 1.$$

Moreover, by re-indexing again if necessary, we may assume that

$$d_{\mathbb{H}}(\zeta_1, \zeta_2) \leq d_{\mathbb{H}}(\zeta_2, \zeta_3). \tag{12.4}$$

(Here, we have abused notation if $m = 2$ by redefining ζ_3 to mean ζ_1.) However, because $\phi(\partial U) \cap \overline{U} = \varnothing$, the points $\phi(\zeta_1)$ and $\phi(\zeta_2)$ must lie in the interiors of D_2 and D_3, respectively. Thus, the closed interval $[\zeta_2, \zeta_3]$ must be properly contained in the open interval $(\phi(\zeta_1), \phi(\zeta_2))$. By inequality (12.4), then, we have

$$d_{\mathbb{H}}\big(\zeta_1, \zeta_2\big) < d_{\mathbb{H}}\big(\phi(\zeta_1), \phi(\zeta_2)\big). \tag{12.5}$$

If we had $\deg_\zeta(\phi) = 1$ for all $\zeta \in (\zeta_1, \zeta_2)$, then it would follow from Theorem 7.22 and a simple compactness argument (see Exercise 12.4) that

$$d_{\mathbb{H}}(\zeta_1, \zeta_2) \geq d_{\mathbb{H}}\big(\phi(\zeta_1), \phi(\zeta_2)\big),$$

contradicting inequality (12.5). Thus, there must be some

$$\zeta \in (\zeta_1, \zeta_2) \subseteq U$$

for which $\deg_\zeta(\phi) \geq 2$. Hence, we are done by Lemma 12.3. □

Proof of Theorem 12.5. The finitely many boundary points $\zeta_1, \ldots, \zeta_m$ of U are necessarily of Type II or III. Because $\phi(\partial U)$ is also a finite set, and disjoint from $\partial U \subseteq \overline{U}$ by hypothesis, we may choose open connected affinoids $U_1, \ldots, U_m$ small enough that

- $\zeta_j \in U_j$ for each $j = 1, \ldots, m$,
- $\phi(\partial U) \cap U_j = \varnothing$ for each $j = 1, \ldots, m$ and
- the U_j's are pairwise disjoint.

Let $V_j := \phi(U) \cap U_j$, which is an open connected affinoid containing ζ_j, since $\zeta_j \in \overline{U} \subseteq \phi(U)$. Define

$$V := U \cup V_1 \cup \cdots \cup V_m.$$

Then V is an open connected affinoid. Moreover,

$$V \cap \phi(\partial U) = \varnothing \quad \text{and} \quad \overline{U} \subseteq V \subseteq \phi(U). \tag{12.6}$$

Pick $\zeta \in \phi^{-1}(V) \cap U$; such a point exists by the second statement of (12.6). Let W be the connected component of $\phi^{-1}(V)$ containing ζ. By Theorem 7.8(b)–(c), W is itself an open connected affinoid, with

$$\phi(W) = V \quad \text{and} \quad \phi(W) \cap \phi(\partial W) = \varnothing.$$

We claim that $W \subseteq U$. Indeed, if there were a point $\xi \in W \smallsetminus U$, then because W is connected, the interval $[\xi, \zeta]$ would be contained in W. Since $\zeta \in U$ but $\xi \notin U$, there would have to be some point

$$\xi_0 \in [\xi, \zeta] \cap \partial U \subseteq W \cap \partial U.$$

Then

$$\phi(\xi_0) \in \phi(W) \cap \phi(\partial U) = V \cap \phi(\partial U),$$

contradicting the first statement of (12.6) and proving our claim.

We have shown that W is an open connected affinoid for which

$$\phi(W) \cap \phi(\partial W) = \varnothing \quad \text{and} \quad \phi(W) = V \supseteq \overline{U} \supseteq \overline{W}.$$

Thus, by Theorem 12.7, $W \subseteq U$ contains a repelling fixed point of Type I or II. □

Remark 12.8. Theorem 12.5 generalizes Theorem 12.7 by relaxing the hypothesis that $\phi(\partial U) \cap \phi(U) = \varnothing$ to $\phi(\partial U) \cap \overline{U} = \varnothing$. (Recall that both results also assume the hypothesis that $\phi(U) \supseteq \overline{U}$.) This condition cannot be removed, however.

For example, for $\lambda \in \mathbb{C}_v$ with $|\lambda| > 1$, the map $\phi(z) = z^2 + \lambda z$ of Example 5.29 has Type I repelling fixed points at 0 and $1 - \lambda$, as well as a superattracting fixed point at ∞, but the reader can check that ϕ has no periodic points in $\mathbb{H}$; see Exercise 8.17. Thus, for $r > 0$ small enough, the open connected affinoid

$$U := \mathbb{P}^1_{\text{an}} \smallsetminus \big(\overline{D}_{\text{an}}(0, r) \cup \overline{D}_{\text{an}}(1 - \lambda, r)\big)$$

satisfies $\phi(U) = \mathbb{P}^1_{\text{an}} \supseteq \overline{U}$, but U contains no repelling fixed points, although it *does* contain the attracting fixed point at ∞. (Incidentally, choosing $R > |\lambda|$, the connected open affinoid

$$V := D_{\text{an}}(0, R) \smallsetminus \big(\overline{D}_{\text{an}}(0, r) \cup \overline{D}_{\text{an}}(1 - \lambda, r)\big)$$

satisfies $\phi(V) \supseteq \overline{V}$ but contains no fixed points at all.) The problem, of course, is that

$$\zeta(0, r) \in \partial U \quad \text{and} \quad \phi\big(\zeta(0, r)\big) = \zeta(0, |\lambda| r) \in U.$$

It is also possible for Theorem 12.5 to fail if we weaken the hypothesis $\phi(\partial U) \cap \overline{U} = \varnothing$ further to $\phi(\partial U) \cap U = \varnothing$ but allow $\phi(\partial U)$ to intersect ∂U; see Exercise 12.10.

Theorem 12.5 allows us to sharpen Theorem 8.15(a) as follows.

Proposition 12.9. *Let $\phi \in \mathbb{C}_v(z)$ be a rational function of degree at least 2, let E_ϕ be the exceptional set of ϕ in $\mathbb{P}^1(\mathbb{C}_v)$, and let $U \subseteq \mathbb{P}^1_{\text{an}}$ be a connected open set that is not dynamically stable under ϕ. Then for any open set V containing E_ϕ, there is an integer $N \geq 0$ such that*

$$\phi^N(U) \supseteq \mathbb{P}^1_{\text{an}} \smallsetminus V.$$

In particular if $E_\phi = \varnothing$, then $\phi^N(U) = \mathbb{P}^1_{\text{an}}$ for some $N \geq 0$.

Proof. Let $S := \mathbb{P}^1_{\text{an}} \smallsetminus \bigcup_{n\geq 0} \phi^n(U)$. Then $\phi^{-1}(S) \subseteq S$, but at the same time, S is finite because U is not dynamically stable. Thus, all elements of S have finite backward orbit. In addition, by Theorem 12.5, ϕ has a repelling fixed point $\zeta \in \mathbb{P}^1_{\text{an}}$ of Type I or II. We now consider these two cases.

Case 1. The repelling fixed point ζ is of Type I. If $\zeta \in S$, then being of Type I with finite backward orbit, ζ would belong to E_ϕ and hence, by Corollary 5.16, to the Fatou set. But ζ is repelling; therefore, $\zeta \notin S$. Thus, there is some $M \geq 0$ such that $\zeta \in \phi^M(U)$.

After a change of coordinates, $\zeta = 0$. Shrinking U if necessary, we may assume that $\phi^M(U) = D_{\text{an}}(0, r)$, where $r > 0$ is small enough that

$$\phi^{M+1}(U) = D_{\text{an}}\big(0, |\phi'(0)|r\big).$$

Since 0 is repelling, we have $\phi^M(U) \subsetneq \phi^{M+1}(U)$, and hence

$$\phi^n(U) \subseteq \phi^{n+1}(U) \quad \text{for all } n \geq M.$$

Each iterate $\phi^n(U)$ is either a disk or all of $\mathbb{P}^1_{\text{an}}$, for $n \geq M$. If the latter ever occurs, we are done. Otherwise, the union $\bigcup_{n\geq M} \phi^n(U)$ is of the form $\mathbb{P}^1_{\text{an}} \smallsetminus \{a\}$, with $a \in E_\phi$. After a change of coordinates, $a = \infty$ (and ϕ is then a polynomial, although we do not need that fact here). Thus, the open set V containing E_ϕ contains a disk of the form $\mathbb{P}^1_{\text{an}} \smallsetminus \overline{D}_{\text{an}}(0, R)$. Therefore, for N large enough, we must have

$$\phi^N(U) \supseteq \overline{D}_{\text{an}}(0, R),$$

and we are done.

Case 2. The repelling fixed point ζ is of Type II. After a change of coordinates, we may assume that $\zeta = \zeta(0, 1)$ is the Gauss point. We claim once again that $\zeta \notin S$. Indeed, if $\zeta \in S$, then by Proposition 8.13, ϕ would have explicit good reduction. Since $\zeta(0, 1) \in S$, we would have $\zeta \notin U$, and therefore, because U is connected, U must be contained in a single residue class $\boldsymbol{v}$. But then, just as in the proof of Proposition 8.12, U would be dynamically stable, contradicting our hypotheses. After all, if $\boldsymbol{v}$ is periodic, then $\bigcup_{n\geq 0} \phi^n(U)$ would be contained in the union of finitely many residue classes; and if not, then $\bigcup_{n\geq 0} \phi^n(U)$ would fail to intersect any preimage $\boldsymbol{w}$ of $\boldsymbol{v}$. Thus, we have proven our claim: $\zeta \notin S$.

Therefore, just as in Case 1, there is some $M \geq 0$ such that $\zeta \in \phi^M(U)$. Then because $\phi^M(U)$ contains an open connected affinoid containing ζ (see Theorem 6.17 and Corollary 7.9), $\phi^M(U)$ must contain all but finitely many residue classes $\boldsymbol{w}_1, \ldots, \boldsymbol{w}_m$ at ζ. Moreover, for each $\boldsymbol{w}_j$, $\phi^M(U)$ must contain an open annulus extending from ζ into $\boldsymbol{w}_j$.

By Lemma 8.11, $\bigcup_{n\geq M} \phi^n(U)$ contains all of $\mathbb{P}^1_{\text{an}}$ except for at most one Type I point a_j of each $\boldsymbol{w}_j$. Since $S' := \{a_1, \ldots, a_m\}$

- is finite,
- consists only of Type I points, and
- satisfies $\phi^{-1}(S') \subseteq S'$,

we have $S \subseteq E_\phi$. Thus, the open set V containing E_ϕ contains a small disk around each a_j, and by Lemma 8.11(a)–(b), some large enough iterate $\phi^N(U)$ must contain $\mathbb{P}^1_{\text{an}} \smallsetminus V$. □

12.3. Repelling density

In complex dynamics, it is a fundamental (and relatively easy) theorem that the Julia set of a rational function (of degree at least 2) is equal to the closure of the set of repelling periodic points. Fatou and Julia each gave proofs of this fact, but by very different strategies.

Fatou's strategy (see the proof of [**39**, Theorem III.3.1]) was as follows. Given an arbitrary open disk U intersecting the Julia set of $\phi \in \mathbb{C}(z)$, shrink U if necessary so that it contains no poles or critical values of ϕ; then we can define $d := \deg(\phi) \geq 2$ local analytic inverses $f_1, \ldots, f_d : U \to \mathbb{C}$ to ϕ, with disjoint images. By writing down an appropriate family $\mathcal{G}$ involving f_1, f_2, and the iterates ϕ^n, Fatou used the failure of equicontinuity to prove that U must contain a periodic point. Since he had already proven the far deeper bound of at most $6d - 6$ nonrepelling periodic cycles (see the discussion in Section 1.6.3), the repelling density result followed.

Julia, by contrast, starts from the existence (in complex dynamics) of a repelling or rationally indifferent fixed point x_0, from Exercise 1.27. Given an open set U intersecting the complex Julia set, he then constructs a sequence of points $\{x_n\}_{n\geq 0}$ such that

- $\phi(x_{n+1}) = x_n$ for all $n \geq 0$,
- $x_m \in U$ for some $m \geq 0$, and
- $\lim_{n\to\infty} x_n = x_0$.

(We say a particular backward orbit $\cdots x_2 \mapsto x_1 \mapsto x_0$ is *homoclinic* if it satisfies the third property.) He uses the fact that U intersects the Julia set to obtain the second property, and he uses the repulsion at x_0 (at least in one direction, in the case of multiplier 1) to obtain the third. By considering preimages of a small enough neighborhood of x_0, he shows that there must be a *repelling* periodic point in U. See [**104**, Section 14] for Julia's proof, as well as another presentation of Fatou's proof.

Here is the analogous result for dynamics on $\mathbb{P}^1_{\text{an}}$.

Theorem 12.10. *Let $\phi \in \mathbb{C}_v(z)$ be a rational function of degree at least 2 with Berkovich Julia set $\mathcal{J}_{\phi,\text{an}}$. Then $\mathcal{J}_{\phi,\text{an}}$ is contained in the closure of the set of repelling periodic points of ϕ in $\mathbb{P}^1_{\text{an}}$.*

Rivera-Letelier gave a proof of Theorem 12.10 in private communication with Baker and Rumely, who published it in [**8**, Theorem 10.88]. His proof essentially follows Julia's complex proof, by constructing a homoclinic orbit from a repelling fixed point. However, the argument is rather complicated, due to some difficult technical obstacles, especially in the case that the repelling fixed point is of Type II. We have therefore opted for a different proof, more along the lines of Fatou's strategy. First, we need some technical lemmas to deal with preimages.

Lemma 12.11. *Let $\phi \in \mathbb{C}_v(z)$ be a rational function, let $W \subseteq \mathbb{P}^1_{\text{an}}$ be an open connected affinoid, and let $\zeta \in W$. Let $1 \leq n \leq \deg(\phi)$ be the number of distinct preimages of ζ in $\mathbb{P}^1_{\text{an}}$, and let $1 \leq \ell \leq n$ be the number of connected components of $\phi^{-1}(W)$. Then there is an open connected affinoid $W' \subseteq W$ such that*

(a) *each connected component of $\phi^{-1}(W')$ contains exactly one element of $\phi^{-1}(\zeta)$, and*
(b) $\#\partial W' \leq n - \ell + \#\partial W$.

Proof. For any open connected affinoid Y, define

$$N(Y) := \#\big(\partial Y\big) - \#\big(\text{components of } \phi^{-1}(Y)\big) \in \mathbb{Z}.$$

Our strategy in this proof is to pare down W to the new affinoid W' so as to make condition (a) hold while at the same time guaranteeing that $N(W') \leq N(W)$. Condition (b) then holds, since $\phi^{-1}(W')$ has exactly n components.

If two distinct points ξ_1 and ξ_2 of $\phi^{-1}(\zeta)$ lie in the same component of $\phi^{-1}(W)$, then without loss we may assume that no other points of $\phi^{-1}(\zeta)$ lie between them. Let $\boldsymbol{u}$ be the direction at ξ_1 containing ξ_2, and let $\boldsymbol{v}$ be the direction at ξ_2 containing ξ_1. Since the interval (ξ_1, ξ_2) is connected but does not intersect $\phi^{-1}(\zeta)$, it must be that $\phi((\xi_1, \xi_2))$ is contained in a single direction $\boldsymbol{w}$ at ζ. That is, ϕ folds the interval $[\xi_1, \xi_2]$ onto itself somehow, with both endpoints mapping to ζ, and with the rest of the interval mapping into $\boldsymbol{w}$. In particular, we must have $\phi_{\#}(\boldsymbol{u}) = \phi_{\#}(\boldsymbol{v}) = \boldsymbol{w}$.

Moreover, by Definition 7.18 and Theorem 7.22, there must be distinct points $\eta_1, \eta_2 \in (\xi_1, \xi_2)$ lying close enough to ξ_1 and ξ_2, respectively, that if we define U_i to be the annulus or punctured disk with boundary points ξ_i and η_i, then

- $\overline{U}_1 \cap \overline{U}_2 = \varnothing$,
- $\phi(\eta_1) = \phi(\eta_2) \in \boldsymbol{w}$, and
- $\phi(U_1) = \phi(U_2)$ is an annulus (or punctured disk) extending from ζ in the direction $\boldsymbol{w}$.

Hence, if we replace W by the connected component of $W \smallsetminus \{\phi(\eta_1)\}$ containing ζ, then the new W has at most one more boundary point, namely $\phi(\eta_1)$. (In fact, it may have the same number or even fewer, if there were boundary points of W lying in the direction $\boldsymbol{w}$.) At the same time, the number of connected components of $\phi^{-1}(W)$ has increased by at least one (possibly by more), since ξ_1 and ξ_2 now lie in separate components of $\phi^{-1}(W)$. Thus, in making this change, $N(W)$ may have decreased, but it has not increased.

Repeat this process until we obtain an open connected affinoid $W' \subseteq W$ containing ζ such that all of the n points of $\phi^{-1}(\zeta)$ lie in separate components of $\phi^{-1}(W')$. Thus, condition (a) holds. By construction, we also have $N(W') \leq N(W)$, and therefore condition (b) holds as well, as we noted at the start of this proof. □

Lemma 12.12. *Let $\phi \in \mathbb{C}_v(z)$ be a rational function, let V, W be open connected Berkovich affinoids, let $\xi \in V$ and $\zeta \in W$ be points of Type II or III, and let $\boldsymbol{u}$ be a direction at ξ. Suppose that*

- *V is a connected component of $\phi^{-1}(W)$,*
- *$\phi^{-1}(\zeta) \cap V = \{\xi\}$, and*
- *$\boldsymbol{u} \not\subseteq V$.*

Then there is a point $\eta_0 \in V \cap \boldsymbol{u}$ such that for any $\eta \in (\eta_0, \xi)$, the connected component W' of $W \smallsetminus \{\phi(\eta)\}$ containing ζ has the properties

(a) *$\#\partial W' \leq \#\partial W$, and*

(b) *for every direction $\boldsymbol{w}$ at ζ, we have $\boldsymbol{w} \subseteq W'$ if and only if $\boldsymbol{w} \subseteq W$.*

Proof. After a change of coordinates on the domain and range, we may assume that $\xi = \zeta(0, r)$ for some $r > 0$, that $\boldsymbol{u}$ is the direction at ξ containing 0, that $\zeta = \zeta(0, s)$ for some $s > 0$, and that $\boldsymbol{v} := \phi_{\#}(\boldsymbol{u})$ is the direction at ζ containing 0. Let $m := \deg_{\xi,\boldsymbol{u}}(\phi)$, and choose $\lambda < 1$ sufficiently large, in the sense of Theorem 7.12, for ϕ to have inner and outer Weierstrass degree m on the annulus

$$U_\lambda := \{\lambda r < |\zeta| < r\}.$$

Increase λ if necessary to ensure that $U_\lambda \subseteq V$, and let $\eta_0 := \zeta(0, \lambda r)$.

Given a point $\eta \in (\eta_0, \xi)$, we may write $\eta = \zeta(0, \mu r)$ for $\lambda < \mu < 1$. By Theorem 7.12, we have $\phi(\eta) = \zeta(0, \mu^m s)$, and

$$\{\mu^m s < |\zeta| < s\} = \phi\big(\{\mu r < |\zeta| < r\}\big) \subseteq \phi(V) = W.$$

Thus, $W' := W \smallsetminus \overline{D}_{\mathrm{an}}(0, \mu^m s)$ is the connected component of $W \smallsetminus \{\phi(\eta)\}$ containing ζ. Moreover, W' contains exactly the directions at ζ that W does, with the possible exception of $\boldsymbol{v} = \phi_\#(\boldsymbol{u})$ itself. In addition, W' has exactly the boundary points outside of $\boldsymbol{v}$ that W does, and exactly one boundary point in $\boldsymbol{v}$, namely $\zeta(0, \mu^m s)$. Therefore, it suffices to show that W had at least one boundary point in the direction $\boldsymbol{v}$.

However, $\phi(V \cap \boldsymbol{u})$ is a connected subset of $\mathbb{P}^1_{\mathrm{an}}$ that intersects the direction $\boldsymbol{v}$ but, because $\phi^{-1}(\zeta) \cap V = \{\xi\}$, cannot contain ζ. Therefore,

$$\phi(V \cap \boldsymbol{u}) \subseteq \boldsymbol{v}.$$

Furthermore, because $\boldsymbol{u} \not\subseteq V$, there must be at least one boundary point ξ' of V in $\boldsymbol{u}$. By continuity, $\phi(\xi')$ must lie in the closure $\boldsymbol{v} \cup \{\zeta\}$ of $\boldsymbol{v}$. On the other hand, by Corollary 7.10, $\phi(\xi')$ must be a boundary point of W; in particular, $\phi(\xi') \neq \zeta$. Hence, $\phi(\xi')$ is the desired element of $\boldsymbol{v} \cap \partial W$. $\square$

The heart of the proof of Theorem 12.10 is the next lemma, which is a weak version of the theorem itself. It says that any open set U intersecting the Julia set either contains a repelling periodic point or else satisfies a certain technical condition. As we will see in the proof of Theorem 12.10, it turns out that the technical condition *also* implies the existence of a repelling periodic point in U.

Lemma 12.13. *Let $\phi \in \mathbb{C}_v(z)$ be a rational function of degree at least 2. Let $\zeta \in \mathcal{J}_{\phi,\mathrm{an}}$ be a nonperiodic Julia point, and let $U \subseteq \mathbb{P}^1_{\mathrm{an}}$ be an open set containing ζ. Then either U contains a repelling periodic point, or else there exist open connected affinoids $V_0, V_1, \ldots, V_M \subseteq \mathbb{P}^1_{\mathrm{an}}$ and integers $n_1, \ldots, n_M \geq 1$ such that*

(a) $\zeta \in V_0 \subseteq U$,

(b) *V_i is a connected component of $\phi^{-n_i}(V_0)$ for each $i = 1, \ldots, M$,*

(c) *the affinoids $V_0, \ldots, V_M$ are pairwise disjoint, and*

(d) *∂V_0 consists of at most M points.*

Proof. By Theorem 6.17, we may assume without loss that U is an open connected affinoid. We now consider three cases.

Case 1. Suppose ζ is of Type I or IV. Choose $\xi \in \phi^{-1}(\zeta)$. Then ξ is a point of the same type, and $\xi \neq \zeta$ since ζ is not periodic. We may therefore choose a sufficiently small open disk V_1 containing ξ such that $V_0 := \phi(V_1)$ is a disk contained in U. Shrinking V_1 (and hence also V_0) further if necessary, we may assume that $V_0 \cap V_1 = \varnothing$. By Corollary 7.10, V_1 is a connected component of $\phi^{-1}(V_0)$. With $m = 1$ and $n_1 = 1$, the affinoids V_0 and V_1 satisfy all the desired conditions.

Case 2. Suppose ζ is of Type III. Choose $\xi \in \phi^{-1}(\zeta)$ and $\eta \in \phi^{-1}(\xi)$, which are also of Type III; and the three points are all distinct, since ζ is not periodic. Using Proposition 3.32 and Theorem 3.33, we may choose an open annulus V_2 containing η and sufficiently thin that

- $V_1 := \phi(V_2)$ and $V_0 := \phi(V_1)$ are open annuli,
- $V_0 \subseteq U$,
- $\phi(\partial V_2) = \partial V_1$ and $\phi(\partial V_1) = \partial V_0$, and
- V_0, V_1, and V_2 are pairwise disjoint.

The first and third conditions above, together with Corollary 7.10, imply that V_0 is a connected component of $\phi^{-1}(V_1)$, and that V_1 is a connected component of $\phi^{-1}(V_2)$. With $M = 2$, $n_1 = 1$, and $n_2 = 2$, the affinoids V_0, V_1, and V_2 satisfy all the desired conditions.

Case 3. Finally, suppose ζ is of Type II. Since U is Hausdorff and ζ is not periodic, we may shrink the connected affinoid U if necessary to guarantee that $\phi^{-1}(U) \cap U = \varnothing$. (We can do this by choosing disjoint open connected affinoids about ζ and about the points of $\phi^{-1}(\zeta)$. Take the forward images of the latter ones, and finally intersect all of the resulting neighborhoods of ζ with U.)

Let $W_0 := U$ and $m := \#\partial W_0$, so that $1 \leq m < \infty$. Let

$$\ell_i := \#\phi^{-i}(\zeta) \quad \text{for each } i = 1, \dots, m.$$

Shrink W_0 by applying Lemma 12.11 to ϕ^m, W_0, and ζ. Let M be the number of boundary points of the new W_0. According to that lemma, writing $W_i := \phi^{-i}(W_0)$ for $i = 1, \dots, m$, and writing W_i as a disjoint union of open connected affinoids $V_{i,1}, \dots, V_{i,\ell_i}$, we know that each $V_{i,j}$ contains exactly one element $\zeta_{i,j}$ of $\phi^{-i}(\zeta)$. In addition, condition (b) of Lemma 12.11 guarantees that $M \leq \ell_m - 1 + m$. Hence,

$$\ell_1 + \cdots + \ell_m \geq m - 1 + \ell_m \geq M,$$

since each ℓ_i is a positive integer. That is, the total number of all these connected affinoids $V_{i,j}$ is at least M.

Our next goal is to shrink W_0, and thus every W_i and $V_{i,j}$ as well, to guarantee that $W_i \cap W_j = \varnothing$ for $i \neq j$. We do so by sliding existing boundary points of W_0 closer to ζ, as follows.

Step 1. For each direction $\boldsymbol{w}$ at ζ with the properties that

- $\boldsymbol{w} \not\subseteq W_0$, and
- $\zeta_{i,j} \in \boldsymbol{w}$ for some $1 \leq i \leq m$ and $1 \leq j \leq \ell_i$,

choose an open annulus $U_{\boldsymbol{w}}$ extending from ζ in the direction $\boldsymbol{w}$ and with second boundary point some $\xi \in \boldsymbol{w}$. Ensure that $U_{\boldsymbol{w}}$ is sufficiently thin that $U_{\boldsymbol{w}} \cup \{\xi\} \subseteq \boldsymbol{w} \cap W_0$, but that none of the preimages $\zeta_{i,j}$ lie in $\overline{U}_{\boldsymbol{w}}$. Redefine W_0 to be the connected component of $W_0 \smallsetminus \{\xi\}$ containing ζ. Note that this new W_0 still has at most M boundary points (since the old W_0 had at least one boundary point in $\boldsymbol{w}$, and the new W_0 has only ξ as a boundary point in $\boldsymbol{w}$), but it no longer contains any of the points $\zeta_{i,j}$ that lay in $\boldsymbol{w}$, and neither does its boundary.

Thus, after having repeated the process of the previous paragraph for each of the finitely many such directions $\boldsymbol{w}$, we may assume that the only points $\zeta_{i,j}$ lying in $\overline{W}_0$ lie in directions $\boldsymbol{w}$ at ζ for which $\boldsymbol{w} \subseteq W_0$. Now redefine each W_i (for $1 \leq i \leq m$) as $\phi^{-i}(W_0)$ as before, and all the $V_{i,j}$ as the various components of W_i. Note that the number ℓ_i of such components is unchanged, because we still have $\ell_i = \#\phi^{-i}(\zeta)$.

Step 2. For each of the finitely many points $\zeta_{i,j}$, let $\boldsymbol{u}$ be the direction at $\zeta_{i,j}$ containing ζ, and let $\boldsymbol{w}$ be the direction at ζ containing $\zeta_{i,j}$. We claim that $\boldsymbol{u} \not\subseteq V_{i,j}$.

To prove the claim, suppose that $\boldsymbol{u} \subseteq V_{i,j}$. If $\boldsymbol{w} \not\subseteq W_0$, then by the conclusions of Step 1 above, we know that $W_0 \subseteq \boldsymbol{u} \subseteq V_{i,j}$. But $W_0 = \phi^i(V_{i,j})$, and hence $\phi^{ni}(V_{i,j}) \subseteq V_{i,j}$ for all $n \geq 0$, implying that $V_{i,j}$ is dynamically stable under ϕ^i. Invoking Proposition 8.2, it would follow that

$$\zeta_{i,j} \in \mathcal{F}_{\phi^i,\mathrm{an}} = \mathcal{F}_{\phi,\mathrm{an}},$$

and hence that $\zeta = \phi^i(\zeta_{i,j}) \in \mathcal{F}_{\phi,\mathrm{an}}$, contradicting the hypotheses.

On the other hand, if $\boldsymbol{w} \subseteq W_0$, then since we opened Case 3 by guaranteeing that $\phi^{-1}(W_0) \cap W_0 = \varnothing$, we must have $i \geq 2$, since $\zeta_{1,j} \in \phi^{-1}(W_0)$. Then

$$\phi^{i-1}(\zeta_{i,j}) = \zeta_{1,k} \quad \text{for some } 1 \leq k \leq \ell_1.$$

Again because $\phi^{-1}(W_0) \cap W_0 = \varnothing$, we must have $\zeta_{1,k} \notin \boldsymbol{w}$. Thus, $\zeta_{1,k}$, and hence the whole connected component $V_{1,k}$ of $\phi^{-1}(W_0)$ containing $\zeta_{1,k}$, must lie in a direction $\boldsymbol{v} \neq \boldsymbol{w}$ at ζ. Hence,

$$\phi^{i-1}(V_{i,j}) = V_{1,k} \subseteq \boldsymbol{v} \subseteq \boldsymbol{u} \subseteq V_{i,j},$$

and just as in the previous paragraph, it follows (since $i-1\geq 1$) that $\zeta\in\mathcal{F}_{\phi,\mathrm{an}}$, a contradiction. The claim is proven: $\boldsymbol{u}\not\subseteq V_{i,j}$.

Step 3. For each $\zeta_{i,j}$, let $\boldsymbol{u}$ be the direction at $\zeta_{i,j}$ containing ζ, and let $\boldsymbol{w}$ be the direction at ζ containing $\zeta_{i,j}$, as in Step 2. Apply Lemma 12.12 to

- the map ϕ^i,
- the affinoids $V=V_{i,j}$ and $W=W_0$,
- the points $\xi=\zeta_{i,j}$ and ζ, and
- the direction $\boldsymbol{u}$ at $\zeta_{i,j}$.

Specifically, choose a point $\eta\in V_{i,j}\cap\boldsymbol{u}$ sufficiently close to $\zeta_{i,j}$ in the sense of Lemma 12.12, lying between $\zeta_{i,j}$ and ζ. If $\boldsymbol{w}\not\subseteq W_0$, then $\zeta_{i,j}\notin\overline{W}_0$ by the conclusion of Step 1, and hence we can ensure that η is close enough to $\zeta_{i,j}$ that $\eta\notin\overline{W}_0$. (That is, if ξ' is the unique boundary point of W_0 lying in $\boldsymbol{w}$, we can ensure that η lies between $\zeta_{i,j}$ and ξ', by moving η even closer to $\zeta_{i,j}$.) Replace W_0 by the affinoid W' of Lemma 12.12, i.e., the component of $W_0\smallsetminus\{\phi^i(\eta)\}$ containing ζ, so that the new W_0 still has at most M boundary points. As in Step 1, shrink each successive $W_i:=\phi^{-i}(W_0)$ correspondingly. Note that this shrinks the particular component $V_{i,j}$ we have just been considering so that it now lies entirely in the direction at η containing $\zeta_{i,j}$. As a result, $\overline{V}_{i,j}$ lies entirely inside $\boldsymbol{w}$.

Step 4. Having shrunk W_0 in Steps 1 and 3, consider once again each $\zeta_{i,j}$, with the associated directions $\boldsymbol{u}$ at $\zeta_{i,j}$ and $\boldsymbol{w}$ at ζ.

If some $\zeta_{i,j}$ has $\boldsymbol{w}\subseteq W_0$ (a condition unaffected by our changes to W_0, according to Lemma 12.12(b)), then by the conclusion of Step 3,

$$\overline{V}_{i,j}\subseteq\boldsymbol{w}\subseteq W_0=\phi^i(V_{i,j}).$$

Since $V_{i,j}$ is a component of $\phi^{-i}(W_0)$, by Corollary 7.10, the hypotheses of Theorem 12.5 apply to ϕ^i and $V_{i,j}$. Thus,

$$V_{i,j}\subseteq W_0\subseteq U$$

contains a repelling fixed point of ϕ^i, and we are done.

Finally, if $\boldsymbol{w}\not\subseteq W_0$ for each of the finitely many points $\zeta_{i,j}$, then by the conclusions of Step 3, we have $W_0\cap V_{i,j}=\varnothing$ for each i,j. Hence,

$$W_0\cap W_i=\varnothing\quad\text{for each } i=1,\ldots,m.$$

Since $W_i=\phi^{-i}(W_0)$, it follows immediately that $W_i\cap W_j=\varnothing$ for any $i\neq j$ between 0 and m.

Thus, setting $V_0:=W_0$, letting $V_1,V_2,\ldots$ be the components $V_{i,j}$ for $i=1,\ldots,m$ and $j=1,\ldots,\ell_i$, and letting $n_1,n_2,\ldots$ be the associated iterate numbers i, we have exactly the desired conclusions. Indeed, we have

V_0 along with

$$\ell_1 + \cdots + \ell_m > M \geq \#\partial V_0$$

other connected affinoids $V_1, V_2, \ldots$, all of which (including V_0) are pairwise disjoint. □

Proof of Theorem 12.10. Given $\zeta \in \mathcal{J}_{\phi,\mathrm{an}}$ and an open set $U \subseteq \mathbb{P}^1_{\mathrm{an}}$ containing ζ, we must find a repelling periodic point in U. If ζ has finite grand orbit and is of Type II, III, or IV, then by Proposition 8.13, it is a Type II fixed point of multiplier d. That is, $\zeta \in U$ itself is the desired repelling periodic point. If ζ has finite grand orbit and is of Type I, then by Corollary 5.16, it belongs to the Fatou set, a contradiction.

We may therefore assume that ζ has infinite grand orbit. In particular, there is some nonperiodic $\zeta' \in \mathcal{J}_{\phi,\mathrm{an}}$ with $\phi^n(\zeta') = \zeta$ for some $n \geq 0$. Let $U' = \phi^{-n}(U)$, which is an open set containing ζ'. If U' contains a repelling periodic point ξ, then $\phi^n(\xi) \in U$ is also a repelling periodic point. Thus, we may assume for the remainder of the proof that ζ is nonperiodic.

Let $S := \mathbb{P}^1_{\mathrm{an}} \setminus \bigcup_{n\geq 0} \phi^n(U)$, which is a finite set of Type I exceptional points by Theorem 8.15(a). Those points are all superattracting periodic points, and hence we can cover S with a finite union K of small closed Berkovich disks such that

$$\zeta \notin K \quad \text{and} \quad \phi(K) \subseteq \operatorname{int} K,$$

where $\operatorname{int} K$ denotes the interior of K. Because K is compact and $\mathbb{P}^1_{\mathrm{an}}$ is Hausdorff, we may shrink U if necessary and assume that $U \cap K = \varnothing$.

By Lemma 12.13, either U contains a repelling periodic point (and we are done), or else there are open connected affinoids

$$V_0, \ldots, V_M \subseteq \mathbb{P}^1_{\mathrm{an}}$$

and integers $n_1, \ldots, n_M$ with the four properties listed in Lemma 12.13. Shrink V_0 by moving each of its boundary points slightly back toward $\zeta \in V_0$, and shrink each of $V_1, \ldots, V_M$, which are components of preimages of V_0, correspondingly. This guarantees that the closures $\overline{V}_0, \ldots, \overline{V}_M$ are pairwise disjoint.

Since V_0 contains $\zeta \in \mathcal{J}_{\phi,\mathrm{an}}$, Proposition 12.9 gives an integer $N \geq 0$ such that

$$\phi^N(V_0) \supseteq \mathbb{P}^1_{\mathrm{an}} \setminus \operatorname{int} K.$$

In addition, because $\phi(K) \subseteq \operatorname{int} K \subseteq K$, and because

$$\phi^{n_i}(V_i) = V_0 \quad \text{for each } i = 1, \ldots, M,$$

we have $\overline{V}_i \cap \operatorname{int} K = \varnothing$ for $i = 0, \ldots, M$. Thus,

$$\overline{V}_i \subseteq \phi^N(V_0) \quad \text{for every } i = 0, \ldots, M.$$

On the other hand, because ϕ^N is an open map, the boundary $\partial(\phi^N(V_0))$ is contained in $\phi^N(\partial V_0)$ and hence has at most M elements. Thus, at least one of the $M+1$ disjoint sets

$$\overline{V}_0, \ldots, \overline{V}_M \subseteq \phi^N(V_0)$$

does not intersect $\partial(\phi^N(V_0))$. If V_i is one of the sets for which this happens, note also that $\phi^{n_i}(\partial V_i) = \partial V_0$ by Corollary 7.10, where we write $n_0 = 0$ if $i = 0$. Thus, V_i is an open connected affinoid such that

$$\phi^{N+n_i}(V_i) \supseteq \overline{V}_i \quad \text{and} \quad \phi^{N+n_i}(\partial V_i) \cap \overline{V}_i = \varnothing.$$

Therefore, by Theorem 12.5, ϕ^{N+n_i} has a repelling fixed point ξ in V_i. Hence, $\phi^{n_i}(\xi) \in V_0 \subseteq U$ is a repelling periodic point of ϕ. □

12.4. Type I repelling density: Hsia's theorem

In light of Theorem 12.10 and the complex dynamics repelling density theorem of Fatou and Julia, it is reasonable to ask whether the *classical* Julia set $\mathcal{J}_{\phi,\mathrm{I}}$ of a rational function $\phi \in \mathbb{C}_v(z)$ is contained in the closure of the set of *classical* (i.e., Type I) repelling periodic points of ϕ. The first result addressing this question is due to Hsia [**77**], who proved the following theorem by adapting Fatou's strategy (sketched at the start of Section 12.3) to the non-archimedean setting. (See [**57**, Proposition 6.4] for a similar result for non-archimedean *entire* functions in residue characteristic 0.) The proof we give below is a slight modification of Hsia's.

Theorem 12.14 (Hsia). *Let $\phi \in \mathbb{C}_v(z)$ be a rational function of degree at least 2 with classical Julia set $\mathcal{J}_{\phi,\mathrm{I}}$. Then $\mathcal{J}_{\phi,\mathrm{I}}$ is contained in the closure of the set of* all *Type I periodic points of ϕ.*

Proof. Given a point $y_0 \in \mathcal{J}_{\phi,\mathrm{I}}$ and an open set $U \subseteq \mathbb{P}^1(\mathbb{C}_v)$ containing y_0, we know that y_0 does not belong to the exceptional set $E_\phi \subseteq \mathbb{P}^1(\mathbb{C}_v)$, since all Type I exceptional points are Fatou. In particular, there is some point $x_0 \in \phi^{-1}(y_0)$ distinct from y_0. After a change of coordinates if necessary, we may assume that $x_0, y_0 \neq \infty$.

Choose a radius $r > 0$ small enough that ϕ may be written as a convergent power series on $D(x_0, r)$, necessarily of finite Weierstrass degree. By Theorem 3.15, $\phi(D(x_0, r)) = D(y_0, s)$ for some $s > 0$. Shrinking r if necessary, and hence s with it, we may assume that $D(x_0, r) \cap D(y_0, s) = \varnothing$, and that $D(y_0, s) \subseteq U$.

For each $n \geq 1$, define

$$\psi_n(z) := \frac{\phi^{n+1}(z) - \phi(z)}{\phi^{n+1}(z) - z},$$

and consider ψ_n acting on $D(x_0, r)$. Since $\phi(D(x_0, r)) = D(y_0, s)$, the map ϕ has no poles in $D(x_0, r)$, and therefore $\psi_n(x) = 1$ for any pole x of ϕ^{n+1} in $D(x_0, r)$. Thus, ψ_n has a pole at $x \in D(x_0, r)$ if and only if x is a fixed point of ϕ^{n+1}. Similarly, ψ_n has a zero at $x \in D(x_0, r)$ if and only if $\phi(x)$ is a fixed point of ϕ^n. (Recall that $D(x_0, r)$ and $\phi(D(x_0, r))$ are disjoint, and therefore no point $x \in D(x_0, r)$ can be a root of both $\phi^{n+1}(z) - \phi(z)$ and $\phi^{n+1}(z) - z$.) We now show that there is some $n \geq 1$ such that ψ_n has a zero or pole in $D(x_0, r)$.

For any $n \geq 1$ and $x \in D(x_0, r)$ with $\phi^{n+1}(x) = x_0$, observe that because $\phi(x) \in D(y_0, s)$, we have $|\phi(x) - x_0| > r$, and hence

$$\left|\psi_n(x)\right| = \frac{|x_0 - \phi(x)|}{|x_0 - x|} > \frac{r}{r} = 1.$$

Similarly, if $\phi^{n+1}(x) = y_0$, then $|\psi_n(x)| < 1$. In addition, as we noted in the previous paragraph, if $\phi^{n+1}(x) = \infty$, then $\psi_n(x) = 1$.

Because $y_0 \in \mathcal{J}_{\phi,\mathrm{I}}$, there is some $n \geq 1$ such that $\phi^{n+1}(D(x_0, r))$ contains at least two of the three points x_0, y_0, and ∞; see Exercise 5.6. By the previous paragraph, there are points $y_1, y_2 \in \mathbb{P}^1(\mathbb{C}_v)$ with

$$|y_1| < 1 < |y_2|,$$

and such that $\psi_n(D(x_0, r))$ contains at least two of the points y_1, y_2, and 1.

However, $\psi_n(D(x_0, r))$ is either all of $\mathbb{P}^1(\mathbb{C}_v)$ or else a $\mathbb{P}^1(\mathbb{C}_v)$-disk. Since it contains at least two of y_1, y_2, and 1, it must contain either 0 or ∞. Indeed, if it does not contain ∞, then it is a disk in $\mathbb{C}_v$ that contains two points of different absolute values; therefore, it must contain 0.

By our observations a few paragraphs above, it follows that ϕ either has an n-periodic point in $D(y_0, s) \subseteq U$ or else an $(n+1)$-periodic point x_1 in $D(x_0, r)$. In the former case, we are done immediately; in the latter case, $\phi(x_1) \in D(y_0, x) \subseteq U$ is also periodic, as desired. □

The conclusion of Hsia's theorem is weaker than the repelling density proven by Fatou and Julia over $\mathbb{C}$, because there is no analogue of Fatou's $6d - 6$ bound over a non-archimedean field. Indeed, one can easily force the number of Type I nonrepelling cycles to be infinite simply by choosing ϕ to have good reduction. Worse, as we have seen in Example 10.20 and Exercise 10.17, there can be nonrepelling Type I periodic points accumulating at Type I Julia points, at least when the residue characteristic is positive.

In fact, when $\mathbb{C}_v$ itself has positive characteristic, the Type I Julia set $\mathcal{J}_{\phi,\mathrm{I}}$ is *not* always the closure of the Type I repelling periodic points. The following example, first suggested to the author by Rivera-Letelier [**120**], illustrates this failure of Type I repelling density.

Example 12.15. Fix a prime $p \geq 2$, and assume $\operatorname{char} \mathbb{C}_v = p$; for example, $\mathbb{C}_v$ could be the completion of an algebraic closure of $\overline{\mathbb{F}}_p((t))$. Fix $b \in \mathbb{C}_v$ with $|b| > 1$, and define

$$\phi(z) := bz(1-z)^p = bz - bz^{p+1}.$$

Then $\phi'(z) = b(1-z)^p$. In other words,

$$\phi'(z) = \begin{cases} \phi(z)/z & \text{if } z \in \mathbb{C}_v \smallsetminus \{0\}, \\ b & \text{if } z = 0. \end{cases}$$

Clearly, ∞ is a superattracting fixed point, and 0 is a repelling fixed point. However, for any periodic point $x \in \mathbb{C}_v \smallsetminus \{0\}$ of some minimal period $n \geq 1$, the multiplier of x is

$$(\phi^n)'(x) = \prod_{i=0}^{n-1} \phi'\big(\phi^i(x)\big) = \prod_{i=0}^{n-1} \frac{\phi^{i+1}(x)}{\phi^i(x)} = 1,$$

since the finite product is telescoping. In particular, there are *no* Type I repelling periodic points besides the fixed point at 0.

On the other hand, because 0 is a repelling fixed and $\phi(1) = 0$, we have $0, 1 \in \mathcal{J}_{\phi,\mathrm{I}}$. Thus, the closure of the set of Type I repelling periodic points is simply $\{0\}$, which does not contain the Type I Julia point 1.

The reader may explore other maps sharing the pathological behavior of Example 12.15 in Exercises 12.12 and 12.13. After excluding such examples, we are led to the following conjecture.

Conjecture 12.16. *Let $\phi \in \mathbb{C}_v(z)$ be a rational function of degree at least 2 with classical Julia set $\mathcal{J}_{\phi,\mathrm{I}}$. If* $\operatorname{char} \mathbb{C}_v = p > 0$*, assume that ϕ is not conjugate to a map of the form*

$$z\big(\psi(z)\big)^p \quad \textit{with} \quad 1 < |\psi(0)| < \infty$$

or of the form

$$z^{-1}\big(\psi(z)\big)^p \quad \textit{with} \quad 0 < |\psi(0)| < |\psi(\infty)| < \infty$$

for some $\psi(z) \in \mathbb{C}_v(z)$.

Then $\mathcal{J}_{\phi,\mathrm{I}}$ is the closure of the set of Type I repelling periodic points of ϕ in $\mathbb{P}^1(\mathbb{C}_v)$.

12.5. Type I repelling density: Bézivin's theorem

Part of what makes the strange behavior of Example 12.15 possible is the fact that the only strict preimage of the repelling fixed point $x_0 = 0$ is the wildly ramified critical point $x_1 = 1$. This observation motivates the following definition.

Definition 12.17. Let $\phi(z) \in \mathbb{C}_v(z)$ be a rational function, and let $x \in \mathbb{P}^1(\mathbb{C}_v)$ be a periodic point of ϕ. We say that x is *backward wild* if x is not in the exceptional set of ϕ, and if for every sequence $\{x_n\}_{n\geq 0} \subseteq \mathbb{P}^1(\mathbb{C}_v)$ with

- $x_0 = x$,
- $\phi(x_{n+1}) = x_n$ for all $n \geq 0$, and
- x_n is not in the forward orbit of x for any $n \geq 1$,

there is some $m \geq 1$ such that x_m is a critical point of ϕ with ramification index divisible by $\operatorname{char} \mathbb{C}_v$.

Observe that if $\operatorname{char} \mathbb{C}_v = 0$, as is the case for $\mathbb{C}_p$ and for the field of Puiseux series, then a nonconstant map $\phi \in \mathbb{C}_v(z)$ has no backward wild points at all, as the conclusion of Definition 12.17 can never hold. The same is true in positive characteristic if $\deg(\phi) < \operatorname{char} \mathbb{C}_v$. On the other hand, if $\operatorname{char} \mathbb{C}_v > 0$ and ϕ is an inseparable map, then *every* nonexceptional periodic point of ϕ is backward wild.

Thus, Definition 12.17 is a nontrivial condition only in positive characteristic, provided that $\deg(\phi) \geq \operatorname{char} \mathbb{C}_v$ and that ϕ is separable. In that case, ϕ has only finitely many critical points in $\mathbb{P}^1(\mathbb{C}_v)$. Thus, for a periodic point x to be backward wild, there must be some $n \geq 1$ such that for *every* $y \in \phi^{-n}(x)$, at least one of the n points

$$y, \phi(y), \ldots, \phi^{n-1}(y)$$

either lies in the forward orbit of x or is one of the finitely many critical points with ramification index divisible by $\operatorname{char} \mathbb{C}_v$.

Theorem 12.18. *Let $\phi(z) \in \mathbb{C}_v(z)$ be a rational function of degree at least 2 with a Type I repelling periodic point $x_0 \in \mathbb{P}^1(\mathbb{C}_v)$ that is* not *backward wild. Then $\mathcal{J}_{\phi,\mathrm{I}}$ is the closure in $\mathbb{P}^1(\mathbb{C}_v)$ of the set of Type I repelling periodic points of ϕ.*

Corollary 12.19. *Let $\phi(z) \in \mathbb{C}_v(z)$ be a rational function, and suppose that at least one of the following conditions holds:*

(a) *ϕ has a Type I repelling periodic point but no Type I points with local degree divisible by* $\operatorname{char} \mathbb{C}_v$.

(b) *ϕ has at least three distinct Type I repelling periodic points.*

Then $\mathcal{J}_{\phi,\mathrm{I}}$ is the closure in $\mathbb{P}^1(\mathbb{C}_v)$ of the set of Type I repelling periodic points of ϕ.

Bézivin proved Corollary 12.19(a) for $\mathbb{C}_p$ in [**27**, Théorème 3], by essentially the same method we use below to prove Theorem 12.18. Note that if $\operatorname{char} \mathbb{C}_v = 0$ (as is the case for $\mathbb{C}_v = \mathbb{C}_p$) or if $\operatorname{char} \mathbb{C}_v \geq \deg(\phi) + 1$, then the condition that ϕ has no wildly ramified Type I critical points is immediate.

Note also that part (b) specifies only that ϕ has at least three distinct Type I repelling periodic *points*, not cycles. Thus, if ϕ has even a single Type I repelling cycle of minimal period 3 or more, then the Type I Julia set is the closure of the set of Type I repelling periodic points.

Proof of Theorem 12.18. The set of Type I repelling periodic points is contained in $\mathcal{J}_{\phi,\mathrm{I}}$, and hence its closure is also contained in $\mathcal{J}_{\phi,\mathrm{I}}$. Given an open set $U \subseteq \mathbb{P}^1(\mathbb{C}_v)$ intersecting $\mathcal{J}_{\phi,\mathrm{I}}$, then, it suffices to show that U contains a Type I repelling periodic point.

Step 1. Our first goal is to construct a homoclinic orbit at the repelling periodic point x_0 specified in the hypotheses. Changing coordinates if necessary, we may assume that $x_0 \neq \infty$. In addition, replacing ϕ by an iterate, we may assume that x_0 is fixed. (By Exercise 12.14, this does not affect the hypothesis that x_0 is not backward wild.)

Let

$$\lambda := \phi'(x_0) \in \mathbb{C}_v$$

be the multiplier of x_0, so that $|\lambda| > 1$. Pick a small enough radius $R > 0$ that

$$\phi(z) - x_0 = \lambda(z - x_0) + a_2(z - x_0)^2 + \cdots$$

has Weierstrass degree 1 on $D(x_0, R)$, and hence

$$|\phi(x) - x_0| = |\lambda| \cdot |x - x_0| \quad \text{for all } x \in D(x_0, R). \tag{12.7}$$

Let

$$h : D\big(x_0, |\lambda|^{-1}R\big) \to D(x_0, R)$$

denote the inverse to the map $\phi : D(x_0, R) \to D(x_0, |\lambda|^{-1}R)$. By Proposition 3.22, h is given by a power series $h(z) \in \mathbb{C}_v[[z - x_0]]$ converging on $D(x_0, |\lambda|^{-1}R)$.

The map ϕ must be separable, because x_0 is repelling, and hence ϕ' is nontrivial. Thus, ϕ has only finitely many (Type I) critical points. Since x_0 is not backward wild, there must be some $q \geq 1$ and some $x_q \in \phi^{-q}(x_0)$ such that the forward orbit of x_q contains no critical points with ramification index divisible by $\operatorname{char} \mathbb{C}_v$, and the backward orbit of x_q contains no critical points at all.

Note that $x_q \notin E_\phi$, where $E_\phi \subseteq \mathbb{P}^1(\mathbb{C}_v)$ is the exceptional set, since E_ϕ is contained in the Fatou set, but x_q is Julia. Applying Theorem 8.15(a) to the open set U, there exist some $r \geq q$ and $x_r \in U$ such that $\phi^{r-q}(x_r) = x_q$. Applying the same theorem to $D(x_0, R)$, there is some $s \geq r$ and $x_s \in D(x_0, R)$ such that $\phi^{s-r}(x_s) = x_r$; see Figure 12.4.

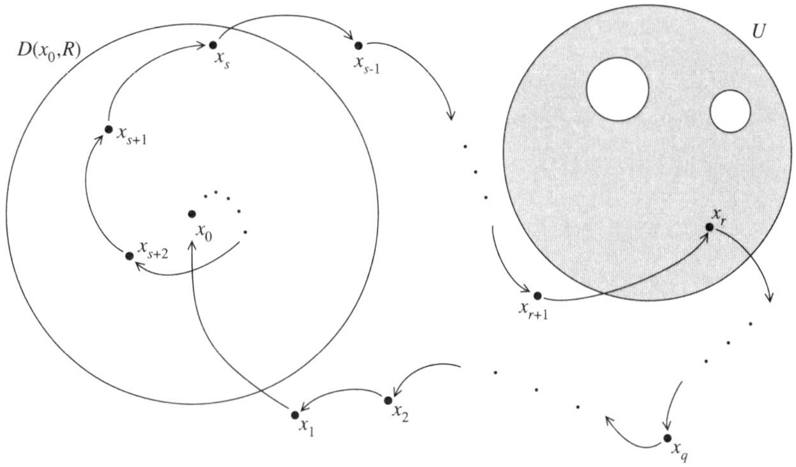

Figure 12.4. The homoclinic orbit in Step 1 of the proof of Theorem 12.18.

We can now define a sequence of points $\{x_n\}_{n\geq 0} \in \mathbb{P}^1(\mathbb{C}_v)$ by

$$x_n := \begin{cases} h^{n-s}(x_s) & \text{if } n > s, \\ \phi^{s-n}(x_s) & \text{if } 0 \leq n \leq s. \end{cases}$$

Note that this definition is consistent with our previous definitions of x_0, x_q, x_r, and x_s. In addition, we have

$$\phi(x_{n+1}) = x_n \text{ for all } n \geq 0 \quad \text{and} \quad \lim_{n\to\infty} x_n = x_0.$$

Note also if $x_s = x_0$, then because x_0 is fixed, we see that

$$x_0 = \phi^{s-r}(x_0) = \phi^{s-r}(x_s) = x_r \in U$$

is the desired repelling periodic point, and we would be done. Thus, we may assume for the remainder of the proof that $x_s \neq x_0$.

Step 2. Our next goal is to choose a small disk about x_s and produce a periodic point in it. Expand $\phi^s(z)$ as a power series centered at x_s, noting that $\phi^s(x_s) = x_0$:

$$\phi^s(z) = x_0 + \sum_{j=d}^{\infty} c_j(z - x_s)^j, \tag{12.8}$$

where $d \geq 1$ is the smallest positive integer such that $c_d \neq 0$. Because we chose the x_n's to avoid wild critical points, the integer d is not divisible by $\operatorname{char} \mathbb{C}_v$. Thus, $0 < |d| \leq 1$.

Since the power series (12.8) converges on some disk containing x_s, we can choose a radius $\sigma > 0$ small enough that

(i) $|dc_d|\sigma^d > |c_j|\sigma^j$ for all $j > d$,

(ii) $\phi^{s-r}\big(D(x_s,\sigma)\big) \subseteq U$ (since $x_r \in U$),

(iii) $|c_d|\sigma^d < |x_s - x_0|$ (since $|x_s - x_0| > 0$), and

(iv) $0 < \sigma < |d| \cdot |x_s - x_0|$.

In particular, by properties (i) and (iii),

$$\phi^s\big(D(x_s,\sigma)\big) = D\big(x_0, |c_d|\sigma^d\big) \subseteq D(x_0, |x_s - x_0|) \subseteq D(x_0, R).$$

Let $m \geq 1$ be the smallest integer such that $|\lambda^m c_d|\sigma^d > |x_s - x_0|$. Then by equation (12.7),

$$\phi^{s+m}\big(D(x_s,\sigma)\big) = D(x_0, |\lambda^m c_d|\sigma^d) \supsetneq D(x_0, |x_s - x_0|) \supseteq D(x_s,\sigma).$$

By Proposition 4.17(a), ϕ^{s+m} has a fixed point y in $D(x_s,\sigma)$. Note that $y \neq x_s$, since $\phi^{s+m}(x_s) = x_0$. In fact, by property (i) above, we have

$$|\phi^s(y) - x_0| = |c_d(y - x_s)^d|,$$

and hence

$$|y - x_0| = \big|\phi^{s+m}(y) - x_0\big| = \big|c_d(y - x_s)^d \lambda^m\big|. \tag{12.9}$$

Step 3. It remains to show that the periodic point $y \in D(x_s,\sigma)$ from Step 2 is repelling. Let $p = \operatorname{char} k$ be the residue characteristic of $\mathbb{C}_v$. Define

$$\rho := \begin{cases} |p|^{1/(p-1)}|y - x_s| & \text{if } p|d, \\ |y - x_s| & \text{if } p \nmid d. \end{cases}$$

Recall from Step 2 that $|d| > 0$; therefore, $0 < \rho \leq |y - x_s|$. By property (i) above, we may apply Lemma 11.5 to $\overline{D}(x_s,\sigma)$, and hence

$$\phi^s : D(y,\rho) \to D\Big(\phi^s(y), \big|dc_d(y - x_s)^{d-1}\big|\rho\Big) \quad \text{is one-to-one and onto.}$$

By equation (12.7) and the fact that $\phi^{s+m}(y) = y$, we have

$$\phi^{s+m} : D(y,\rho) \to D\Big(y, \big|dc_d(y - x_s)^{d-1}\lambda^m\big|\rho\Big) \quad \text{is one-to-one and onto.}$$

However, by equation (12.9) and property (iv) above,

$$\big|dc_d(y - x_s)^{d-1}\lambda^m\big|\rho = |d| \cdot \frac{|y - x_0|}{|y - x_s|}\rho = |d| \cdot \frac{|x_s - x_0|}{|y - x_s|}\rho > \frac{\sigma\rho}{|y - x_s|} > \rho,$$

where we have used the fact that $|y - x_s| < |y - x_0| < \sigma$, and hence also $|x_s - x_0| = |y - x_0|$.

Thus, ϕ^{s+m} maps $D(y,\rho)$ bijectively onto a strictly larger disk. By Theorem 4.18(c), y must be a repelling fixed point of ϕ^{s+m}, and hence a repelling periodic point of ϕ. Its forward iterate $\phi^{s-r}(y)$, which belongs to U by property (ii) above, is therefore the desired repelling periodic point. $\square$

Proof of Corollary 12.19. Part (a) is immediate from Theorem 12.18, since in those cases, ϕ has no backward wild points at all. To prove part (b), first note that because ϕ has a repelling point, it must be separable, just as in Step 1 of the proof of Theorem 12.18.

Our main goal is to show that for any backward wild periodic point $x \in \mathbb{P}^1(\mathbb{C}_v)$ of ϕ, we have

$$\sum_{y \in S(x)} \rho_y \geq d - 1. \tag{12.10}$$

Here, $S(x)$ denotes the subset of the backward orbit $\mathrm{Orb}^-(x)$ consisting of all $y \in \mathbb{P}^1(\mathbb{C}_v)$ for which there is some $n \geq 1$ with $\phi^n(y) = x$, but for which none of $y, \phi(y), \ldots, \phi^{n-1}(y)$ is in the forward orbit of x. In addition, ρ_y is the quantity we called the order of the critical point y. (As the reader may recall from Section 1.4, if $y, \phi(y) \neq \infty$, then ρ_y is the order of vanishing of ϕ' at y.)

Observe that for any two distinct periodic points $x_1 \neq x_2$ of ϕ, we must have $S(x_1) \cap S(x_2) = \varnothing$, even if x_1 and x_2 belong to the same periodic cycle. Thus, inequality (12.10) and the Riemann–Hurwitz formula (Theorem 1.9, which we can apply because ϕ is separable) together imply that at least one of the three repelling periodic points of our hypotheses is not backward wild. Hence, to prove part (b), it suffices to prove inequality (12.10).

Let $d := \deg(\phi)$ and $p := \operatorname{char} \mathbb{C}_v$. Given a periodic point x that is backward wild, let $\ell \geq 1$ be the minimal period of x, and let $w := \phi^{\ell-1}(x)$, which is the immediate preimage of x in the periodic cycle. Note that by definition of backward wild, we have $p \geq 2$. We consider two cases.

Case 1. Suppose every preimage $y \in \phi^{-1}(x)$ other than w is ramified with index divisible by p. For each such y, let ρ_y be the associated order, as above, and let e_y be the ramification index of ϕ at y. By assumption, we have $p | e_y$. As discussed in Section 1.4 (see also Exercise 1.20(b)), we also have $\rho_y \geq e_y$. Thus,

$$\begin{aligned} \sum_{y \in S(x)} \rho_y \geq \sum_{y \in \phi^{-1}(x) \smallsetminus \{w\}} \rho_y &\geq \sum_{y \in \phi^{-1}(x) \smallsetminus \{w\}} e_y \\ &= -1 + \sum_{y \in \phi^{-1}(x)} e_y = d - 1, \end{aligned}$$

where the first equality is because ϕ cannot be ramified at the repelling periodic point $w = \phi^{\ell-1}(x)$, and hence $\deg_w(\phi) = 1$; and the second is by the identity $\sum_{y \in \phi^{-1}(x)} e_y = d$.

Case 2. The only other possibility is that ϕ has ramfication index prime to p at some point $y_0 \in \phi^{-1}(x) \smallsetminus \{w\}$. Since ϕ is separable, it has only finitely many critical points, and therefore the assumption that x is backward wild implies that there is some $n \geq 0$ and $y_n \in \phi^{-n}(y_0)$ such that ϕ has ramification index divisible by p at each point $y \in \phi^{-1}(y_n)$. Just as in Case 1, then, we have $\rho_y \geq e_y$ for all such y, and hence

$$\sum_{y\in S(x)} \rho_y \geq \sum_{y\in\phi^{-1}(y_n)} \rho_y \geq \sum_{y\in\phi^{-1}(y_n)} e_y = d > d-1. \qquad \square$$

If Conjecture 12.16 is valid, then it should be possible to remove the hypothesis of the existence of a Type I repelling periodic point from Theorem 12.18 or Corollary 12.19. Unfortunately, there are significant obstacles to that goal, as the following example illustrates.

Example 12.20. Assume that $\mathbb{C}_v$ has positive residue characteristic $p > 0$, but that $\operatorname{char}\mathbb{C}_v = 0$. For any $a \in \mathbb{C}_v$, with $|a| > 1$, define

$$\phi_a(z) := z^{p+1}(z-a)^{2p}.$$

Let $U_0 =: \overline{D}(0, |a|^{-2})$ and $V_0 := \overline{D}(0, |a|)$. Then all points of $\mathbb{P}^1(\mathbb{C}_v) \smallsetminus V_0$ are attracted to ∞ under iteration, while $\phi_a(U_0) = U_0$. Thus, the Julia set, and hence any repelling periodic points of ϕ_a, must lie in the annulus $W_0 := V_0 \smallsetminus U_0$, and in fact in $\phi^{-1}(W_0) \subsetneq W_0$.

The reader can check that $\phi^{-1}(W_0)$ is a disjoint union of two thin annuli, one centered at 0, and the other at a. By a careful analysis of the orbits of points in these two annuli, it can be shown that for any $n \geq 1$, there is a choice of the parameter a (in fact, all that matters is the choice of $|a|$) such that ϕ_a has no Type I repelling periodic points of period n or smaller, but ϕ_a *does* have a Type I repelling periodic point of period $n+1$. See Exercise 12.16 for details.

The pathological maps of Example 10.20, Exercise 10.17, and Example 12.20 all require $\mathbb{C}_v$ to have positive residue characteristic. Example 12.20, in fact, requires that $\mathbb{C}_v$ has mixed characteristic. It would be interesting to know whether there is an analogue of Example 12.20 when $\operatorname{char}\mathbb{C}_v > 0$; and it would be even more interesting to know whether there are counterexamples to Conjecture 12.16. In light of Theorem 12.14 and Corollary 12.19, such a map ϕ would have to have $\mathcal{J}_{\phi,\mathrm{I}}$ nonempty but not containing a single periodic point, and yet *every* point of $\mathcal{J}_{\phi,\mathrm{I}}$ is an accumulation point of Type I nonrepelling periodic points. No such example has been found, and it is hard to imagine how such a situation could exist; but as yet, we have no proof that it cannot.

Exercises for Chapter 12

12.1. With notation as in Step 1 of the proof of Lemma 12.1, use Zorn's lemma to prove that an interval $[\zeta_0, \zeta_\infty) \subseteq X$ exists that is maximal under $\subseteq$ among all intervals of the form $[\zeta_0, \zeta) \subseteq X$.

(In particular, show that any nested union of such intervals $[\zeta_0, \zeta) \subseteq X$ is again such an interval; see Exercise 6.25.)

12.2. With notation as in Step 1 of the proof of Lemma 12.1, recall our observation that for any $\zeta \in X$, the original point ζ_0 lies between ζ and $\phi(\zeta_0)$. Prove that ζ_0 lies *strictly* between ζ_∞ and $\phi(\zeta_0)$; cf. Figure 12.1.

(*Warning*: $\zeta_\infty \notin X$.)

12.3. By supplementing its proof with an extra argument, show that the conclusion of Lemma 12.1 can be strengthened to say that the repelling fixed point ζ_∞ also has a repelling fixed direction, namely the direction containing the original point ζ_0.

12.4. Let $\phi \in \mathbb{C}_v(z)$ be a rational function, and let $\zeta_1, \zeta_2 \in \mathbb{H}$. Suppose that $\deg_\zeta(\phi) = 1$ for all $\zeta \in (\zeta_1, \zeta_2)$.

(a) Use Theorem 7.22 and the compactness of the interval $[\zeta_1, \zeta_2]$ to show that

$$d_{\mathbb{H}}(\zeta_1, \zeta_2) \geq d_{\mathbb{H}}\big(\phi(\zeta_1), \phi(\zeta_2)\big),$$

as claimed late in the proof of Theorem 12.7.

(b) Prove that, in fact, $d_{\mathbb{H}}(\zeta_1, \zeta_2) = d_{\mathbb{H}}\big(\phi(\zeta_1), \phi(\zeta_2)\big)$.

12.5. Let $\phi \in \mathbb{C}_v(z)$ be a rational function of degree at least 2. Let $x \in \mathbb{P}^1(\mathbb{C}_v)$ be a Type I fixed point of ϕ, and let $\zeta \in \mathbb{P}^1_{\text{an}}$ be a point that lies strictly between x and its own image $\phi(\zeta)$. Prove that there is a repelling fixed point in the interval $[x, \zeta)$.

(*Suggestion*: Change coordinates so that $x = \infty$ and $\zeta = \zeta(0, R)$, with $\phi(\zeta) \in D_{\text{an}}(0, R)$. Consider the supremum of the set of radii $r \geq R$ for which $\phi(\zeta(0, r)) \in D_{\text{an}}(0, r)$. This result first appeared as [**115**, Lemme 6.1].)

12.6. Generalize Exercise 12.5 to the case that $x \in \mathbb{P}^1_{\text{an}}$ is a fixed point of any type.

12.7. Give an example of a rational function with infinitely many Type II repelling periodic points but no Type I repelling periodic points.

(There are many ways to do this; one such function appeared in an example of an earlier section.)

12.8. Give an example of a rational function with infinitely many Type II repelling periodic points and infinitely many Type I repelling periodic points.

(*Suggestion*: Work in positive residue characteristic, and use a variant of Example 5.31 or 5.32.)

12.9. For any integer $n \geq 1$, give an example of a rational function with exactly n Type II repelling periodic points and infinitely many Type I repelling periodic points.

(*Suggestion*: Again, there are many ways to do this, including variations on Example 5.31 or 9.12.)

12.10. Let $a, b \in \mathbb{C}_v$ and $s \in \mathbb{R}$. Assume that $0 < |a| < |b| < 1$ and $0 < s < |a||b|^{1/2}$. Define

$$\phi(z) := \frac{z^3 - bz^2}{(z-a)^2}$$

and

$$U := D_{\text{an}}(0,1) \setminus \left(\overline{D}_{\text{an}}(0, |a|^2|b|^{-1}) \cup \overline{D}_{\text{an}}(a,s)\right).$$

Prove that

$$\phi(U) \supseteq \overline{U}, \quad \phi(\partial U) \cap U = \varnothing, \quad \text{and} \quad \phi(\partial U) \cap \partial U \neq \varnothing,$$

but that ϕ has no repelling fixed points in U.

Thus, as noted in Remark 12.8, the hypothesis $\phi(\partial U) \cap \overline{U} = \varnothing$ of Theorem 12.5 cannot be weakened to $\phi(\partial U) \cap U = \varnothing$.

(*Suggestion*: To show there are no repelling fixed points in U, partitition U into pieces, such as

$$U_1 := \{|b| \le |z| < 1\}, \qquad U_2 := \{|a| \le |z| < |b| \text{ and } |a| \le |z-a|\},$$
$$U_3 := \{s < |z-a| < |a|\}, \quad U_4 := \{|a|^2|b|^{-1} < |z| < |a|\},$$

then work with each piece separately.)

12.11. Assume $\mathbb{C}_v$ is of mixed characteristic, and let $\phi \in \mathbb{C}_v(z)$ be a rational function of degree $d \ge 2$. Prove the following strengthening of Hsia's theorem: $\mathcal{J}_{\phi,\text{I}}$ is precisely the set of accumulation points in $\mathbb{P}^1(\mathbb{C}_v)$ of Type I periodic points.

(*Suggestion*: Use the Classification Theorem and Theorem 10.10. For $\mathbb{C}_v = \mathbb{C}_p$, this result appeared as [**114**, Corollaire 4.28]. It is currently unknown whether this result holds in equal characteristic.)

12.12. Let $\phi(z) \in \mathbb{C}_v(z)$ be a rational function of degree at least 2. Let us say that ϕ is *globally unitary* if there is a nonconstant rational function $F(z) \in \mathbb{C}_v(z)$ and a constant $\mu \in \mathbb{C}_v^\times$ such that

$$\phi'(z) = \mu \frac{F(\phi(z))}{F(z)}.$$

For such ϕ, F, and μ, prove the following statements.

(a) If $x \in \mathbb{C}_v$ is a periodic point of ϕ of minimal period n, then either the multiplier of x is μ^n, or else some iterate of x is ∞, a pole of F, or a zero of F.

(b) For any $h \in \text{PGL}(2, \mathbb{C}_v)$, the conjugate $h \circ \phi \circ h^{-1}$ is also globally unitary.

(c) For any $b \in \mathbb{P}^1(\mathbb{C}_v)$, define

$$\operatorname{val}_b(F) := \begin{cases} 1 + \operatorname{ord}_b(F) & \text{if } b = \infty, \\ -1 + \operatorname{ord}_b(F) & \text{if } b \neq \infty. \end{cases}$$

Then

$$\operatorname{val}_b(F) \leq \operatorname{val}_{\phi(b)}(F) \cdot \deg_b(\phi) \quad \text{for all } b \in \mathbb{P}^1(\mathbb{C}_v),$$

where $\deg_b(\phi)$ is the ramification index of ϕ at the point b.

(d) For all $b \in \mathbb{P}^1(\mathbb{C}_v)$, we have $\operatorname{val}_b(F) \geq -1$.
(*Suggestion*: If $\operatorname{val}_b(F) \leq -2$, consider the backward orbit of b under ϕ, using the previous part.)

(e) F must be a polynomial of degree 1 or 2.
(*Suggestion*: Use part (d).)

(f) After a change of coordinates, we may assume $F(z) = z$.
(*Suggestion*: Move a critical value of ϕ to ∞.)

(g) If $F(z) = z$, then $\phi(z) = z^j(\psi(z))^p$ for some $\psi \in \mathbb{C}_v(z)$ and $j \in \mathbb{Z}$ with $p \nmid j$, where $p := \operatorname{char} \mathbb{C}_v$. Moreover, $\mu = j$ when viewed as an element of $\mathbb{C}_v$.

(h) If $\operatorname{char} \mathbb{C}_v = 0$, then ϕ is conjugate to a monomial z^j.

12.13. Assume $\operatorname{char} \mathbb{C}_v = p > 0$, and let $\phi(z) \in \mathbb{C}_v(z)$ be globally unitary, in the sense of Exercise 12.12. Prove the following.

(a) ϕ has a Type I repelling fixed point if and only if ϕ is conjugate to a map of the form

$$\tilde{\phi}(z) = z(\psi(z))^p$$

for some nonconstant $\psi \in \mathbb{C}_v(z)$ with $1 < |\psi(0)| < \infty$. The repelling fixed point of $\tilde{\phi}$ is at 0, and ∞ either is also fixed or else maps to 0.

(b) ϕ has a Type I repelling 2-cycle if and only if ϕ is conjugate to a map of the form

$$\tilde{\phi}(z) = z^{-1}(\psi(z))^p$$

for some nonconstant $\psi \in \mathbb{C}_v(z)$ with $0 < |\psi(0)| < |\psi(\infty)| < \infty$. The repelling 2-cycle of $\tilde{\phi}$ consists of 0 and ∞.

(c) All Type I repelling periodic points of ϕ are backward wild.

(d) ϕ has at most two Type I repelling periodic points.

12.14. Let $\phi \in \mathbb{C}_v(z)$ be a rational function of degree $d \geq 2$, let $x \in \mathbb{P}^1(\mathbb{C}_v)$ be a periodic point of ϕ, and let $n \geq 1$. Prove that x is backward wild for ϕ^n if and only if x is backward wild for ϕ.

12.15. Assume $\operatorname{char} \mathbb{C}_v = p > 0$, let $\phi(z) \in \mathbb{C}_v(z)$ be a rational function of degree p, and let x be a periodic point of ϕ of nonzero multiplier $a \in \mathbb{C}_v^\times$. If x is backward wild, prove that x is fixed and that ϕ is conjugate to the globally unitary map

$$\phi_a(z) := \frac{az}{(1-z)^p} \in \mathbb{C}_v(z).$$

12.16. Assume $\operatorname{char} \mathbb{C}_v = 0$ but $\mathbb{C}_v$ has positive residue characteristic $p > 0$. Let

$$\phi_a(z) := z^{p+1}(z-a)^{2p}$$

as in Example 12.20. Prove that for any $n \geq 1$, there exists $a \in \mathbb{C}_v$ with $|a| > 1$ such that ϕ_a has no Type I periodic points of period up to n, but ϕ_a has Type I periodic points of period $n+1$.

(*Suggestion*: Find precise formulas for each of $|\phi(x)|$, $|\phi(x) - a|$, and $|\phi'(x)|$ for all x in the two thin annuli described in Example 12.20. Use these formulas to compute the absolute values of multipliers of periodic points of various periods in these annuli.)

12.17. Assume $\operatorname{char} \mathbb{C}_v = 0$ but $\mathbb{C}_v$ has *odd* positive residue characteristic $p > 0$. Prove that for any $n \geq 1$, there exists $a \in \mathbb{C}_v$ with $|a| > 1$ such that

$$\psi_a(z) := z^2(z-a)^p$$

has no Type I periodic points of period up to n, but ψ_a has Type I periodic points of period $n+1$.

(*Suggestion*: Follow the same strategy as in Exercise 12.16. This example first appeared as [**15**, Example 3]. The question of what the smallest degree is of a map for which this phenomenon is possible for a given field $\mathbb{C}_v$ is currently open.)

Chapter 13

The Equilibrium Measure

As we noted in Section 1.6.4, a complex rational function $\phi \in \mathbb{C}(z)$ of degree $d \geq 2$ has an associated probability measure μ_ϕ supported on its Julia set $\mathcal{J}_\phi$. Around 2004, non-archimedean analogues of μ_ϕ on $\mathbb{P}^1_{\mathrm{an}}$, together with associated equidistribution theorems, were discovered independently by three different groups: Chambert-Loir [**40**] and Thuillier [**138**] using Arakelov theory; Baker and Rumely using a non-archimedean potential theory inspired by capacity theory [**7**, **8**]; and Favre and Rivera-Letelier [**61**, **63**] using a very similar non-archimedean potential theory related to Favre's work with Jonsson on $\mathbb{R}$-trees [**58**]. In this chapter, we develop the relevant non-archimedean potential theory of the Baker–Rumely and Favre–Rivera–Letelier approaches, leading to the construction of μ_ϕ on $\mathbb{P}^1_{\mathrm{an}}$. Our discussion mostly follows the expositions by Favre and Jonsson [**58**] and by Favre and Rivera-Letelier [**61**, **63**], but we also draw on that of Baker and Rumely [**8**].

13.1. Some measure theory

13.1.1. Review of the basics. We begin by recalling a few concepts from a standard real variable course. For more details, see any text covering abstract measure theory, such as [**64**, **123**, **124**].

A *σ-algebra* $\mathcal{A}$ on a set X is a subset of the power set of X that is closed under the operations of complements and countable unions, and such that $X \in \mathcal{A}$. Elements of $\mathcal{A}$ are called *$\mathcal{A}$-measurable sets* or simply *measurable* *sets.* If X is a topological space, then the *Borel σ-algebra* of X is the σ-algebra on X generated by the set of open subsets of X. Elements of the Borel σ-algebra are called *Borel measurable sets* or simply *Borel sets.*

If $\mathcal{A}$ and $\mathcal{B}$ are σ-algebras on sets X and Y, respectively, then a function $f : X \to Y$ is said to be *measurable* if $f^{-1}(V) \in \mathcal{A}$ for all $V \in \mathcal{B}$. If $\mathcal{B}$ is the Borel σ-algebra on Y, this is equivalent to specifying that $f^{-1}(V) \in \mathcal{A}$ for all open $V \subseteq Y$. If f is continuous, then it is measurable with respect to the Borel σ-algebras on X and Y.

Let X be a set with σ-algebra $\mathcal{A}$. A *positive measure* on X is a function $\mu : \mathcal{A} \to [0, +\infty]$ with $\mu(\varnothing) = 0$ that is *countably additive.* That is, for any countable set $\{E_n\}_{n\geq 1}$ of pairwise disjoint elements of $\mathcal{A}$, we have

$$\mu\bigl(E_1 \cup E_2 \cup \cdots\bigr) = \sum_{n\geq 1} \mu(E_n).$$

If $\mu(E) \in [0, \infty)$ for all $E \in \mathcal{A}$, we say that μ is a *finite positive measure.* If the positive measure μ satisfies $\mu(X) = 1$, we say that μ is a *probability measure.*

A (finite) *signed measure* on X is a function $\mu : \mathcal{A} \to \mathbb{R}$ with $\mu(\varnothing) = 0$ that is countably additive. Some texts allow a signed measure to take on the value $+\infty$ or $-\infty$ (but not both), but we follow the convention that

all signed measures are finite.

If $\mathcal{A}$ is the Borel σ-algebra on a topological space X, a (positive or signed) measure on X is a *Borel measure.*

Every signed measure μ has a unique *Jordan decomposition*

$$\mu = \mu_+ - \mu_-,$$

where μ_+ and μ_- are finite positive measures on X that are *mutually singular.* (That is, there exists $A \in \mathcal{A}$ such that for all $E \in \mathcal{A}$, we have $\mu_+(E) = 0$ if $E \cap A = \varnothing$, and we have $\mu_-(E) = 0$ if $E \subseteq A$.) The *total variation measure* $|\mu|$ of μ is the finite positive measure $|\mu| := \mu_+ + \mu_-$.

If X is a topological space, the *support* $\operatorname{supp}(\mu)$ of a positive Borel measure μ is the set of all $x \in X$ for which every open neighborhood U of x has $\mu(U) > 0$. The support $\operatorname{supp}(\nu)$ of a signed Borel measure ν is the support of the total variation measure $|\nu|$. The support of any (positive or signed) Borel measure is a closed subset of X.

Because of the overuse of the letter δ, we denote the delta-measure (or Dirac mass) at a point $a \in X$ by $[a]$. That is, $[a]$ is the probability measure on X given by

$$[a](E) := \begin{cases} 1 & \text{if } a \in E, \\ 0 & \text{if } a \notin E. \end{cases}$$

An *atomic measure* is a finite linear combination of delta-measures.

Given a signed measure μ on a set X, a measurable subset $E \subseteq X$, and a Borel-measurable function $f : X \to \mathbb{R}$, we assume the reader is familiar with the definition and properties of the integral $\int_E f\,d\mu$. We also assume the reader is familiar with the usual Lebesgue measure on $\mathbb{R}$ (which we consider only with the Borel σ-algebra, not the larger Lebesgue σ-algebra).

13.1.2. The Riesz representation theorem. We need the Riesz representation theorem to construct certain measures on a locally compact Hausdorff space X. To state it, we must first define Radon measures.

Definition 13.1. Let X be a locally compact Hausdorff space. A positive Borel measure μ on X is said to be

- *outer regular* on a Borel set E if
$$\mu(E) = \inf\{\mu(U) : U \subseteq X \text{ open, with } E \subseteq U\},$$
- *inner regular* on a Borel set E if
$$\mu(E) = \sup\{\mu(U) : C \subseteq X \text{ compact, with } E \supseteq C\},$$
- a *Radon measure* if it is finite on all compact sets, outer regular on all Borel sets, and inner regular on all open sets.

A (finite) signed Borel measure ν is a *signed Radon measure* if its Jordan decomposition components ν_+ and ν_- are both Radon.

The set $M_{\mathrm{Rad}}(X)$ of all signed Radon measures on X forms a Banach space, with norm $\|\mu\| := |\mu|(X)$, where $|\mu|$ is the total variation measure of μ. For convenience, we denote $M_{\mathrm{Rad}}(\mathbb{P}^1_{\mathrm{an}})$ simply by M_{Rad}.

The Riesz representation theorem, stated below, says that $M_{\mathrm{Rad}}(X)$ is dual to the Banach space $C_c(X)$ of continuous, compactly supported functions $f : X \to \mathbb{R}$. Here, $C_c(X)$ is equipped with the L^∞-norm $\|f\|_\infty := \sup\{f(x) : x \in X\}$, and the dual $B^\vee$ of a Banach space B is the Banach space of all *continuous* linear functionals $T : B \to \mathbb{R}$, with norm given by
$$\|T\| := \sup\{|Tf| : f \in B \text{ with } \|f\| \leq 1\}.$$

Theorem 13.2 (Riesz representation theorem). *Let X be a locally compact Hausdorff space. Then the map*
$$\mu \mapsto \left[F \mapsto \int_X F\,d\mu\right] \tag{13.1}$$
is an isometric isomorphism from the space $M_{\mathrm{Rad}}(X)$ of signed Radon measures on X to the dual $C_c(X)^\vee$ of the space of continuous, compactly supported functions on X.

Proof. See, for example, [**64**, Theorem 7.17], [**123**, Section 21.5], [**124**, Section 6.19]. □

In light of the Riesz representation theorem, some texts define Radon measures on a locally compact Hausdorff space to be elements of $C_c(X)$, i.e., continuous linear functionals $\mu : C_c(X) \to \mathbb{R}$. See, for example, [**11**, Definition 7.3.5], [**33**, Definition III.1.3.2].

Corollary 13.3. *The map of equation* (13.1) *is an isometric isomorphism from* M_{Rad} *to* $C(\mathbb{P}^1_{\mathrm{an}})^\vee$*, where* $M_{\mathrm{Rad}} := M_{\mathrm{Rad}}(\mathbb{P}^1_{\mathrm{an}})$*, and* $C(\mathbb{P}^1_{\mathrm{an}})$ *is the Banach space of continuous, real-valued functions on* $\mathbb{P}^1_{\mathrm{an}}$*.*

Proof. This is immediate from Theorem 13.2, since $\mathbb{P}^1_{\mathrm{an}}$ is compact. □

Remark 13.4. Both [**58**, Proposition 7.14] and [**8**, Lemma 5.6(A)] state that every signed Borel measure on $\mathbb{P}^1_{\mathrm{an}}$ is Radon, but the proofs have a gap. In fact, as noted in [**58**, Errata 5], it is possible to construct counterexamples if one assumes the existence of real-valued measurable cardinals, a question independent of the standard ZFC axioms of set theory. Thus, although many results we quote from [**8**, **58**] in this chapter were originally stated for signed Borel measures on $\mathbb{P}^1_{\mathrm{an}}$, we state them here for signed Radon measures. This subtlety was first pointed out to the author by Matt Baker [**5**].

13.1.3. Push-forward and pull-back measures.

Definition 13.5. Let X be a topological space, let $\phi : X \to X$ be a continuous function, and let μ be a signed Borel measure on X. Then the *push-forward measure* $\phi_*\mu$ is the signed Borel measure on X given by

$$\phi_*\mu(E) := \mu(\phi^{-1}(E)) \quad \text{for all Borel sets } E \subseteq \mathbb{P}^1_{\mathrm{an}}.$$

Proposition 13.6. *Let* X *be a topological space, and let* $M(X)$ *be the space of (finite) signed Borel measures on* X*. Let* $\phi : X \to X$ *be a continuous function, and let* $\mu \in M(X)$*. Then the following hold.*

(a) ϕ_* *is a linear map from* $M(X)$ *to itself.*

(b) *If* μ *is a positive measure, then so is* $\phi_*\mu$*.*

(c) $\phi_*\mu(X) = \mu(X)$*; in particular, if* μ *is a probability measure, then so is* $\phi_*\mu$*.*

(d) *If* X *is locally compact and Hausdorff, and if* μ *is Radon, then* $\phi_*\mu$ *is also Radon.*

(e) $\int_X f\, d(\phi_*\mu) = \int_X f \circ \phi \, d\mu$ *for every measurable* $f : X \to \mathbb{R}$ *for which* $\int_{\mathbb{P}^1_{\mathrm{an}}} |f \circ \phi| \, d|\mu| < \infty$*.*

(f) $(\phi \circ \psi)_*(\mu) = \phi_*(\psi_*\mu)$ *for continuous functions* $\psi : X \to X$*.*

Proof. This is left to the reader; see Exercise 13.2. □

Definition 13.7. Let $\phi \in \mathbb{C}_v(z)$ be a nonconstant rational function. For any Borel measurable function $f : \mathbb{P}^1_{\text{an}} \to \mathbb{R}$, the *pushforward* of f by ϕ is $\phi_* f : \mathbb{P}^1_{\text{an}} \to \mathbb{R}$, given by

$$\phi_* f(\zeta) := \sum_{\xi \in \phi^{-1}(\zeta)} \deg_\xi(\phi) f(\xi).$$

For any signed Radon measure $\mu \in M_{\text{Rad}}$, the *pull-back measure* is the unique measure in $\phi^* \mu \in M_{\text{Rad}}$ such that

$$\int_{\mathbb{P}^1_{\text{an}}} f \, d(\phi^* \mu) = \int_{\mathbb{P}^1_{\text{an}}} (\phi_* f) \, d\mu \quad \text{for all continuous } f : \mathbb{P}^1_{\text{an}} \to \mathbb{R}.$$

Proposition 13.8. *Let $\phi \in \mathbb{C}_v(z)$ be a nonconstant rational function, let $f : \mathbb{P}^1_{\text{an}} \to \mathbb{R}$ be Borel measurable, and let $\mu \in M_{\text{Rad}}$ be a signed Radon measure on $\mathbb{P}^1_{\text{an}}$. Then the following hold.*

(a) *The pushforward $\phi_* f : \mathbb{P}^1_{\text{an}} \to \mathbb{R}$ is a Borel measurable function. If f is continuous, then $\phi_* f$ is also continuous.*

(b) *ϕ^* is a well-defined linear map from M_{Rad} to itself.*

(c) *If μ is a positive measure, then so is $\phi^* \mu$.*

(d) *$\phi^* \mu(\mathbb{P}^1_{\text{an}}) = \deg(\phi) \cdot \mu(\mathbb{P}^1_{\text{an}})$; in particular, if μ is a probability measure, then so is $(\deg(\phi))^{-1} \phi^* \mu$.*

(e) *If μ is positive, then $\operatorname{supp}(\phi^* \mu) = \phi^{-1}(\operatorname{supp}(\mu))$.*

(f) *$(\phi \circ \psi)^*(\mu) = \psi^*(\phi^* \mu)$ for nonconstant $\psi \in \mathbb{C}_v(z)$.*

(g) *$\phi_*(\phi^* \mu) = \deg(\phi) \mu$.*

(h) *If $\phi^* \mu = \deg(\phi) \mu$, then $\phi_* \mu = \mu$.*

Proof. This is left to the reader; see Exercise 13.3. □

Example 13.9. Let $\zeta \in \mathbb{P}^1_{\text{an}}$, and let $\phi \in \mathbb{C}_v(z)$ be a nonconstant rational function. Recall that $[\zeta]$ denotes the delta-measure at ζ. Then

$$\phi_*[\zeta] = [\phi(\zeta)] \quad \text{and} \quad \phi^*[\zeta] = \sum_{\xi \in \phi^{-1}(\zeta)} \big(\deg_\xi(\phi)\big)[\xi]; \tag{13.2}$$

see Exercise 13.4.

13.2. Weak convergence

To construct the equilibrium measure, we need an appropriate notion of the limit of measures. A naive notion of convergence of a sequence $\{\mu_n\}_{n \geq 1}$ of measures might be to define the limiting measure μ on our space X by

$$\mu(U) = \lim_{n \to \infty} \mu_n(U) \quad \text{for any measurable set } U \subseteq X.$$

Unfortunately, this definition is too strong for our purposes; many reasonable sequences of measures would not converge at all. (See Example 13.12 below.) Instead, what we need is the following notion of convergence.

Definition 13.10. Let X be a topological space, let $\{\mu_n\}_{n\geq 1}$ be a sequence of signed Borel measures on X, and let μ be a signed Borel measure on X. We say that $\{\mu_n\}$ *converges weakly* to μ as $n \to \infty$ if for every $f \in C_c(X)$ we have

$$\lim_{n\to\infty} \int_X f(x)\, d\mu_n(x) = \int_X f(x)\, d\mu(x).$$

Remark 13.11. Definition 13.10 is an abuse of terminology; the convergence $\mu_n \to \mu$ it describes should more accurately be called weak* convergence. However, because this abuse of terminology is common in the theory of probability measures, and because it allows simpler phrasing ("converges weakly" instead of "converges in the weak* topology"), we reluctantly use it here. In French, the same property is sometimes called *vague* convergence.

Example 13.12. Let $X = \mathbb{R}$. For each $a \in \mathbb{R}$, recall that $[a]$ denotes the delta-measure at a. We claim that the sequence $\{[1/n]\}_{n\geq 1}$ converges weakly to $[0]$. Indeed, for any $f \in C_c(\mathbb{R})$, we have

$$\lim_{n\to\infty} \int_{\mathbb{R}} f(x)\, d[1/n](x) = \lim_{n\to\infty} f\left(\frac{1}{n}\right) = f(0) = \int_{\mathbb{R}} f(x)\, d[0](x). \tag{13.3}$$

Note that the continuity of f is essential for the second equality of equation (13.3) to hold. On the other hand, if we were to define μ on Borel subsets of $\mathbb{R}$ by

$$\mu(U) := \lim_{n\to\infty} [1/n](U),$$

then μ would *not* be a measure on $\mathbb{R}$; see Exercise 13.6.

The weak limit of a sequence of signed Radon measures on a locally compact Hausdorff space X is also a signed Radon measure; see Exercise 13.7(a).

13.3. Potential functions

Definition 13.13. Let $\xi \in \mathbb{H}$. The *Gromov product* based at ξ is the function $\langle \cdot, \cdot \rangle_\xi : \mathbb{P}^1_{\mathrm{an}} \times \mathbb{P}^1_{\mathrm{an}} \to [0, +\infty]$ given by

$$\langle \zeta_1, \zeta_2 \rangle_\xi := d_{\mathbb{H}}(\xi, \zeta'),$$

where $\zeta' \in \mathbb{P}^1_{\mathrm{an}}$ is the unique point lying between ξ, ζ_1, and ζ_2, and where we understand $d_{\mathbb{H}}(\xi, \zeta')$ to be $+\infty$ if ζ' is of Type I; see Figure 13.1.

Note that $\langle \zeta_1, \zeta_2 \rangle_\xi = +\infty$ if and only if ζ_1 and ζ_2 are both the same Type I point. See Exercises 13.10, 13.11, and 13.12 for alternate descriptions of the Gromov product in various cases.

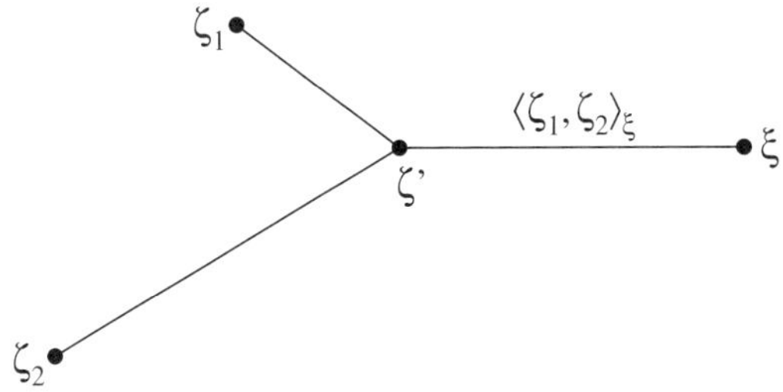

Figure 13.1. The Gromov product $\langle \zeta_1, \zeta_2 \rangle_\xi$ of Definition 13.13.

If we fix ξ and ζ_1, the function $G(\zeta) := \langle \zeta_1, \zeta \rangle_\xi$ has the following intuitive description. Define

$$G_0 : [\xi, \zeta_1] \to [0, +\infty] \quad \text{by} \quad G_0(\zeta) := d_{\mathbb{H}}(\zeta, \xi).$$

That is, as ζ moves along the Berkovich interval $[\xi, \zeta_1]$, starting from ξ, G_0 increases linearly with slope 1 from $G_0(\xi) = 0$. Then $G : \mathbb{P}^1_{\text{an}} \to [0, +\infty]$ is the unique continuous extension of G_0 to $\mathbb{P}^1_{\text{an}}$ that is constant on every connected component of $\mathbb{P}^1_{\text{an}} \smallsetminus [\xi, \zeta_1]$.

For example, for fixed $r > 1$, the function $\zeta \mapsto \langle \zeta(0, r), \zeta \rangle_{\zeta(0,1)}$ is

$$\zeta \mapsto \begin{cases} 0 & \text{if } \zeta \in \overline{D}_{\text{an}}(0, 1), \\ \log |\zeta| & \text{if } 1 \le |\zeta| \le r, \\ \log r & \text{if } |\zeta| \ge r. \end{cases}$$

Definition 13.14. Let $\mu \in M_{\text{Rad}}$ be a signed Radon measure on $\mathbb{P}^1_{\text{an}}$, and let $\xi \in \mathbb{H}$. Then the *potential of* μ *based at* ξ is the function $u_{\mu,\xi} : \mathbb{H} \to \mathbb{R}$ given by

$$u_{\mu,\xi}(\zeta) := -\mu(\mathbb{P}^1_{\text{an}}) - \int_{\mathbb{P}^1_{\text{an}}} \langle \zeta, \zeta' \rangle_\xi \, d\mu(\zeta').$$

Note that $u_{\mu,\xi}(\xi) = -\mu(\mathbb{P}^1_{\text{an}})$. In addition, note that the potential of a delta-measure $[\zeta']$ is

$$u_{[\zeta'],\xi}(\zeta) = -1 - \langle \zeta, \zeta' \rangle_\xi. \tag{13.4}$$

In particular, the potential of the delta-measure at the basepoint ξ is simply the constant function $u_{[\xi],\xi} = -1$.

Remark 13.15. The constant term $-\mu(\mathbb{P}^1_{\text{an}})$ in Definition 13.14 comes from [**61**, Section 4.1] and [**63**, Section 2.4], and ultimately from a normalization chosen in [**58**, Section 7.6]. It is not especially important for our limited purposes, but one of its benefits is that it ensures that all constant functions are potentials. (Specifically, the constant function c is the potential function based at $\xi \in \mathbb{H}$ of the point-mass $-c[\xi]$.)

Definition 13.16. Let $\xi \in \mathbb{H}$. The *space of potentials* is

$$\mathcal{P} := \{f : \mathbb{H} \to \mathbb{R} \,:\, f = u_{\mu,\xi} \text{ for some } \mu \in M_{\mathrm{Rad}}\}.$$

Proposition 13.17. *The space $\mathcal{P}$ of potentials is a vector space and is independent of the choice of $\xi \in \mathbb{H}$. In addition, $\mathcal{P}$ contains all functions $f : \mathbb{H} \to \mathbb{R}$ that are*

- *of the form $\zeta \mapsto \langle \zeta, \zeta_0 \rangle_\eta$ for fixed $\zeta_0 \in \mathbb{P}^1_{\mathrm{an}}$ and $\eta \in \mathbb{H}$,*
- *of the form $\zeta \mapsto \log \max\{r, \|z - a\|_\zeta\}$ for fixed $a \in \mathbb{C}_v$ and $r > 0$,*
- *of the form $\zeta \mapsto \log \|z - a\|_\zeta$ for fixed $a \in \mathbb{C}_v$, or*
- *constant.*

Proof. This is left to the reader; see Exercise 13.16. □

Proposition 13.18. *Let $\mu \in M_{\mathrm{Rad}}$ be a signed Radon measure on $\mathbb{P}^1_{\mathrm{an}}$, and let $\xi \in \mathbb{H}$. Then the potential function $u_{\mu,\xi}$*

(a) *is continuous on $\mathbb{H}$ with respect to the hyperbolic metric,*
(b) *is continuous on $\mathbb{H} \smallsetminus \mathrm{supp}(\mu)$ with respect to the Gel'fand topology,*
(c) *has a unique extension to a function $\tilde{u} : \mathbb{P}^1_{\mathrm{an}} \smallsetminus \mathrm{supp}(\mu) \to \mathbb{R}$ that is continuous with respect to the Gel'fand topology.*

Proof. This is left to the reader; see Exercise 13.17 and also [**8**, Proposition 6.12]. □

Warning 13.19. Many potential functions $u_{\mu,\xi}$ are continuous even on $\mathrm{supp}(\mu)$, but many others are not. For example, the potential $\zeta \mapsto \log \|z\|_\zeta$ of the atomic measure $[0] - [\infty]$ (and based at $\xi := \zeta(0,1)$) approaches $+\infty$ as $\zeta \to \infty$, and it approaches $-\infty$ as $\zeta \to 0$.

It is also possible for $u_{\mu,\xi}$ to be discontinuous at $\zeta \in \mathrm{supp}(\mu) \cap \mathbb{H}$; see Exercise 13.20 for an example.

Definition 13.20. Let $\mu \in M_{\mathrm{Rad}}$. We say that μ has *bounded potentials* (respectively, *continuous potentials*) if the associated potential function $u_{\mu,\xi}$ is bounded (respectively, extends to a function on $\mathbb{P}^1_{\mathrm{an}}$ that is continuous with respect to the Gel'fand topology) for some $\xi \in \mathbb{H}$.

For fixed $\mu \in M_{\mathrm{Rad}}$, if the potential $u_{\mu,\xi}$ is continuous (respectively, bounded) for one choice of basepoint $\xi \in \mathbb{H}$, then it is continuous (respectively, bounded) for all $\xi \in \mathbb{H}$; see Exercise 13.18.

Proposition 13.21. *Let $\mu \in M_{\mathrm{Rad}}$ be a finite linear combination of delta-measures at points of $\mathbb{H}$. Then μ has continuous potentials.*

Proof. This is left to the reader; see Exercise 13.19. □

Remark 13.22. In [**8**, Section 3.5,5.4], Baker and Rumely characterize the space $\mathcal{P}$ intrinsically. They define a space $\mathrm{BDV}(\mathbb{P}^1_{\mathrm{an}})$ of functions $f : \mathbb{H} \to \mathbb{R}$ that are of *bounded differential variation*, and they show that $\mathrm{BDV}(\mathbb{P}^1_{\mathrm{an}})$ coincides with the set of potential functions. (See Proposition 5.28 and Section 5.8 of [**8**]. More precisely, their functions are defined on most Type I points, but not necessarily all; the space $\mathcal{P}$ consists of restrictions of $f \in \mathrm{BDV}(\mathbb{P}^1_{\mathrm{an}})$ to $\mathbb{H}$.) Loosely speaking, BDV functions have a derivative defined on the tree $\mathbb{H}$ that does not wiggle too much. Less loosely, one can meaningfully take the Laplacian of a BDV function; see Section 13.4.3.

Baker and Rumely also define the potential of a measure slightly differently. First, after fixing $\xi \in \mathbb{P}^1_{\mathrm{an}}$, they define the *generalized Hsia kernel* based at ξ to be the real-valued function $\delta(\cdot,\cdot)_\xi$ given by

$$\delta(\zeta_1,\zeta_2)_\xi := \exp\big(\langle\zeta_1,\xi\rangle_{\zeta(0,1)} + \langle\zeta_2,\xi\rangle_{\zeta(0,1)} - \langle\zeta_1,\zeta_2\rangle_{\zeta(0,1)}\big), \tag{13.5}$$

which is defined on $(\mathbb{P}^1_{\mathrm{an}})^2$ if $\xi \in \mathbb{H}$, or on $(\mathbb{P}^1_{\mathrm{an}} \smallsetminus \{\xi\})^2$ if ξ is of Type I; see [**8**, Section 4.4]. Then, given a measure $\mu \in M_{\mathrm{Rad}}$ and assuming that ξ lies either in $\mathbb{H}$ or in $\mathbb{P}^1(\mathbb{C}_v) \smallsetminus \mathrm{supp}(\mu)$, they define

$$u_{\mu,\xi}(\zeta) := -\int_{\mathbb{P}^1_{\mathrm{an}}} \log\big(\delta(\zeta,\zeta')_\xi\big)\, d\mu(\zeta').$$

(See [**8**, Example 5.21]. The negative sign is because the Baker–Rumely Laplacian operator is the negative of that in [**58**, **61**, **63**].) If $\xi \in \mathbb{H}$, then as the reader can check in Exercise 13.22, $-\log\delta$ differs from the Gromov product by a constant, and hence the negative of the Baker–Rumely potential function differs from that of Definition 13.14 by a constant. One advantage of the Baker–Rumely approach is that $u_{\mu,\xi}(\zeta)$ can be defined even when ξ is of Type I, provided $\xi \notin \mathrm{supp}(\mu)$.

13.4. The Laplacian operator

13.4.1. Definition and examples.

The following is the key result that allows us to define a Laplacian operator on $\mathbb{P}^1_{\mathrm{an}}$.

Theorem 13.23. *Fix $\xi \in \mathbb{H}$. The function $M_{\mathrm{Rad}} \to \mathcal{P}$ given by $\mu \mapsto u_{\mu,\xi}$ is a bijective linear map.*

Proof. See [**58**, Theorem 7.50]. □

It turns out that the correct definition of the Laplacian is not as the inverse of the map $\mu \mapsto u_{\mu,\xi}$, but rather the following closely related function.

Definition 13.24. Fix $\xi \in \mathbb{H}$. The function $\Delta : \mathcal{P} \to M_{\mathrm{Rad}}$ given by

$$\Delta\big(u_{\mu,\xi}\big) := \mu - \big(\mu(\mathbb{P}^1_{\mathrm{an}})\big)\cdot[\xi] \tag{13.6}$$

is called the *Laplacian* on $\mathbb{P}^1_{\mathrm{an}}$.

Theorem 13.25. *The Laplacian* $\Delta : \mathcal{P} \to M_{\text{Rad}}$ *of Definition* 13.24 *is linear and is independent of the choice of basepoint* $\xi \in \mathbb{H}$. *Its image is*

$$\Delta(\mathcal{P}) = \{\mu \in M_{\text{Rad}} : \mu(\mathbb{P}^1_{\text{an}}) = 0\},$$

and its kernel is the set of constant real-valued functions on $\mathbb{H}$.

Proof. This is left to the reader; see Exercise 13.23. □

Example 13.26. Fix $\xi \in \mathbb{H}$, $\zeta_0 \in \mathbb{P}^1_{\text{an}}$, and $\zeta_1 \in \mathbb{A}^1_{\text{an}}$. We claim that the following functions G_1, G_2, G_3 belong to $\mathcal{P}$, with the following Laplacians:

$G_1(\zeta) := \langle\zeta, \zeta_0\rangle_\xi$	has only	$\Delta G_1 = [\xi] - [\zeta_0]$,
$G_2(\zeta) := \log\max\{\lvert\zeta\rvert, 1\}$	has only	$\Delta G_2 = [\zeta(0,1)] - [\infty]$,
$G_3(\zeta) := \log\operatorname{diam}(\zeta \vee \zeta_1)$	has only	$\Delta G_3 = [\zeta_1] - [\infty]$.

To prove the claim for G_1, let $\mu := [\xi] - [\zeta_0]$, and observe that

$$u_{\mu,\xi}(\zeta) = u_{[\xi],\xi}(\zeta) - u_{[\xi],\zeta_0}(\zeta) = -1 - \big(-1 - \langle\zeta, \zeta_0\rangle_\xi\big) = G_1(\zeta),$$

by the linearity of u and by equation (13.4). Since $\mu(\mathbb{P}^1_{\text{an}}) = 0$, our claim follows for G_1. The claim for G_2 is then immediate by Exercise 13.12(d), using $\xi := \zeta(0,1)$ and $\zeta_0 := \infty$ in the formula for G_1. For G_3, let $\nu := [\zeta_1] - [\infty]$ and $\xi := \zeta(0,1)$, and observe that

$$\begin{aligned} u_{\nu,\xi}(\zeta) &= u_{[\zeta_1],\xi}(\zeta) - u_{[\infty],\zeta_0}(\zeta) = -\langle\zeta, \zeta_1\rangle_\xi + \langle\zeta, \infty\rangle_\xi \\ &= G_3(\zeta) - \log\max\{\lvert\zeta_1\rvert, 1\}, \end{aligned}$$

where the third equality is by Exercise 13.12(a), (d). Since $\log\max\{\lvert\zeta_1\rvert, 1\}$ is constant and $\nu(\mathbb{P}^1_{\text{an}}) = 0$, we have $\Delta(G_3) = \nu$, as claimed.

Example 13.27. Let $\phi \in \mathbb{C}_v(z)$ be a nonzero rational function. Write the divisor of ϕ as $\operatorname{div}(\phi) = \sum_{i=1}^m n_i(a_i)$, with $n_i \in \mathbb{Z}$ and $a_i \in \mathbb{P}^1(\mathbb{C}_v)$. (That is, ϕ has a zero of multiplicity n_i at a_i, where $n_i < 0$ means that ϕ has a pole of multiplicity $-n_i$ at a_i.) Then the reader can check (see Exercise 13.24) that the function $\zeta \mapsto \log\|\phi(z)\|_\zeta$ belongs to $\mathcal{P}$, and

$$\Delta\big(\zeta \mapsto \log\|\phi(z)\|_\zeta\big) = \sum_{i=1}^m n_i[a_i]. \tag{13.7}$$

13.4.2. Properties of the Laplacian.

Proposition 13.28. *Let* $\{f_n\}_{n\geq 1}$ *be a sequence of potentials in* $\mathcal{P}$, *and let* $f : \mathbb{H} \to \mathbb{R}$. *Suppose that*

(a) $f_n \to f$ *uniformly on every closed interval* $[\zeta_1, \zeta_2]$ *in* $\mathbb{H}$, *and*

(b) *there exists* $B > 0$ *such that the each of the total variation measures* $\lvert\Delta f_n\rvert$ *has* $\lvert\Delta f_n\rvert(\mathbb{P}^1_{\text{an}}) \leq B$.

Then $f \in \mathcal{P}$, *and the measures* Δf_n *converge weakly to* Δf.

Proof. See [**8**, Proposition 5.32]. □

Definition 13.29. Let $\xi \in \mathbb{H}$, and let $\mu, \nu \in M_{\text{Rad}}$. The *Dirichlet pairing* of μ and ν is

$$\langle \mu, \nu \rangle_{\text{Dir}} := \iint_{\mathbb{P}^1_{\text{an}} \times \mathbb{P}^1_{\text{an}}} \langle \zeta_1, \zeta_2 \rangle_\xi \, d\mu(\zeta_1)\, d\nu(\zeta_2). \tag{13.8}$$

If $\mu := \Delta f$ and $\nu := \Delta g$, the Dirichlet pairing of f and g is

$$\langle f, g \rangle_{\text{Dir}} := \langle \mu, \nu \rangle_{\text{Dir}}.$$

The integral of equation (13.8) is not always defined, but it *is* defined for a large class of measures $\mu, \nu \in M_{\text{Rad}}$. In addition, it is independent of the choice of the basepoint $\xi \in \mathbb{H}$. The following result makes these claims precise.

Proposition 13.30. *Let $\mu, \nu \in M_{\text{Rad}}$ with $\mu(\mathbb{P}^1_{\text{an}}) = \nu(\mathbb{P}^1_{\text{an}}) = 0$ and with Jordan decompositions $\mu = \mu_1 - \mu_2$ and $\nu = \nu_1 - \nu_2$. Suppose that μ_2 and ν_2 have bounded potentials. Then the Dirichlet pairing $\langle \mu, \nu \rangle_{\text{Dir}}$ is defined, belongs to $\mathbb{R} \cup \{\infty\}$, and is independent of the choice of $\xi \in \mathbb{H}$ in equation* (13.8).

Proof. See [**8**, Proposition 5.41]. □

Remark 13.31. The Dirichlet pairing is defined and independent of ξ under various other hypotheses as well. For example, this happens if $\text{supp}(\mu)$ is finite and $|\nu|$ has continuous potentials. It also happens if both $\text{supp}(\mu)$ and $\text{supp}(\nu)$ are finite and we restrict the integral of equation (13.8) to the complement $\mathbb{P}^1_{\text{an}} \smallsetminus \text{Diag}_{\text{I}}$ of the Type I diagonal

$$\text{Diag}_{\text{I}} := \{(x, x) : x \in \mathbb{P}^1(\mathbb{C}_v)\};$$

see [**61**, Section 4.4].

Proposition 13.32. *Let $f \in \mathcal{P} \cap C(\mathbb{P}^1_{\text{an}})$ be a continuous potential function. Then $\langle f, f \rangle_{\text{Dir}} \geq 0$.*

Proof. See [**61**, Proposition 4.5]; see also [**8**, Proposition 5.37]. □

Proposition 13.33. *Let $f, g \in \mathcal{P}$, and suppose that f is continuous on $\mathbb{P}^1_{\text{an}}$. Then the Dirichlet pairing $\langle f, g \rangle_{\text{Dir}}$ is defined and finite, and*

$$\langle f, g \rangle_{\text{Dir}} = \int_{\mathbb{P}^1_{\text{an}}} f \, d(\Delta g).$$

Proof. See [**8**, Proposition 5.35]. □

Corollary 13.34. *Let $f, g \in \mathcal{P} \cap C(\mathbb{P}^1_{\mathrm{an}})$ be continuous potential functions. Then the Dirichlet pairing $\langle f, g\rangle_{\mathrm{Dir}}$ is defined and finite, and*

$$\int_{\mathbb{P}^1_{\mathrm{an}}} f\, d(\Delta g) = \int_{\mathbb{P}^1_{\mathrm{an}}} g\, d(\Delta f).$$

Proof. This is immediate from Proposition 13.33 and the symmetry of $\langle \cdot, \cdot\rangle_{\mathrm{Dir}}$. □

Theorem 13.35. *Let $f \in \mathcal{P}$ be a potential function, and let $\phi \in \mathbb{C}_v(z)$ be a nonconstant rational function. Then*

$$\phi^*(\Delta f) = \Delta(f \circ \phi) \quad \textit{and} \quad \phi_*(\Delta f) = \Delta(\phi_* f).$$

Proof. For the first identity, see [**8**, Proposition 9.56] or [**81**, Proposition 4.15]; for the second, see Exercise 13.25. □

13.4.3. The Baker–Rumely Laplacian. Definition 13.24 unfortunately does not provide much insight into what the Laplacian operator Δ actually does. In [**8**], Baker and Rumely define Δ more concretely. (See [**8**, Definition 5.11], which leans heavily on [**8**, Section 3.5].) To convey some intuition into the action of Δ, we now present an informal description of their rather involved definition, adjusted with a sign change to bring their sign convention into alignment with ours.

Let $\zeta \in \mathbb{H}$, let $\boldsymbol{v}$ be a direction at ζ, and let $f : \mathbb{H} \to \mathbb{R}$. The *directional derivative* of f at ζ in the direction of $\boldsymbol{v}$ is

$$\partial_{\boldsymbol{v}} f(\zeta) := \lim_{t\to 0^+} \frac{f(h(t)) - f(\zeta)}{t}, \tag{13.9}$$

where $h : [0, 1] \to \mathbb{H}$ is a distance-preserving map with $h(0) = \zeta$ and $h(t) \in \boldsymbol{v}$ for $t > 1$. The limit of equation (13.9) does not necessarily converge, but it is independent of the choice of h; see Exercise 13.26. We only consider maps f for which $\partial_{\boldsymbol{v}} f(\zeta)$ is defined for every $\zeta \in \mathbb{H}$ and every direction $\boldsymbol{v}$ at ζ; indeed, this is one of the several defining properties of the space $\mathrm{BDV}(\mathbb{P}^1_{\mathrm{an}})$.

Let U be an open connected Berkovich affinoid with boundary points $\zeta_1, \ldots, \zeta_m$. At each boundary point ζ_i, let $\boldsymbol{v}_i$ be the unique direction at ζ_i intersecting U. Then the measure Δf assigns U a mass of

$$(\Delta f)(U) := -\sum_{i=1}^{m} \partial_{\boldsymbol{v}_i} f(\zeta_i).$$

Similarly, if V is a closed connected Berkovich affinoid with boundary points $\zeta_1, \ldots, \zeta_m$, then at each boundary point ζ_i, let $\boldsymbol{w}_i$ be the unique direction at ζ_i *not* intersecting U. Then the measure Δf assigns V a mass of

$$(\Delta f)(V) := \sum_{i=1}^{m} \partial_{\boldsymbol{v}_i} f(\zeta_i).$$

In addition, for $\zeta \in \mathbb{H}$, the singleton set $\{\zeta\}$ has measure

$$(\Delta f)(\{\zeta\}) := \sum_{\boldsymbol{v}} \partial_{\boldsymbol{v}} f(\zeta),$$

where the sum is over all directions $\boldsymbol{v}$ at ζ. Using the defining properties of $\mathrm{BDV}(\mathbb{P}^1_{\mathrm{an}})$ (not just the the convergence of every limit of equation (13.9)), Baker and Rumely show that this definition can be extended to all Borel sets, yielding $\Delta f \in M_{\mathrm{Rad}}$. (See [**8**, Sections 3.5, 5.3, and 5.4].)

In fact, Baker and Rumely first define Δf on finite trees $\Gamma \subseteq \mathbb{H}$ of finite hyperbolic diameter. They then show that the resulting measures on all such trees Γ agree with one another in an appropriate sense (they "form a coherent system of measures"). Next, they use this coherent system of measures to define a bounded linear functional on $C(\mathbb{P}^1_{\mathrm{an}})$, the space of continuous functions on $\mathbb{P}^1_{\mathrm{an}}$. Finally, they apply the Riesz representation theorem, producing the measure Δf.

Intuitively then, Δf has a point mass at any $\zeta \in \mathbb{H}$ for which the net rate of increase of f moving away from ζ is nonzero. More precisely, the mass at ζ is the total rate at which f increases in directions from ζ minus the total rate that it decreases in the other directions. For $x \in \mathbb{P}^1(\mathbb{C}_v)$, if f increases or decreases linearly (with respect to the hyperbolic metric) along an interval emanating from x, then Δf has a point mass equal to the slope of f along this interval.

Example 13.36. The function $G_2(\zeta) := \log\max\{|\zeta|, 1\}$ of Example 13.26 is locally constant away from the interval $[\zeta(0,1), \infty]$ and hence has directional derivative 0 in every direction at every point $\zeta \in \mathbb{H} \smallsetminus [\zeta(0,1), \infty)$. However, at $\zeta(0,1)$, the directional derivatives are

$$\partial_{\vec{v}(\infty)}\big(\zeta(0,1)\big) = 1 \quad \text{and} \quad \partial_{\boldsymbol{v}}\big(\zeta(0,1)\big) = 0 \text{ for } \boldsymbol{v} \neq \vec{v}(\infty),$$

where $\vec{v}(\infty)$ is the direction at ζ containing ∞. Indeed, G_2 is identically zero on $\overline{D}_{\mathrm{an}}(0,1)$, but $G_2(\zeta(0,r)) = \log r$ for $r > 1$. Thus, ΔG_2 has a point mass of measure 1 at the Type II point $\zeta(0,1)$.

In addition, at every $\zeta \in (\zeta(0,1), \infty)$, we have

$$\partial_{\vec{v}(\infty)}(\zeta) = 1 \quad \text{and} \quad \partial_{\vec{v}(0)}(\zeta) = -1,$$

with $\partial_{\boldsymbol{v}}(\zeta) = 0$ for $\boldsymbol{v} \neq \vec{v}(\infty), \vec{v}(0)$. Thus, $\Delta G_2(\{\zeta\}) = 0$ for every such ζ. Finally, moving along the interval $(\infty, \zeta(0,1))$ away from ∞, $f(\zeta)$ decreases linearly with slope -1, and hence ΔG_2 has a point mass of measure -1 at ∞.

Having computed all directional derivatives $\partial_{\boldsymbol{v}} f(\zeta)$ in $\mathbb{H}$, the reader can easily check that for any connected affinoid U, $\Delta G_2(U)$ is 1 if U contains $\zeta(0,1)$ but not ∞, -1 if U contains ∞ but not $\zeta(0,1)$, or 0 if U contains neither or both of these points. Thus, we have confirmed the claim of Example 13.26 that $\Delta G_2 = [\zeta(0,1)] - [\infty]$.

In most of our examples so far, the Laplacian Δf has turned out to be an atomic measure. However, in general it can be any signed Radon measure of total mass zero; see Theorem 13.25. In particular, if f varies smoothly along an interval $[\zeta, \xi] \subseteq \mathbb{H}$ of length R and is constant on each component of $\mathbb{H} \smallsetminus [\zeta, \xi]$, then Δf is just an appropriately scaled second derivative of f times the Lebesgue measure on $[\zeta, \xi] \cong [0, R]$. See Exercise 13.27 for a more precise statement, and see Exercise 13.28 for a characterization of $f \in \mathcal{P}$ for which Δf is atomic.

Indeed, just like its complex counterpart, the Laplacian on $\mathbb{P}^1_{\mathrm{an}}$ should be thought of as a sort of second derivative. If f varies smoothly along an interval I in $\mathbb{P}^1_{\mathrm{an}}$, then Δf is essentially f'' times Lebesgue measure λ on I. If f has a sharp corner at a point ζ, as can happen if f is piecewise affine, then Δf has a point mass at ζ. Moreover, the formula of Corollary 13.34 can be thought of as arising from two integrations by parts, changing the integral of fg'' into that of $g''f$. Indeed, if f and g both vary smoothly along a Berkovich interval I, then this is exactly what happens: both integrals equal $-\int_I f'g'\, d\lambda$. This idea can be generalized to arbitrary potential functions; see [**8**, Proposition 5.37] or [**61**, Proposition 4.5]. Of course, integrating by parts requires declaring an orientation on the interval I, but the double integration by parts means that any resulting sign cancels itself.

Ultimately, our main interest is in the equilibrium measure μ_ϕ of a rational function $\phi \in \mathbb{C}_v(z)$. As we will see in the next section, the support $\mathrm{supp}(\mu_\phi)$ turns out to be precisely the Julia set of ϕ, and hence μ is generally neither atomic nor a multiple of the Lebesgue measure on an interval.

13.5. Construction of the equilibrium measure

Theorem 13.37. *Let $\phi \in \mathbb{C}_v(z)$ be a rational function of degree $d \geq 2$, and let $\xi \in \mathbb{H}$. For each $n \geq 0$, let*

$$\mu_n := d^{-n}(\phi^n)^*[\xi].$$

Then the sequence of Radon probability measures $\{\mu_n\}_{n\geq 0}$ converges weakly to a Radon probability measure μ_ϕ. Moreover,

(a) *μ_ϕ is independent of the choice of the basepoint $\xi \in \mathbb{H}$,*

(b) *$\phi^*(\mu_\phi) = d \cdot \mu_\phi$,*

(c) *$\phi_*(\mu_\phi) = \mu_\phi$, and*

(d) *for any $n \geq 1$, $\mu_{\phi^n} = \mu_\phi$.*

Proof. By Example 13.9, $(\phi^n)^*[\xi]$ is a finite sum of delta-measures with positive integer coefficients, and it has total mass $\deg(\phi^n) = d^n$. Thus, μ_n is a finite sum of delta-measures with positive coefficients, and it has total mass 1. In particular, μ_n is indeed a Radon probability measure.

Let $\nu := \mu_1 - [\xi]$, which is a signed Radon measure of total mass 0. Applying Theorem 13.25 to ν, there is a potential function $g \in \mathcal{P}$ such that $\Delta g = \nu$. For each $m \geq 0$, Theorem 13.35 yields

$$\begin{aligned}\Delta(g \circ \phi^m) &= (\phi^m)^*(\Delta g) = (\phi^m)^*\big(\mu_1 - [\xi]\big)\\ &= (\phi^m)^*\big(d^{-1}\phi^*[\xi] - [\xi]\big) = d^{-1}(\phi^{m+1})^*[\xi] - (\phi^m)^*[\xi],\end{aligned}$$

where the last equality is by Proposition 13.8(f). Therefore,

$$\mu_n - [\xi] = \Delta G_n, \quad \text{where} \quad G_n := \sum_{m=0}^{n-1} d^{-m}\big(g \circ \phi^m\big).$$

Since ξ and all its preimages lie in $\mathbb{H}$, Proposition 13.21 implies that g has a continuous extension to $\mathbb{P}^1_{\text{an}}$. Since $\mathbb{P}^1_{\text{an}}$ is compact, it follows that g is bounded. Therefore, the sequence $\{G_n\}_{n\geq 1}$ converges uniformly to the function

$$G_\infty := \sum_{n=0}^{\infty} d^{-n}\big(g \circ \phi^n\big) \ : \mathbb{H} \to \mathbb{R}. \tag{13.10}$$

In addition, each ΔG_n is the difference of two probability measures, and hence has total variation $|\Delta G_n|$ bounded by 2. Thus, by Proposition 13.28, we have $G_\infty \in \mathcal{P}$, and the sequence $\{\mu_n\}_{n\geq 1}$ converges weakly to

$$\mu_\phi := [\xi] + \Delta G_\infty.$$

By Exercise 13.7, μ_ϕ is a Radon probability measure.

To prove statement (a), given any other choice of basepoint $\eta \in \mathbb{H}$, let h be the potential function of the measure $[\eta]$ with respect to the original basepoint ξ. Thus, $[\eta] = [\xi] + \Delta h$. By Proposition 13.21 again, h has a continuous (and hence bounded) extension to $\mathbb{P}^1_{\text{an}}$. Define $\mu'_n := d^{-n}(\phi^n)^*[\eta]$ for each $n \geq 0$. Then

$$\mu'_n - \mu_n = d^{-n}(\phi^n)^*\big([\eta] - [\xi]\big) = \Delta\big(d^{-n}(h \circ \phi^n)\big).$$

By the boundedness of h, the potential functions $d^{-n}(h \circ \phi^n)$ approach 0 uniformly. Moreover, as the difference of two probability measures, each $\mu'_n - \mu_n$ has total variation bounded by 2. Therefore, by Proposition 13.28 again, the sequence of measures $\{\mu'_n - \mu_n\}_{n\geq 1}$ converges weakly to the zero measure. That is, $\{\mu'_n\}_{n\geq 1}$ converges weakly to the same limit measure μ_ϕ as before.

To prove (b) and (c), observe that for every $n \geq 1$, we have

$$\phi^*\mu_n = \phi^*\big(d^{-n}(\phi^n)^*[\xi]\big) = d^{-n}(\phi^{n+1})^*[\xi] = d \cdot \mu_{n+1}.$$

Taking the weak limit of both sides yields (b). Proposition 13.8(h) then immediately yields (c).

Finally, to prove (d), fix $n \geq 1$. The sequence $\{\mu_{mn}\}_{m\geq 1}$ is a subsequence of the original sequence of measures $\{\mu_m\}_{m\geq 1}$. Since the latter has weak limit μ_ϕ, so does the former. That is, $\mu_{\phi^n} = \mu_\phi$. □

Definition 13.38. Let $\phi \in \mathbb{C}_v(z)$ be a rational function of degree $d \geq 2$. The measure μ_ϕ of Theorem 13.37 is the *equilibrium measure* of ϕ.

Baker and Rumely call μ_ϕ the *canonical measure* of ϕ; see [**8**, Section 10.1]. The equilibrium measure is also sometimes called the *invariant measure* of ϕ, because in ergodic theory, a probability measure satisfying property (c) of Theorem 13.37 is called invariant for ϕ. In general, however, many probability measures are invariant for ϕ, so the wording "the" invariant measure is a bit misleading. Instead, the stronger property of Theorem 13.37(b) is more important for characterizing μ_ϕ.

Theorem 13.39. *Let $\phi \in \mathbb{C}_v(z)$ be a rational function of degree $d \geq 2$, let E_ϕ be the Type I exceptional set of ϕ, and let μ_ϕ be its equilibrium measure. Then the following hold.*

(a) $\text{supp}(\mu_\phi) = \mathcal{J}_{\phi,\text{an}}$.

(b) *For all $\zeta \in \mathbb{P}^1_{\text{an}}$, we have $\mu_\phi(\{\zeta\}) = 0$ unless ζ is an Type II point and $\mathcal{J}_{\phi,\text{an}} = \{\zeta\}$, so that ϕ has potential good reduction.*

(c) *Let ν be a probability measure on $\mathbb{P}^1_{\text{an}}$. Then the sequence of measures $\{d^{-n}(\phi^n)^*\nu\}_{n\geq 1}$ has weak limit μ_ϕ if and only if $\nu(E_\phi) = 0$.*

(d) *μ_ϕ is the unique probability measure μ on $\mathbb{P}^1_{\text{an}}$ such that $\phi^*\mu = d \cdot \mu$ and $\mu(E_\phi) = 0$.*

(e) *μ_ϕ is mixing, and hence ergodic, for $\phi : \mathbb{P}^1_{\text{an}} \to \mathbb{P}^1_{\text{an}}$.*

To explain Theorem 13.39(e), recall that if $T : X \to X$ is a continuous function and μ is a probability measure on X, we say that μ is *invariant* for T if $T_*\mu = \mu$. Furthermore, we say that a T-invariant measure μ is

- *ergodic* for T if for any measurable set $E \subseteq X$ which is T-invariant (meaning that $T^{-1}(E) = E$), we have either $\mu(E) = 0$ or $\mu(E) = 1$;
- *mixing* (or *strong-mixing*) for T if for any two measurable sets $E_1, E_2 \subseteq X$,

$$\lim_{n\to\infty} \mu\big(E_1 \cap T^{-n}(E_2)\big) = \mu(E_1) \cdot \mu(E_2).$$

Choosing $E_1 = E_2 = E$, it is clear that mixing implies ergodic. For further details, see, for example, [**140**, Section 1.7] or [**50**, Section 2.7].

Proof of Theorem 13.39.

(a) We begin by showing that the Fatou set is disjoint from $\operatorname{supp}(\mu_\phi)$. Given $\zeta \in \mathcal{F}_{\phi,\mathrm{an}}$, we first claim that there exist a point $\xi \in \mathbb{H}$ and an open connected affinoid $U \subseteq \mathbb{P}^1_{\mathrm{an}}$ containing ζ such that the backward orbit of ξ is disjoint from U.

To prove the claim, let W be the Fatou component of ζ. If there is another Fatou component V not in the forward orbit $\{\phi^n(W)\}_{n\geq 0}$ of W, we may simply pick $\xi \in \mathbb{H} \cap V$, and choose U to be an open connected affinoid containing ζ and contained in W.

Otherwise, W must be a periodic component of some period $m \geq 1$, such that ϕ^m maps W onto itself d^m-to-1. Since $d^m > 1$, W is an attracting component. Let $a \in W \cap \mathbb{P}^1(\mathbb{C}_v)$ be the unique attracting periodic point in W, and choose a small open disk $D \subsetneq W$ containing a, with $\phi^m(D) \subsetneq D$. Let $\ell \geq 1$ be the smallest positive integer such that $\phi^{\ell m}(\zeta) \in D$, and let U be the component of $\phi^{-\ell m}(D)$ containing a. Then the component V of $\phi^{-m}(U)$ containing a is a connected open affinoid satisfying $U \subsetneq V$. We may therefore pick $\xi \in V \smallsetminus U$. Since $\phi(U) \subseteq U$, the backward orbit of ξ is disjoint from U, proving the claim.

Because U is an intersection of finitely many open disks, we may shrink each of those disks slightly to obtain another connected open affinoid U' such that

$$\zeta \in U' \subsetneq \overline{U'} \subsetneq U,$$

where $\overline{U'}$ is the closure of U' in $\mathbb{P}^1_{\mathrm{an}}$. Since $\mathbb{P}^1_{\mathrm{an}}$ is normal topological space (cf. Exercise 6.20), there is a continuous function $F : \mathbb{P}^1_{\mathrm{an}} \to [0,1]$ that is 1 on U' and 0 on $\mathbb{P}^1_{\mathrm{an}} \smallsetminus U$. Thus,

$$0 \leq \mu_\phi(U') \leq \int_{\mathbb{P}^1_{\mathrm{an}}} F \, d\mu = \lim_{n\to\infty} \int_{\mathbb{P}^1_{\mathrm{an}}} F \, d\Big(d^{-n} (\phi^n)^*[\xi]\Big) = \lim_{n\to\infty} 0 = 0.$$

Therefore, $\mu_\phi(U') = 0$, and hence $\zeta \notin \operatorname{supp}(\mu_\phi)$, as desired.

We have shown that $\operatorname{supp}(\mu_\phi) \subseteq \mathcal{J}_{\phi,\mathrm{an}}$. To prove the reverse inclusion, observe that $\operatorname{supp}(\mu_\phi)$ is a closed set, and it is nonempty because

$$\mu_\phi(\mathbb{P}^1_{\mathrm{an}}) = 1 \neq 0.$$

Since $E_\phi \subseteq \mathcal{F}_{\phi,\mathrm{an}}$ is disjoint from $\operatorname{supp}(\mu_\phi)$, then, $\operatorname{supp}(\mu_\phi)$ is not contained in E_ϕ. In addition, by Proposition 13.8(e) and Theorem 13.37(b), we have

$$\phi^{-1}(\operatorname{supp}(\mu_\phi)) = \operatorname{supp}(\phi^* \mu_\phi) = \operatorname{supp}(\mu_\phi).$$

Thus, by Theorem 8.15(d), we have $\operatorname{supp}(\mu_\phi) \supseteq \mathcal{J}_{\phi,\mathrm{an}}$, proving (a).

(b) Suppose there is some $\zeta \in \mathbb{P}^1_{\mathrm{an}}$ with $\mu_\phi(\{\zeta\}) > 0$. Then $\zeta \notin \mathcal{F}_{\phi,\mathrm{an}}$ by part (a), and hence $\zeta \notin E_\phi$. If ζ has finite backward orbit, then by Proposition 8.13, ϕ has potential good reduction, since $\zeta \notin E_\phi$. Therefore, $\mathcal{J}_{\phi,\mathrm{an}} = \{\zeta\}$, and ζ is of Type II.

Otherwise, there is some $n \geq 0$ and $\zeta' \in \phi^{-n}(\zeta)$ such that ζ' is not periodic. By Exercise 13.5 and Theorem 13.37(b), we have

$$\mu_\phi(\{\zeta'\}) = d^{-n}\big((\phi^n)^*\mu_\phi\big)(\{\zeta'\}) = \frac{\deg_{\zeta'}(\phi^n)}{d^n}\mu_\phi(\zeta) > 0.$$

Replacing ζ by ζ', we may assume that ζ is not periodic. In particular, the sets $\phi^{-1}(\zeta), \phi^{-2}(\zeta), \phi^{-3}(\zeta), \ldots$ are pairwise disjoint.

However, it is immediate from Theorem 13.37(c) and the definition of push-forward measures that for each $n \geq 1$,

$$\mu_\phi\big(\phi^{-n}(\zeta)\big) = \phi^n_*\mu_\phi\big(\{\zeta\}\big) = \mu_\phi\big(\{\zeta\}\big).$$

Thus,

$$1 = \mu_\phi(\mathbb{P}^1_{\text{an}}) \geq \sum_{n\geq 1} \mu_\phi\big(\phi^{-n}(\zeta)\big) = \sum_{n\geq 1} \mu_\phi(\{\zeta\}) = \infty.$$

This contradiction completes the proof of (b).

(c) We refer the reader to [**63**, Théorème A] and its proof in [**63**, Section 3.2]. One can prove (c) in the special case that ν has bounded potentials by an argument very similar to the proof of Theorem 13.37(a). Thus, the more significant content of (c) is that it holds even when ν has unbounded potentials. The idea of the proof is essentially the same—to show that the potentials of the measures $d^{-n}(\phi^n)^*\nu$ approach the potential of μ_ϕ—but the analysis is more delicate, because this time the convergence is pointwise rather than uniform; see also [**8**, Section 10.4].

(d) This is left to the reader; see Exercise 13.29.

(e) See [**63**, Section 3.3]. □

Before presenting some examples, we note some consequences of Theorem 13.39.

Corollary 13.40 (Equidistribution of preimages). *Let $\phi \in \mathbb{C}_v(z)$ be a rational function of degree $d \geq 2$, let E_ϕ be the Type I exceptional set of ϕ, and let μ_ϕ be its equilibirum measure. For any $\zeta \in \mathbb{P}^1_{\text{an}} \smallsetminus E_\phi$ and any $n \geq 0$, let*

$$\nu_n = \frac{1}{d^n} \sum_{\eta\in\phi^{-n}(\zeta)} \deg_\eta(\phi^n)[\eta].$$

Then each ν_n is a probability measure, and $\nu_n \to \mu_\phi$ weakly.

Proof. For each $n \geq 0$, equation (13.2) applied to ϕ^n yields

$$\nu_n = \frac{1}{d^n}(\phi^n)^*[\zeta].$$

Because $\zeta \notin E_\phi$, we have $\nu_n(E_\phi) = 0$. Therefore, by Theorem 13.39(c), $\{\nu_n\}$ converges weakly to μ_ϕ. □

Proposition 13.41. *Let $\phi \in \mathbb{C}_v(z)$ be a rational function of degree $d \geq 2$. Then the Julia set $\mathcal{J}_{\phi,\mathrm{an}}$ is separable and metrizable.*

Proof. According to [**58**, Lemma 7.15] or [**8**, Lemma 5.7], the support of any (finite) signed Radon measure on $\mathbb{P}^1_{\mathrm{an}}$ is a separable, metrizable space. The desired result is therefore immediate by considering the equilibrium measure μ_ϕ and applying Theorem 13.39(a). ☐

Proposition 13.42. *Let $\phi \in \mathbb{C}_v(z)$ be a rational function of degree $d \geq 2$, with Type I exceptional set E_ϕ and with equilibrium measure μ_ϕ. Let $U \subseteq \mathbb{P}^1_{\mathrm{an}}$ be an affinoid. Suppose that either*

- *ϕ does not have potential good reduction, or*
- *ϕ has potential good reduction, and the unique point ξ in $\mathcal{J}_{\phi,\mathrm{an}}$ does not lie in ∂U.*

Then for any point $\zeta \in \mathbb{P}^1_{\mathrm{an}} \smallsetminus E_\phi$, we have

$$\mu_\phi(U) = \lim_{n\to\infty} \frac{1}{d^n} \sum_{\eta \in \phi^{-n}(\zeta)\cap U} \deg_\eta(\phi^n).$$

That is, $\mu_\phi(U)$ is approximated by the proportion of nth preimages of ζ in U, counted with multiplicity.

Proof. For each $n \geq 0$, let

$$\nu_n := \frac{1}{d^n} \sum_{\eta \in \phi^{-n}(\zeta)} \deg_\eta(\phi^n)[\eta].$$

By Corollary 13.40, $\nu_n \to \mu_\phi$ weakly. In addition, our hypotheses combined with Theorem 13.39(b) imply that $\mu(\partial U) = 0$. The Portmanteau Theorem in probability states that in this situation—that a sequence $\{\nu_n\}$ of probabiliity measures converges weakly to μ_ϕ, with $\mu_\phi(\partial U) = 0$—we have $\mu_\phi(U) = \lim_{n\to\infty} \nu_n(U)$. (See any of [**31**, Theorem 2.1], [**128**, Theorem III.1.1], or [**8**, Theorem A.13].) Since

$$\nu_n(U) = \frac{1}{d^n} \sum_{\eta \in \phi^{-n}(\zeta)\cap U} \deg_\eta(\phi^n),$$

we are done. ☐

Example 13.43. Let $\phi(z) = z^2 + \lambda z \in \mathbb{C}_v[z]$, as in Examples 5.29 and 8.16. If $|\lambda| \leq 1$, then ϕ has good reduction, and hence $\mathcal{J}_{\phi,\mathrm{an}} = \{\zeta(0,1)\}$. Therefore, the equilibrium measure μ_ϕ is the delta-measure $[\zeta(0,1)]$.

If $|\lambda| > 1$, recall that $\mathcal{J}_{\phi,\mathrm{an}}$ is a Cantor set consisting only of Type I points contained in $U_0 = \overline{D}_{\mathrm{an}}(0, |\lambda|)$. More precisely, setting $\xi := \zeta(0, |\lambda|)$,

$\phi^{-n}(\xi)$ consists of 2^n points of the form $\zeta(a_{n,i}, |\lambda|^{1-n})$. Thus,

$$(\phi^n)^*[\xi] = \sum_{i=1}^{2^n} \frac{1}{2^n}\Big[\zeta\big(a_{n,i}, |\lambda|^{1-n}\big)\Big]$$

is a sum of 2^n distinct delta-measures, each of mass 2^{-n}. The equilibrium measure μ_ϕ therefore assigns a mass of $1/2^n$ to each of the 2^n subsets of $\mathcal{J}_{\phi,\mathrm{an}}$ of the form $\mathcal{J}_{\phi,\mathrm{an}} \cap \overline{D}_{\mathrm{an}}(a_{n,i}, |\lambda|^{1-n})$.

Example 13.44. Let $c \in \mathbb{C}_v$ with $0 < |c| < 1$, and let

$$\phi(z) = \frac{z^4 - 8c^4 z - c^4}{4z^3 + z^2 + 4c^4} \in \mathbb{C}_v(z)$$

be the Lattès map of Example 5.33. As observed in Example 8.20, the Julia set $\mathcal{J}_{\phi,\mathrm{an}}$ is the interval $J = [\zeta(0,1), \zeta(0,|c|^2)]$. In fact, the equilibrium measure μ_ϕ is Lebesgue measure on J, divided by the length

$$d_{\mathbb{H}}(\zeta(0,1), \zeta(0,|c|^2)) = -2\log|c|$$

of J; see Exercise 13.30.

13.6. Local heights

The potential function G_∞ of equation (13.10) played an essential role in the construction of the equilibrium measure μ_ϕ given in the proof of Theorem 13.37. It is also of interest in its own right, in the theory of Weil local height functions.

Definition 13.45. Fix $a \in \mathbb{P}^1(\mathbb{C}_v)$ and $\xi \in \mathbb{H}$. A (*Weil*) *local height* on $\mathbb{P}^1_{\mathrm{an}}$ relative to a is a continuous function $\lambda_{v,a} : \mathbb{P}^1_{\mathrm{an}} \setminus \{a\} \to \mathbb{R}$ such that

$$\lambda_{v,a}(\zeta) - \langle \zeta, a\rangle_\xi$$

is bounded as a function of $\zeta \in \mathbb{P}^1_{\mathrm{an}} \setminus \{a\}$.

For $\xi, \eta \in \mathbb{H}$, the Gromov products $\langle \zeta, a\rangle_\xi$ and $\langle \zeta, a\rangle_\eta$ differ by a bounded amount, by Exercise 13.13(b). Thus, although the conditions of Definition 13.45 depend on the choice of $a \in \mathbb{P}^1(\mathbb{C}_v)$, they do not actually depend on the choice of $\xi \in \mathbb{H}$.

Example 13.46. Let $a = \infty$ and $\xi = \zeta(0,1)$. Then

$$\lambda_{v,\infty,\mathrm{std}} : \mathbb{A}^1_{\mathrm{an}} \to \mathbb{R} \quad \text{by} \quad \lambda_{v,\infty,\mathrm{std}}(\zeta) = \log\max\{1, \|z\|_\zeta\}$$

is a local height relative to ∞, since $\lambda_{v,\infty,\mathrm{std}}(\zeta) = \langle \zeta, \infty\rangle_{\zeta(0,1)}$; see Exercise 13.12(d). This function, which on $\mathbb{C}_v$ is given by

$$\lambda_{v,\infty,\mathrm{std}}(x) = \log\max\{1, |x|\},$$

is called the *standard local height* relative to ∞. Thus, a local height relative to ∞ is a continuous real-valued function on $\mathbb{P}^1_{\mathrm{an}}$ that differs on $\mathbb{P}^1(\mathbb{C}_v)$ by a bounded amount from the standard local height $x \mapsto \log\max\{1,|x|\}$.

In the literature, local heights are often defined only on Type I points. See, for example, [**75**, Section B.8] and [**88**, Sections 10.1–10.2].

For any rational function $\phi \in \mathbb{C}_v(z)$ of degree $d \geq 1$ with denominator $L(z) \in \mathbb{C}_v[z]$, it is a fundamental fact that

$$\lambda_{v,\infty,\mathrm{std}}\big(\phi(\zeta)\big) - d \cdot \lambda_{v,\infty,\mathrm{std}}(\zeta) + \log\|L\|_\zeta \tag{13.11}$$

is a bounded function of $\zeta \in \mathbb{P}^1_{\mathrm{an}} \smallsetminus (\phi^{-1}(\infty) \cup \{\infty\})$. (See Exercise 13.31. For Type I points only, see also Theorem B.8.1(c) and Remark B.8.2 of [**75**], or [**88**, Proposition 2.6].) More generally, if we fix $a \in \mathbb{P}^1(\mathbb{C}_v)$, and if $L \in \mathbb{C}_v(z)$ is a rational function with divisor $\mathrm{div}(L) = \phi^*((a)) - d \cdot (a)$, then for any local height $\lambda_{v,a}$ relative to a, the function

$$\zeta \mapsto \lambda_{v,a}\big(\phi(\zeta)\big) - d \cdot \lambda_{v,a}(\zeta) + \log\|L\|_\zeta, \tag{13.12}$$

is bounded for $\zeta \in \mathbb{P}^1_{\mathrm{an}} \smallsetminus (\phi^{-1}(a) \cup \{a\})$.

However, it can be useful to have a local height for which this bounded error is in fact identically zero, as follows.

Definition 13.47. Fix $a \in \mathbb{P}^1(\mathbb{C}_v)$. Let $\phi \in \mathbb{C}_v(z)$ be a rational function of degree $d \geq 2$, and let $L \in \mathbb{C}_v(z)$ with $\mathrm{div}(L) = \phi^*((a)) - d \cdot (a)$. A *canonical local height* for ϕ, or *Call–Silverman local height* for ϕ, is a local height $\hat{\lambda}_\phi = \hat{\lambda}_{v,a,\phi}$ relative to a such that

$$\hat{\lambda}_{v,a,\phi}\big(\phi(\zeta)\big) = d \cdot \hat{\lambda}_{v,a,\phi}(\zeta) - \log\|L\|_\zeta \tag{13.13}$$

for all $\zeta \in \mathbb{P}^1_{\mathrm{an}} \smallsetminus (\phi^{-1}(a) \cup \{a\})$.

Call and Silverman [**37**, Theorem 2.1] proved that for a given choice of a, ϕ, and L, there is a unique associated canonical local height $\hat{\lambda}_{v,a,\phi}$. (See also [**131**, Theorem 3.27] for the case of $\mathbb{P}^1$ with $a = \infty$.) From the uniqueness, it is straightforward to show that $\hat{\lambda}_{v,a,\phi}$ is also a canonical local height for ϕ^n, for any $n \geq 1$. We now show that this canonical local height is essentially given by the potential function G_∞ of equation (13.10).

Theorem 13.48. *Let $\phi \in \mathbb{C}_v(z)$ be a rational function of degree $d \geq 2$, let $a \in \mathbb{P}^1(\mathbb{C}_v)$, and let $L \in \mathbb{C}_v(z)$ with divisor $\mathrm{div}(L) = \phi^*((a)) - d \cdot (a)$. Let $\xi \in \mathbb{H}$, and let $g, h \in \mathcal{P}$ be potential functions such that*

$$\Delta g = d^{-1}\phi^*[\xi] - [\xi] \quad \textit{and} \quad \Delta h = [\xi] - [a].$$

Define $G_\infty := \sum_{n=0}^{\infty} d^{-n}(g \circ \phi^n)$. Then there is some $c \in \mathbb{R}$ such that

$$\hat{\lambda}_{v,a,\phi} := G_\infty + h + c \in \mathcal{P}$$

is a canonical local height for ϕ relative to a and L.

Proof. We saw in the proof of Theorem 13.37 that the series defining G_∞ converges, and that $G_\infty \in \mathcal{P}$. We compute

$$G_\infty \circ \phi = \sum_{n\geq 0} d^{-n}\bigl(g \circ \phi^{n+1}\bigr) = d\sum_{n\geq 1} d^{-n}\bigl(g \circ \phi^{n}\bigr) = d \cdot \bigl(G_\infty - g\bigr),$$

and hence

$$\hat{\lambda}_{v,a,\phi} \circ \phi - d \cdot \hat{\lambda}_{v,a,\phi} = (1-d)c + h \circ \phi - d \cdot h - d \cdot g. \tag{13.14}$$

Taking the Laplacian yields

$$\begin{aligned}\Delta\bigl(\hat{\lambda}_{v,a,\phi} \circ \phi - d \cdot \hat{\lambda}_{v,a,\phi}\bigr) &= \phi^*(\Delta h) - d\Delta h - d\Delta g \\ &= \phi^*[\xi] - \phi^*[a] - d[\xi] + d[a] - \phi^*[\xi] + d[\xi] \\ &= -\phi^*[a] + d[a] = -\Delta\bigl(\zeta \mapsto \log \|L\|_\zeta\bigr),\end{aligned}$$

where the last equality is by Example 13.27. Since $\ker(\Delta)$ is the set of constant functions, the desired identity (13.13) holds up to the addition of a real constant. Because c appears in equation (13.14) with a multiple of $(1-d) \neq 0$, we may choose $c \in \mathbb{R}$ so that the identity holds exactly. □

In [**8**, Section 10.2], Baker and Rumely defined a homogeneous version of the canonical local height and used it to define a pairing called Arakelov–Green's function. Before presenting their definitions, we note that for $\phi(z) \in \mathbb{C}_v(z)$ of degree $d \geq 2$, there are homogeneous polynomials $F, G \in \mathbb{C}_v[X,Y]$ of degree d such that $\phi(z) = F(z,1)/G(z,1)$. (Simply multiply the numerator and denominator of $\phi(X/Y)$ by Y^d.) Moreover, the pair (F,G), which we call a *lift* of ϕ to $\mathbb{C}_v^2$, is unique up to multiplication by a nonzero constant. In addition, a *lift* of $z \in \mathbb{P}^1(\mathbb{C}_v)$ to $\mathbb{C}_v^2$ is a pair $(x,y) \in \mathbb{C}_v^2 \smallsetminus \{(0,0)\}$ such that $z = x/y$, where we understand $x/0$ to be ∞ if $x \neq 0$. Finally, for $(x,y) \in \mathbb{C}_v^2$, define

$$\bigl|(x,y)\bigr|_{\max} := \max\{|x|, |y|\}.$$

Definition 13.49. Let $\phi \in \mathbb{C}_v(z)$ be a rational function of degree $d \geq 2$, and let $\Phi = (F,G)$ be a lift of ϕ to $\mathbb{C}_v^2$. The *homogeneous canonical local height* for Φ is the function $\hat{\Lambda}_\Phi : \mathbb{C}_v^2 \smallsetminus \{(0,0)\} \to \mathbb{R}$ given by

$$\hat{\Lambda}_\Phi(x,y) := \lim_{n\to\infty} \frac{1}{d^n} \log \bigl|\Phi^n(x,y)\bigr|_{\max}.$$

Arakelov–Green's function for ϕ is $AG_\phi : \mathbb{P}^1(\mathbb{C}_v)^2 \to \mathbb{R} \cup \{\infty\}$ given by

$$\begin{aligned}AG_\phi(z_1,z_2) := &- \log|x_1y_2 - x_2y_1| \\ &+ \hat{\Lambda}_\Phi(x_1,y_1) + \hat{\Lambda}_\Phi(x_2,y_2) - \frac{1}{d(d-1)} \log|\operatorname{Res}(F,G)|,\end{aligned}$$

where $(x_1,y_1),(x_2,y_2) \in \mathbb{C}_v^2 \smallsetminus \{(0,0)\}$ are lifts of $z_1, z_2 \in \mathbb{P}^1(\mathbb{C}_v)$, respectively, and $\operatorname{Res}(F,G)$ is the homogeneous resultant of Remark 4.11.

Proposition 13.50. *Let $\phi \in \mathbb{C}_v(z)$ be a rational function of degree $d \geq 2$, and let AG_ϕ be the associated Arakelov–Green's function for ϕ. Then for any $z_1, z_2 \in \mathbb{P}^1(\mathbb{C}_v)$, the following hold.*

(a) *$AG_\phi(z_1, z_2)$ is independent of the choice of lifts of ϕ to $\Phi = (F, G)$ and of z_1, z_2 to $(x_1, y_1), (x_2, y_2)$.*

(b) *$AG_\phi(z_1, z_2) = AG_\phi(z_2, z_1)$.*

(c) *$AG_\phi(z_1, z_2) = \infty$ if and only if $z_1 = z_2$.*

In addition, for any $n \geq 1$, $AG_{\phi^n} = AG_\phi$.

Proof. This is left to the reader; see Exercise 13.32. □

Proposition 13.51. *Let $\phi \in \mathbb{C}_v(z)$ be a rational function of degree $d \geq 2$, let AG_ϕ be Arakelov–Green's function for ϕ, and let $\Phi = (F, G)$ be a lift of ϕ to $\mathbb{C}_v^2$. Define*

- *$\hat{\lambda}_{v,\infty,\phi}$ to be the canonical local height for ϕ relative to ∞ and L_∞, where $L_\infty(z) := G(z, 1)$, and*
- *$\hat{\lambda}_{v,0,\phi}$ to be the canonical local height for ϕ relative to 0 and L_0, where $L_0(z) := F(1, 1/z)$.*

Then for all $z_1, z_2 \in \mathbb{C}_v$, the following hold.

(a) $$AG_\phi(z_1, z_2) = -\log\left|z_1 - z_2\right| + \hat{\lambda}_{v,\infty,\phi}(z_1) + \hat{\lambda}_{v,\infty,\phi}(z_2) - \tfrac{1}{d(d-1)} \log|\operatorname{Res}(F, G)|.$$

(b) $$AG_\phi(z_1, \infty) = \hat{\lambda}_{v,\infty,\phi}(z_1) + \hat{\lambda}_{v,0,\phi}(\infty) - \tfrac{1}{d(d-1)} \log|\operatorname{Res}(F, G)|.$$

In addition, AG_ϕ extends to a function

$$AG_\phi : (\mathbb{P}^1_{\text{an}})^2 \to \mathbb{R} \cup \{\infty\}$$

which is continuous and finite-valued on $(\mathbb{P}^1_{\text{an}})^2 \smallsetminus \mathrm{Diag}_{\mathrm{I}}$, where $\mathrm{Diag}_{\mathrm{I}}$ is the Type I diagonal

$$\mathrm{Diag}_{\mathrm{I}} := \{(x, x) : x \in \mathbb{P}^1(\mathbb{C}_v)\}.$$

Proof. Parts (a) and (b) can be proven by direct computation using various properties of canonical local heights. For the final statement, rewrite the formula in part (a) using Gromov products, and replace each $z_i \in \mathbb{C}_v$ in the formulas of parts (a) and (b) by $\zeta_i \in \mathbb{A}^1_{\text{an}}$ in order to define $AG_\phi(\zeta_1, \zeta_2)$ for $\zeta_1, \zeta_2 \in \mathbb{P}^1_{\text{an}}$. We leave the details to the reader; see Exercise 13.33. □

We close our discussion of Arakelov–Green's functions by noting two more useful properties.

Theorem 13.52. *Let $\phi \in \mathbb{C}_v(z)$ be a rational function of degree $d \geq 2$, and let AG_ϕ be the associated Arakelov–Green's function.*

(a) *For any $\zeta_1, \zeta_2 \in \mathbb{P}^1_{\mathrm{an}}$, we have*

$$AG_\phi\big(\phi(\zeta_1), \zeta_2\big) = \sum_{\xi \in \phi^{-1}(\zeta_2)} \deg_\xi(\phi) \cdot AG_\phi(\zeta_1, \xi).$$

(b) *Fix $a \in \mathbb{P}^1(\mathbb{C}_v)$. Then the function*

$$\hat{\lambda} : \mathbb{P}^1_{\mathrm{an}} \smallsetminus \{a\} \to \mathbb{R} \quad by \quad \hat{\lambda}(\zeta) := AG_\phi(\zeta, a)$$

is a canonical local height for ϕ relative to a.

Proof. This is left to the reader; see Exercise 13.35. □

Remark 13.53. The function AG_ϕ arises as a special case of Baker and Rumely's broader theory of Arakelov–Green's functions. A *subharmonic function* on an open set $U \subseteq \mathbb{P}^1_{\mathrm{an}}$ is a potential $f \in \mathcal{P}$ for which the restriction of Δf to U is a positive measure. (Actually, Baker and Rumely define a Laplacian on affinoids, and so they do not require f to extend to a potential on all of $\mathbb{H}$, but only on U.) In particular, if $\Delta f = \mu - [\xi]$, where μ is a probability measure and $\xi \in \mathbb{H}$, then f is subharmonic on $\mathbb{P}^1_{\mathrm{an}} \smallsetminus \{\xi\}$. Baker and Rumely [**8**, Section 8.6] denote the set of all such $f \in \mathrm{BDV}(\mathbb{P}^1_{\mathrm{an}})$ by $\mathcal{AG}[\xi]$; when $\xi = \zeta(0,1)$, Favre and Rivera-Letelier [**63**, Section 2.4], following Favre and Jonsson [**58**, Section 7.9], denote essentially the same space by $\mathcal{P}_+$.

Of course, $\mathcal{AG}$ stands for Arakelov–Green. In [**8**, Section 8.10], given a Radon probability measure $\mu \in M_{\mathrm{Rad}}$ with *continuous* potentials, Baker and Rumely define the *normalized Arakelov–Green's function* of μ to be

$$AG_\mu(\zeta_1, \zeta_2) := \int_{\mathbb{P}^1_{\mathrm{an}}} -\log\left(\delta(\zeta_1, \zeta_2)_\xi\right) d\mu(\xi) + C, \tag{13.15}$$

where δ is the generalized Hsia kernel of equation (13.5), and where the constant C is chosen so that

$$\iint_{(\mathbb{P}^1_{\mathrm{an}})^2} AG_\mu(\zeta_1, \zeta_2)\, d\mu(\zeta_1)\, d\mu(\zeta_2) = 0.$$

More generally, for another Radon probability measure ν, they consider

$$\iint AG_\mu(\zeta_1, \zeta_2)\, d\nu(\zeta_1)\, d\nu(\zeta_2) \tag{13.16}$$

to be an "energy integral" of μ against ν; the special choice of the constant C guarantees that the energy has minimum value 0, attained for $\nu = \mu$; see [**8**, Theorem 8.67].

Baker and Rumely further observe that substituting equation (13.5) into (13.15) yields

$$AG_{\mu}(\zeta_1, \zeta_2) = \langle \zeta_1, \zeta_2 \rangle_{\zeta(0,1)} - u_{\mu,\zeta(0,1)}(\zeta_1) - u_{\mu,\zeta(0,1)}(\zeta_2) + C,$$

which is suggestive of the formulas for $AG_{\phi}(\zeta_1, \zeta_2)$ in Proposition 13.51. In particular, they prove in [**8**, Theorem 10.21] that AG_{ϕ} concides with $AG_{\mu_{\phi}}$. Thus, the constant term involving the resultant is not just a fudge factor to make AG_{ϕ} independent of the choice of lift $\Phi = (F, G)$; it is also the correct normalization factor to make the energy of (13.16) equal to zero when $\nu = \mu$.

13.7. Equidistribution of points of small canonical height

Throughout this section, K denotes a global field, i.e., either a number field (a finite extension of $\mathbb{Q}$) or a global function field (a finite extension of $\mathbb{F}_q(T)$, where $\mathbb{F}_q$ is a finite field). We assume some familiarity with basic algebraic number theory: the set of extensions of such fields, associated sets of primes, and some Galois theory.

Let K^{sep} denote the separable closure of K, and let $\text{Gal}(K^{\text{sep}}/K)$ be the Galois group of K^{sep} over K. Let M_K denote the set of places (or primes) of K, including the archimedean places. For each $v \in M_K$, let $|\cdot|_v$ denote the associated absolute value on K, let K_v denote the completion of K with respect to v, and let $\mathbb{C}_v$ denote the completion of an algebraic closure of K_v. (Recall that $|\cdot|_v$ extends uniquely to $\mathbb{C}_v$.) For each $x \in K^{\times}$, we have $|x|_v = 1$ for all but finitely many $v \in M_K$.

There are integers $N_v \geq 1$ such that

$$\sum_{v \in M_K} N_v \log |x|_v = 0 \quad \text{for all} \quad x \in K^{\times}.$$

This identity is known as the *product formula* for K. More generally, for $x \in K^{\text{sep}}$, let G_x be the $\text{Gal}(K^{\text{sep}}/K)$-orbit of x. Then

$$\sum_{v \in M_K} \sum_{y \in G_x} \frac{N_v}{\# G_x} \log |y|_v = 0 \quad \text{for all} \quad x \in K^{\text{sep}} \setminus \{0\}.$$

A *system of local heights* over K relative to $a \in \mathbb{P}^1(K)$ is a choice of a local height $\lambda_{v,a}$ relative to a (see Definition 13.45) at each place $v \in M_K$, such that for all but finitely many v, we have

$$\lambda_{v,a}(x) = \langle x, a \rangle_{\zeta(0,1)} = \begin{cases} \log \max \{1, |x - a|_v^{-1}\} & \text{if } a \neq \infty, \\ \log \max \{1, |x|_v\} & \text{if } a = \infty. \end{cases}$$

Equivalently, after a $\text{PGL}(2, K)$-coordinate change to move a to ∞, $\lambda_{v,a}$ is the standard local height $\lambda_{v,\infty,\text{std}}$ for all but finitely many v.

For any system of local heights, the function $h : \mathbb{P}^1(K^{\text{sep}}) \smallsetminus \{a\} \to \mathbb{R}$ given by

$$h(x) := \sum_{v \in M_K} \sum_{y \in G_x} \frac{N_v}{\#G_x} \lambda_{v,a}(y) \tag{13.17}$$

is a *global height* over K (or a *Weil height*, or simply a *height*, over K). We often declare $h(a)$ to be an arbitrarily chosen real number, so that h is defined on all of $\mathbb{P}^1(K^{\text{sep}})$.

If $a = \infty$ and $\lambda_{v,\infty}$ is the standard local height for *all* $v \in M_K$, the resulting global height $h_{\text{std}} : K^{\text{sep}} \to \mathbb{R}$ given by

$$h_{\text{std}}(x) := \sum_{v \in M_K} \sum_{y \in G_x} \frac{N_v}{\#G_x} \log \max\{1, |y|_v\}, \quad h_{\text{std}}(\infty) := 0 \tag{13.18}$$

is called the *standard (global/Weil) height.* Any global height over K differs from h_{std} by a bounded amount; see Exercise 13.37.

Global fields have the *Northcott property* that for any bounds $B \in \mathbb{R}$ and $D \geq 1$,

$$\{x \in K^{\text{sep}} : [K(x) : K] \leq D \text{ and } h_{\text{std}}(x) \leq B\} \quad \text{is finite.}$$

Since an arbitrary Weil height h differs from h_{std} by a bounded amount, the same finiteness property applies to h as well.

Definition 13.54. Let K be a global field, and let $\phi \in K(z)$ be a rational function of degree $d \geq 2$. The *canonical height* or *Call–Silverman height* for ϕ is the function $\hat{h}_\phi : \mathbb{P}^1(K^{\text{sep}}) \to \mathbb{R}$ given by

$$\hat{h}_\phi(x) = \lim_{n \to \infty} d^{-n} h_{\text{std}}(\phi^n(x)).$$

Proposition 13.55. *Let K be a global field, and let $\phi \in K(z)$ be a rational function of degree $d \geq 2$. Let $a \in \mathbb{P}^1(K)$, and let $L \in K(z)$ with divisor $\operatorname{div}(L) = \phi^*((a)) - d \cdot (a)$. For each $v \in M_K$, let $\hat{\lambda}_{v,a,\phi}$ be the canonical local height for ϕ relative to a and L. Then the following hold.*

(a) *The limit defining $\hat{h}_\phi$ converges.*

(b) *$\hat{h}_\phi$ is a Weil height over K.*

(c) *$\hat{h}_\phi(\phi(x)) = d \cdot \hat{h}_\phi(x)$ for all $x \in \mathbb{P}^1(K^{\text{sep}})$.*

(d) *For any Weil height h over K such that $h(\phi(x)) = d \cdot h(x)$ for all $x \in \mathbb{P}^1(K^{\text{sep}})$, we have $h = \hat{h}_\phi$.*

(e) *For all $x \in \mathbb{P}^1(K^{\text{sep}}) \smallsetminus \{a\}$,*

$$\hat{h}_\phi(x) = \sum_{v \in M_K} \sum_{y \in G_x} \frac{N_v}{\#G_x} \hat{\lambda}_{v,a,\phi}(y).$$

Proof. This is left to the reader; see Exercise 13.38. Proposition 13.55(a)–(d) are special cases of the far more general [**37**, Theorem 1.1], and part (e) is the corresponding special case of [**37**, Theorem 2.3]. □

Recall that a subset $S \subseteq \mathbb{P}^1(K^{\mathrm{sep}})$ is said to be *Galois-invariant* if for all $x \in S$ and all $\sigma \in \mathrm{Gal}(K^{\mathrm{sep}}/K)$, we have $\sigma(x) \in S$. Observe that for a finite Galois-invariant set S and a Weil height h over K arising from a system $\{\lambda_{v,a} : v \in M_K\}$ of local heights, we have

$$\sum_{x\in S} h(x) = \sum_{x\in S}\sum_{v\in M_K} N_v \lambda_{v,a}(x). \tag{13.19}$$

The principal dynamical result in each of [**7**, **61**, **138**] is the following equidistribution theorem.

Theorem 13.56. *Let K be a global field, let $v \in M_K$, and for each $n \geq 1$, let S_n be a finite Galois-invariant subset of $\mathbb{P}^1(K^{\mathrm{sep}})$. Suppose that $S_m \cap S_n = \varnothing$ for $m \neq n$, and that*

$$\lim_{n\to\infty} \frac{1}{\#S_n} \sum_{x\in S_n} \hat{h}_\phi(x) = 0. \tag{13.20}$$

For any place $v \in M_K$, define a sequence of atomic probability measures $\{\nu_n\}_{n\geq 1}$ on $\mathbb{P}^1_{\mathrm{an}}$ by

$$\nu_n := \frac{1}{\#S_n} \sum_{x\in S_n} [x].$$

Then the sequence $\{\nu_n\}_{n\geq1}$ converges weakly to $\mu_{\phi,v}$, where $\mu_{\phi,v}$ is the equilibrium measure for ϕ at v.

We sketch the proof of Theorem 13.56 below; full proofs can be found in [**8**, Theorem 10.24] and [**61**, Théorème 2]. It is a partial generalization of Corollary 13.40, in the following sense. If ϕ is separable, and if we let $S_n := \phi^{-n}(x_0)$ for some $x_0 \in \mathbb{P}^1(K)$, then S_n is a Galois-invariant finite set satisfying the small height condition (13.20) (by Proposition 13.55(c)), and the measures ν_n of Corollary 13.40 and Theorem 13.56 coincide.

Proof of Theorem 13.56 (Sketch). We mostly follow the argument presented in [**61**, Sections 4.8, 5.2, 5.4, and 6]. In [**61**, Théorème 8], Favre and Rivera-Letelier prove that for any finite Galois-invariant set S, we have

$$\frac{1}{\#S}\sum_{x\in S} \hat{h}_\phi(x) = \frac{1}{2}\sum_{v\in M_K} N_v \langle \nu - \mu_{\phi,v}, \nu - \mu_{\phi,v}\rangle_{\mathrm{Dir}}, \tag{13.21}$$

where $\nu = (\#S)^{-1}\sum_{x\in S}[x]$. Note that the Dirichlet pairing here is defined by Proposition 13.30 and Remark 13.31. (Actually, [**61**] uses a version of the Dirichlet pairing based at ∞ rather than at a point $\xi \in \mathbb{H}$, but the principle is the same.)

Specifically, Favre and Rivera-Letelier prove equation (13.21) as follows. First, in [**61**, Section 5.2], they use equation (13.19) and direct expansion of the Dirichlet pairing

$$\langle \nu - \mu_{\phi,v}, \nu - \mu_{\phi,v} \rangle_{\mathrm{Dir}}$$

to show that the expression on the right-hand side of (13.21) differs by a bounded amount from

$$\frac{1}{\#S} \sum_{x \in S} h_{\mathrm{std}}(x).$$

They deduce that the right-hand side of (13.21) is a Weil height over K. Moreover, a further computation using log cross ratios (see Exercise 13.11), shows that the same expression satisfies the canonical height identity $h(\phi(x)) = d \cdot h(x)$. Equation (13.21) then follows from Proposition 13.55(d).

Next, they use the Northcott property to show that $\#S_n \to \infty$, since the sets S_n are pairwise disjoint and of bounded height. They also show that for each $v \in M_K$, we have

$$\limsup_{n \to \infty} \langle \nu_n - \mu_{\phi,v}, \nu_n - \mu_{\phi,v} \rangle_{\mathrm{Dir}} \geq 0. \tag{13.22}$$

For each of the (all but finitely many) places v at which ϕ has explicit good reduction, we have $\mu_{\phi,v} = [\zeta(0,1)]$, and direct computation shows that

$$\langle \nu_n - [\zeta(0,1)], \nu_n - [\zeta(0,1)] \rangle_{\mathrm{Dir}} = \frac{1}{(\#S_n)^2} \sum_{x \neq y \in S_n} \langle x, y \rangle_{\zeta(0,1)} \geq 0.$$

For the finitely many other places v, Favre and Rivera-Letelier approximate ν_n by another atomic measure $\tilde{\nu}_{n,\varepsilon}$ with support contained in $\mathbb{H}$. Therefore,

$$\langle \tilde{\nu}_{n,\varepsilon} - \mu_{\phi,v}, \tilde{\nu}_{n,\varepsilon} - \mu_{\phi,v} \rangle_{\mathrm{Dir}} \geq 0$$

by Proposition 13.32, since $\tilde{\nu}_{n,\varepsilon}$ and $\mu_{\phi,v}$ both have continuous potentials. Letting $\varepsilon \to 0$ and $n \to \infty$ at appropriate relative speeds, they deduce inequality (13.22). (*Note*: When v is an archimedean place, this step requires arguments in complex potential theory; see [**61**, Section 2].)

Combining hypothesis (13.20) with equations (13.21) and (13.22), they show that in fact, we have

$$\lim_{n \to \infty} \langle \nu_n - \mu_{\phi,v}, \nu_n - \mu_{\phi,v} \rangle_{\mathrm{Dir}} = 0 \quad \text{for all} \quad v \in M_K.$$

In [**61**, Proposition 4.12] (or [**61**, Proposition 2.11] when v is archimedean), they use this fact, combined with $\#S_n \to \infty$, to deduce that ν_n converges weakly to $\mu_{\phi,v}$. They do so using the approximation of ν_n by $\tilde{\nu}_{n,\varepsilon}$ again, allowing the application of Proposition 13.32. A closer analysis shows that the directional derivatives of the potential functions f_n of $\nu_n - \mu_{\phi,v}$ approach zero at all points and all directions of $\mathbb{P}^1_{\mathrm{an}}$. Thus, f_n approaches a constant, and hence $\Delta(\nu_n - \mu_{\phi,v})$ converges weakly to zero, as desired. □

Remark 13.57. Baker and Rumely's argument to prove Theorem 13.56, which they state as [**8**, Theorem 10.24], has the same broad outlines. In place of equation (13.21), they observe that for a finite Galois-invariant set $S \subseteq \mathbb{P}^1(K^{\mathrm{sep}})$, we have

$$\frac{1}{\#S}\sum_{x\in S}\hat{h}_\phi(x) = \frac{1}{2}\sum_{v\in M_K} N_v \kappa_{\phi,v}(S), \tag{13.23}$$

where $\kappa_{\phi,v}$ is the the *discrepancy* of the set S, which they define to be

$$\kappa_{\phi,v}(z_1,\ldots,z_N) := \frac{1}{N(N-1)}\sum_{i\neq j} AG_{\phi,v}(z_i,z_j).$$

(Baker and Rumely use the notation D_ϕ for the discrepancy, but we have opted for different notation, given the heavy use of D in this book.)

Equation (13.23) is straightforward to prove: Using the same lifts Φ and (x_i, y_i) for each $v \in M_K$, sum the terms of Definition 13.49 separately over all v. The $\hat{\Lambda}_\Phi$ terms sum to the $\hat{h}_\phi(x)$ terms, and by the product formula, the $\log|x_1y_2 - x_2y_1|_v$ and $\log|\operatorname{Res}(\Phi)|_v$ terms sum to zero.

Next, for fixed v, they prove that there is a constant $C_{\phi,v} > 0$ such that for all $N \geq 1$ and all sets $\{z_1,\ldots,z_N\}$ of N distinct points in $\mathbb{P}^1_{\mathrm{an}}$, we have

$$\kappa_{\phi,v}(z_1,\ldots,z_N) \geq -\frac{C_{\phi,v}\log N}{N-1}; \tag{13.24}$$

see [**8**, Theorem 10.23]. In addition, by the same reasoning as in Favre and Rivera-Letelier's proof, we have $\lim_{n\to\infty}\#S_n = \infty$. Thus for each $v \in M_K$, we have

$$\limsup_{n\to\infty}\kappa_{\phi,v}(S_n) \geq 0.$$

Applying hypothesis (13.20) yields $\lim_{n\to\infty}\kappa_{\phi,v}(S_n) = 0$.

To prove inequality (13.24), Baker and Rumely write

$$\sum_{i\neq j}\log|x_iy_j - x_jy_i|_v = \log|\det(V_N)|,$$

where V_N is a homogeneous Vandermonde matrix. They then do row operations to replace high powers of x_i and y_i with lower powers of $\Phi^n(x_i, y_i)$. Theorem 13.52 then allows them to bound the original sum $\kappa_{\phi,v}$ of $N(N-1)$ terms using only $O(N\log N)$ terms. Alternatively, to reach the weaker lim sup statement above more directly, they analyze the energy integral of (13.16) when $\mu = \mu_{\phi,v}$ and $\nu = \nu_n$; see [**8**, Theorem 10.22].

Next, Baker and Rumely use the formula $\lim_{n\to\infty}\kappa_{\phi,v}(S_n) = 0$ to show that the same energy integral approaches zero as $n \to \infty$. They then invoke Prokhorov's theorem, that a sequence of probability measures on a suitable space X has a weakly convergent subsequence. See [**31**, Section 1.6] or [**128**, Theorem III.2.1] for the standard statement of Prokhorov's theorem (when X is separable and metrizable, with a mild restriction on the sequence if X is not compact), or [**8**, Theorem A.11] for a modified statement when X is compact and Hausdorff.

By Prokhorov's theorem, it remains to show that any such limit measure is $\mu_{\phi,v}$ itself. Finally, [**8**, Theorem 8.67] is Baker and Rumely's analogue of Favre and Rivera-Letelier's [**61**, Proposition 4.12]: if μ is a probability measure with continuous potentials, then the energy integral of (13.16) is nonnegative, and it is zero if and only if $\nu = \mu$. Theorem 13.56 is then immediate.

Exercises for Chapter 13

13.1. Prove that $\mathbb{H}$ and $\mathbb{P}^1(\mathbb{C}_v)$ are Borel subsets of $\mathbb{P}^1_{\text{an}}$.

(*Suggestion*: Prove that $\{\zeta \in \mathbb{H} : d_{\mathbb{H}}(\zeta, \zeta(0,1)) \leq R\}$ is closed.)

13.2. Prove Proposition 13.6.

(*Suggestion*: For part (d), use the Riesz representation theorem and the fact that compact Hausdorff spaces are normal, and hence the Tietze extension theorem holds for such spaces.)

13.3. Prove Proposition 13.8.

(*Suggestion*: Use the Riesz representation theorem, in the form of Corollary 13.3, to prove the existence and uniqueness of $\phi^*\mu$.)

13.4. Prove the equations (13.2) of Example 13.9.

13.5. Let $\phi \in \mathbb{C}_v(z)$ be a nonconstant rational function, and let $\mu \in M_{\text{Rad}}$ be a signed Radon measure on $\mathbb{P}^1_{\text{an}}$. Prove that

$$(\phi^*\mu)\big(\{\zeta\}\big) = \deg_\zeta(\phi) \cdot \mu\big(\{\phi(\zeta)\}\big)$$

for any $\zeta \in \mathbb{P}^1_{\text{an}}$.

(*Suggestion*: First assume μ is positive; because μ is Radon, choose an open set U containing ζ so that $|\mu(U) - \mu(\{\zeta\})|$ and $|\mu(\phi(U)) - \mu(\{\phi(\zeta)\})|$ are both very small. Then there is a continuous $f : \mathbb{P}^1_{\text{an}} \to [0,1]$ with $f(\zeta) = 1$ and $f = 0$ outside U. Integrate f with respect to $\phi^*\mu$.)

13.6. As in Example 13.12, let $X = \mathbb{R}$, and define a real-valued function μ on the Borel subsets of X by

$$\mu(U) := \lim_{n\to\infty} \big[1/n\big](U).$$

Prove that μ is *not* a measure on X.

(*Suggestion*: Write $(0,1]$ as a union of sets of the form $(1/(n+1), 1/n]$. What is the value of μ on all these sets?)

13.7. Let X be a locally compact Hausdorff space, and let $\{\mu_n\}_{n\geq 1}$ be a sequence of (finite) signed measures on X converging weakly to a (finite) signed measure μ.

(a) If each μ_n is a signed Radon measure, prove that μ is also a signed Radon measure.
(*Suggestion*: Use the Uniform Boundedness Principle from functional analysis, together with the Riesz representation theorem.)

(b) If X is compact and each μ_n is a probability measure, prove that μ is also a probability measure.

13.8. Let $X = \mathbb{R}$, and for each $n \geq 1$, let μ_n be the delta-measure $[n]$. Prove that $\{\mu_n\}_{n\geq 1}$ converges weakly to the trivial measure.

(In particular, Exercise 13.7(b) can fail if the locally compact Hausdorff space X is not compact.)

13.9. Let $X = [0,1]$, with the Borel σ-algebra, and let λ denote the usual Lebesgue measure on X. For each integer $n \geq 1$, define

$$\mu_n = \frac{1}{n}\sum_{i=1}^{n}[i/n],$$

where $[a]$ is the delta-measure at a. Prove that μ_n converges weakly to λ as $n \to \infty$.

13.10. Let $\xi \in \mathbb{H}$. Prove that for $\zeta_1, \zeta_2 \in \mathbb{H}$, the Gromov product $\langle \zeta_1, \zeta_2\rangle_\xi$ satisfies

$$\langle \zeta_1, \zeta_2\rangle_\xi = \frac{1}{2}\big(d_{\mathbb{H}}(\zeta_1,\xi) + d_{\mathbb{H}}(\zeta_2,\xi) - d_{\mathbb{H}}(\zeta_1,\zeta_2)\big).$$

13.11. For any field K, the cross ratio of $(a,b,c,d)_{\mathrm{CR}}$ of $a,b,c,d \in \mathbb{P}^1(K)$ is

$$(a,b,c,d)_{\mathrm{CR}} := \frac{(a-b)(c-d)}{(a-d)(c-b)},$$

or $(a-b)/(c-b)$ if $d=\infty$, or $(a-b)/(a-d)$ if $c = \infty$, and similarly if either of a,b is ∞.

(a) Define the *log cross ratio* of $a,b,c,d \in \mathbb{A}^1_{\mathrm{an}}$ to be

$$(a,b,c,d)_{\mathrm{LCR}} := \log\frac{\mathrm{diam}(a\vee b)\,\mathrm{diam}(c\vee d)}{\mathrm{diam}(a\vee d)\,\mathrm{diam}(c\vee b)},$$

or $\log\mathrm{diam}(a\vee b) - \log\mathrm{diam}(c\vee b)$ if $d = \infty$, and similarly if any of a,b,c is ∞. Prove that for $a,b,c,d \in \mathbb{P}^1(\mathbb{C}_v)$, we have

$$(a,b,c,d)_{\mathrm{LCR}} = \log\big|(a,b,c,d)_{\mathrm{CR}}\big|.$$

(b) Let $\xi \in \mathbb{H}$ and $\zeta_1,\zeta_2 \in \mathbb{P}^1_{\mathrm{an}}$. Prove that

$$\langle \zeta_1,\zeta_2\rangle_\xi = (\zeta_1,\xi,\xi,\zeta_2)_{\mathrm{LCR}}.$$

13.12. In this exercise, we give explicit formulas for the Gromov product when the basepoint is the Gauss point.

(a) Let $\zeta_1, \zeta_2 \in \mathbb{A}^1_{\mathrm{an}}$. Prove that

$$\langle \zeta_1, \zeta_2 \rangle_{\zeta(0,1)} = -\log \operatorname{diam}(\zeta_1 \vee \zeta_2) + \log \max\{|\zeta_1|, 1\} + \log \max\{|\zeta_2|, 1\}.$$

(b) For any $a \in \mathbb{C}_v$, $r > 0$, and $\zeta \in \mathbb{A}^1_{\mathrm{an}}$, prove that

$$\langle \zeta, \zeta(a,r) \rangle_{\zeta(0,1)} = -\log \max\{r, \|z-a\|_\zeta\} + \log \max\{|a|, r, 1\} + \log \max\{\|z\|_\zeta, 1\}.$$

(*Suggestion*: Use part (a) and Exercise 6.22.)

(c) For any $a \in \mathbb{C}_v$ and $\zeta \in \mathbb{A}^1_{\mathrm{an}}$, prove that

$$\langle \zeta, a \rangle_{\zeta(0,1)} = -\log \|z-a\|_\zeta + \log \max\{|a|, 1\} + \log \max\{\|z\|_\zeta, 1\}.$$

(d) For any $\zeta \in \mathbb{A}^1_{\mathrm{an}}$, prove that

$$\langle \zeta, \infty \rangle_{\zeta(0,1)} = \log \max\{|\zeta|, 1\}.$$

13.13. Let $\zeta_1, \zeta_2 \in \mathbb{P}^1_{\mathrm{an}}$ and $\xi_1, \xi_2 \in \mathbb{H}$.

(a) Prove that $\langle \zeta_1, \zeta_2 \rangle_{\xi_1} - \langle \zeta_1, \xi_2 \rangle_{\xi_1} = \langle \zeta_1, \zeta_2 \rangle_{\xi_2} - \langle \zeta_2, \xi_1 \rangle_{\xi_2}$.
(*Suggestion*: First use Exercise 13.10 to prove the identity when all four points are in $\mathbb{H}$.)

(b) Prove that

$$\langle \zeta, \zeta_1 \rangle_{\xi_1} - \langle \zeta, \zeta_1 \rangle_{\xi_2}$$

is bounded as a function of $\zeta \in \mathbb{P}^1_{\mathrm{an}}$.

13.14. Prove that the Gromov product $\langle \zeta_1, \zeta_2 \rangle_\xi$ is

(a) continuous in $\xi \in \mathbb{H}$ with respect to the hyperbolic metric $d_{\mathbb{H}}$,

(b) continuous in $(\zeta_1, \zeta_2) \in (\mathbb{P}^1_{\mathrm{an}} \times \mathbb{P}^1_{\mathrm{an}}) \smallsetminus \mathrm{Diag}$ with respect to the Gel'fand topology, where $\mathrm{Diag} := \{(\zeta, \zeta) : \zeta \in \mathbb{P}^1_{\mathrm{an}}\}$ is the diagonal,

(c) continuous in $\zeta_1 \in \mathbb{P}^1_{\mathrm{an}}$ with respect to the Gel'fand topology, if $\zeta_2 \in \mathbb{H}$ (even at $\zeta_1 = \zeta_2$),

(d) discontinuous at every $\xi \in \mathbb{H}$ with respect to the Gel'fand topology,

(e) discontinuous in (ζ_1, ζ_2) with respect to the Gel'fand topology at every $(\zeta_1, \zeta_2) \in \mathrm{Diag}$.

(*Note*: Part (c) also holds with the roles of ζ_1 and ζ_2 reversed, by symmetry.)

13.15. Let $\theta \in \mathrm{PGL}(2, \mathbb{C}_v)$, let $\xi \in \mathbb{H}$, and let $\zeta, \zeta' \in \mathbb{P}^1_{\mathrm{an}}$. Prove that

$$\big\langle \theta(\zeta), \theta(\zeta') \big\rangle_{\theta(\xi)} = \langle \zeta, \zeta' \rangle_\xi.$$

13.16. Prove Proposition 13.17.

(*Suggestion*: For proving the independence of ξ, Exercise 13.13 may be helpful; cf. [**61**, Proposition 4.1].)

13.17. Prove Proposition 13.18.

13.18. Prove the claim made after Definition 13.20: Let $\mu \in M_{\text{Rad}}$ and $\xi, \eta \in \mathbb{H}$. Then the potential function $u_{\mu,\xi}$ is continuous (respectively, bounded) if and only if $u_{\mu,\eta}$ is continuous (respectively, bounded).

(*Suggestion*: Use Exercise 13.13.)

13.19. Prove Proposition 13.21.

13.20. Let $\{a_n\}_{n\geq 1}$ be a sequence of (Type I) points in $\overline{D}(0,1)$ that all lie in different directions at $\zeta(0,1)$. Define

$$\mu := \sum_{n\geq 1} \frac{1}{2^n}[a_n].$$

(a) Prove that μ is a probability measure on $\mathbb{P}^1_{\text{an}}$, and that $\zeta(0,1) \in \text{supp}(\mu)$.

(b) For any $\xi \in \mathbb{H}$, prove that the potential function $u_{\mu,\xi}$ is discontinuous at $\zeta(0,1)$.

(This example comes from [**8**, Remark 6.14].)

13.21. Let $\mu \in M_{\text{Rad}}$ be a finite *positive* Radon measure on $\mathbb{P}^1_{\text{an}}$, let $\xi \in \mathbb{H}$, and let $u_{\mu,\xi}$ be the associated potential function.

(a) Prove that $u_{\mu,\xi}(\zeta)$ achieves a maximum at $\zeta = \xi$.

(b) Prove that for every Type I point $a \in \mathbb{P}^1(\mathbb{C}_v)$, the limit

$$\lim_{\substack{\zeta\to a \\ \zeta\in(a,\xi]}} u_{\mu,\xi}(\zeta)$$

either converges or approaches $-\infty$, where the limit is as ζ approaches a along the interval $[a,\xi]$. Define $u_{\mu,\xi}(a)$ to be this limit value in $\mathbb{R}\cup\{-\infty\}$.

(c) Prove that the extension of $u_{\mu,\xi}$ to $u_{\mu,\xi} : \mathbb{P}^1_{\text{an}} \to \mathbb{R}\cup\{-\infty\}$ from part (b) is upper semicontinuous on $\mathbb{P}^1_{\text{an}}$.

(This is a portion of the statement of [**8**, Proposition 6.12].)

13.22. Fix $\xi \in \mathbb{H}$. Prove that the generalized Hsia kernel $\delta(\cdot,\cdot)_\xi$ of (13.5) satisfies

$$-\log\big(\delta(\zeta_1,\zeta_2)_\xi\big) = \langle\zeta_1,\zeta_2\rangle_\xi - d_{\mathbb{H}}\big(\xi,\zeta(0,1)\big)$$

for all $\zeta_1,\zeta_2 \in \mathbb{P}^1_{\text{an}}$.

(*Suggestion*: Exercise 13.13 may be helpful.)

13.23. Prove Theorem 13.25.

13.24. Let $\phi \in \mathbb{C}_v(z)$ be a nonzero rational function. Prove the claims of Example 13.27, that $(\zeta \mapsto \log\|\phi(z)\|_\zeta) \in \mathcal{P}$, with Laplacian given by equation (13.7).

13.25. Prove the second identity of Theorem 13.35.

(*Suggestion*: Use the first identity, Corollary 13.34, and the results of Section 13.1.3.)

13.26. Let $\zeta \in \mathbb{H}$, let $\boldsymbol{v}$ be a direction at ζ, and let $f : \mathbb{H} \to \mathbb{R}$. Let $g, h : [0,1] \to \mathbb{H}$ be distance-preserving maps with $g(0) = h(0) = \zeta$ and $g(t), h(t) \in \boldsymbol{v}$ for $t > 0$. Prove that

$$\lim_{t\to 0^+} \frac{f(g(t)) - f(\zeta)}{t} = \lim_{t\to 0^+} \frac{f(h(t)) - f(\zeta)}{t},$$

in the sense that if either limit converges, then both converge and are equal.

13.27. Let $\zeta, \xi \in \mathbb{H}$, let $R := d_{\mathbb{H}}(\zeta, \xi)$, and let $i : [\zeta, \xi] \to [0, R]$ be the isometry $i(\eta) := d_{\mathbb{H}}(\zeta, \eta)$. Let $g : [0, R] \to \mathbb{R}$ be a continuously differentiable function for which g'' is integrable on $[0, R]$, and for which

$$g'(0) = g'(R) = 0.$$

Define $f : \mathbb{H} \to \mathbb{R}$ to be the unique continuous function such that

$$f(\eta) = g \circ i(\eta) \quad \text{for all} \quad \eta \in [\zeta, \xi]$$

and that is constant on each connected component of $\mathbb{H} \smallsetminus [\zeta, \xi]$. Prove that Δf is supported on $[\zeta, \xi]$, and that for any Borel set $E \subseteq [\zeta, \xi]$,

$$\Delta f(E) = \int_{i(E)\cap[\zeta,\eta]} g''(t)\, d\lambda(t),$$

where λ is Lebesgue measure on $[0, R]$.

13.28. Let $f \in \mathcal{P}$. Prove that Δf is atomic (i.e., a finite linear combination of delta-measures) if and only if there is a tree $\Gamma \subseteq \mathbb{H}$ with only finitely many edges and vertices, such that

(a) f is constant on every connected component of $\mathbb{H} \smallsetminus \Gamma$, and

(b) for any two adjacent vertices ζ, ξ of Γ, f is affine linear on the edge $[\zeta, \xi]$. That is, there are real constants $a, b \in \mathbb{R}$ such that

$$f(\eta) = a \cdot d_{\mathbb{H}}(\eta, \zeta) + b \quad \text{for all } \eta \in (\zeta, \xi).$$

(*Note*: The tree Γ may have infinite hyperbolic diameter; that is, it is allowed to have vertices that are points of Type I.)

13.29. Prove Theorem 13.39(d).

(*Suggestion*: Use part (c) of the same theorem.)

13.30. Prove that the equilibrium measure μ_ϕ of the Lattès map ϕ of Example 13.44 is a Lebesgue measure on the interval J divided by the (hyperbolic) length of J.

(*Suggestion*: Use $\xi = \zeta(0,1)$, and study the preimages of ξ under the action of ϕ^n via the description in Example 7.31.)

13.31. Let $\phi \in \mathbb{C}_v(z)$ be a rational function of degree $d \geq 2$, and write $\phi = f/g$ for polynomials $f, g \in \mathbb{C}_v[z]$ with no common roots and with $\max\{\|f\|_{\zeta(0,1)}, \|g\|_{\zeta(0,1)}\} = 1$.

(a) Prove that there is a bound $B \geq 0$ such that for all $x \in \mathbb{C}_v$,

$$-B \leq \log\max\bigl\{|f(x)|, |g(x)|\bigr\} - d \cdot \log\max\bigl\{|x|, 1\bigr\} \leq B.$$

(*Suggestion*: For the lower bound, make use of the resultant $\operatorname{Res}(f,g)$ of equation (4.1).)

(b) Prove that expression (13.11) is bounded for $\zeta \in \mathbb{A}^1_{\text{an}} \smallsetminus \phi^{-1}(\infty)$.

(c) Let $a \in \mathbb{P}^1(\mathbb{C}_v)$, and let $L \in \mathbb{C}_v(z)$ be a rational function with divisor $\operatorname{div}(L) = \phi^*((a)) - d \cdot (a)$. Prove that expression (13.12) is a bounded function of $\zeta \in \mathbb{P}^1_{\text{an}} \smallsetminus (\phi^{-1}(a) \cup \{a\})$.

(d) If ϕ and L have good reduction, prove that

$$\bigl\langle \phi(\zeta), a \bigr\rangle_{\zeta(0,1)} = d \cdot \langle \zeta, a \rangle_{\zeta(0,1)} - \log\|L\|_\zeta \quad \text{for all } \zeta \in \mathbb{P}^1_{\text{an}} \smallsetminus \{a\}.$$

13.32. Prove Proposition 13.50.

13.33. Let $\phi \in \mathbb{C}_v(z)$ be a rational function of degree $d \geq 2$ with Arakelov–Green's function AG_ϕ. Let $\Phi = (F, G)$ be a lift of ϕ to $\mathbb{C}_v^2$, with associated homogeneous canonical local height $\hat{\Lambda}_\Phi$. As in Proposition 13.51, define $\hat{\lambda}_{v,\infty,\phi}$ to be the canonical local height for ϕ relative to ∞ and L_∞, and $\hat{\lambda}_{v,0,\phi}$ to be the canonical local height for ϕ relative to 0 and L_0, where $L_\infty(z) := G(z,1)$, and $L_0(z) := F(1, 1/z)$.

(a) Prove that $\hat{\Lambda}_\Phi(\Phi(x,y)) = d \cdot \hat{\Lambda}_\Phi(x,y)$ for all $x, y \in \mathbb{C}_v$.

(b) Prove that $z \mapsto \hat{\Lambda}_\Phi(z, 1)$ is a canonical local height for ϕ relative to ∞ and L_∞.

(c) Prove that $z \mapsto \hat{\Lambda}_\Phi(1, 1/z)$ is a canonical local height for ϕ relative to 0 and L_0.

(d) Use the uniqueness of canonical local heights (discussed immediately after Definition 13.47) to prove Proposition 13.51, parts (a) and (b).

(e) For all $z_1, z_2 \in \mathbb{C}_v$, prove that

$$-\log|z_1 - z_2| = \langle z_1, z_2 \rangle_{\zeta(0,1)} - \log\max\{1, |z_1|\} - \log\max\{1, |z_2|\}.$$

(*Note*: See Exercise 13.12(a).)

(f) Prove the final statement of Proposition 13.51.
(*Suggestion*: Define $AG_\phi(\zeta_1, \zeta_2)$ for $\zeta_1, \zeta_2 \in \mathbb{P}^1_{\text{an}}$ by substituting $\zeta_i \in \mathbb{A}^1_{\text{an}}$ for $z_i \in \mathbb{C}_v$ in the formulas of parts (a) and (b) of the proposition, after adjusting with part (e) of this exercise. Then use properties of Gromov products (from earlier exercises in this section) and of local heights to prove the desired properties.)

13.34. Let $\phi \in \mathbb{C}_v(z)$ be a rational function of degree $d \geq 2$ with Arakelov–Green's function AG_ϕ. Let $\Phi = (F, G)$ be a lift of ϕ to $\mathbb{C}_v^2$, with associated homogeneous canonical local height $\hat{\Lambda}_\Phi$.

(a) Let $A \in \mathrm{GL}(2, \mathbb{C}_v)$, and define $\Psi := A^{-1} \circ \Phi \circ A$, with associated homogeneous canonical local height $\hat{\Lambda}_\Psi$. Prove that

$$\hat{\Lambda}_\Phi\big(A(x,y)\big) = \hat{\Lambda}_\Psi(x,y) \quad \text{for all } (x,y) \in \mathbb{C}_v^2.$$

(See also [**8**, Lemma 10.13].)

(b) Let $\theta \in \mathrm{PGL}(2, \mathbb{C}_v)$, and define $\psi := \theta^{-1} \circ \phi \circ \theta$. Prove that

$$AG_\phi\big(\theta(z_1), \theta(z_2)\big) = AG_\psi(z_1, z_2) \quad \text{for all } (z_1, z_2) \in \mathbb{P}^1(\mathbb{C}_v).$$

(*Suggestion*: Use part (a) and Exercise 4.12. See also [**8**, Corollary 10.14].)

(c) For θ and ψ as in part (b), prove that

$$AG_\phi\big(\theta(\zeta_1), \theta(\zeta_2)\big) = AG_\psi(\zeta_1, \zeta_2) \quad \text{for all } (\zeta_1, \zeta_2) \in \mathbb{P}^1_{\mathrm{an}}.$$

(*Suggestion*: Use part (b) and the continuity of Proposition 13.51.)

13.35. Prove Theorem 13.52.

(*Suggestion*: For part (a) (cf. [**8**, Theorem 10.18]), prove it for $\zeta_1 \in \mathbb{C}_v$ and $\zeta_2 = \infty$, and generalize with Exercise 13.34(b) and the continuity of Proposition 13.51. For part (b), use Exercises 13.33(b) and 13.34(c).)

13.36. Let h be a Weil height over a global field K, as in equation (13.17). Let $\phi \in K(z)$ be a rational function of degree $d \geq 1$. Prove that there is a bound $B \geq 0$ so that for every $x \in \mathbb{P}^1(K) \smallsetminus \{a\}$, we have

$$-B \leq h\big(\phi(x)\big) - d \cdot h(x) \leq B.$$

(*Suggestion*: Use Exercise 13.31(c), (d) and the product formula.)

13.37. Let K be a global field, let h_{std} be the standard Weil height of equation (13.18), and let $h : \mathbb{P}^1(K^{\mathrm{sep}}) \to \mathbb{R}$ be a Weil height over K.

(a) Prove that there is a bound $C \geq 0$ so that

$$-C \leq h(x) - h_{\mathrm{std}}(x) \leq C \quad \text{for all } x \in \mathbb{P}^1(K) \smallsetminus \{a\}.$$

(*Suggestion*: Use Exercise 13.36 to move a to ∞ via some $\theta \in \mathrm{PGL}(2, K)$, and otherwise work directly from the definitions.)

(b) Let $\phi \in K(z)$ be a rational function of degree $d \geq 1$, and let B be the associated bound of Exercise 13.36. Prove that for any integers $n \geq m \geq 0$ and any $x \in \mathbb{P}^1(K)$, we have

$$-\frac{B}{d^m(d-1)} \leq d^{-n} h\big(\phi^n(x)\big) - d^{-m} h\big(\phi^m(x)\big) \leq \frac{B}{d^m(d-1)}.$$

13.38. Use Exercise 13.37 and properties of local heights and Weil heights to prove Proposition 13.55.

Part 5

Proofs from Non-Archimedean Analysis

Chapter 14

Proofs of Results from Non-Archimedean Analysis

In this chapter, we supply the missing proofs from Chapter 3 on the analysis of non-archimedean power series. We also present a number of supplementary results; some are necessary lemmas for our main results, and some are simply interesting in their own right.

14.1. Basic power series proofs

The following proposition shows that like absolutely convergent series in real or complex analysis, any convergent non-archimedean series can be rearranged arbitrarily.

Proposition 14.1. *Let $\{a_n\}_{n\geq 0} \subseteq K$ be a sequence, and let $\sigma : \mathbb{N} \to \mathbb{N}$ be any permutation of the natural numbers. If $\sum_{n\geq 0} a_n$ converges, then so does $\sum_{n\geq 0} a_{\sigma(n)}$, and the two series have the same sum.*

Proof. The fact that $\sum_{n\geq 0} a_{\sigma(n)}$ converges is an immediate corollary of Proposition 3.1. However, the fact that the two sums are the same requires a little more work, and recovers the convergence as a byproduct, as follows.

Let $A = \sum_{n\geq 0} a_n$. Given $\varepsilon > 0$, there is an integer $N_1 \geq 0$ such that

$$\left| \sum_{n=0}^{m} a_n - A \right| < \varepsilon \quad \text{for all } m \geq N_1.$$

By Proposition 3.1, there is an integer $N_2 \geq 0$ such that that $|a_n| < \varepsilon$ for all $n \geq N_2$. (In fact, one can take $N_2 = N_1$, but we do not need that here.) Define

$$N' := \max\{N_1, N_2\} \quad \text{and} \quad N := \max\{\sigma^{-1}(n) : 0 \leq n \leq N'\} \geq N'.$$

Then for any $m \geq N$, the set

$$T := \{\sigma^{-1}(0), \sigma^{-1}(1), \ldots, \sigma^{-1}(N')\}$$

is contained in the set $S := \{0, 1, \ldots, m\}$, and $|a_{\sigma(n)}| < \varepsilon$ for all $n \in S \smallsetminus T$. Thus,

$$\begin{aligned} \left|\sum_{n=0}^{m} a_{\sigma(n)} - A\right| &= \left|\left(\sum_{n\in T} a_{\sigma(n)} - A\right) + \sum_{n\in S\smallsetminus T} a_{\sigma(n)}\right| \\ &\leq \max\left\{\left|\sum_{n=0}^{N'} a_n - A\right|, \max_{n\in S\smallsetminus T}\{|a_{\sigma(n)}|\}\right\} < \varepsilon, \end{aligned}$$

as desired. □

Proof of Theorem 3.2. Part (a) is immediate from Proposition 3.1, and part (b) follows from the same proposition by choosing $x \in \mathbb{C}_v$ with $|x| = r$ and considering $f(a + x)$. For part (c), let

$$X := \{x \in \mathbb{C}_v : f(x) \text{ converges}\}$$

and

$$R := \sup\{r \in |\mathbb{C}_v^\times| : f \text{ converges on } \overline{D}(a, r)\}.$$

We now show that X is either $D(a, R)$ or $\overline{D}(a, R)$, where, for the purposes of this proof, we consider $D(a, \infty)$ to be $\mathbb{C}_v$ and $\overline{D}(a, 0)$ to be $\{a\}$. Since

$$D(a, R) = \bigcup_{0<r<R} \overline{D}(a, r),$$

we have $X \supseteq D(a, R)$. For any $b \in \mathbb{C}_v$ with $|b| > R$, part (b) shows that $f(a+b)$ does not converge; hence, $X \subseteq \overline{D}(a, R)$. Thus, it remains to consider what happens to $a + b$ for $|b| = R$.

If $f(a + b)$ converges for even one such b, then f converges on $\overline{D}(a, R)$ by part (b), and hence $X = \overline{D}(a, R)$. Otherwise, $X = D(a, R)$. Either way, we are done. □

Proof of Proposition 3.3. Write $f(z) = \sum_{n\geq 0} c_n(z - a)^n$, and define

$$g(z) := \sum_{n\geq 0} c'_n(z - b)^n,$$

where

$$c'_n := \sum_{m=n}^{\infty} c_m \binom{m}{n} (b-a)^{m-n}. \tag{14.1}$$

We chose formula (14.1) by writing the expression $(z-a)^n$ from the definition of $f(z)$ as

$$((z-b)+(b-a))^n = \sum_{k=0}^{n} \binom{n}{k} (b-a)^{n-k}(z-b)^k,$$

and formally switching the order of summation in the expression for f.

The sum in formula (14.1) converges, for the following reason. Since $\binom{m}{n}$ is an integer, we have $|\binom{m}{n}| \leq 1$. Therefore, the absolute value of the mth term in (14.1) is

$$\left| c_m \binom{m}{n} (b-a)^{m-n} \right| \leq |b-a|^{-n} \cdot \left| c_m (b-a)^m \right|. \tag{14.2}$$

Since $c_m(b-a)^m$ is the mth term in the sum describing $f(b)$, and since f converges at b, Proposition 3.1 shows that $|c_m(b-a)^m| \to 0$ as $m \to \infty$. Hence, expression (14.2) also approaches 0 as $m \to \infty$. Thus, the sum (14.1) also converges, again by Proposition 3.1.

Finally, the two series (for f and g) have the same region of convergence and the same sum by Proposition 14.1. □

Proof of Theorem 3.5. Pick $a \in D$. We wish to show that $\mathcal{A}_D$ is a subring of the formal power series ring $\mathbb{C}_v[[z-a]]$. Clearly, $0, 1 \in \mathcal{A}_D$. It remains to show that $\mathcal{A}_D$ is closed under addition, subtraction, and multiplication.

Given $f, g \in \mathcal{A}_D$, write

$$f(z) = \sum_{n \geq 0} b_n (z-a)^n \quad \text{and} \quad g(z) = \sum_{n \geq 0} c_n (z-a)^n.$$

Then for any $x \in D$, we have

$$\lim_{n\to\infty} \big(a_n \pm b_n\big)(x-a)^n = \lim_{n\to\infty} a_n (x-a)^n \pm \lim_{n\to\infty} b_n (x-a)^n = 0,$$

and hence the series $f \pm g$ converges on D, by Proposition 3.1.

We also have

$$fg(z) = \sum_{n \geq 0} \big(b_0 c_n + \cdots + b_n c_0\big)(z-a)^n.$$

For any $x \in D$, we have

$$\lim_{n\to\infty} |b_n||x-a|^n = \lim_{n\to\infty} |c_n||x-a|^n = 0.$$

Thus, for any $m \geq 0$, we may define

$$B_m := \max_{n \geq m} \{|b_n||x-a|^n\} \quad \text{and} \quad C_m := \max_{n \geq m} \{|c_n||x-a|^n\}.$$

For any $n \geq 0$, let $m := n/2$ if n is even, or $m := (n-1)/2$ if n is odd. Then

$$\begin{aligned}
&\big|(b_0c_n + \cdots + b_nc_0)(x-a)^n\big| \\
&\qquad \leq \max\{|b_j||x-a|^j \cdot |c_{n-j}||x-a|^{n-j} : 0 \leq j \leq n\} \\
&\qquad \leq \max\{B_0C_m, C_0B_m\}.
\end{aligned}$$

Since $B_m, C_m \to 0$ as $n \to \infty$, it follows from Proposition 3.1 that the series fg converges at x, as desired. □

Proof of Proposition 3.7. Choose $a \in E$, and let $r > 0$ be the radius of E. Properties (a) and (b) are immediate from Definition 3.6, using the fact that the radius of D is at least r. Property (c) is also clear from the definition, using the fact that the conditions on E imply that the supremum in Definition 3.6 is attained.

To prove the other two properties, given $f, g \in \mathcal{A}_D$, write

$$f = \sum_{n \geq 0} b_n(z-a)^n \quad \text{and} \quad g = \sum_{n \geq 0} c_n(z-a)^n.$$

Property (d) is immediate from the fact that $|b_n + c_n| \leq \max\{|b_n|, |c_n|\}$. As for property (e), the $(z-a)^n$-coefficient of fg is

$$c'_n := b_0c_n + b_1c_{n-1} + \cdots + b_nc_0, \tag{14.3}$$

and hence $\|fg\|_E \leq \|f\|_E \cdot \|g\|_E$, again by the non-archimedean inequality. Thus, it suffices to prove that $\|fg\|_E \geq \|f\|_E \cdot \|g\|_E$.

Let $M, N \geq 0$ be the smallest integers such that

$$|b_M|r^M = \|f\|_E \quad \text{and} \quad |c_N|r^N = \|g\|_E.$$

Then the term b_Mc_N appears in the expression (14.3) for c'_{M+N}, and b_Mc_N has strictly larger absolute value than any of the other terms in that expression. Thus,

$$|c'_{M+N}|r^{M+N} = |b_Mc_N|r^{M+N} = \|f\|_E\|g\|_E,$$

and hence $\|fg\|_E \geq \|f\|_E \cdot \|g\|_E$, as desired. □

14.2. The Weierstrass preparation theorem

The term *Weierstrass degree* of Section 3.2 comes from the following theorem, which is central to much of non-archimedean analysis.

Theorem 14.2 (Weierstrass preparation theorem). *Let $D := \overline{D}(a, r)$, where $a \in K$ and $r \in |\mathbb{C}_v^\times|$. Let $f \in \mathcal{A}_D$ be a nonzero power series defined over K and converging on D. Then there exist a monic polynomial $g \in \mathbb{C}_v[z]$ and a power series $h \in \mathcal{A}_D$ converging on D such that*

(a) *$f = gh$,*

(b) *all zeros of g lie in $\overline{D}(a, r)$, and*

(c) *$h(x) \neq 0$ for all $x \in \overline{D}(a, r)$.*

Moreover, this factorization of f is unique, $\deg(g) = \operatorname{wdeg}_D(f)$, *and g and h are both defined over K.*

Theorem 14.2 is the dimension-1 case of the more general Weierstrass preparation theorem; see [**30**, Sections 5.2.1–5.2.2], for example. The theorem is sometimes stated only for $r = 1$; the version above for arbitrary $r \in |\mathbb{C}_v^\times|$ can then be obtained by considering the function $f(bz)$, where $|b| = r$. However, more work is then required to show that g and h are defined over K if $r \in |\mathbb{C}_v^\times| \smallsetminus |K^\times|$. On the other hand, the direct proof of Theorem 14.2 is essentially the same as that of the $r = 1$ statement, and hence we opt for the proof for general $r \in |\mathbb{C}_v^\times|$ below. First, however, we need the following generalization of Euclidean division.

Lemma 14.3 (Weierstrass division theorem for rational closed disks). *Let D, a, and f be as in Theorem 14.2. Let $F \in \mathcal{A}_D$ be a power series converging on D and defined over K. Then F can be written as $F = Qf + R$, where $Q \in \mathcal{A}_D$ is a power series converging on D and defined over K, and where $R \in K[z]$ is a polynomial of degree strictly less than $d :=$ $\operatorname{wdeg}_D(f)$ such that $\|R\|_D \leq \|F\|_D$.*

Proof. Without loss, assume that $a = 0$. Write $f = \sum_{n \geq 0} c_n z^n$, and define $f_d(z)$ to be the degree-d truncation

$$f_d(z) := c_0 + \cdots + c_d z^d \in K[z]$$

of $f(z)$. Note that $\|f - f_d\|_D < \|f\|_D = \|f_d\|_D$ by definition of Weierstrass degree.

Step 1. We first prove Lemma 14.3 for the polynomial f_d in place of the power series f. Write $F(z) = \sum_{n \geq 0} C_n z^n$. For every $n \geq 0$, Euclidean division tells us that the term $C_n z^n$ may be written as $C_n z^n = q_n f_d + g_n$, where $q_n, g_n \in K[z]$ with $\deg(g_n) < d$. Moreover, because

$$\|f\|_D = |c_d| r^d = \max_{n \geq 0} \{|c_n| r^n\},$$

the reader can check that

$$\|q_n\|_D \cdot \|f\|_D \leq |C_n| r^n \quad \text{and} \quad \|g_n\|_D \leq |C_n| r^n; \tag{14.4}$$

see Exercise 14.1. By Proposition 3.1, since $\lim_{n\to\infty} |C_n|r^n = 0$, it makes sense to define

$$Q := \sum_{n\geq 0} q_n \quad \text{and} \quad R := \sum_{n\geq 0} g_n,$$

yielding $F = Qf_d + R$ as desired. Moreover, by inequalities (14.4), we obtain the additional fact that

$$\|Q\|_D \|f\|_D \leq \|F\|_D \quad \text{and} \quad \|R\|_D \leq \|F\|_D.$$

Step 2. We now apply Step 1 inductively. Given $F \in K[[z]]$ converging on $\overline{D}(0,r)$, set $F_0 := F$, and invoke Step 1 to define

$$Q_0 \in K[[z]] \quad \text{and} \quad R_0 \in K[z] \quad \text{so that} \quad F_0 = Q_0 f_d + R_0,$$

with all the associated properties. Next, define $F_1 := Q_0 \cdot (f_d - f)$, and obtain

$$Q_1 \in K[[z]] \quad \text{and} \quad R_1 \in K[z] \quad \text{so that} \quad F_1 = Q_1 f_d + R_1,$$

In general, given Q_{i-1} and R_{i-1}, set $F_i := Q_{i-1} \cdot (f_d - f)$, and obtain Q_i and R_i such that $F_i = Q_i f_d + R_i$.

Thus, we have a sequence $\{Q_i\}_{i\geq 0}$ of power series in $K[[z]]$, each of which converges on $\overline{D}(0,r)$, and a sequence $\{R_i\}_{i\geq 0}$ of polynomials in $K[z]$ of degree at most $d-1$. Inductively applying the above bounds, we obtain

$$\big\|F_i\big\|_D \leq \|F\|_D \cdot \left(\frac{\|f - f_d\|_D}{\|f\|_D}\right)^i,$$

as well as

$$\big\|Q_i\big\|_D \leq \frac{\|F\|_D \cdot \|f - f_d\|_D^i}{\|f\|_D^{i+1}} \quad \text{and} \quad \big\|R_i\big\|_D \leq \frac{\|F\|_D \cdot \|f - f_d\|_D^i}{\|f\|_D^{i+1}}.$$

Because $\lim_{i\to\infty}(\|f - f_d\|_D / \|f\|_D)^i = 0$, the sums

$$Q := \sum_{i\geq 0} Q_i \quad \text{and} \quad R := \sum_{i\geq 0} R_i \tag{14.5}$$

converge, by Proposition 3.1. The reader can check that $F = Qf + R$, that Q converges on $\overline{D}(0,r)$, and that R is a polynomial of degree at most $d-1$ with $\|R\|_D \leq \|F\|_D$; see Exercise 14.2. □

Proof of Theorem 14.2. Once again, assume without loss that $a = 0$. Write $f = \sum_{n\geq 0} c_n z^n$, and let $d := \operatorname{wdeg}_D(f)$.

Write $z^d = Q(z)f(z) + R(z)$ according to Lemma 14.3, and set

$$g(z) := Q(z)f(z) = z^d - R(z),$$

which is a monic polynomial of degree d with $\|g\|_D = r^d$. Thus, writing

$$g = z^d + B_{d-1}z^{d-1} + \cdots + B_0,$$

we have $|B_i| \le r^{d-i}$. For any $x \in \mathbb{C}_v$ with $|x| > r$, then,

$$|g(x)| = |x|^d \neq 0,$$

since the lead term dominates the others. Hence, all zeros of g lie in $\overline{D}(0, r)$.

Write $Q(z) = \sum_{n\ge 0} A_n z^n$. We claim that $\operatorname{wdeg}_D(Q) = 0$. That is, we claim that $|A_n|r^n < |A_0|$ for all $n \ge 1$.

To prove the claim, let $M := \operatorname{wdeg}_D(Q) \ge 0$, and consider the z^{M+d} term of $Q(z)f(z)$. Its coefficient is a sum of terms of the form $A_m c_n$ with $m + n = M + d$. Moreover, each such term satisfies $|A_m c_n| \le |A_M c_d|$, with strict inequality unless $m = M$ and $n = d$, since $M = \operatorname{wdeg}_D(Q)$ and $d = \operatorname{wdeg}_D(f)$. Thus, the z^{M+d}-coefficient of $Q(z)f(z)$ has absolute value $|A_M c_d| > 0$. However,

$$Q(z)f(z) = z^d - R(z)$$

is a polynomial of degree d, and therefore $M + d \le d$; that is, $M = 0$, as claimed.

Define $h := 1/Q$. By the claim and by Exercise 14.4, we have $h \in \mathcal{A}_D$, with h defined over K. Because $g = Qf$, it follows that $f = gh$. In addition, for any $x \in D$, we have $h(x) = 1/Q(x) \neq 0$, as desired.

It remains only to show that the factorization is unique. Suppose that $\tilde{g} \in \mathbb{C}_v[z]$ is a monic polynomial with all its zeros in D, and that $\tilde{h} \in \mathcal{A}_D$ is a convergent power series with no zeros in D, such that

$$\tilde{g}\tilde{h} = f = gh.$$

Let $q := \gcd(g, \tilde{g}) \in \mathbb{C}_v[z]$. After dividing both g and $\tilde{g}$ by q, we may assume that $gh = \tilde{g}\tilde{h}$, where g and $\tilde{g}$ are monic polynomials with no common zeros. If $x \in \mathbb{C}_v$ is a zero of $\tilde{g}$, then by assumption $x \in D$, but we have

$$0 = \tilde{g}(x)\tilde{h}(x) = g(x)h(x) \neq 0,$$

a contradiction. Thus, $\tilde{g}$ must have degree 0; similarly, so must g. Therefore, since they are monic polynomials, we have $\tilde{g} = 1 = g$, and hence $\tilde{h} = h$ as well. □

Remark 14.4. Let $D = D(a, r)$ be an open disk, either rational or irrational, and let $f(z) = \sum_{n\ge 0} c_n(z - a)^n$ be a power series converging on D. Recall from Definition 3.8(b) that the Weierstrass degree of f on D is the *smallest* integer $d \ge 0$ such that $|c_d|r^d = \sup_{n\ge 0} |c_n|r^n$, or $d = \infty$ if there is no such integer. There may indeed be no such integer, because the supremum may not be attained even though f converges on D.

Note that if $D = D(a, r) = \overline{D}(a, r)$ is irrational, i.e., if $r \notin |\mathbb{C}_v^\times|$, and if the supremum is attained, then it is attained at a *unique* integer. However, if $D = D(a, r)$ is rational, there may be more than one, or even infinitely many, such integers.

If $d := \mathrm{wdeg}_D(f)$ is finite, one can prove an analogue of the Weierstrass preparation theorem, that f factors as $f = gh$, where h is a convergent power series with no zeros in D, and g is a monic polynomial of degree d with all its zeros in D. The proof involves writing D as an infinite union of nested rational closed disks; see Exercise 14.5.

The Weierstrass preparation theorem now allows us to prove some other results from Chapter 3.

Proof of Proposition 3.10. Subtracting g from both f and g, we may assume without loss that $g = 0$. If $f \neq 0$ as a power series, then f has some Weierstrass degree $d < \infty$. By Theorem 14.2, $f(x) = 0$ for at most d points in $\overline{D}(a, r)$, contradicting the hypotheses and hence proving that $f = 0$. □

Proof of Proposition 3.11. This proposition says that three sets coincide. The last two sets are equal by Theorem 3.13(a). Next, if $f \in \mathcal{A}_D$ has multiplicative inverse $g \in \mathcal{A}_D$, then $f(x)g(x) = 1$ for all $x \in D$, and hence f cannot have any zeros in D. Conversely, if $f \in \mathcal{A}_D$ has Weierstrass degree 0 on D, then by Exercise 14.4 it has a multiplicative inverse $g \in \mathcal{A}_D$. □

14.3. Proofs related to Newton polygons

Proof of Theorem 3.13.

(a) If D is a rational closed disk, then statement (a) is immediate from Theorem 14.2. If D is open, we may write D as an increasing union of rational closed disks

$$D_1 \subseteq D_2 \subseteq \cdots .$$

We leave it to the reader to check that the Weierstrass degrees d_i of f on D_i approach $d := \mathrm{wdeg}_D(f)$ as $i \to \infty$; see Exercise 14.7(a). Since f has d_i zeros on D_i, f must also have d zeros on D.

(b) If D is a rational closed disk, then statement (b) is immediate from Theorem 14.2. If D is open with $d := \mathrm{wdeg}_D(f) < \infty$, then again we leave it to the reader to check that there is some slightly smaller rational closed subdisk $E := \overline{D}(a, r) \subsetneq D$ such that $\mathrm{wdeg}_E(f) = d$; see Exercise 14.7(b). By part (a), f has exactly d zeros in each of E and D (counting multiplicity), and therefore all of the zeros of f in D lie in E. Since part (b) holds for E, then, it also holds for D.

(c) If $\overline{D}(a, r)$ is rational closed, then the corresponding Weierstrass degree j is finite. Similarly, even if $r \notin |\mathbb{C}_v^\times|$, then the hypothesis that $\overline{D}(a, r) \subsetneq D$ implies that j is finite, as we leave the reader to check in Exercise 14.7(c).

Statement (i), i.e., the fact that $i \le j$, is immediate from Definition 3.8. Statement (ii) is immediate from part (b) applied to both of the disks $D(a,r)$ and $\overline{D}(a,r)$. Finally, statement (iii) follows by letting $g_o, g_c \in K[z]$ be the polynomials for $D(a,r)$ and $\overline{D}(a,r)$, respectively. Since every zero of g_o is a zero of g_c with the same multiplicity, the quotient $g_r := g_c/g_o \in K[z]$ is indeed a polynomial and has the other desired properties. □

Another key result on zeros of non-archimedean polynomials is Hensel's lemma, which we state in two versions. We recall that $\mathcal{O}_K$ denotes the ring of integers of K, $\mathcal{M}_K$ denotes the unique maximal ideal, and $k = \mathcal{O}_K/\mathcal{M}_K$ denotes the residue field.

Theorem 14.5 (Hensel's lemma, version 1). *Let $f(z) \in \mathcal{O}_K[z]$ be a monic polynomial. Suppose that when we reduce the coefficients of f modulo $\mathcal{M}_v$, the resulting polynomial $\overline{f} \in k[z]$ factors as*

$$\overline{f} = \overline{g} \cdot \overline{h},$$

where $\overline{g}, \overline{h} \in k[z]$ are relatively prime polynomials. Then there are polynomials $g, h \in \mathcal{O}_K[z]$ satisfying

$$\deg(g) = \deg(\overline{g}), \quad \deg(h) = \deg\left(\overline{h}\right), \quad f = gh,$$

and reducing the coefficients of g and h modulo $\mathcal{M}_v$ gives $\overline{g}$ and $\overline{h}$, respectively.

Proof. See [**30**, Section 3.3.4]. □

Theorem 14.6 (Hensel's lemma, version 2). *Let $f(z) \in \mathcal{O}_K[z]$ be a polynomial, and let $a \in \mathcal{O}_K$ be a point such that $|f(a)| < |f'(a)|^2$. Then f has a unique zero in $\overline{D}(a, |f(a)|/|f'(a)|)$, and this zero is K-rational.*

Sketch of Proof. Set $a_0 = a$, and inductively produce a sequence of points $a_0, a_1, a_2, \ldots \in K$ using Newton's method. Prove that the sequence is Cauchy (using the $|f(a)| < |f'(a)|^2$ hypothesis), and observe that the resulting limit must be a root of f. Then prove uniqueness, again using $|f(a)| < |f'(a)|^2$; see Exercise 14.8. For a more detailed proof over $\mathbb{Q}_p$, see [**122**, Section II.1.5] or [**70**, Theorem 3.4.1]. □

Both versions of Hensel's lemma can be generalized to power series. They can also be used in combination with the Newton polygon to give more precise descriptions of the zeros of a power series. That is, after the Newton polygon detects the absolute value of zeros, Hensel's lemma can help one decide whether the $j-i$ zeros of a given absolute value lie in a degree $j-i$ extension of K, or whether they actually lie in lower-degree extensions.

14.4. Proofs related to mapping properties of disks

Proof of Proposition 3.17. Consider the case that the disk $D = D(a,r)$ is open; the rational closed case is similar. Write

$$f = \sum_{n\geq 0} c_n(z-a)^n.$$

The hypothesis that $\mathrm{wdeg}_D(f) < \infty$ implies that

$$t := \max_{n\geq 0} \{|c_n|r^n\}$$

is attained, at some minimal integer $n = d := \mathrm{wdeg}_D(f)$.

If $d = 0$, then $\|f\|_D = |c_0|$, and in addition,

$$\left|c_n(x-a)^n\right| < |c_n|r^n \leq |c_0|$$

for all $n \geq 1$ and all $x \in D$. Thus, $|f(x)| = |c_0|$ for all $x \in D$, and hence

$$\|f\|_D = |c_0| = \sup\{|f(x)| : x \in D\}.$$

If $d \geq 1$, then we also have $d = \mathrm{wdeg}_D(f - c_0)$. Therefore, by Theorem 3.15, we have

$$f\big(D(a,r)\big) = D(c_0, t) = D(0,t).$$

Thus,

$$\sup\{|f(x)| : x \in D\} = \sup\{|y| : y \in D(0,t)\} = t = \|f\|_D. \qquad \square$$

Remark 14.7 (The valuation polygon)**.** If we fix $a \in \mathbb{C}_v$ and $f \in K[[z-a]]$, then in light of Corollary 3.18, it is natural to ask how fast the image disk $f(\overline{D}(a,s))$ grows as s grows—or perhaps, to simplify things a bit, how fast $\|f\|_{\zeta(a,s)}$ grows. (That is, consider the norm of f, rather than the norm of $f - f(a)$.) It turns out that the best way to understand this growth is by taking the logarithm of both radii.

More precisely, if $f \in \mathcal{A}(a,r)$, define $N_{f,a} : (-\infty, \log r) \to \mathbb{R}$ by

$$N_{f,a}(\log s) := \log \|f\|_{\zeta(a,s)}.$$

By the definition of the norm, we have

$$N_{f,a}(\log s) = \max_{n\geq 0}\{|c_n| + n\log s\},$$

from which it is clear that $N_{f,a}$, as the maximum of various linear functions, is convex (i.e., concave up), continuous, and piecewise linear. In addition, our formula for $N_{f,a}$ tells us that at a particular radius $s = s_0$, the image radius $\|f\|_{\zeta(a,s)}$ is growing like a degree-d monomial (specifically, $|c_d|s^d$), where d is the number of zeros of f in $\overline{D}(a,s)$.

The graph of the function $N_{f,a}$ is called the *valuation polygon* or the *Newton copolygon* of f at a. As the latter name suggests, it is in some sense dual to the Newton polygon. Indeed, each vertex of the copolygon corresponds to a pivotal radius, and hence to a line segment of the polygon. Moreover, each segment of the copolygon corresponds to an interval of radii $[s_i, s_j)$ on which the number of zeros is constant, equal to some integer d. Hence, the copolygon segment, which has slope equal to d, corresponds to the vertex with x-coordinate d on the polygon. For more on valuation polygons and Newton polygons, see [**2**, Section 4.3], [**121**, Section 3], or for a broader and more accessible exposition, see [**122**, Section VI.1.6].

Proof of Proposition 3.22. Consider the case that the disk $D = D(a, r)$ is open; the rational closed case is similar. Write

$$f(z) = \sum_{n\geq 0} c_n(z-a)^n \quad \text{with} \quad |c_n|r^n \leq |c_1|r \quad \text{for all } n \geq 1.$$

Then $f(D) = D(c_0, s)$, where $s := |c_1|r$, by Theorem 3.15. Note that $c_1 \neq 0$, because the Weierstrass degree is 1 and not 0.

We first find a formal power series that is the inverse to f under composition of formal power series. Write an arbitrary formal power series $g \in K[[z - c_0]]$ with constant term a as

$$g(z) = a + \sum_{n\geq 1} b_n(z-c_0)^n.$$

Then $g \circ f \in K[[z - a]]$ is of the form

$$g \circ f(z) = a + b_1c_1(z-a) + \sum_{n\geq 2}\left(b_n c_1^n + \sum_{i=1}^{n-1} b_i F_{n,i}(c_1, \ldots, c_{n-1})\right)(z-a)^n,$$

where each $F_{n,i}$ is a polynomial, and each monomial of $F_{n,i}$ is an integer multiple of $\prod_{j=1}^{n-1} c_j^{e_j}$, for some integers e_j satisfying

$$\sum_{j=1}^{n-1} je_j = n \quad \text{and} \quad \sum_{j=1}^{n-1} e_j = i.$$

In particular, since $|c_j| \leq |c_1|r^{1-j}$, it is easy to compute that

$$\big|F_{n,i}(c_1, \ldots, c_{n-1})\big| \leq |c_1|^i r^{i-n} = s^i r^{-n}. \tag{14.6}$$

Thus, we may set $b_1 := c_1^{-1}$, and for each $n \geq 2$,

$$b_n := -c_1^{-n} \sum_{i=1}^{n-1} b_i F_{n,i}(c_1, \ldots, c_{n-1}) \in K,$$

yielding $g \circ f(z) = z$ as formal power series.

We claim that $|b_n|s^n \le r$ for all $n \ge 1$, proceeding by induction on n. The claim is obvious for $n = 1$. For $n \ge 2$, by inequality (14.6) and the inductive assumption of the claim for all $1 \le i < n$, we have

$$\left| \sum_{i=1}^{n-1} b_i F_{n,i}(c_1, \dots, c_{n-1}) \right| \le \max_{1 \le i \le n-1} \left\{ |b_i| s^i r^{-n} \right\} \le r^{1-n}.$$

Therefore, $|b_n|s^n \le |c_1|^{-n} r^{1-n} s^n = r$, proving the claim.

By the claim, together with Theorem 3.15, g is a convergent power series on $D(c_0, s)$ of Weierstrass degree 1, with $g(D(c_0, s)) = D(a, r)$. As previously noted, we have $g \circ f(z) = z$ on $D(a, r)$.

It remains to show that $f \circ g(z) = z$ on $D(c_0, s)$. Given any $w \in D(c_0, s)$, we may write $w = f(z)$ for some $z \in D(a, r)$, and we compute

$$f \circ g(w) = f\Big(g\big(f(z)\big)\Big) = f(z) = w. \qquad \square$$

Before proving Theorem 3.23, we need to state and prove the following lemma, suggested by Remark 14.7.

Lemma 14.8. *Let $D, E \subseteq \mathbb{C}_v$ be disks, let $a \in D$, let $f \in \mathcal{A}_D$ be a power series converging on D, and suppose that $f(a) \in E \subseteq f(D)$. Then there is a unique disk $U \subseteq D$ with $a \in U$ such that $f(U) = E$.*

Proof. Write $f(z) = \sum_{n \ge 0} c_n (z-a)^n$. Note that f is nonconstant because $E \subseteq f(D)$; in particular, we have $c_n \ne 0$ for some $n \ge 1$. Write D as either $D = D(a, r)$ or $D = \overline{D}(a, r)$. Similarly, write E as either $\overline{D}(f(a), R)$ or $D(f(a), R)$.

Define $T : [0, r] \to \mathbb{R} \cup [0, \infty]$ by $T(s) := \max_{n \ge 1}\{|c_n| s^n\}$, which is continuous and, because $c_n \ne 0$ for some $n \ge 1$, strictly increasing. Clearly, $T(0) = 0$, and by hypothesis, $T(r) \ge R$. Thus, by the Intermediate Value Theorem, there is some radius $s_0 \in (0, r]$ such that $T(s_0) = R$. Moreover, because T is strictly increasing, s_0 is unique.

If D is open and $s_0 = r$, then $\phi(D) = D(f(a), R)$, by Theorem 3.15. In that case, the assumption that $E \subseteq f(D)$ forces $E = D(f(a), R)$, and we may set $U := D$.

Otherwise, $\overline{D}(f(a), R) \subseteq f(D)$, and again by Theorem 3.15, we have

$$f\big(D(a, s_0)\big) = D\big(f(a), R\big) \quad \text{and} \quad f\Big(\overline{D}(a, s_0)\Big) = \overline{D}\big(f(a), R\big).$$

Thus, we have proven the existence of the desired disk U.

Finally, if V is another such disk, then it must be either $D(a, s_0)$ or $\overline{D}(a, s_0)$, by Theorem 3.15 and the uniqueness of s_0. Without loss we then have

$$U = D(a, s_0) \quad \text{and} \quad V = \overline{D}(a, s_0).$$

By the same theorem again, it follows that

$$D(f(a), R) = f(U) = f(V) = \overline{D}(f(a), R),$$

which implies $R \notin |\mathbb{C}_v^\times|$. Yet again by the same theorem, $s_0 \notin |\mathbb{C}_v^\times|$ as well, and hence

$$U = D(a, s_0) = \overline{D}(a, s_0) = V,$$

proving uniqueness. □

Proof of Theorem 3.23. Every point $x \in \mathbb{C}_v$ has exactly d preimages in $\mathbb{C}_v$, counting multiplicity. We construct the desired disks $D_1, \ldots, D_m$ inductively. We start with $j = 0$ disks.

Given $j \geq 1$, suppose we already have disjoint disks $D_1, \ldots, D_{j-1}$, and positive integers $d_1, \ldots, d_{j-1} \geq 1$, such that each D_i maps d_i-to-1 onto D, and such that

$$\phi^{-1}(D) \supseteq D_1 \cup \cdots \cup D_{j-1}. \tag{14.7}$$

If the inclusion (14.7) is equality, then we must have $d_1 + \cdots + d_{j-1} = d$ and $1 \leq j - 1 \leq d$, and therefore we are done by setting $m = j - 1$.

Otherwise, choose

$$a \in \phi^{-1}(D) \smallsetminus (D_1 \cup \cdots \cup D_{j-1}),$$

and view $\phi(z)$ as a power series on some disk large enough to contain $\phi^{-1}(D)$. (Such a disk exists because $\phi^{-1}(D)$ is bounded. Indeed, writing $\phi(z) = c_d z^d + \cdots + c_0$, we have $|\phi(z)| = |c_d z^d|$ for $|z|$ sufficiently large.) By Lemma 14.8, there is a unique disk D_j containing a with image $\phi(D_j) = D$, and by Theorem 3.15, there is an integer $d_j \geq 1$ such that

$$\phi : D_j \to D \quad \text{is} \quad d_j\text{-to-1}.$$

Finally, D_j cannot intersect any of the other disks. Indeed, if $D_j \cap D_i \neq \varnothing$ for some $i < j$, then one disk would contain the other. Therefore, choosing any point b in the smaller disk, the uniqueness statement of Lemma 14.8 would imply that $D_j = D_i$, contradicting our original choice of $a \notin D_i$. □

Proof of Proposition 3.25. If $\phi(D) = \mathbb{P}^1(\mathbb{C}_v)$, we are done. Otherwise, choose a point $a \in \mathbb{P}^1(\mathbb{C}_v) \smallsetminus \phi(D)$, and choose a linear fractional transformation $\eta \in \mathrm{PGL}(2, \mathbb{C}_v)$ such that $\eta(a) = \infty$. Replacing ϕ by $\eta \circ \phi$, we may assume that ϕ has no poles in D. Similarly, replacing ϕ by $\phi \circ \theta$ for some appropriately chosen $\theta \in \mathrm{PGL}(2, \mathbb{C}_v)$, we may also assume that D does not contain the point ∞.

Write $\phi = f/g$, and view both polynomials f and g as convergent power series on D. Because g has no zeros in D, Theorem 3.5 and Proposition 3.11 tell us that ϕ can be written as a convergent power series on D. Moreover, by Theorem 3.15(b), the Weierstrass degree $\mathrm{wdeg}_D(\phi)$ is finite, because every point of $\phi(D)$ has at most $\deg(\phi) < \infty$ preimages in all of $\mathbb{P}^1(\mathbb{C}_v)$, let alone in D. Thus, by Theorem 3.15, $\phi(D)$ is a disk of the same type as D, and $\phi : D \to \phi(D)$ is everywhere d-to-1, for some integer $d \geq 1$. □

Proof of Proposition 3.29 (Sketch). This result requires the deeper theory of rigid analysis (see Section 14.7), and thus we refer most of it to other sources. (See also [**8**, Corollary 9.11], for an alternate proof that the image $\phi(U)$ of an *open* connected affinoid under $\phi \in \mathbb{C}_v(z)$ is again an open connected affinoid.)

In the case that U is a rational closed connected affinoid, the fact that $\phi(U)$ is also a rational closed connected affinoid is the content of [**41**, Lemma A5.6]. The fact that $\phi^{-1}(U)$ is a finite union of rational closed connected affinoids is a little easier but still requires the machinery of rigid analysis; see [**30**, Proposition 7.2.4]. However, the fact that each V_i maps d_i-to-1 onto U requires the discussion in [**41**] following Lemma A5.6; see [**12**, Lemma 2.5.4], for example.

A more general connected affinoid U can be written as an infinite increasing union of rational closed connected affinoids. Thus, $\phi(U)$ is also such an infinite increasing union of connected affinoids, and $\phi^{-1}(U)$ is a finite union of infinite increasing unions of connected affinoids. We leave it to the reader to show that these unions are in fact still connected affinoids, and of the same type (if any) as U; see Exercise 14.9. □

14.5. Weierstrass preparation theorem for open annuli

To prove the results of Sections 3.6 and 3.7 on Laurent series on annuli, we first need the following statement.

Proposition 14.9. *Let $a \in \mathbb{C}_v$, let $0 < r < R$, and let U be the open annulus $U = \{r < |z - a| < R\}$. Let $f, g \in \mathcal{A}_U$, and write*

$$f(z) = \sum_{n \in \mathbb{Z}} b_n (z-a)^n \quad \text{and} \quad g(z) = \sum_{n \in \mathbb{Z}} c_n (z-a)^n.$$

If $f(x) = g(x)$ for all $x \in U$, then $b_n = c_n$ for all $n \in \mathbb{Z}$.

Proof. This is left to the reader; see Exercise 14.11. □

Many of our results on the mapping properties of annuli rely on the following analogue of Theorem 14.2 (the Weierstrass preparation theorem for disks). Although it can be generalized to arbitrary annuli or even connected affinoids, we only need it for open annuli.

Theorem 14.10 (Weierstrass preparation theorem for open annuli). *Let $a \in \mathbb{C}_v$, let $0 < r < R$ be real numbers, and let U be the open annulus*

$$U := \{r < |z - a| < R\} = D(a, R) \smallsetminus \overline{D}(a, r).$$

Let $f(z) \in \mathcal{A}_U$ be a nonzero Laurent series converging on U of finite inner Weierstrass degree $M \in \mathbb{Z}$ and finite outer Weierstrass degree $N \in \mathbb{Z}$. Then there are convergent power series $g \in \mathcal{A}(a, R)$ and $h \in \mathcal{A}(0, 1/r)$ and a monic polynomial $P \in \mathbb{C}_v[z]$ such that

(a) $f(z) = (z - a)^M P(z)g(z)h(1/(z - a))$,

(b) $h(0) = 1$,

(c) *all zeros of P lie in U,*

(d) $g(x) \neq 0$ *for all* $x \in D(a, R)$, *and*

(e) $h(y) \neq 0$ *for all* $y \in D(0, 1/r)$.

Moreover, this factorization of f is unique, and $\deg(P) = N - M$. In particular, f has exactly $N - M$ zeros in U, counting multiplicity.

Our proof requires the following lemma.

Lemma 14.11. *Let $0 < r < 1 < R$ be real numbers, and let U be the open annulus $U := \{r < |z| < R\}$. Let $f(z) = \sum_{n \in \mathbb{Z}} c_n z^n \in \mathcal{A}_U$ be a convergent Laurent series on U of inner Weierstrass degree 0 and finite outer Weierstrass degree $N \geq 0$. Suppose that $|c_n| < |c_0|$ for all $n \in \mathbb{Z} \smallsetminus \{0\}$. Then there are convergent power series $G \in \mathcal{A}(0, R)$ and $H \in \mathcal{A}(0, 1/r)$ such that*

(a) $f(z) = G(z)H(1/z)$,

(b) $H(0) = 1$,

(c) *G has Weierstrass degree N on $D(0, R)$,*

(d) *G has Weierstrass degree 0 on $\overline{D}(0, r)$, and*

(e) *H has Weierstrass degree 0 on $D(0, 1/r)$.*

Proof. Because f has inner Weierstrass degree 0, we have $c_0 \neq 0$. Dividing f by c_0, we may assume that $c_0 = 1$. By our hypotheses on the inner and outer Weierstrass degrees of f, we have

- $|c_n| \leq r^{-n}$ for all $n \leq 0$,
- $|c_n| < \min\{r^{-n}, |c_N| R^{N-n}\}$ for all $1 \leq n \leq N - 1$,
- $|c_n| \leq |c_N| R^{N-n}$ for all $n \geq N$, and
- if $N \geq 1$, then $|c_N| < 1$ but $|c_N| R^N > 1$.

We construct G and H by inductively choosing certain sequences

$$\{g_j\}_{j \geq 0} \subseteq \mathcal{A}_U \quad \text{and} \quad \{h_j\}_{j \geq 0} \subseteq \mathcal{A}(0, 1/r)^\times$$

of the forms

$$g_j(z) = \sum_{n\in\mathbb{Z}} c_{j,n} z^n \quad \text{and} \quad h_j(z) = 1 + c_{j,0}^{-1} \sum_{n\geq 1} c_{j,-n} z^n$$

and which have coefficients satisfying

$$\begin{aligned}
&|c_{j,n}| \leq r^{j-n} < 1 && \text{for } n \leq -1,\\
&|c_{j,0}| = 1,\\
&|c_{j,n}| < \min\{1, |c_N| R^{N-n}\} && \text{for } 1 \leq n \leq N-1,\\
&|c_{j,N}| = |c_N|,\\
&|c_{j,n}| \leq |c_N| R^{N-n} < 1 && \text{for } n \geq N+1.
\end{aligned} \tag{14.8}$$

Note that $c_{j,0}h_j(1/z)$ is simply the truncation of $g_j(z)$ consisting of the terms of nonpositive power. In addition, conditions (14.8) say that

- $g_j \in \mathcal{A}_U$,
- $h_j \in \mathcal{A}(0, 1/r)^\times$,
- g_j has inner Weierstrass degree 0 on both the annuli

$$\{r < |z| < R\} \quad \text{and} \quad \{1 < |z| < R\},$$

- g_j has outer Weierstrass degree N on the same two annuli, and
- for each $n \leq -1$, we have $\lim_{j\to\infty} c_{j,n} = 0$.

We begin our construction by setting $g_0 := f$ and letting $h_0(1/z)$ be the corresponding truncation. The coefficients of g_0 and h_0 satisfy conditions (14.8) by hypothesis, and hence we have $g_0 \in \mathcal{A}_U$ and $h_0 \in \mathcal{A}(0, 1/r)$. In fact, h_0 has Weierstrass degree 0 and therefore no roots in $D(0, 1/r)$. Thus, by Proposition 3.11, $h_0 \in \mathcal{A}(0, 1/r)^\times$.

Proceeding inductively, for any $j \geq 0$, suppose we have already constructed $g_j \in \mathcal{A}_U$ and $h_j \in \mathcal{A}(0, 1/r)^\times$. Let

$$q_j(z) := \frac{1}{h_j(z)} = 1 + \sum_{n\geq 1} b_{j,n} z^n \in \mathcal{A}(0, 1/r)^\times$$

denote the multiplicative inverse of h_j. The reader may check, in Exercise 14.12, that the coefficients of q_j satisfy

$$\left|b_{j,n}\right| \leq r^{j+n} < 1 \text{ for all } n \geq 1. \tag{14.9}$$

Define

$$g_{j+1}(z) := g_j(z) \cdot q_j(1/z)$$

so that

$$g_{j+1}(z) = c_{j,0} + \big(g_j(z) - c_{j,0} \cdot h_j(1/z)\big) \cdot q_j(1/z) = \sum_{n\in\mathbb{Z}} c_{j+1,n} z^n,$$

where

$$c_{j+1,n} := \begin{cases} \sum_{m\geq 1-n} b_{j,m}c_{j,m+n} & \text{for } n \leq -1, \\ \\ c_{j,n} + \sum_{m\geq 1} b_{j,m}c_{j,m+n} & \text{for } n \geq 0. \end{cases}$$

It is straightforward to check, using the bounds of (14.8) and (14.9), that the coefficients $c_{j+1,n}$ satisfy inequality (14.8) with $j+1$ in the role of j. In particular, $g_{j+1} \in \mathcal{A}_U$. Set h_{j+1} to be the truncation of $c_{j+1,0}^{-1} g_{j+1}(1/z)$ and observe, as we did with h_0, that h_{j+1} has Weierstrass degree 0 on $D(0,1/r)$. Hence, $h_{j+1} \in \mathcal{A}(0,1/r)^\times$, by Proposition 3.11, completing our inductive construction.

Define

$$H_n := \prod_{j=0}^{n} h_j \in \mathcal{A}(0,1/r)^\times,$$

which has multiplicative inverse $Q_n := \prod_{j=0}^{n} q_j$. We leave it to the reader to show that the infinite products

$$H := \prod_{j\geq 0} h_j = \lim_{n\to\infty} H_n \quad \text{and} \quad Q := \prod_{j\geq 0} q_j = \lim_{n\to\infty} Q_n \tag{14.10}$$

converge to elements of $\mathcal{A}(0,1/r)$ and are multiplicative inverses; see Exercise 14.13. In particular, the Weierstrass degree of H on $D(0,1/r)$ is 0. Moreover, $H(0) = 1$, because $h_j(0) = 1$ for all $j \geq 0$.

Next, define

$$G(z) := Q(1/z)f(z) = \lim_{j\to\infty} g_0(z)\prod_{j=0}^{n} q_j(1/z) = \lim_{j\to\infty} g_j(z).$$

By the bounds of (14.8), G is a power series converging on $D(0,R)$. Moreover, it has Weierstrass degree N on $D(0,R)$ but Weierstrass degree 0 on $\overline{D}(0,r)$. Finally, we have $f(z) = G(z)H(1/z)$ by construction. □

Proof of Theorem 14.10. We may assume that $a = 0$; and replacing $f(z)$ by $z^{-M}f(z)$, we may also assume that $M = 0$.

Uniqueness. Suppose

$$P_1(z)g_1(z)h_1(1/z) = P_2(z)g_2(z)h_2(1/z)$$

for some monic polynomials P_i and series g_i, h_i as specified in the statement of the theorem. Write $P_1 = P\tilde{P}_1$ and $P_2 = P\tilde{P}_2$, where $P := \gcd(P_1, P_2)$. Then

$$\tilde{P}_1(z)g_1(z)h_1(1/z) = \tilde{P}_2(z)g_2(z)h_2(1/z), \tag{14.11}$$

but $\tilde{P}_1$ and $\tilde{P}_2$ have no common zeros. If $x \in \mathbb{C}_v$ is a zero of $\tilde{P}_1$, then $x \in U$ by hypothesis (c). Hence $g_2(x), h_2(1/x) \neq 0$ by hypothesis (d), and

$\tilde{P}_2(x) \neq 0$ by assumption, contradicting equation (14.11). Thus, $\tilde{P}_1$ has no zeros; similarly, neither does $\tilde{P}_2$. Being monic polynomials, then, they must be $\tilde{P}_1 = \tilde{P}_2 = 1$, and therefore $P_1 = P_2$.

We now have $g_1(z)h_1(1/z) = g_2(z)h_2(1/z)$ as an equality of functions on U. By hypotheses (d) and (e), by Theorem 3.5, and by Proposition 3.11, we may define

$$\tilde{g}(z) := \frac{g_2(z)}{g_1(z)} \in \mathcal{A}(0, R) \quad \text{and} \quad \tilde{h}(z) := \frac{h_1(z)}{h_2(z)} \in \mathcal{A}(0, 1/r).$$

Thus, we have $\tilde{g}(z) = \tilde{h}(1/z)$, as an equality of convergent Laurent series in the ring $\mathcal{A}_U$. However, there are no negative-power terms on the left-hand side of this equation, and no positive-power terms on the right. Therefore, by Proposition 14.9, $\tilde{g}$ and $\tilde{h}$ are both some constant c. In addition,

$$\tilde{h}(0) = h_1(0)/h_2(0) = 1/1 = 1$$

by hypothesis (b), and hence $c = 1$. Thus, $h_1 = h_2$ and $g_1 = g_2$.

Existence. Write $f(z) = \sum_{n\in\mathbb{Z}} c_n z^n$. Because $|\mathbb{C}_v^\times|$ is dense in $(0, \infty)$, and because $|c_n|r^n < |c_0|$ for all $n \geq 1$, with

$$|c_n|R^n \leq |c_N|R^N \quad \text{for all } n \geq N,$$

there is some radius $s \in (r, R) \cap |\mathbb{C}_v^\times|$ such that

$$|c_n|s^n < |c_0| \quad \text{for all} \quad n \geq 1.$$

Moreover, since $|c_n|r^n \leq |c_0|$ for all $n \leq -1$, we in fact have $|c_n|s^n < |c_0|$ for all $n \in \mathbb{Z} \setminus \{0\}$.

Choose $\alpha \in \mathbb{C}_v^\times$ with $|\alpha| = s$. Then $f(\alpha z)$ is a convergent Laurent series on the open annulus $\{r/s < |z| < R/s\}$ satisfying the conditions of Lemma 14.11. Thus, there exist $G \in \mathcal{A}(0, R/s)$ and $H \in \mathcal{A}(0, s/r)$ satisfying the conclusions of that lemma for the disks $D(0, R/s)$ and $D(0, s/r)$.

Define

$$h(z) := H(\alpha z) \in \mathcal{A}(0, 1/r) \quad \text{and} \quad \tilde{G}(z) := G(z/\alpha) \in \mathcal{A}(0, R).$$

Then $f(z) = \tilde{G}(z)h(1/z)$ and $h(0) = H(0) = 1$. Moreover, h has Weierstrass degree 0 on $D(0, 1/r)$, while $\tilde{G}$ has Weierstrass degree N on $D(0, R)$ but Weierstrass degree 0 on $\overline{D}(0, r)$. By Theorem 14.2, h has no zeros on $D(0, 1/r)$, whereas $\tilde{G}$ factors as $\tilde{G} = Pg$ for some monic polynomial $P \in \mathbb{C}_v[z]$ of degree N and some power series $g \in \mathcal{A}(0, R)$ with no zeros in $D(0, R)$. Finally, we note that none of the N zeros of P can lie in $\overline{D}(0, r)$; for if any did, then $\tilde{G}$ would have positive Weierstrass degree on $\overline{D}(0, r)$. □

14.6. Proofs about Laurent series on annuli

Having proven the Weierstrass preparation theorem for open annuli (Theorem 14.10), we can now prove the remaining results from Chapter 3.

Proof of Proposition 3.31. We leave it to the reader to prove that $\mathcal{A}_U$ forms a commutative ring with unity; see Exercise 14.14. It remains to show that the unit group $(\mathcal{A}_U)^\times$ is given by equation (3.3).

The forward inclusion of equation (3.3) is clear. Conversely, given $f \in \mathcal{A}_U$ with no zeros in U, we wish to show that f is a unit. By Exercise 14.15, f has finite inner and outer Weierstrass degrees. Factor

$$f(z) = (z-a)^M P(z) g(z) h(1/(z-a))$$

as in Theorem 14.10. Since f has no zeros in U, the monic polynomial P must be the constant 1. By Proposition 3.11, g has a multiplicative inverse $1/g \in \mathcal{A}(a, R)$, and h has a multiplicative inverse $1/h \in \mathcal{A}(0, 1/r)$. Thus, each of the Laurent series $(z-a)^{-M}$, $1/g(z)$, and $1/h(1/(z-a))$ belongs to $\mathcal{A}_U$. Since $\mathcal{A}_U$ is a commutative ring, their product is a multiplicative inverse of f in $\mathcal{A}_U$, as desired. □

Proof of Proposition 3.32.

(a) Write $\phi = f/g$ with $f, g \in \mathbb{C}_v[z]$ relatively prime. By hypothesis, g has no roots in U. Therefore, by Proposition 3.31, we have $1/g \in \mathcal{A}_U$. Since $\mathcal{A}_U$ is a ring, and $f \in \mathbb{C}_v[z] \subseteq \mathcal{A}_U$, it follows that $\phi = f \cdot (1/g)$ also belongs to $\mathcal{A}_U$.

(b) We first claim that both the inner and outer Weierstrass degrees of ϕ must be finite. If, say, the outer Weierstrass degree were ∞, then after decreasing the outer radius slightly, the outer Weierstrass degree would be a very large positive integer. Similarly, increasing the inner radius slightly would guarantee that the inner Weierstrass degree is finite but much smaller. By Theorem 14.10, U would contain far more than $\deg(\phi)$ zeros of ϕ; this contradiction proves the claim.

Finally, if U has inner and outer Weierstrass degrees $M \leq N \in \mathbb{Z}$, then Theorem 14.10 says that ϕ has $N - M$ zeros in U. Thus, if ϕ has no zeros in U, then $N = M$, as desired. □

Proof of Theorem 3.33. By considering $f(z+a) - c_0$ instead of $f(z)$, and on the annulus $U := \{r < |z| < R\}$, we may assume without loss that $a = c_0 = 0$. Note that the Weierstrass degrees M, N are both nonzero. We now examine the three cases.

(a) If $M < N$, then for any $x \in U$ and $n \in \mathbb{Z}$, we have

$$|c_n x^n| < \begin{cases} |c_n| r^n \le s & \text{if } n \le 0, \\ |c_n| R^n \le t & \text{if } n \ge 0. \end{cases}$$

Thus, $f(U) \subseteq D(0, \max\{s,t\})$.

Conversely, given $y \in D(0, \max\{s,t\})$, we have either $|y| < s$ or $|y| < t$ (or both). Therefore, although either the inner or outer Weierstrass degree of $f(z) - y$ may be zero, the other remains at M or N, respectively. Since $M, N \neq 0$, it follows that the inner and outer Weierstrass degrees of $f(z) - y$ on U are distinct. Therefore, by Theorem 14.10, $f(z) - y$ has at least one zero in U, as desired.

(b) If $M = N \ge 1$, we begin by noting that $|c_M x^M| > |c_n x^n|$ for all $x \in U$ and all $n \in \mathbb{Z} \smallsetminus \{M\}$, by definition of inner and outer Weierstrass degree. Thus, $|f(z)| = |c_M||x|^M \in (s,t)$ for all $x \in U$, proving that

$$f(U) \subseteq \{s < |z| < t\},$$

and also verifying the final statement of Theorem 3.33.

Conversely, given $y \in \{s < |z| < t\}$, the Laurent series $f(z) - y$ has outer Weierstrass degree M on U, since $|y| < t = |c_M| R^M$, but it has inner Weierstrass degree 0, since $|y| > s = |c_M| r^M$. Thus, by Theorem 14.10, $f(z) - y$ has $M - 0 = M \ge 1$ zeros in U, counting multiplicity.

(c) If $M = N \le -1$, then let $g(z) = f(1/z) \in \mathcal{A}_V$, where V is the annulus $V = \{1/R < |z| < 1/r\}$. By applying part (b) to g on V, we obtain the desired conclusion for f on U. □

14.7. An overview of rigid analysis (optional)

Connected affinoids are a key object of study in the theory of rigid analysis, which was developed by Tate [**135**] in the 1960s but not published until 1971. The impetus for the theory is the following problem. In complex analysis, it is appropriate to define an analytic function on a domain D to be a function that satisfies a certain local condition at every point of D. However, any analogous definition on a non-archimedean field $\mathbb{C}_v$ would be too general, because $\mathbb{C}_v$ is totally disconnected. For example, as we noted in Section 3.2, the function

$$f(z) = \begin{cases} 0 & \text{if } |z| < 1, \\ 1 & \text{if } |z| \ge 1 \end{cases}$$

is *locally* analytic. Thus, the set of locally analytic functions in $\mathbb{C}_v$ is too large to permit many of the useful consequences of complex analysis.

Tate's solution to this problem was to consider only functions given by (convergent) power series, also called *rigid* analytic functions, giving the subject its name. The "rigidity" here is the fact that the function does not merely satisfy a local condition at each point, but instead is already given by a power series known to converge on a large set. In addition, Tate stipulated that one should only attempt to do analytic continuation by gluing series expansions on *overlapping* domains. However, because two disks can overlap only if one contains the other, he needed to work with domains more general than disks, and hence he developed affinoids.

The rigorous definition of affinoid domain, and especially affinoid subdomains, is rather abstract, involving maximal ideal spectra, representable functors, and other technicalities. The idea, however, is that one starts with a ring of (convergent) power series in several variables and takes a quotient by an appropriate ideal, to obtain an *affinoid algebra*. Then the set of maximal ideals of this algebra, equipped an appropriate topology, is an *affinoid domain*. Intuitively, the affinoid domain is the set of points at which all the series in the ring converge; conversely, the affinoid algebra is some appropriate ring of power series converging on the domain.

For example, let $\mathcal{A} \subseteq \mathbb{C}_v[[x,y]]$ consist of those two-variable power series whose coefficients approach zero. The corresponding affinoid is the bidisk $\overline{D}(0,1) \times \overline{D}(0,1) \subseteq \mathbb{C}_v^2$. If we take the quotient of $\mathcal{A}$ by the principal ideal $I = \langle xy - a \rangle$, for some fixed $a \in \mathcal{O} \smallsetminus \{0\}$, then the affinoid corresponding to the quotient algebra $\mathcal{A}/I$ is the rational closed annulus

$$\overline{D}(0,1) \smallsetminus D(0,|a|) = \big\{|a| \leq |z| \leq 1\big\} \subseteq \mathbb{C}_v.$$

After all, we are effectively setting $y = a/x$, and we still have the stipulation that $|x|, |y| \leq 1$, which means that we are considering all $x \in \mathbb{C}_v$ such that $|a| \leq |x| \leq 1$.

Similarly, for any fixed $a, b \in \mathbb{C}_v$ with $0 < |a|, |b| \leq 1$, the ring

$$\mathbb{C}_v[[x,y,z]]/\big\langle xy - a, (x-1)z - b\big\rangle,$$

consists of all convergent series in x, a/x, and $b/(x-1)$. Its corresponding (rational closed) affinoid is

$$\overline{D}(0,1) \smallsetminus \big(D(0,|a|) \cup D(1,|b|)\big).$$

Thus, in rigid analysis, as in algebraic geometry, one considers not just the set of points in a given domain, but also a ring of functions attached to that set. An affinoid is now said to be *connected* if its associated affinoid algebra cannot be written as the direct product of two nontrivial affinoid algebras. (Compare this condition with the complex case: if $U, V \subseteq \mathbb{C}$ are disjoint open sets, then the ring of analytic functions on $U \cup V$ is the direct product of the ring for U and the ring for V.) It is a deep theorem that any connected affinoid domain of $\mathbb{P}^1(\mathbb{C}_v)$ is simply a rational closed $\mathbb{P}^1(\mathbb{C}_v)$-disk with finitely many open $\mathbb{P}^1(\mathbb{C}_v)$-disks removed, and an arbitrary affinoid domain in $\mathbb{P}^1$ is simply a finite union of such connected affinoids.

We should also mention, as described in the comments following Definition 3.27, that our definition of "connected affinoid" is more general than the rigid analysis definition in two ways. First, we consider $\mathbb{P}^1(\mathbb{C}_v)$ itself to be a connected affinoid, whereas this is not the case in rigid analysis, as the root "affine" of the word "affinoid" suggests. Second, the connected affinoids of rigid analysis are all rational closed connected affinoids, in our terminology. For example, an open disk would *not* be considered an affinoid domain in the rigid analytic sense.

Still, we opt for the broader definition because our aims are somewhat different from those of rigid analysis. The goal of rigid analysis is to do non-archimedean analysis on algebraic varieties, and especially to do analytic continuation. By contrast, our main interest is in overcoming the totally disconnected topology on $\mathbb{P}^1(\mathbb{C}_v)$ by working with regions that behave like connected sets. Hence, we allow a broader library of fundamental connected sets, even if the associated rings of power series are a little messier.

For a more detailed development of rigid analysis, the reader may consult Tate's published notes [**135**] or the exposition of the same material by Goss [**69**]. For a longer, more detailed, but still very readable exposition with more examples, especially in dimension 1, see the excellent text [**66**]. Various other useful results in the theory appear in Coleman's paper [**41**]. Finally, the serious rigid analyst must consult [**30**], the utterly thorough, very general, and highly intimidating bible on the subject.

Exercises for Chapter 14

14.1. Prove inequality (14.4) in the proof of Lemma 14.3.

14.2. Consider the power series Q and polynomial R defined near the end of the proof of Lemma 14.3, in equation (14.5). Prove that $F = Qf + R$, that Q converges on $\overline{D}(0, r)$, and that R is a polynomial of degree at most $d-1$ with $\|R\|_D \leq \|F\|_D$.

14.3. Prove that the functions Q, R in Lemma 14.3 are unique.

14.4. Let $D \subseteq \mathbb{C}_v$ be a disk, and let $f \in \mathcal{A}_D$. If $\operatorname{wdeg}_D(f) = 0$, prove that the function $g := 1/f$ is given by a power series in $\mathcal{A}_D$. If f is defined over a complete subfield K of $\mathbb{C}_v$, also prove that g is also defined over K.

(*Suggestion*: Use the familiar power series for $1/(1-z)$ to describe the formal power series g for $1/f$. Then show that g converges on D.)

14.5. Let D be an open disk, and let f be a power series converging on D *and of finite Weierstrass degree on* D. Prove the analogue of Theorem 14.2 for f on D, as described in Remark 14.4.

(*Suggestion*: Write D as an increasing union of rational closed disks, and apply Theorem 14.2 to f on each such disk.)

14.6. Let $D := D(a, r) \subseteq \mathbb{C}_v$ be an open disk. Prove that there are power series $f, g \in \mathbb{C}_v[[z - a]]$ converging on D, each of finite Weierstrass degree on D, for which $f + g$ does *not* have finite Weierstrass degree on D.

14.7. In this exercise, you will fill in some details from the proof of Theorem 3.13.

(a) Let $D \subseteq \mathbb{C}_v$ be an open disk, and write D as an increasing union of disks $D_1 \subseteq D_2 \subseteq \cdots$. Let $f \in \mathcal{A}_D$, and suppose that f has Weierstrass degree d_i on D_i. Prove that

$$d_i \nearrow d := \mathrm{wdeg}_D(f).$$

(b) Let $D = D(a, s) \subseteq \mathbb{C}_v$ be an open disk, and let $f \in \mathcal{A}_D$ with $\mathrm{wdeg}_D(f) < \infty$. Prove that there is some $r < s$ with $r \in |\mathbb{C}_v^\times|$ such that the disk $E := \overline{D}(a, s)$ satisfies $\mathrm{wdeg}_E(f) = \mathrm{wdeg}_D(f)$.

(c) Let $D \subseteq \mathbb{C}_v$ be an arbitrary disk, let $E \subsetneq D$ be a rational closed proper subdisk, and let $f \in \mathcal{A}_D$. Prove that $\mathrm{wdeg}_E(f) < \infty$.

14.8. Give a full proof of Theorem 14.6 (Hensel's lemma, version 2).

14.9. In this exercise, you will extend the proof of Proposition 3.29 to arbitrary connected affinoids.

Let U be a connected affinoid. Write U as a union of rational closed connected affinoids $U_1 \subseteq U_2 \subseteq \cdots$, and use the truth of Proposition 3.29 for each U_i to prove it for U.

14.10. Let $D \subseteq \mathbb{C}_v$ be a disk, let $a \in D$, and let $f, g \in \mathcal{A}_D$. Suppose that $f = g$ as functions on D. Prove that the coefficients of f and g, when expanded as power series centered at a, also coincide.

14.11. Prove Proposition 14.9. Do *not* assume the Weierstrass preparation theorem for open annuli (Theorem 14.10) or any other results from Sections 14.5 or 14.6.

(*Suggestion*: Reduce to the case that $g = 0$ and $U = \{r < |z| < R\}$, and write $f = \sum c_n z^n$. If f is not the trivial series, prove that there is some $s \in (r, R)$ and some $N \in \mathbb{Z}$ such that $|c_n| s^n < |c_N| s^N$ for all $n \neq N$.)

14.12. Prove the bounds $|b_{j,n}| \leq r^{j+n} < 1$ of inequality (14.9) from the proof of Lemma 14.11, using the facts that the coefficients of the power series h_j satisfy conditions (14.8).

14.13. Use the bounds of inequalities (14.8) and (14.9) from the proof of Lemma 14.11 to show that the infinite products of equation (14.10) converge to elements of $\mathcal{A}(0, 1/r)$ that are multiplicative inverses of one another.

14.14. Prove the first half of Proposition 3.31: if $U \subseteq \mathbb{C}_v$ is an open annulus, then $\mathcal{A}_U$ is a commutative ring with unity.

(*Suggestion*: Mimic the proof of Theorem 3.5.)

14.15. Let $U \subseteq \mathbb{C}_v$ be an open annulus, and let $f \in \mathcal{A}_U$. If either the inner or outer Weierstrass degree of f on U is infinite, prove that f has infinitely many zeros in U.

(*Suggestion*: Write U as an increasing union of strictly smaller open annuli $U_1 \subseteq U_2 \subseteq \cdots$.)

14.16. State and prove the analogue of Theorem 14.10 for rational closed annuli $U = \{r \leq |z - a| \leq R\}$, with $r, R \in |\mathbb{C}_v^\times|$.

This is a matter of making appropriate modifications to the (very long) proof of Theorem 14.10. It is essential to define inner and outer Weierstrass degrees appropriately; modify their desciption in Definition 3.30 according to the contrast between parts (a) and (b) of Definition 3.8.

14.17. Use the result of Exercise 14.16 to prove Proposition 3.32 in the case that $U = \{r \leq |z - a| \leq R\}$ is a rational closed annulus.

14.18. Use the result of Exercise 14.16 to state and prove the analogue of Theorem 3.33 in the case that $U = \{r \leq |z - a| \leq R\}$ is a rational closed annulus.

Chapter 15

Proofs of Berkovich Space Results

In this chapter, we supply the missing proofs of various results that we stated in Chapter 6 on the Berkovich projective line. As in Chapter 14, we also state and prove a number of supplementary results.

15.1. Basic results on seminorms

We begin with four fundamental lemmas on seminorms.

Lemma 15.1. *Let A be a ring with unity, and let $\|\cdot\|_\zeta$ and $\|\cdot\|_\xi$ be multiplicative seminorms on A.*

(a) *$\|\cdot\|_\zeta$ is bounded with respect to $\|\cdot\|_\xi$ if and only if $\|f\|_\zeta \leq \|f\|_\xi$ for all $f \in A$.*

(b) *If $\|\cdot\|_\zeta$ is bounded with respect to $\|\cdot\|_\xi$, then $\|u\|_\zeta = \|u\|_\xi$ for all units $u \in A^\times$.*

(c) *If $\|\cdot\|_\zeta$ is bounded with respect to $\|\cdot\|_\xi$ and if $\|\cdot\|_\xi$ is non-archimedean, then $\|\cdot\|_\zeta$ is also non-archimedean.*

Proof. We leave parts (a) and (b) to the reader; see Exercise 15.1. For part (c), we present the following proof from [**8**, Lemma 1.1].

Given $f, g \in A$, we wish to show that $\|f+g\|_\zeta \leq \max\{\|f\|_\zeta, \|g\|_\zeta\}$. For any integer $n \geq 1$, observe that

$$\|f+g\|_\zeta^n = \|(f+g)^n\|_\zeta = \left\| \sum_{i=0}^{n} \binom{n}{i} f^i g^{n-i} \right\|_\zeta .$$

Therefore, by the (standard) triangle equality, we have

$$\|f+g\|_\zeta^n \leq \sum_{i=0}^{n} \left\| \binom{n}{i} \right\|_\zeta \|f\|_\zeta^i \|g\|_\zeta^{n-i} \leq \left[\sum_{i=0}^{n} \left\| \binom{n}{i} \right\|_\zeta \right] \max\left\{ \|f\|_\zeta, \|g\|_\zeta \right\}^n .$$

Since $\|\cdot\|_\xi$ is non-archimedean and bounds $\|\cdot\|_\zeta$, we have $\|m\|_\zeta \leq \|m\|_\xi \leq 1$ for any $m \in \mathbb{Z}$, by part (a). Thus, taking nth roots in the inequality above, we have

$$\|f+g\|_\zeta \leq (n+1)^{1/n} \max\left\{ \|f\|_\zeta, \|g\|_\zeta \right\} \tag{15.1}$$

for every $n \geq 1$. Taking the limit as $n \to \infty$ yields the desired inequality $\|f+g\|_\zeta \leq \max\{\|f\|_\zeta, \|g\|_\zeta\}$. □

Lemma 15.2. *Let $\zeta, \xi \in \mathbb{A}^1_{\mathrm{an}}$. Then the following hold.*

(a) *$\|\cdot\|_\zeta$ is non-archimedean.*

(b) *For any $a \in \mathbb{C}_v$, there is some $r \in |\mathbb{C}_v^\times|$ such that $\|\cdot\|_\zeta$ is bounded by $\|\cdot\|_{\zeta(a,r)}$.*

(c) *$\|\cdot\|_\zeta$ is bounded by $\|\cdot\|_\xi$ if and only if $\|z-a\|_\zeta \leq \|z-a\|_\xi$ for all $a \in \mathbb{C}_v$.*

(d) *$\zeta = \xi$ if and only if $\|z-a\|_\zeta = \|z-a\|_\xi$ for all $a \in \mathbb{C}_v$.*

Proof.

(a) Given $f, g \in \mathbb{C}_v[z]$ and $n \geq 1$, the derivation of inequality (15.1) in the proof of Lemma 15.1 is still valid, because $\|m\|_\zeta = |m| \leq 1$ for every $m \in \mathbb{Z}$. Taking the limit as $n \to \infty$, we are done.

(b) Let $s = \|z-a\|_\zeta$, and pick $r \geq s$ with $r \in |\mathbb{C}_v^\times|$. For any $f \in \mathbb{C}_v[z]$, write

$$f = c_0 + c_1(z-a) + \cdots + c_n(z-a)^n .$$

Then because $\|\cdot\|_\zeta$ is non-archimedean, by part (a) we have

$$\|f\|_\zeta \leq \max_{0 \leq i \leq n} \left\{ |c_i| \cdot \|z-a\|_\zeta^i \right\} \leq \max_{0 \leq i \leq n} \left\{ |c_i| r^i \right\} = \|f\|_{\zeta(a,r)} .$$

(c) The forward implication is immediate from Lemma 15.1(a). Conversely, given $f \in \mathbb{C}_v[z]$, write $f = c\prod_{i=1}^n (z-a_i)$. Then

$$\|f\|_\zeta = |c| \prod_{i=1}^{n} \|z-a_i\|_\zeta \leq |c| \prod_{i=1}^{n} \|z-a_i\|_\xi = \|f\|_\xi .$$

(d) This is immediate from part (c) and Lemma 15.1(a). □

Lemma 15.3. *Let $a \in \mathbb{C}_v$, let $r > 0$, and let $\zeta, \xi \in \overline{D}_{\mathrm{an}}(a, r)$.*

(a) $\|\cdot\|_\zeta$ *is bounded by* $\|\cdot\|_\xi$ *if and only if* $\|z - b\|_\zeta \le \|z - b\|_\xi$ *for all* $b \in \overline{D}(a, r)$.

(b) $\zeta = \xi$ *if and only if* $\|z - b\|_\zeta = \|z - b\|_\xi$ *for all* $b \in \overline{D}(a, r)$.

Proof.

(a) The forward implication is immediate from Lemma 15.1(a). For the converse, given any $b \in \mathbb{C}_v \smallsetminus \overline{D}(a, r)$, we have

$$\|z - a\|_\zeta \le \|z - a\|_{\zeta(a,r)} = r < |a - b|,$$

and hence by the non-archimedean property of Lemma 15.2(a),

$$\|z - b\|_\zeta = \max\left\{\|z - a\|_\zeta, |a - b|\right\} = |a - b|.$$

By the same reasoning applied to ξ, we have

$$\|z - b\|_\zeta = |a - b| = \|z - b\|_\xi \quad \text{for all} \quad b \in \mathbb{C}_v \smallsetminus \overline{D}(a, r).$$

Combined with the hypothesis, we have $\|z - b\|_\zeta \le \|z - b\|_\xi$ for all $b \in \mathbb{C}_v$. Therefore, by Lemma 15.2(c), ζ is bounded by ξ.

(b) Immediate from part (a). □

Lemma 15.4. *Let $D_{\mathrm{an}}, E_{\mathrm{an}} \subseteq \mathbb{A}^1_{\mathrm{an}}$ be Berkovich disks. If*

$$D_{\mathrm{an}} \cap E_{\mathrm{an}} \neq \varnothing,$$

then either $D_{\mathrm{an}} \subseteq E_{\mathrm{an}}$, or $D_{\mathrm{an}} \supseteq E_{\mathrm{an}}$.

Proof. Write $D_{\mathrm{an}} = \overline{D}_{\mathrm{an}}(a, r)$ or $D_{\mathrm{an}} = D_{\mathrm{an}}(a, r)$, and similarly write $E_{\mathrm{an}} = \overline{D}_{\mathrm{an}}(b, s)$ or $E_{\mathrm{an}} = D_{\mathrm{an}}(b, s)$. Without loss, assume that $r \ge s$; if $r = s$ and at least one of the disks is closed, assume that D_{an} is closed.

Given $\zeta \in D_{\mathrm{an}} \cap E_{\mathrm{an}}$, Lemma 15.2(a) implies that

$$|a - b| \le \max\left\{\|z - a\|_\zeta, \|z - b\|_\zeta\right\} \le \max\{r, s\} = r,$$

where the second inequality is strict if D_{an} is open, by our assumptions. Thus, $b \in D_{\mathrm{an}}$, and we may therefore write D_{an} as $\overline{D}_{\mathrm{an}}(b, r)$ or $D_{\mathrm{an}}(b, r)$. In either case, by our assumptions, D_{an} contains E_{an}. □

15.2. Proofs on Berkovich points

In this section, we prove results from Sections 6.2 and 6.3.

Proof of Proposition 6.6. For any $\zeta \in \overline{D}_{\mathrm{an}}(a,r) \setminus \{\zeta(a,r)\}$, we claim that there must be some $c \in \overline{D}(a,r)$ such that $\|z-c\|_\zeta < r$. Indeed, for any $c \in \overline{D}(a,r)$, we have

$$\|z-c\|_\zeta \le \|z-c\|_{\zeta(a,r)} = r.$$

If the claim were false, then $\|z-c\|_\zeta = r$ for all $c \in \overline{D}(a,r)$, and hence

$$\|z-c\|_\zeta = \max\{r, |a-c|\} = \|z-c\|_{\zeta(a,r)} \quad \text{for all } c \in \mathbb{C}_v.$$

But then $\zeta = \zeta(a,r)$ by Lemma 15.3(b), contradicting our choice of ζ and proving the claim.

The claim gives $\zeta \in D_{\mathrm{an}}(c,r)$, proving the forward inclusion. The reverse inclusion follows from the previously noted facts that

$$D_{\mathrm{an}}(c,r) \subseteq \overline{D}_{\mathrm{an}}(c,r) = \overline{D}_{\mathrm{an}}(a,r).$$

Next, we show that the open disks are in fact pairwise disjoint. Given $c_1, c_2 \in \overline{D}(a,r)$ and a point $\zeta \in D_{\mathrm{an}}(c_1,r) \cap D_{\mathrm{an}}(c_2,r)$, we have $\|z-c_i\|_\zeta < r$ for $i = 1,2$. Then by Lemma 15.2(a),

$$|c_1 - c_2| \le \max\{\|z-c_1\|_\zeta, \|z-c_2\|_\zeta\} < r.$$

Hence, $D(c_1,r) = D(c_2,r)$, but we still need to prove $D_{\mathrm{an}}(c_1,r) = D_{\mathrm{an}}(c_2,r)$. Given $\xi \in D_{\mathrm{an}}(c_1,r)$, then because $\|\cdot\|_\xi$ is non-archimedean, we have

$$\|z-c_2\|_\xi \le \max\{\|z-c_1\|_\xi, |c_1-c_2|\} < r,$$

and therefore $\xi \in D_{\mathrm{an}}(c_2,r)$. Thus, $D_{\mathrm{an}}(c_1,r) \subseteq D_{\mathrm{an}}(c_2,r)$. The reverse inclusion holds similarly, proving that $D_{\mathrm{an}}(c_1,r) = D_{\mathrm{an}}(c_2,r)$, as desired.

Finally, we note that the union is indeed a disjoint union in the topological sense, because each open disk $D_{\mathrm{an}}(c,r)$ is open in the Gel'fand topology by Proposition 6.5. □

The following technical lemma is the key to the proof of Berkovich's classification (Theorem 6.9).

Lemma 15.5. *Let $\zeta \in \mathbb{A}^1_{\mathrm{an}}$. Then there is a decreasing sequence of closed disks $D_1 \supseteq D_2 \supseteq \cdots$ in $\mathbb{C}_v$ such that for all $f \in \mathbb{C}_v[z]$,*

$$\|f\|_\zeta = \lim_{n\to\infty} \|f\|_{\zeta_n},$$

where ζ_n is the Berkovich point associated to the disk D_n.

Proof. For each $a \in \mathbb{C}_v$, define E_a to be the disk

$$E_a := \overline{D}\big(a, \|z-a\|_\zeta\big) \subseteq \mathbb{C}_v.$$

Since $\|\cdot\|_\zeta$ is non-archimedean and extends $|\cdot|$, it is immediate that for any $a, b \in \mathbb{C}_v$, we have

$$E_a \subseteq E_b \quad \text{if and only if} \quad \|z-a\|_\zeta \le \|z-b\|_\zeta. \tag{15.2}$$

Let $r := \mathrm{diam}(\zeta) \geq 0$, and choose a sequence $\{a_n\}_{n\geq 1} \subseteq \mathbb{C}_v$ such that $\{\|z-a_n\|_\zeta\}_{n\geq 1}$ decreases to r. (The existence of such a sequence is immediate from Definition 6.3.) Set

$$D_n := E_{a_n} = \overline{D}\big(a_n, \|z - a_n\|_\zeta\big),$$

and define $\zeta_n \in \mathbb{A}^1_{\mathrm{an}}$ to be the norm corresponding to the disk D_n. Observe, by fact (15.2) above, that $D_1 \supseteq D_2 \supseteq \cdots$.

Define $\|\cdot\|_\xi : \mathbb{C}_v[z] \to [0, \infty)$ by

$$\|f\|_\xi := \lim_{n\to\infty} \|f\|_{\zeta_n}. \tag{15.3}$$

We leave it to the reader (see Exercise 15.2) to verify that the limit defining $\|f\|_\xi$ converges and that $\|\cdot\|_\xi$ is indeed an element of $\mathbb{A}^1_{\mathrm{an}}$. It suffices to show that $\zeta = \xi$. That is, by Lemma 15.2(d), it suffices to show that

$$\|z - b\|_\zeta = \|z - b\|_\xi \quad \text{for all} \quad b \in \mathbb{C}_v.$$

Given $b \in \mathbb{C}_v$, we consider two cases. In the first case, suppose $b \in D_n$ for all $n \geq 1$. Then

$$\|z - b\|_{\zeta_n} = \|z - a_n\|_\zeta \searrow r \quad \text{as} \quad n \to \infty,$$

where the equality is because $\|z - a_n\|_\zeta$ is the radius of D_n, and the limit is by definition of the sequence $\{a_n\}_{n\geq 1}$. Thus, $\|z - b\|_\xi = r$.

On the other hand, for all $n \geq 1$, we have

$$\|a_n - b\|_\zeta = |a_n - b| \leq \|z - a_n\|_\zeta, \tag{15.4}$$

again because $\|z - a_n\|_\zeta$ is the radius of D_n. Therefore,

$$r \leq \|z - b\|_\zeta \leq \max\big\{\|z - a_n\|_\zeta, \|a_n - b\|_\zeta\big\} = \|z - a_n\|_\zeta \searrow r \quad \text{as} \quad n \to \infty,$$

where the first inequality is by definition of r, the second inequality is by Lemma 15.2(a), and the equality is by (15.4). Hence,

$$\|z - b\|_\zeta = r = \|z - b\|_\xi,$$

as desired.

In the second case, suppose $b \notin D_N$ for some $N \geq 1$. Then $b \notin D_n$ for all $n \geq N$. That is, $|b - a_n| > \|z - a_n\|_\zeta$, and hence

$$\|z - b\|_{\zeta_n} = |b - a_n| = \|z - b\|_\zeta \quad \text{for all} \quad n \geq N, \tag{15.5}$$

where the first equality is by definition of ζ_n and the fact that $b \notin D_n$, and the second is by the non-archimedean property of Lemma 15.2(a). Taking the limit of equation (15.5) as $n \to \infty$, we have

$$\|z - b\|_\xi = \|z - b\|_\zeta. \qquad \square$$

Proof of Theorem 6.9. By Lemma 15.5, there is a decreasing sequence of closed disks $D_1 \supseteq D_2 \supseteq \cdots$ in $\mathbb{C}_v$ with corresponding Berkovich points $\zeta_1, \zeta_2, \ldots \in \mathbb{A}^1_{\mathrm{an}}$ such that

$$\|\cdot\|_\zeta = \lim_{n\to\infty} \|\cdot\|_{\zeta_n}.$$

Write $D_n = \overline{D}(a_n, r_n)$; note that the sequence $\{r_n\}_{n\geq 1}$ is decreasing.

Since $r_n > 0$, the limit $r := \lim_{n\to\infty} r_n \geq 0$ converges. We have

$$\|z - b\|_{\zeta_n} \geq r_n \geq r \quad \text{for all } b \in \mathbb{C}_v.$$

In addition,

$$\|z - a_n\|_{\zeta_m} = \max\left\{r_m, |a_m - a_n|\right\} \leq r_n \quad \text{for all} \quad m \geq n \geq 1,$$

and hence

$$r = \inf\left\{\|z - b\|_\zeta : b \in \mathbb{C}_v\right\} = \operatorname{diam}(\zeta). \tag{15.6}$$

We distinguish the various types of Berkovich points by determining whether this infimum is attained, and if so, what its value is.

Consider the intersection

$$D_\infty := \bigcap_{n\geq 1} D_n.$$

If $D_\infty = \varnothing$, then by definition, ζ is a point of Type IV. In addition, by Lemma 2.7(a), we have $r_n > r > 0$ for all $n \geq 1$. However, for any $b \in \mathbb{C}_v$, we must have $b \notin D_N$ for some $N \geq 1$, and therefore

$$\|z - b\|_{\zeta_n} = |a_n - b| > r_N > r \quad \text{for all} \quad n \geq N.$$

As a result, $\|z-b\|_\zeta > r$ for all $b \in \mathbb{C}_v$; that is, the infimum in equation (15.6) is not attained.

On the other hand, if there is some point $a \in D_\infty$, then each disk D_n may be written as $D_n = \overline{D}(a, r_n)$, and hence $\zeta_n = \zeta(a, r_n)$. Note that for any $b \in \mathbb{C}_v$, we have

$$\begin{aligned} \|z - b\|_\zeta &= \lim_{n\to\infty} \|z - b\|_{\zeta(a,r_n)} = \lim_{n\to\infty} \max\left\{|a - b|, r_n\right\} \\ &= \max\left\{|a - b|, r\right\} = \begin{cases} \|z - b\|_a & \text{if } r = 0, \\ \|z - b\|_{\zeta(a,r)} & \text{if } r > 0. \end{cases} \end{aligned}$$

Thus, by Lemma 15.2(d), $\zeta = a$ is a point of Type I if $r = 0$; and $\zeta = \zeta(a, r)$ is a point of Type II or III if $r > 0$, depending on whether or not $r \in |\mathbb{C}_v^\times|$.

Combining the previous two paragraphs, then, we see that the infimum of equation (15.6) is

- attained at $r = 0$ if and only if ζ is a point of Type I,
- attained at $r \in |\mathbb{C}_v^\times|$ if and only if ζ is a point of Type II,

- attained at $r \in (0,\infty) \smallsetminus |\mathbb{C}_v^\times|$ if and only if ζ is a point of Type III,
- unattained if and only if ζ is a point of Type IV.

Since these four cases cover all possibilities and are mutually exclusive, we have shown that ζ is of exactly one of the four types.

We now show that ζ determines its underlying geometric data. For Type I points, suppose $\|\cdot\|_x = \|\cdot\|_y$ for some $x, y \in D$. Then

$$|x-y| = \|z-y\|_x = \|z-y\|_y = 0,$$

and hence $x = y$. For Type II and III points, if $\|\cdot\|_{\zeta(a,r)} = \|\cdot\|_{\zeta(b,s)}$, then

$$r = \|z-a\|_{\zeta(a,r)} = \|z-a\|_{\zeta(b,s)} = \max\{|a-b|, s\}$$

and

$$s = \|z-b\|_{\zeta(b,s)} = \|z-b\|_{\zeta(a,r)} = \max\{|a-b|, r\}.$$

It follows that $r = s \geq |a-b|$, and hence $\overline{D}(a,r) = \overline{D}(b,s)$, as desired.

Finally, suppose that

$$D_1 \supseteq D_2 \supseteq \cdots \quad \text{and} \quad E_1 \supseteq E_2 \supseteq \cdots$$

are decreasing sequences of closed disks in D with empty intersection. Write

$$D_n = \overline{D}(a_n, r_n) \quad \text{and} \quad E_n = \overline{D}(b_n, s_n),$$

and let $\zeta_n = \zeta(a_n, r_n)$ and $\xi_n = \zeta(b_n, s_n)$. Suppose also that the two sequences $\{\zeta_n\}_{n\geq 1}$ and $\{\xi_n\}_{n\geq 1}$ both converge to the same Type IV point $\zeta \in \mathbb{A}^1_{\mathrm{an}}$. We must show that the two sequences of disks are equivalent.

Without loss, the sequences $\{r_n\}_{n\geq 1}$ and $\{s_n\}_{n\geq 1}$ are strictly decreasing. By the discussion surrounding equation (15.6), they have the same limit $r = \operatorname{diam}(\zeta) > 0$. For any integer $N \geq 1$, we need to find $M \geq N$ such that $E_M \subseteq D_N$.

For any $n \geq N+1$, we have

$$\|z - a_{N+1}\|_{\zeta(a_n,r_n)} = \max\left\{|a_n - a_{N+1}|, r_n\right\} \leq r_{N+1},$$

and therefore

$$\lim_{n\to\infty} \|z - a_{N+1}\|_{\zeta(b_n,s_n)} = \lim_{n\to\infty} \|z - a_{N+1}\|_{\zeta(a_n,r_n)} \leq r_{N+1} < r_N.$$

Thus, there is some $M \geq N$ such that

$$\max\left\{|b_M - a_{N+1}|, s_M\right\} = \|z - a_{N+1}\|_{\zeta(b_M,s_M)} < r_N.$$

In other words, since $a_{N+1} \in \overline{D}(a_N, r_N)$, we have

$$E_M = \overline{D}(b_M, s_M) \subseteq \overline{D}(a_{N+1}, r_N) = \overline{D}(a_N, r_N) = D_N,$$

as desired. Applying the same arguments with the roles of $\{D_n\}$ and $\{E_n\}$ reversed and taking the maximum of the two values of M, we have proven that the two sequences of disks are equivalent. □

15.3. Basic proofs on the Berkovich projective line

In this section, we prove results from Section 6.4 and from the first half of Section 6.5.

Proof of Proposition 6.13. Throughout this proof, we abbreviate notation and write D for $D_{\mathrm{an}}(0,1)$, and $\overline{D}$ for $\overline{D}_{\mathrm{an}}(0,1)$.

First, it is immediate from the definitions of $|\cdot|$ on $\mathbb{P}^1_{\mathrm{an}}$ and of Berkovich disks in Definition 6.4 that

$$D = \{\zeta \in \mathbb{A}^1_{\mathrm{an}} : |\zeta| < 1\} \quad \text{and} \quad \overline{D} = \{\zeta \in \mathbb{A}^1_{\mathrm{an}} : |\zeta| \le 1\}. \tag{15.7}$$

Thus, as implicitly claimed in the statement of the proposition,

$$\overline{D} \setminus D = \{\zeta \in \mathbb{A}^1_{\mathrm{an}} : |\zeta| = 1\}.$$

Consider the quotient map

$$q : \mathbb{A}^1_{\mathrm{an}} \amalg \mathbb{A}^1_{\mathrm{an}} \to \mathbb{P}^1_{\mathrm{an}}$$

from two disjoint copies of $\mathbb{A}^1_{\mathrm{an}}$, glued via $\zeta \mapsto 1/\zeta$, that defines $\mathbb{P}^1_{\mathrm{an}}$. Let

$$i : \overline{D} \amalg \overline{D} \hookrightarrow \mathbb{A}^1_{\mathrm{an}} \amalg \mathbb{A}^1_{\mathrm{an}}$$

be the embedding induced by $\overline{D} \subseteq \mathbb{A}^1_{\mathrm{an}}$.

We claim that $q \circ i$ is a quotient map. Indeed, observe that $q \circ i$ takes the first copy of $\overline{D}$ to $\overline{D} \subseteq \mathbb{P}^1_{\mathrm{an}}$, and the second to $\mathbb{P}^1_{\mathrm{an}} \setminus D$. Since both of these images are closed in $\mathbb{P}^1_{\mathrm{an}}$, it follows immediately that $q \circ i$ is a closed map, and hence a quotient map, as claimed.

Therefore, $q \circ i$ induces an equivalence relation on $\overline{D} \amalg \overline{D}$, given by

$$\zeta \sim \xi \quad \text{if and only if} \quad q \circ i(\zeta) = q \circ i(\xi).$$

On the other hand, the gluing of the two copies of $\overline{D}$ along $\overline{D} \setminus D$ via $\zeta \mapsto 1/\zeta$ also induces an equivalence relation on $\overline{D} \amalg \overline{D}$. Using characterization (15.7) of D and $\overline{D}$ in terms of $|\cdot|$, as well as the fact that $|1/\zeta| = 1/|\zeta|$, the reader can check that these two equivalence relations are precisely the same; see Exercise 15.3. Thus, the quotient space described in the statement of the proposition is homeomorphic to $\mathbb{P}^1_{\mathrm{an}}$, as desired. □

Proof of Theorem 6.17.

Step 1. As noted above in the discussion following Definition 6.16, all open Berkovich affinoids are indeed open in the Gel'fand topology.

Step 2. Given any $f \in \mathbb{C}_v[z]$ and any $R > 0$, define

$$U_{f,R} := \{\zeta \in \mathbb{A}^1_{\mathrm{an}} : \|f\|_\zeta < R\}$$

and

$$V_{f,R} := \{\zeta \in \mathbb{A}^1_{\mathrm{an}} : \|f\|_\zeta > R\}.$$

We wish to show that both $U_{f,R}$ and $V_{f,R}$ are unions of open connected Berkovich affinoids intersected with $\mathbb{A}^1_{\mathrm{an}}$.

If $f = c$ is constant, then $U_{f,R}$ is either empty or all of $\mathbb{A}^1_{\mathrm{an}}$, depending on whether $R \leq |c|$ or $R > |c|$, respectively. Similarly, $V_{f,R}$ is also either empty or all of $\mathbb{A}^1_{\mathrm{an}}$. Since $\mathbb{A}^1_{\mathrm{an}} = \bigcup_{r>0} D_{\mathrm{an}}(a, r)$ is a union of open Berkovich disks, we are done.

We may therefore assume f is nonconstant. Write

$$f(z) = c \prod_{j=1}^{\ell} (z - \alpha_i)^{e_i},$$

where $\ell \geq 1$, the roots $\alpha_1, \ldots, \alpha_\ell \in \mathbb{C}_v$ are distinct, $e_1, \ldots, e_\ell \geq 1$, and $c \in \mathbb{C}_v^\times$. For each $j = 1, \ldots, \ell$, the real-valued function

$$t \mapsto |c| \prod_{i=1}^{\ell} \max\{t, |\alpha_j - \alpha_i|\}^{e_i}$$

is continuous, maps $0 \mapsto 0$, and increases to ∞ as $t \to \infty$. (In fact, this function is strictly increasing.) Thus, there is some (unique) radius $r_j > 0$ such that

$$|c| \prod_{i=1}^{\ell} \max\left\{r_j, |\alpha_j - \alpha_i|\right\}^{e_i} = R.$$

Note that r_j is indeed strictly positive, because we assumed $R > 0$.

We claim that

$$U_{f,R} = \bigcup_{j=1}^{\ell} D_{\mathrm{an}}(\alpha_j, r_j) \tag{15.8}$$

and

$$V_{f,R} = \mathbb{A}^1_{\mathrm{an}} \smallsetminus \bigcup_{j=1}^{\ell} \overline{D}_{\mathrm{an}}(\alpha_j, r_j). \tag{15.9}$$

To prove equation (15.8), given any $\zeta \in U_{f,R}$, choose $j \in \{1, \ldots, \ell\}$ minimizing $\|z - \alpha_j\|_\zeta$. Then

$$|\alpha_j - \alpha_i| \leq \max\{\|z - \alpha_j\|_\zeta, \|z - \alpha_i\|_\zeta\} \leq \|z - \alpha_i\|_\zeta$$

for all $i = 1 \ldots, \ell$. Thus, if $\|z - \alpha_j\|_\zeta \geq r_j$, then

$$\begin{aligned} R = |c| \prod_{i=1}^{\ell} \max\left\{r_j, |\alpha_j - \alpha_i|\right\}^{e_i} &\leq |c| \prod_{i=1}^{\ell} \max\left\{\|z - \alpha_j\|_\zeta, \|z - \alpha_i\|_\zeta\right\}^{e_i} \\ = |c| \prod_{i=1}^{\ell} \|z - \alpha_i\|_\zeta^{e_i} = \|f\|_\zeta, \end{aligned}$$

contradicting the assumption that $\zeta \in U_{f,R}$. Hence, we have $\|z-\alpha_j\|_\zeta < r_j$, and therefore $\zeta \in D_{\text{an}}(\alpha_j, r_j)$.

Conversely, given $\zeta \in D_{\text{an}}(\alpha_j, r_j)$, we have $\|z - \alpha_j\|_\zeta < r_j$, while at the same time

$$\|z - \alpha_i\|_\zeta \leq \max\left\{\|z - \alpha_j\|_\zeta, |\alpha_j - \alpha_i|\right\}$$

for all $i = 1, \ldots, \ell$. Thus,

$$\begin{aligned}\|f\|_\zeta = |c| \prod_{i=1}^{\ell} \|z - \alpha_i\|_\zeta^{e_i} &\leq |c| \prod_{i=1}^{\ell} \max\left\{\|z - \alpha_j\|_\zeta, |\alpha_j - \alpha_i|\right\}^{e_i} \\ &< |c| \prod_{i=1}^{\ell} \max\left\{r_j, |\alpha_j - \alpha_i|\right\}^{e_i} = R,\end{aligned}$$

where the strict inequality is due to the $i = j$ term. Therefore, $\zeta \in U_{f,R}$, proving equation (15.8).

We leave it to the reader to verify equation (15.9); see Exercise 15.5. Equations (15.8) and (15.9) show that $U_{f,R}$ is a union of open connected Berkovich affinoids (in fact, disks) and that $V_{f,R}$ is a single open connected Berkovich affinoid intersected with $\mathbb{A}^1_{\text{an}}$, as desired.

Step 3. In Exercise 15.6, we leave it to the reader to use the result of Step 2 above to show that every open subset of $\mathbb{P}^1_{\text{an}}$ is a union of open connected Berkovich affinoids. □

The following useful lemma is a variation on Theorem 6.17, focusing on neighborhoods of Berkovich points.

Lemma 15.6. *Let $\zeta \in \mathbb{P}^1_{\text{an}}$, and let $U \subseteq \mathbb{P}^1_{\text{an}}$ be an open subset. Then there is an open connected affinoid V such that $\zeta \in V \subseteq U$. More precisely, the following hold.*

(a) *If $\zeta = x$ is of Type I, then we can choose*

$$V = \begin{cases} D_{\text{an}}(x, \varepsilon), & \text{for some } \varepsilon > 0, \quad \text{if } x \in \mathbb{C}_v, \text{ or} \\ \mathbb{P}^1_{\text{an}} \smallsetminus \overline{D}_{\text{an}}(0, R), & \text{for some } R > 0, \quad \text{if } \zeta = \infty. \end{cases}$$

(b) *If $\zeta = \zeta(a, r)$ is of Type II, then we can choose*

$$V = D_{\text{an}}(a, \lambda r) \smallsetminus \bigcup_{i=1}^{\ell} \overline{D}_{\text{an}}\bigl(b_i, \lambda^{-1} r\bigr),$$

where $\lambda > 1$ is a real number and $b_1, \ldots, b_\ell$ are Type I points in $\overline{D}(a, r)$.

(c) *If $\zeta = \zeta(a, r)$ is of Type III, then we can choose*

$$V = \left\{\lambda^{-1} r < |\zeta - a| < \lambda r\right\} = D_{\text{an}}\bigl(a, \lambda r\bigr) \smallsetminus \overline{D}_{\text{an}}\bigl(a, \lambda^{-1} r\bigr),$$

for some real number $\lambda > 1$.

(d) *If ζ is of Type IV, represented by a decreasing sequence of closed disks $D_1 \supsetneq D_2 \supsetneq \cdots$ in $\mathbb{C}_v$, then we can choose $V = D_{\text{an}}(a, r)$, where $D(a, r) \subseteq \mathbb{C}_v$ is some open disk containing all but finitely many of the disks D_n.*

Proof. The case that $\zeta = \infty$ is covered by the case that $\zeta = 0$, by reversing the roles of the two copies of $\mathbb{A}^1_{\text{an}}$ that are glued to form $\mathbb{P}^1_{\text{an}}$. We may therefore assume that $\zeta \in \mathbb{A}^1_{\text{an}}$.

By equation (6.1), we may choose $b \in \mathbb{C}_v$ and $s > 0$ so that $\zeta \in D_{\text{an}}(b, s)$. In addition, by Theorem 6.17, we may assume that U is an open connected Berkovich affinoid. Replacing U by its intersection with $D_{\text{an}}(b, s)$, then, we may assume that U is of the form

$$U = D_{\text{an}}(b, s) \smallsetminus \left[\overline{D}_{\text{an}}(c_1, t_1) \cup \cdots \cup \overline{D}_{\text{an}}(c_m, t_m)\right].$$

If $\zeta = x \in \mathbb{C}_v$ is of Type I, then $|x - b| < s$ and $|x - c_i| > t_i$ for all i. Set

$$\varepsilon := \min\{s, t_1, \ldots, t_m\} > 0.$$

Then $D_{\text{an}}(x, \varepsilon) \subseteq D_{\text{an}}(b, s)$, and

$$D_{\text{an}}(x, \varepsilon) \cap \overline{D}_{\text{an}}(c_i, t_i) = \varnothing \quad \text{for all } i = 1, \ldots, m.$$

Hence, $x \in D_{\text{an}}(x, \varepsilon) \subseteq U$.

If $\zeta = \zeta(a, r)$ is of Type II or III, then

$$\overline{D}(a, r) \subseteq D(b, s) \quad \text{and} \quad \overline{D}(a, r) \not\subseteq \overline{D}(c_i, t_i),$$

by Remark 6.24. In particular, we must have

$$r < s, \quad \text{and} \quad t_i < \max\left\{|a - c_i|, r\right\} \quad \text{for each } i.$$

We may therefore choose $\lambda > 1$ small enough that

- $\lambda r \leq s$,
- $\lambda r \leq |a - c_i|$ for all i such that $|a - c_i| > r$, and
- $\lambda^{-1} r \geq t_i$ for all i such that $|a - c_i| \leq r$.

Let $b_1, \ldots, b_\ell$ be precisely those c_i's lying in $\overline{D}(a, r)$, and define

$$V := D_{\text{an}}(a, \lambda r) \smallsetminus \bigcup_{i=1}^{\ell} \overline{D}_{\text{an}}(b_i, \lambda^{-1} r).$$

Then $\zeta \in V \subseteq U$, as the reader can confirm by carefully checking each disk involved against Definition 6.4.

If ζ is of Type III, then we have $|a - b_i| < r$ for each $i = 1, \ldots, \ell$, and hence we can shrink $\lambda > 1$ further to guarantee that

- $\lambda^{-1} r \geq |a - b_i|$ for all $i = 1, \ldots, \ell$.

Then, redefining

$$V := D_{\mathrm{an}}(a, \lambda r) \setminus \overline{D}_{\mathrm{an}}\big(a, \lambda^{-1} r\big),$$

we still have $\zeta \in V \subseteq U$.

Finally, suppose ζ is of Type IV, corresponding to a decreasing sequence of disks $D(a_1, r_1) \supsetneq D(a_2, r_2) \supsetneq \cdots$ in $\mathbb{C}_v$ with empty intersection. Since the intersection is empty, we may choose N large enough that $b, c_1, \ldots, c_m \notin D_N$.

In particular,

$$s > \|z - b\|_\zeta = \lim_{n\to\infty} \|z - b\|_{\zeta(a_n, r_n)} = \lim_{n\to\infty} \max\big\{|a_n - b|, r_n\big\}, \tag{15.10}$$

where the inequality is by the fact that $\zeta \in U \subseteq D_{\mathrm{an}}(b, s)$, the first equality is the definition of $\|\cdot\|_\zeta$ at a Type IV point, and the second is because $b \notin D_n$ for $n \geq N$. By inequality (15.10), we may increase N if necessary to guarantee that

$$|a_n - b| < s \quad \text{and} \quad r_n < s \quad \text{for all } n \geq N.$$

By a similar argument, using the fact that

$$\|z - c_i\|_\zeta > t_i \quad \text{but} \quad c_i \notin D_N,$$

we can further increase N to guarantee that

$$|a_n - c_i| > t_i \quad \text{for all } i = 1, \ldots, m \text{ and all } n \geq N.$$

Set $V := D_{\mathrm{an}}(a_{N+1}, r_N)$, which contains all but finitely many of the disks $\overline{D}(a_n, r_n)$. Then $\zeta \in V$, because

$$\begin{aligned}\|z - a_{N+1}\| &= \lim_{n\to\infty} \|z - a_{N+1}\|_{\zeta(a_n, r_n)} \\ &= \lim_{n\to\infty} \max\big\{|a_n - a_{N+1}|, r_n\big\} \leq r_{N+1} < r_N.\end{aligned}$$

On the other hand, using Definition 6.4 and the results of the previous two paragraphs, the reader can easily check that $V \subseteq U$. □

15.4. Proving compactness

In this section, we prove the remaining results from Section 6.5, showing that $\mathbb{P}^1_{\mathrm{an}}$ is indeed a compactification of $\mathbb{P}^1(\mathbb{C}_v)$.

Proof of Theorem 6.21. By Theorem 6.9, the map i is injective and maps $\mathbb{P}^1(\mathbb{C}_v)$ onto the set of Type I points of $\mathbb{P}^1_{\mathrm{an}}$.

By Theorem 6.17, to prove that i is continuous, we need only consider the inverse images of open connected Berkovich affinoids. Given any such Berkovich affinoid $W \subseteq \mathbb{P}^1_{\mathrm{an}}$, we may write W as an intersection of finitely many open Berkovich disks. For any such open Berkovich disk U, let

$$U_{\mathrm{I}} := U \cap \mathbb{P}^1(\mathbb{C}_v)$$

be the corresponding open $\mathbb{P}^1(\mathbb{C}_v)$-disk. Then $i^{-1}(U) = U_{\mathrm{I}}$ is an open subset of $\mathbb{P}^1(\mathbb{C}_v)$. Thus, $i^{-1}(W)$, which is the intersection of these finitely many corresponding open $\mathbb{P}^1(\mathbb{C}_v)$-disks, is open in $\mathbb{P}^1(\mathbb{C}_v)$. Hence, i is continuous.

Conversely, given any open disk $U \subseteq \mathbb{P}^1(\mathbb{C}_v)$, let U_{an} be the corresponding open Berkovich disk. That is,

$$U_{\mathrm{an}} := \begin{cases} D_{\mathrm{an}}(a,r) & \text{if } U = D(a,r), \text{ or} \\ \mathbb{P}^1_{\mathrm{an}} \smallsetminus \overline{D}_{\mathrm{an}}(a,r) & \text{if } U = \mathbb{P}^1(\mathbb{C}_v) \smallsetminus \overline{D}(a,r). \end{cases}$$

Then

$$i(U) = U_{\mathrm{an}} \cap i(\mathbb{P}^1(\mathbb{C}_v)),$$

which is an open subset of the image $i(\mathbb{P}^1(\mathbb{C}_v)) \subseteq \mathbb{P}^1_{\mathrm{an}}$. Since the open disks U form a basis for the topology on $\mathbb{P}^1(\mathbb{C}_v)$, the map $i^{-1} : i(\mathbb{P}^1(\mathbb{C}_v)) \to \mathbb{P}^1(\mathbb{C}_v)$ is continuous. □

We now turn to proving Theorem 6.22, that $\mathbb{P}^1_{\mathrm{an}}$ is Hausdorff and compact. In fact, the same is true for far more general Berkovich spaces; Berkovich proved that the (seminorm) spectrum of a commutative Banach ring is a nonempty compact Hausdorff space. Berkovich's compactness proof appears as [**26**, Theorem 1.2.1] and uses Stone–Čech compactification; see also [**8**, Theorem C.3] for a short proof using nets. Given our far more specific setting, we present a more explicit, but less elegant, proof of the compactness of $\overline{D}_{\mathrm{an}}(0,1)$, and hence of $\mathbb{P}^1_{\mathrm{an}}$.

Proof of Theorem 6.22.

Hausdorff. Given $\zeta \neq \xi \in \mathbb{P}^1_{\mathrm{an}}$, then either ζ and ξ are 0 and ∞, or by choosing one of the two copies of $\mathbb{A}^1_{\mathrm{an}}$ that are glued to form $\mathbb{P}^1_{\mathrm{an}}$, we have $\zeta, \xi \in \mathbb{A}^1_{\mathrm{an}}$.

If ζ and ξ are 0 and ∞, then the disks

$$D_{\mathrm{an}}(0,1) \quad \text{and} \quad \mathbb{P}^1_{\mathrm{an}} \smallsetminus \overline{D}_{\mathrm{an}}(0,1)$$

are disjoint open subsets, with one containing ζ and the other containing ξ.

On the other hand, if $\zeta \neq \xi \in \mathbb{A}^1_{\mathrm{an}}$, then by definition of $\mathbb{A}^1_{\mathrm{an}}$, there must be some polynomial $f \in \mathbb{C}_v[z]$ such that $\|f\|_\zeta \neq \|f\|_\xi$. Without loss, $\|f\|_\zeta < \|f\|_\xi$. Choose a real number s with $\|f\|_\zeta < s < \|f\|_\xi$. Define

$$U := \{\eta \in \mathbb{A}^1_{\mathrm{an}} : \|f\|_\eta < s\} \quad \text{and} \quad V := \{\eta \in \mathbb{A}^1_{\mathrm{an}} : \|f\|_\eta > s\},$$

both of which are open in $\mathbb{A}^1_{\mathrm{an}}$ and hence in $\mathbb{P}^1_{\mathrm{an}}$, by definition of the Gel'fand topology. Clearly, we have $\zeta \in U$, $\xi \in V$, and $U \cap V = \varnothing$, as desired.

Compact. By Proposition 6.13, it suffices to show that the closed unit disk $\overline{D}_{\mathrm{an}}(0,1)$ is compact.

Let $\mathcal{U}$ be a family of open sets covering $\overline{D}_{\mathrm{an}}(0,1)$. Define

$$\mathcal{S} := \big\{\overline{D}(b,s) \subseteq \overline{D}(0,1) : \overline{D}_{\mathrm{an}}(b,s) \text{ has no finite subcover in } \mathcal{U}\big\}.$$

Suppose, toward a contradiction, that $\overline{D}(0,1) \in \mathcal{S}$.

Define

$$\mathcal{T} := \big\{\overline{D}(b,s) \subseteq \overline{D}(0,1) : \text{there is some } \overline{D}(c,t) \subsetneq \overline{D}(b,s) \text{ with } \overline{D}(c,t) \in \mathcal{S}\big\}.$$

We claim that $\mathcal{T} = \mathcal{S}$. The forward inclusion is immediate; if $\overline{D}_{\mathrm{an}}(c,t)$ fails to have a finite subcover, the same must be true of the larger set $\overline{D}_{\mathrm{an}}(b,s)$. Conversely, suppose $\overline{D}(b,s) \subseteq \overline{D}(0,1)$ is a closed disk *not* in $\mathcal{T}$. Let $U_0 \in \mathcal{U}$ be an element of the cover containing $\zeta(b,s)$, which is a point of Type II or III. By Lemma 15.6(b)–(c), U_0 must contain an affinoid of the form

$$D_{\mathrm{an}}(b,\lambda s) \smallsetminus \Big[\overline{D}_{\mathrm{an}}\big(c_1,\lambda^{-1}s\big) \cup \cdots \cup \overline{D}_{\mathrm{an}}\big(c_\ell,\lambda^{-1}s\big)\Big]$$

for some $\lambda > 1$ and some $c_1,\ldots,c_\ell \in \overline{D}(b,s)$. Because $\overline{D}(b,s) \notin \mathcal{T}$, each smaller Berkovich disk $\overline{D}_{\mathrm{an}}(c_i,\lambda^{-1}s)$ has a finite subcover in $\mathcal{U}$. Thus, there is a finite set $\{U_1,\ldots,U_m\} \subseteq \mathcal{U}$ such that

$$\bigcup_{i=1}^{\ell} \overline{D}_{\mathrm{an}}\big(c_i,\lambda^{-1}s\big) \subseteq U_1 \cup \cdots \cup U_m.$$

Therefore, $\{U_0,U_1,\ldots,U_m\} \subseteq \mathcal{U}$ is a finite subcover of $\overline{D}_{\mathrm{an}}(b,s)$. That is, $\overline{D}(b,s) \notin \mathcal{S}$, proving our claim.

Our next goal is to inductively construct a decreasing sequence of closed disks $D_1 \supsetneq D_2 \supsetneq \cdots$ in $\mathcal{S}$, with radii $r_1 > r_2 > \cdots$ that decrease rapidly, in a sense that will become clear shortly. First, set

$$D_1 := \overline{D}(0,1) \in \mathcal{S} \quad \text{and} \quad r_1 := 1.$$

Given the nth disk $D_n \in \mathcal{S} = \mathcal{T}$ of radius r_n, there must be proper subdisks of D_n in $\mathcal{S}$, by definition of $\mathcal{T}$. Thus, it makes sense to define

$$\rho_n := \inf\{s > 0 : \text{ there is a subdisk of } D_n \text{ of radius } s \text{ in } \mathcal{S}\}, \tag{15.11}$$

and we have $\rho_n \in [0,r_n)$. In particular, there is a closed disk $D_{n+1} \subsetneq D_n$ with $D_{n+1} \in \mathcal{S}$ of some radius r_{n+1}, where

$$0 < r_{n+1} < \min\left\{\rho_n + \frac{1}{n}, r_n\right\}, \tag{15.12}$$

completing our inductive construction. Define

$$D := \bigcap_{n \geq 1} D_n,$$

which must be either empty, a single point, or a closed disk in $\mathbb{C}_v$. For each $n \geq 1$, define $D_{n,\mathrm{an}}$ to be the closed Berkovich disk corresponding to D_n.

If $D = \varnothing$, then the sequence $D_1 \supsetneq D_2 \supsetneq \cdots$ defines a Type IV point $\zeta \in \overline{D}_{\text{an}}(0,1)$. Write $D_n = \overline{D}(a_n, r_n)$. Choose $U_0 \in \mathcal{U}$ containing ζ; by Lemma 15.6(d), U_0 also contains some disk D_N in the sequence. But then $\{U_0\}$ is a finite cover of $D_{N+1,\text{an}}$, contradicting the fact that $D_{N+1} \in \mathcal{S}$.

Similarly, if $D = \{a\}$ is a single point, then the radii r_n must decrease to zero, and in addition, $a \in \overline{D}_{\text{an}}(0,1)$. Choose $U_0 \in \mathcal{U}$ containing a; by Lemma 15.6(a), U_0 contains $D_{\text{an}}(a, \varepsilon)$ for some $\varepsilon > 0$. However, since $r_n \to 0$, there is some $N \geq 0$ such that $r_N < \min\{\varepsilon, 1\}$, and therefore $D_{N,\text{an}} = \overline{D}_{\text{an}}(a, r_N) \subseteq U_0$. Again, we have a finite cover of $D_{N,\text{an}}$ and hence a contradiction.

Finally, suppose that $D = \overline{D}(a,r)$ is a closed disk, with $r_n \searrow r > 0$ as $n \to \infty$. If $D \in \mathcal{S} = \mathcal{T}$, then there is some strictly smaller disk $\overline{D}(b,s) \subsetneq D$ in $\mathcal{S}$. Choose $N \geq 1$ large enough that $1/N < r - s$. Then the disk $\overline{D}(b,s) \in \mathcal{S}$ is contained in D_N but has radius

$$s < r - \frac{1}{N} < r_{N+1} - \frac{1}{N} < \rho_N,$$

by inequality (15.12). Because this contradicts the definition of ρ_N in equation (15.11), we must actually have $D = \overline{D}(a,r) \notin \mathcal{S}$.

Since $D \notin \mathcal{S}$, the Berkovich disk $\overline{D}_{\text{an}}(a,r)$ has a finite subcover

$$\{U_1, \ldots, U_m\} \subseteq \mathcal{U}.$$

Let $U_0 \in \mathcal{U}$ be an element of the cover containing $\zeta(a,r)$, which is a point of Type II or III. By Lemma 15.6(b)–(c), U_0 must contain an affinoid of the form

$$V = D_{\text{an}}\big(a, \lambda r\big) \smallsetminus \bigcup_{i=1}^{\ell} \overline{D}_{\text{an}}\big(b_i, \lambda^{-1} r\big),$$

where $\lambda > 1$ and $b_1, \ldots, b_\ell \in \overline{D}(a,r)$. In particular,

$$D_{\text{an}}\big(a, \lambda s\big) \smallsetminus \overline{D}_{\text{an}}(a,r) \subseteq U_0.$$

Choose an integer $N \geq 1$ large enough that $r_N < \lambda r$. Then

$$\{U_0, U_1, \ldots, U_m\}$$

is a finite subcover of $\overline{D}_{\text{an}}(a, r_N)$. Thus, $D_N = \overline{D}(a, r_N) \notin \mathcal{S}$, contradicting our construction of D_N and hence proving that $\overline{D}(0,1) \notin \mathcal{S}$ after all. □

15.5. Proving path-connectedness

In this section, we prove most of the results from Section 6.6.

Proof of Proposition 6.27.

(a) The relation $\preceq$ is clearly reflexive and transitive. Moreover, if $\xi \preceq \zeta$ and $\zeta \preceq \xi$, then $\|f\|_\xi = \|f\|_\zeta$ for all $f \in \mathbb{C}_v[z]$, and hence $\xi = \zeta$.

(b) Given a Type I point $a \in \mathbb{C}_v$, suppose that $\zeta \in \mathbb{A}^1_{\text{an}}$ lies below a. Then

$$0 \le \|z-a\|_\zeta \le \|z-a\|_a = 0.$$

That is, $\|z-a\|_\zeta = 0$, and hence by the classification of Theorem 6.9, we must have $\zeta = a$.

Given a Type IV point $\xi \in \mathbb{A}^1_{\text{an}}$ corresponding to a decreasing sequence $D_1 \supseteq D_2 \supseteq \cdots$ of closed disks in $\mathbb{C}_v$ with empty intersection, we may write $D_n = \overline{D}(a_n, r_n)$. Note that $\xi \in \overline{D}_{\text{an}}(a_n, r_n)$ for every $n \ge 1$. Given any point $\zeta \in \mathbb{A}^1_{\text{an}}$ lying below ξ, we must show that $\zeta = \xi$.

For any $b \in \mathbb{C}_v$, there is some $n \ge 1$ such that $b \notin D_n$. That is, $\|b - a_n\|_\zeta = |b - a_n| > r_n$. On the other hand,

$$\|z - a_n\|_\zeta \le \|z - a_n\|_\xi \le r_n,$$

where the first inequality holds because $\zeta \preceq \xi$, and the second holds because $\xi \in \overline{D}_{\text{an}}(a_n, r_n)$. Since $\|\cdot\|_\zeta$ is non-archimedean, we must have

$$\|z - a_n\|_\zeta < |b - a_n| = \|z - b\|_\zeta.$$

However, we also have $\|z - b\|_\xi = |b - a_n|$, by the same reasoning applied to ξ. Thus, $\|z-b\|_\zeta = \|z-b\|_\xi$ for all $b \in \mathbb{C}_v$. By Lemma 15.2(d), $\zeta = \xi$.

(c) The reverse implication is immediate from the definitions. For the forward implication, for any $b \in \mathbb{C}_v$, we have

$$\|z-b\|_\xi \le \max\{\|z-a\|_\xi, |a-b|\} \le \max\{r, |a-b|\} = \|z-b\|_{\zeta(a,r)},$$

and hence $\xi \preceq \zeta(a,r)$, by Lemma 15.2(c).

(d) Since $\|z-b\|_\xi \le \|z-b\|_\zeta$ for all $b \in \mathbb{C}_v$, it is immediate that $\operatorname{diam}(\xi) \le \operatorname{diam}(\zeta)$.

If $\xi \preceq \zeta$ with $\operatorname{diam}(\xi) = \operatorname{diam}(\zeta)$, we consider two cases. First, if ζ is of Type I or IV, then $\xi = \zeta$ by part (b). Second, if ζ is of Type II or III, then write $\zeta = \zeta(a,r)$. By part (c), we have $\xi \in \overline{D}_{\text{an}}(a,r)$, and hence

$$\|z-b\|_\xi \ge \operatorname{diam}(\xi) = \operatorname{diam}(\zeta) = r \quad \text{for all } b \in \mathbb{C}_v.$$

On the other hand,

$$\|z-b\|_\xi \le \|z-b\|_\zeta = r \quad \text{for all } b \in \overline{D}(a,r),$$

since $\xi \preceq \zeta$. Thus,

$$\|z-b\|_\xi = r = \|z-b\|_\zeta \quad \text{for all } b \in \overline{D}(a,r).$$

Therefore, by Lemma 15.3(b), $\xi = \zeta$. □

Proof of Proposition 6.29. The uniqueness statement is immediate: if ξ_1 and ξ_2 both satisfy properties (a)–(b) of Definition 6.28, then $\xi_1 \preceq \xi_2$ and $\xi_2 \preceq \xi_1$. Hence $\xi_1 = \xi_2$ by Proposition 6.27(a).

As for existence, we first consider the case that $\zeta_0 = \zeta_1$. Then this same point ζ_0 fulfills both properties of Definition 6.28, and we are done. Thus, we may assume for the remainder of the proof that $\zeta_0 \neq \zeta_1$.

By Lemma 15.2(d), there is some $a \in \mathbb{C}_v$ such that $\|z-a\|_{\zeta_0} \neq \|z-a\|_{\zeta_1}$. Without loss, $\|z-a\|_{\zeta_0} < \|z-a\|_{\zeta_1}$. Define $r := \|z-a\|_{\zeta_1}$. We wish to show that $\zeta(a,r)$ satisfies both properties of Definition 6.28.

Note that $\zeta_0, \zeta_1 \in \overline{D}_{\mathrm{an}}(a,r)$ (by Definition 6.4(a)), and hence by Proposition 6.27(c), we have $\zeta_0, \zeta_1 \preceq \zeta(a,r)$, verifying the first property. For the second, given $\xi \in \mathbb{A}^1_{\mathrm{an}}$ with $\zeta_0, \zeta_1 \preceq \xi$, we need to show that $\zeta(a,r) \preceq \xi$. By Lemma 15.2(c), it suffices to show that

$$\|z-b\|_{\zeta(a,r)} \leq \|z-b\|_{\xi} \quad \text{for any } b \in \mathbb{C}_v.$$

Given $b \in \mathbb{C}_v$, we consider two cases.

First, if $|b-a| \geq r$, then applying the triangle inequality for ζ_0 to $(z-b)+(b-a) = (z-a)$, we have

$$\|z-b\|_{\zeta_0} = |b-a|, \quad \text{since} \quad \|z-a\|_{\zeta_0} < \|z-a\|_{\zeta_1} = r \leq |b-a|.$$

Therefore,

$$\|z-b\|_{\zeta(a,r)} = |b-a| = \|z-b\|_{\zeta_0} \leq \|z-b\|_{\xi}.$$

Second, if $|b-a| < r$, then by similar reasoning, we have

$$\|z-b\|_{\zeta_1} = \|z-a\|_{\zeta_1} = r, \quad \text{since} \quad |b-a| < r = \|z-a\|_{\zeta_1},$$

and hence

$$\|z-b\|_{\zeta(a,r)} = r = \|z-b\|_{\zeta_1} \leq \|z-b\|_{\xi},$$

as desired. □

Our proof of Theorem 6.32 requires the following two technical lemmas.

Lemma 15.7. *Let $\zeta \preceq \eta \in \overline{D}_{\mathrm{an}}(0,1)$, and define*

$$I := \{\xi \in \overline{D}_{\mathrm{an}}(0,1) : \zeta \preceq \xi \preceq \eta\}$$

and

$$F : I \to \mathbb{R} \quad \text{by} \quad F(\xi) = \mathrm{diam}(\xi).$$

Set $r := \mathrm{diam}(\zeta)$ and $s := \mathrm{diam}(\eta)$. Then $r \leq s$, and F maps I homeomorphically onto the real interval $[r,s]$.

Lemma 15.8. *Let $D \subseteq \mathbb{P}^1_{\mathrm{an}}$ be a Berkovich disk, let $\zeta_0, \zeta_1 \in D$, and let $F : [0,1] \to \mathbb{P}^1_{\mathrm{an}}$ be an injective continuous function such that $F(0) = \zeta_0$ and $F(1) = \zeta_1$. Then $F([0,1]) \subseteq D$.*

Proof of Lemma 15.7. Proposition 6.27(d) shows that $r \le s$ and that F maps I injectively into the interval $[r, s] \subseteq \mathbb{R}$. Thus, it suffices to show that F is onto and continuous, and that I is compact. If $\zeta = \eta$, we are already done; we may therefore assume that $\zeta \ne \eta$.

Compactness of I. By Theorem 6.22, we need only show that I is a closed subset of $\mathbb{P}^1_{\text{an}}$. Given $\xi \in \mathbb{A}^1_{\text{an}} \smallsetminus I$, we have either $\xi \not\preceq \eta$ or $\xi \not\succeq \zeta$.

If $\xi \not\preceq \eta$, then by definition, there is some polynomial $f \in \mathbb{C}_v[z]$ such that $\|f\|_\xi > \|f\|_\eta$. Set $R := \|f\|_\eta$, and set

$$V_{f,R} := \{\nu \in \mathbb{A}^1_{\text{an}} : \|f\|_\nu > R\}.$$

Then $V_{f,R}$ is open, contains ξ, and does not intersect I, as desired. The proof is similar in the case that $\xi \not\succeq \zeta$.

Surjectivity of F. Given $t \in (r, s)$, it suffices to show that there is some $a \in \mathbb{C}_v$ such that $\zeta(a, t) \in I$. If ζ is of Type I, then let $a = \zeta \in \mathbb{C}_v$, and we must have $\eta = \zeta(a, s)$, since η is a point of Type II or III lying above a and of diameter s. Clearly,

$$a \preceq \zeta(a, t) \preceq \zeta(a, s),$$

as desired. Similarly, if ζ is of Type II or III, then write $\zeta = \zeta(a, r)$ for some $a \in \mathbb{C}_v$, and we again obtain

$$\zeta(a, r) \preceq \zeta(a, t) \preceq \zeta(a, s).$$

If ζ is of Type IV, then since $t > r$, we may choose an approximating closed disk $\overline{D}(a, r')$ small enough that $r < r' < t$, and $\zeta \in \overline{D}_{\text{an}}(a, r')$. Yet again, we obtain

$$\zeta \preceq \zeta(a, r') \preceq \zeta(a, t) \preceq \zeta(a, s) = \eta,$$

as desired.

Continuity of F. It suffices to show, for any $t \in (r, s)$, that $F^{-1}([r, t))$ and $F^{-1}((t, s])$ are open subsets of I. The set $F^{-1}([r, t))$ is open in I by the fact that $\operatorname{diam} : \mathbb{A}^1_{\text{an}} \to \mathbb{R}$ is upper semicontinuous; see Exercise 6.9(c).

To show that $F^{-1}((t, s])$ is also open in I, begin by writing the unique point $F^{-1}(t)$ as $\zeta(a, t)$, and note that $\eta = \zeta(a, s)$, as in the "surjectivity" portion of this proof. We claim that

$$F^{-1}\big((t, s]\big) = \big\{\zeta(a, \rho) : t < \rho \le s\big\} = I \smallsetminus \overline{D}_{\text{an}}(a, t). \tag{15.13}$$

Indeed, $J := \{\zeta(a, \rho) : t < \rho \le s\} \subseteq I$, since any $\zeta(a, \rho) \in J$ satisfies

$$\zeta \preceq \zeta(a, t) \preceq \zeta(a, \rho) \preceq \zeta(a, s) = \eta.$$

In addition, clearly $F(J) = (t, s]$. The first equality of equation (15.13) then follows because F is injective. The forward inclusion of the second equality is immediate from the fact that $J \subseteq I$, and the reverse inclusion follows from Proposition 6.27(d), confirming equation (15.13). Since $\mathbb{P}^1_{\mathrm{an}} \smallsetminus \overline{D}_{\mathrm{an}}(a,t)$ is open in $\mathbb{P}^1_{\mathrm{an}}$, we conclude that $F^{-1}((t,s])$ is indeed open in I. □

Proof of Lemma 15.8.

Case 1. Suppose D is closed. Then we may write D as either $\overline{D}_{\mathrm{an}}(a,r)$ or $\mathbb{P}^1_{\mathrm{an}} \smallsetminus D_{\mathrm{an}}(a,r)$. Thus, $D \smallsetminus \{\zeta(a,r)\}$ is the union of all but one of the open disks in the partition of $\mathbb{P}^1_{\mathrm{an}} \smallsetminus \{\zeta(a,r)\}$ from Corollary 6.14.

Thus, we may write $\mathbb{P}^1_{\mathrm{an}} \smallsetminus \{\zeta(a,r)\}$ as a disjoint union of two open sets U and V, where

$$U := D \smallsetminus \{\zeta(a,r)\} \quad \text{and} \quad V := \mathbb{P}^1_{\mathrm{an}} \smallsetminus D.$$

We need to show that $F([0,1])$ is disjoint from V.

If there were some $t \in [0,1]$ such that $F(t) \in V$, we claim that there must be some parameter $s_0 \in [0,t)$ such that $F(s_0) = \zeta(a,r)$. Indeed, if there were no such s_0, then in particular $F(0) \neq \zeta(a,r)$, and hence $F(0) = \zeta_0 \in U$. Moreover, we would have $F([0,t]) \subseteq U \cup V$. Since $F(t) \in V$, we have shown that $F([0,t])$ is disconnected, contradicting the facts that F is continuous and the interval $[0,t]$ is connected. Thus, we have proven the claim.

Similarly, there must also be some $s_1 \in (t,1]$ such that $F(s_1) = \zeta(a,r)$. Then $F(s_0) = F(s_1)$ but $s_0 \neq s_1$, contradicting the injectivity of F.

Hence, we must have $F([0,1]) \cap V = \varnothing$. In other words,

$$F([0,1]) \subseteq \mathbb{P}^1_{\mathrm{an}} \smallsetminus V = D.$$

Case 2. Next, suppose D is open. By Exercise 6.8, we may write D as a nested union of closed disks $D_1 \subseteq D_2 \subseteq \cdots$. Choose n large enough that $\zeta_0, \zeta_1 \in D_n$. Then $F([0,1]) \subseteq D_n \subseteq D$ by Case 1. □

Before proceeding to the proof of Theorem 6.32, we need the following weaker version of the result.

Theorem 15.9. *The Berkovich disk $\overline{D}_{\mathrm{an}}(0,1)$ is uniquely path-connected. Moreover, for any $\zeta_0, \zeta_1 \in \overline{D}_{\mathrm{an}}(0,1)$, all points on the unique arc between them, other than perhaps ζ_0 and ζ_1 themselves, are of Type II or III.*

Proof.

Existence. Given $\zeta_0 \neq \zeta_1 \in \overline{D}_{\mathrm{an}}(0,1)$, let

$$\eta := \zeta_0 \vee \zeta_1 \in \mathbb{A}^1_{\mathrm{an}},$$

which exists by Proposition 6.29. Furthermore, since $\zeta_0, \zeta_1 \preceq \zeta(0,1)$, we must also have $\eta \preceq \zeta(0,1)$ by definition of the least upper bound, and therefore $\eta \in \overline{D}_{\mathrm{an}}(0,1)$. Let

$$I_0 := \{\xi \in \mathbb{P}^1_{\mathrm{an}} : \zeta_0 \preceq \xi \preceq \eta\}$$

and

$$I_1 := \{\xi \in \mathbb{P}^1_{\text{an}} : \zeta_1 \preceq \xi \preceq \eta\}.$$

Then $I_0 \cap I_1 = \{\eta\}$ and $I_0 \cup I_1 \subseteq \overline{D}_{\text{an}}(0,1)$, by definition of the least upper bound, and by the facts that $\eta \preceq \zeta(0,1)$ and that $\preceq$ is a partial order.

Moreover, each of the sets I_0 and I_1 is homeomorphic to a compact interval in $\mathbb{R}$, by Lemma 15.7. Because $I_0 \cap I_1 = \{\eta\}$, it follows that $I_0 \cup I_1$ is also homeomorphic to a compact interval, with ζ_0 and ζ_1 mapping to the endpoints. Thus, we have shown the existence of an arc in $\mathbb{P}^1_{\text{an}}$ between the two points. In addition, any point ξ on this arc $I_0 \cup I_1$ lies above either ζ_0 or ζ_1. If $\xi \neq \zeta_0, \zeta_1$, then by Proposition 6.27(b), ξ must be of Type II or III, proving the final statement of the theorem.

Uniqueness. Suppose that $F : [0,1] \to \overline{D}_{\text{an}}(0,1)$ is a continuous injection with $F(i) = \zeta_i$ for $i = 0, 1$. We need to show that

$$I_0 \cup I_1 = F([0,1]).$$

To prove the forward inclusion, we already have $\zeta_0, \zeta_1 \in F([0,1])$, and therefore we need only consider an arbitrary point

$$\xi \in (I_0 \cup I_1) \smallsetminus \{\zeta_0, \zeta_1\}.$$

Without loss, we may assume that $\xi \in (\zeta_0, \eta]$. Then ξ must be of Type II or III, by Proposition 6.27(b), since $\zeta_0 \precneqq \xi$. Write $\xi = \zeta(a,r)$, and consider the partition of $\mathbb{P}^1_{\text{an}} \smallsetminus \{\xi\}$ into open disks, given by Corollary 6.14.

We have $\zeta_0 \in \overline{D}_{\text{an}}(a,r) \smallsetminus \{\xi\}$ by Proposition 6.27(c); without loss, then, we may assume that $\zeta_0 \in D_{\text{an}}(a,r)$. If $\zeta_1 \in D_{\text{an}}(a,r)$ as well, then setting

$$t := \max\left\{\|z-a\|_{\zeta_0}, \|z-a\|_{\zeta_1}\right\} < s,$$

we have $\zeta_0, \zeta_1 \in \overline{D}_{\text{an}}(a,t)$. But then Proposition 6.27(c)–(d) shows that

$$\zeta_0, \zeta_1 \preceq \zeta(a,t) \precneqq \xi \preceq \eta = \zeta_0 \vee \zeta_1,$$

contradicting the minimality of the least upper bound. Thus, ζ_0 and ζ_1 belong to different open disks in the partition of $\mathbb{P}^1_{\text{an}} \smallsetminus \{\xi\}$.

Let U_0 be the partition element containing ζ_0, and let U_1 be the union of all the other partition elements. Note that $\zeta_1 \in U_1$. If $\xi \notin F([0,1])$, then $F([0,1])$ could be written as the disjoint union of the two nonempty open subsets $F([0,1]) \cap U_0$ and $F([0,1]) \cap U_1$. However, $F([0,1])$ is connected, producing a contradiction and proving that $\xi \in F([0,1])$, as desired.

For the reverse inclusion, define

$$S := F^{-1}(I_0 \cup I_1) \subseteq [0,1].$$

Because $I_0 \cup I_1 \subseteq F([0,1])$, we have $F(S) = I_0 \cup I_1$. Therefore S, like $I_0 \cup I_1$, is a homeomorphic copy of a real interval and hence is connected. Moreover, since $\zeta_0, \zeta_1 \in I_0 \cup I_1$, we also have $0, 1 \in S$. That is, S is a connected subset of $[0,1]$ containing both 0 and 1; hence, $S = [0,1]$. Applying F, we have $F([0,1]) = I_0 \cup I_1$. □

We are now finally prepared to prove that all connected affinoids, including $\mathbb{P}^1_{\text{an}}$ itself, are uniquely path-connected.

Proof of Theorem 6.32.

Case 1. Suppose $U = \mathbb{P}^1_{\text{an}}$. Given two distinct points $\zeta_0 \neq \zeta_1 \in \mathbb{P}^1_{\text{an}}$, there is a path from ζ_0 to ζ_1 by Theorem 15.9 and Proposition 6.13. Moreover, such a path has no points of Type I or IV, except perhaps at the endpoints ζ_0 and ζ_1 themselves, again by Theorem 15.9.

It remains to prove the uniqueness of the path from ζ_0 to ζ_1. Consider the partition of $\mathbb{P}^1_{\text{an}} \setminus \{\zeta(0,1)\}$ of Corollary 6.14. Without loss, we are in one of three cases: that $|\zeta_0|, |\zeta_1| \leq 1$, that $|\zeta_0|, |\zeta_1| \geq 1$, or that $|\zeta_0| < 1 < |\zeta_1|$.

In the first case, if $F : [0,1] \to \mathbb{P}^1_{\text{an}}$ is an injective path from ζ_0 to ζ_1, then since $\zeta_0, \zeta_1 \in \overline{D}_{\text{an}}(0,1)$, we have $F([0,1]) \subseteq \overline{D}_{\text{an}}(0,1)$, by Lemma 15.8. Hence, uniqueness is immediate from Theorem 15.9. The second case follows from the first by exchanging the roles of the two disks in Proposition 6.13.

Finally, if $|\zeta_0| < 1 < |\zeta_1|$, then Corollary 6.14 implies that any arc from ζ_0 to ζ_1 must pass through the Gauss point $\zeta(0,1)$. Thus, the arc is the union of two smaller arcs: one from ζ_0 to $\zeta(0,1)$ and the second from $\zeta(0,1)$ to ζ_1. The first smaller arc is unique by the first case, and the second by the second case; hence, the full arc is also unique.

Case 2. Otherwise, $U \subsetneq \mathbb{P}^1_{\text{an}}$. Write $U = D_1 \cap \cdots \cap D_m$, where $m \geq 1$, and each D_i is a Berkovich disk. Given $\zeta_0, \zeta_1 \in U$, then by Case 1 there is a unique path from ζ_0 to ζ_1 in $\mathbb{P}^1_{\text{an}}$. Moreover, this path consists entirely of points of Type II and III, except perhaps ζ_0 and ζ_1 themselves. The path must also lie in each D_i, and hence in U, by Lemma 15.8. □

15.6. Other Berkovich space proofs

In this section, we prove the remaining results from Chapter 6.

Proof of Proposition 6.35. First, suppose that two of the three points coincide; without loss, $\zeta_0 = \zeta_1$. Then $\xi := \zeta_0$ lies between all three pairs of points, as desired. Moreover, ζ_0 is the only point that lies between ζ_0 and itself, and hence ξ is unique. In addition, ζ_0 trivially lies between $\zeta_1 = \zeta_0$ and ζ_2, and therefore the final statement of the proposition is vacuous.

For the rest of the proof, we may assume that all three point are distinct. Let $\alpha, \beta : [0,1] \to \mathbb{P}^1_{\text{an}}$ be injective parametrizations of the unique arcs from

ζ_0 to ζ_1, and from ζ_0 to ζ_2, respectively. Set

$$s := \sup\big\{t \in [0,1] : \alpha(t) \in \beta\big([0,1]\big)\big\},$$

which exists because $\alpha(0) = \zeta_0 \in \beta([0,1])$. Let $\xi := \alpha(s)$, which lies between ζ_0 and ζ_1 by definition. Note that $\xi \in \beta([0,1])$, by the continuity of α and the compactness of $\beta([0,1])$; hence we may write

$$\xi = \beta(s') \quad \text{for some } s' \in [0,1].$$

In particular, ξ lies between ζ_0 and ζ_2.

To show that ξ also lies between ζ_1 and ζ_2, consider $\alpha([s,1])$ and $\beta([s',1])$, both of which are either intervals or singletons in $\mathbb{P}^1_{\text{an}}$. They intersect at their common starting point $\alpha(s) = \xi = \beta(s')$. However, they intersect nowhere else, because α and β are injective. Thus,

$$I := \alpha\big([s,1]\big) \cup \beta\big([s',1]\big)$$

is an interval in $\mathbb{P}^1_{\text{an}}$ with one endpoint at $\alpha(1) = \zeta_1$, and the other at $\beta(1) = \zeta_2$. Furthermore, this interval contains the point ξ; that is, ξ lies between ζ_1 and ζ_2, as desired.

In addition, the set of points lying between all three of ζ_0, ζ_1, and ζ_2 is, by definition,

$$\alpha\big([0,1]\big) \cap \beta\big([0,1]\big) \cap I = \{\xi\},$$

and hence ξ is indeed unique.

To prove the final statement of the proposition, suppose that none of the three points lies between the other two. Then ξ is distinct from all three, and hence ξ is of Type II or III, by Theorem 6.32. If ξ were of Type III, by Corollary 6.14, $\mathbb{P}^1_{\text{an}} \smallsetminus \{\xi\}$ would consist of two open disks U and V, which together contain all three of ζ_0, ζ_1, and ζ_2. Without loss, then, $\zeta_0, \zeta_1 \in U$. Applying Theorem 6.32 to U, then, we have $\xi \in [\zeta_0, \zeta_1] \subseteq U$, which is a contradiction, since $\xi \notin U$ by definition. Thus, ξ must be of Type II. □

Proof of Theorem 6.42. We begin by proving that $d_{\mathbb{H}}$ takes finite non-negative values. Given $\zeta, \xi \in \mathbb{H} \subseteq \mathbb{A}^1_{\text{an}}$, we have

$$\text{diam}(\zeta \vee \xi) \geq \max\big\{\text{diam}(\zeta), \text{diam}(\xi)\big\} > 0, \tag{15.14}$$

where the first inequality is by Proposition 6.27(d) and the definition of the least upper bound, and the second is by the discussion following Definition 6.3. Thus, $d_{\mathbb{H}}(\zeta, \xi) \in [0, \infty)$.

Clearly, $d_{\mathbb{H}}$ is symmetric, and $d_{\mathbb{H}}(\zeta, \zeta) = 0$, since $\zeta \vee \zeta = \zeta$. In addition, if $d_{\mathbb{H}}(\zeta, \xi) = 0$, then

$$\text{diam}(\zeta \vee \xi) = \text{diam}(\zeta) = \text{diam}(\xi),$$

by inequality (15.14). Therefore, by Proposition 6.27(d), we have

$$\zeta = \zeta \vee \xi = \xi,$$

since $\zeta, \xi \preceq \zeta \vee \xi$.

It remains to show that $d_{\mathbb{H}}$ satisfies the triangle inequality. Given $\zeta, \xi, \eta \in \mathbb{P}^1_{\text{an}}$, we wish to show that

$$d_{\mathbb{H}}(\zeta, \eta) \leq d_{\mathbb{H}}(\zeta, \xi) + d_{\mathbb{H}}(\xi, \eta).$$

We may assume that ζ, ξ, η are all distinct; otherwise, the desired inequality is trivial. By Proposition 6.27(b), each of $\zeta \vee \eta$, $\zeta \vee \xi$, and $\xi \vee \eta$ is a point of Type II or III. Thus, we may choose $a, b, c \in \mathbb{C}_v$ and $r, s, t > 0$ such that

$$\zeta \vee \eta = \zeta(a, r), \quad \zeta \vee \xi = \zeta(b, s), \quad \text{and} \quad \xi \vee \eta = \zeta(c, t).$$

We claim that $r \leq \max\{s, t\}$.

Since $\xi \preceq \zeta \vee \xi$ and $\xi \preceq \xi \vee \eta$, Proposition 6.27(c) implies that

$$\xi \in \overline{D}_{\text{an}}(b, s) \cap \overline{D}_{\text{an}}(c, t).$$

Hence, by Lemma 15.4, we must have either

$$\overline{D}_{\text{an}}(b, s) \subseteq \overline{D}_{\text{an}}(c, t) \quad \text{or} \quad \overline{D}_{\text{an}}(b, s) \supseteq \overline{D}_{\text{an}}(c, t).$$

Without loss, assume the former. Then by Proposition 6.27(c) again,

$$\zeta \preceq \zeta \vee \xi = \zeta(b, s) \preceq \zeta(c, t) = \xi \vee \eta.$$

That is, $\xi \vee \eta$ is an upper bound for ζ. Since it is also an upper bound for η, it is immediate from Definition 6.28 that

$$\zeta(a, r) = \zeta \vee \eta \preceq \xi \vee \eta = \zeta(c, t),$$

and hence by Proposition 6.27(d),

$$r = \text{diam}\big(\zeta(a, r)\big) \leq \text{diam}\big(\zeta(c, t)\big) = t,$$

proving the claim.

By the claim, we may assume without loss that $r \leq t$. Therefore,

$$\begin{aligned}
&d_{\mathbb{H}}(\zeta, \eta)\\
&\quad = 2\log \text{diam}(\zeta \vee \eta) - \log \text{diam}(\zeta) - \log \text{diam}(\eta)\\
&\quad = 2\log r - \log \text{diam}(\zeta) - \log \text{diam}(\eta) \leq 2\log t - \log \text{diam}(\zeta) - \log \text{diam}(\eta)\\
&\quad \leq 2\log \text{diam}(\zeta \vee \xi) + 2\log t - 2\log \text{diam}(\xi) - \log \text{diam}(\zeta) - \log \text{diam}(\eta)\\
&\quad = d_{\mathbb{H}}(\zeta, \xi) + 2\log \text{diam}(\xi \vee \eta) - \log \text{diam}(\xi) - \log \text{diam}(\eta)\\
&\quad = d_{\mathbb{H}}(\zeta, \xi) + d_{\mathbb{H}}(\xi, \eta),
\end{aligned}$$

where the inequality on the third line follows from Proposition 6.27(d), since $\xi \preceq \zeta \vee \xi$. □

The following result shows that the metric $d_{\mathbb{H}}$ is consistent with the notion of lying between.

Proposition 15.10. *Let $\zeta, \xi, \eta \in \mathbb{A}^1_{\mathrm{an}}$ be points such that ξ lies between ζ and η. Then*

$$d_{\mathbb{H}}(\zeta, \eta) = d_{\mathbb{H}}(\zeta, \xi) + d_{\mathbb{H}}(\xi, \eta).$$

Proof. By Exercise 6.26, the least upper bound $\zeta \vee \eta$ lies between ζ and η. Thus, ξ lies either between ζ and $\zeta \vee \eta$, or between η and $\zeta \vee \eta$. Without loss, assume the former. By Lemma 15.7, we have

$$\zeta \preceq \xi \preceq \zeta \vee \eta.$$

Hence $\zeta \vee \xi = \xi$, because $\zeta \preceq \xi$.

In addition, $\zeta \vee \eta$ is an upper bound for both ξ and η, implying that $\xi \vee \eta \preceq \zeta \vee \eta$. On the other hand,

$$\xi \vee \eta \succeq \xi \succeq \zeta$$

is an upper bound for both ζ and η, and hence $\xi \vee \eta \succeq \zeta \vee \eta$. Thus, $\xi \vee \eta = \zeta \vee \eta$.

Combining the conclusions of the previous two paragraphs, then,

$$\begin{aligned} d_{\mathbb{H}}(\zeta, \eta) &= 2 \log \operatorname{diam}(\zeta \vee \eta) - \log \operatorname{diam}(\zeta) - \log \operatorname{diam}(\eta) \\ &= 2 \log \operatorname{diam}(\xi \vee \eta) - \log \operatorname{diam}(\zeta) - \log \operatorname{diam}(\eta) \\ &= d_{\mathbb{H}}(\xi, \eta) + \log \operatorname{diam}(\xi) - \log \operatorname{diam}(\zeta) \\ &= d_{\mathbb{H}}(\xi, \eta) + 2 \log \operatorname{diam}(\zeta \vee \xi) - \log \operatorname{diam}(\xi) - \log \operatorname{diam}(\zeta) \\ &= d_{\mathbb{H}}(\zeta, \xi) + d_{\mathbb{H}}(\xi, \eta). \end{aligned}$$

□

As we noted in Warning 6.43, the metric topology on $\mathbb{H}$ is strictly stronger than the Gel'fand topology. However, as the next result shows, the two topologies coincide when they are restricted to subsets of $\mathbb{H}$ that happen to be intervals.

Proposition 15.11. *Let $\zeta_1, \zeta_2 \in \mathbb{H}$. Then the interval $[\zeta_1, \zeta_2] \subseteq \mathbb{H}$ inherits the same topology from the metric $d_{\mathbb{H}}$ as it does from the Gel'fand topology on $\mathbb{P}^1_{\mathrm{an}}$.*

Proof. By Exercise 6.26, the least upper bound $\zeta_1 \vee \zeta_1$ lies in the interval $[\zeta_1, \zeta_2]$. In either topology, $[\zeta_1, \zeta_2]$ can be formed by gluing the two intervals

$$[\zeta_1, \zeta_1 \vee \zeta_2] \quad \text{and} \quad [\zeta_1 \vee \zeta_2, \zeta_2]$$

at the point $\zeta_1 \vee \zeta_2$. Thus, we may assume without loss that $\zeta_1 \preceq \zeta_2$. Set

$$r := \operatorname{diam}(\zeta_1) \quad \text{and} \quad s := \operatorname{diam}(\zeta_2).$$

By Lemma 15.7, the function

$$G : [\zeta_1, \zeta_2] \to [\log r, \log s] \subseteq \mathbb{R} \quad \text{by} \quad G(\zeta) := \log\big(\operatorname{diam}(\zeta)\big)$$

is a homeomorphism with respect to the Gel'fand topology on $[\zeta_1, \zeta_2]$. On the other hand, for any $\zeta, \xi \in [\zeta_1, \zeta_2]$, we may assume without loss that $\zeta \preceq \xi$, in which case

$$d_{\mathbb{R}}\big(G(\zeta), G(\xi)\big) = \log\big(\operatorname{diam}(\xi)\big) - \log\big(\operatorname{diam}(\zeta)\big) = d_{\mathbb{H}}(\zeta, \xi).$$

Thus, G is also an isometry with respect to the metric $d_{\mathbb{H}}$ on $[\zeta_1, \zeta_2]$, proving that the two topologies do indeed coincide. □

Exercises for Chapter 15

15.1. Prove parts (a) and (b) of Lemma 15.1.

15.2. Prove that the limit in equation (15.3) defining the function $\|\cdot\|_\xi$ in the proof of Lemma 15.5 converges, and that $\|\cdot\|_\xi$ is indeed a multiplicative seminorm on $\mathbb{C}_v[z]$ that extends the absolute value on $\mathbb{C}_v$.

15.3. Prove that the two equivalence relations on $\overline{D} \amalg \overline{D}$ from the end of the proof of Proposition 6.13 are the same.

15.4. Let $U, V \subseteq \mathbb{P}^1_{\mathrm{an}}$ be connected Berkovich affinoids, and suppose that $U \cup V$ is a connected set. Prove that $U \cup V$ is a connected Berkovich affinoid. (*Warning*: U and V may or may not be disjoint.)

15.5. Verify equation (15.9) in the proof of Theorem 6.17.

15.6. Given the result of Step 2 in the proof of Theorem 6.17, prove that every open subset of $\mathbb{P}^1_{\mathrm{an}}$ is a union of open connected Berkovich affinoids.

15.7. Prove Theorem 6.25.

15.8. Prove the following converse to Proposition 15.10: Let $\zeta, \xi, \eta \in \mathbb{H}$ be points such that

$$d_{\mathbb{H}}(\zeta, \eta) = d_{\mathbb{H}}(\zeta, \xi) + d_{\mathbb{H}}(\xi, \eta).$$

Then ξ lies between ζ and η.

15.9. Extend $d_{\mathbb{H}}$ to a function $d_{\mathbb{P}} : \mathbb{P}^1_{\mathrm{an}} \times \mathbb{P}^1_{\mathrm{an}} \to [0, \infty]$ by defining

$$d_{\mathbb{P}}(\zeta, \xi) := d_{\mathbb{H}}(\zeta, \xi) \quad \text{for} \quad \zeta, \xi \in \mathbb{H}$$

and

$$d_{\mathbb{P}}(x, \zeta) = d_{\mathbb{P}}(\zeta, x) := \begin{cases} 0 & \text{if } \zeta = x, \\ \infty & \text{if } \zeta \neq x \end{cases} \quad \text{for} \quad x \in \mathbb{P}^1(\mathbb{C}_v).$$

(a) Prove that $d_{\mathbb{P}}$ is symmetric, nondegenerate (i.e., $d_{\mathbb{P}}(\zeta, \xi) = 0$ if and only if $\zeta = \xi$), and satisfies the triangle inequality.

(Of course, it is not technically a metric, because it takes on the value ∞.)

(b) Extend Proposition 15.10 to $d_{\mathbb{P}}$. That is, for all $\zeta, \xi, \eta \in \mathbb{P}^1_{\text{an}}$ such that ξ lies between ζ and η, prove that

$$d_{\mathbb{P}}(\zeta, \eta) = d_{\mathbb{P}}(\zeta, \xi) + d_{\mathbb{P}}(\xi, \eta).$$

15.10. Let $d_{\mathbb{P}}$ be the function of Exercise 15.9. For any point $\xi \in \mathbb{P}^1_{\text{an}}$, find points $\zeta, \eta \in \mathbb{P}^1_{\text{an}}$ such that

$$d_{\mathbb{P}}(\zeta, \eta) = d_{\mathbb{P}}(\zeta, \xi) + d_{\mathbb{P}}(\xi, \eta),$$

but ξ does *not* lie between ζ and η.

(Thus, the statement of Exercise 15.8 fails for $d_{\mathbb{P}}$.)

15.11. Let $\mathbb{H}'$ denote the set of multiplicative norms on the field $\mathbb{C}_v(z)$ that extend the absolute value on $\mathbb{C}_v$. Equip $\mathbb{H}'$ with the Gel'fand topology, i.e., the weakest topology such that for every $f \in \mathbb{C}_v(z)$, the real-valued function $\zeta \mapsto \|f\|_\zeta$ is continuous.

Prove that $\mathbb{H}'$ is homeomorphic to $\mathbb{H} \subseteq \mathbb{A}^1_{\text{an}}$ via the restriction map

$$\zeta \mapsto \zeta\Big|_{\mathbb{C}_v[z]}.$$

Chapter 16

Proofs of Results on Berkovich Maps

In this chapter, we supply the missing proofs for Chapter 7 on the action of rational functions on $\mathbb{P}^1_{\mathrm{an}}$.

16.1. Basic results on Berkovich maps

Proof of Theorem 7.1. Write $\phi = g/h$, where $g, h \in \mathbb{C}_v[z]$ are relatively prime. The hypothesis that ζ is not a pole of ϕ means that $\|h\|_\zeta \neq 0$.

For any $f(z) = a_n z^n + \cdots + a_0 \in \mathbb{C}_v[z]$, we have

$$\|f \circ \phi\|_\zeta = \frac{\|a_n g^n + a_{n-1} g^{n-1} h + \cdots + a_0 h^n\|_\zeta}{\|h\|_\zeta^n},$$

which is defined and belongs to $[0, \infty)$, since $\|h\|_\zeta \neq 0$. Thus, $\|\cdot\|_{\phi(\zeta)}$ is indeed a function from $\mathbb{C}_v[z]$ to $[0, \infty)$.

We clearly have $\|c\|_{\phi(\zeta)} = \|c\|_\zeta = |c|$ for all $c \in \mathbb{C}_v$. In addition, for any $f_1, f_2 \in \mathbb{C}_v[z]$,

$$\|f_1 f_2\|_{\phi(\zeta)} = \big\|(f_1 \circ \phi) \cdot (f_2 \circ \phi)\big\|_\zeta = \|f_1 \circ \phi\|_\zeta \cdot \|f_2 \circ \phi\|_\zeta = \|f_1\|_{\phi(\zeta)} \cdot \|f_2\|_{\phi(\zeta)}$$

and

$$\begin{aligned}\|f_1 + f_2\|_{\phi(\zeta)} &= \big\|(f_1 \circ \phi) + (f_2 \circ \phi)\big\|_\zeta \leq \max\big\{\|f_1 \circ \phi\|_\zeta, \|f_2 \circ \phi\|_\zeta\big\} \\ &= \max\big\{\|f_1\|_{\phi(\zeta)}, \|f_2\|_{\phi(\zeta)}\big\},\end{aligned}$$

as desired. □

Before proving Theorem 7.4, we note the following alternative characterization of the Gel'fand topology.

Lemma 16.1. *The Gel'fand topology on* $\mathbb{A}^1_{\text{an}}$ *is the weakest topology such that for any real number* $t \in \mathbb{R}$ *and any two polynomials* $f, g \in \mathbb{C}_v[z]$, *the set* $\{\zeta \in \mathbb{A}^1_{\text{an}} : \|f\|_\zeta > t\|g\|_\zeta\}$ *is open.*

Proof. Let $\mathcal{U}$ be the Gel'fand topology on $\mathbb{A}^1_{\text{an}}$, and let $\mathcal{V}$ be the topology described in the statement of the lemma. Since sets of the form $[0,t)$ and (t,∞) together generate the topology on $[0,\infty) \subseteq \mathbb{R}$, sets of the form

$$\{\zeta \in \mathbb{A}^1_{\text{an}} : \|h\|_\zeta < t\} \quad \text{and} \quad \{\zeta \in \mathbb{A}^1_{\text{an}} : \|h\|_\zeta > t\},$$

for $h \in \mathbb{C}_v[z]$ and $t \in \mathbb{R}$, together generate $\mathcal{U}$. All such sets are in $\mathcal{V}$ (simply choose one of f or g to be h, and the other to be 1), and hence $\mathcal{U} \subseteq \mathcal{V}$.

Conversely, given $f, g \in \mathbb{C}_v[z]$ and $t \in \mathbb{R}$, we wish to show that

$$W := \{\zeta \in \mathbb{A}^1_{\text{an}} : \|f\|_\zeta > t\|g\|_\zeta\}$$

is $\mathcal{U}$-open. Given $\xi \in W$, choose open intervals $I, J \subseteq \mathbb{R}$ with $\|f\|_\xi \in I$ and $\|g\|_\xi \in J$ such that $x > ty$ for all $x \in I$ and $y \in J$. Let

$$U := \{\zeta \in \mathbb{A}^1_{\text{an}} : \|f\|_\zeta \in I \text{ and } \|g\|_\zeta \in J\}.$$

Then $U \in \mathcal{U}$, and $\xi \in U \subseteq W$. We can do this for all $\xi \in W$, and therefore $W \in \mathcal{U}$, as desired. □

Proof of Theorem 7.4. Because the Type I points are dense in $\mathbb{P}^1_{\text{an}}$ by Theorem 6.25(a), uniqueness is immediate once we know that Definition 7.2 gives *some* continuous extension of ϕ to $\mathbb{P}^1_{\text{an}}$.

Step 1. First we show $\phi : \mathbb{P}^1_{\text{an}} \to \mathbb{P}^1_{\text{an}}$ agrees with the usual action of ϕ on the Type I points. Given $a \in \mathbb{P}^1(\mathbb{C}_v)$, if a is either ∞ or a pole of ϕ, then the two definitions of $\phi(a)$ agree immediately. Thus, we may assume that $a, \phi(a) \in \mathbb{C}_v$. In that case, the definition of $\phi(a)$ from equation (7.1) gives

$$\|f\|_{\phi(a)} = \|f \circ \phi\|_a = \big|f \circ \phi(a)\big| = \big|f\big(\phi(a)\big)\big|$$

for all $f \in \mathbb{C}_v[z]$, coinciding with the earlier definition of the Type I seminorm $\|\cdot\|_{\phi(a)}$ from Section 6.1.1.

Step 2. Now we show $\phi : \mathbb{P}^1_{\text{an}} \to \mathbb{P}^1_{\text{an}}$ is continuous. By Theorem 6.17, combined with the fact that every connected affinoid is a finite intersection of disks, it suffices to show that $\phi^{-1}(U)$ is an open subset of $\mathbb{P}^1_{\text{an}}$ whenever U is of the form either

$$U = D_{\text{an}}(c,r) \quad \text{or} \quad U = \mathbb{P}^1_{\text{an}} \smallsetminus \overline{D}_{\text{an}}(c,r)$$

for $c \in \mathbb{C}_v$ and $r > 0$. For the remainder of the proof, fix $c \in \mathbb{C}_v$ and $r > 0$, and write

$$\phi(z) = c + \frac{f(z)}{g(z)}$$

for some relatively prime $f, g \in \mathbb{C}_v[z]$. Let $d := \deg(\phi)$.

Case 1. Suppose $U = D_{\mathrm{an}}(c, r)$. Then

$$\phi^{-1}(U) \cap \mathbb{A}^1_{\mathrm{an}} = \{\zeta \in \mathbb{A}^1_{\mathrm{an}} : \|f\|_\zeta < r\|g\|_\zeta\},$$

which is open in $\mathbb{A}^1_{\mathrm{an}}$, and hence in $\mathbb{P}^1_{\mathrm{an}}$, by Lemma 16.1. In addition, if $\phi(\infty) \in U$, then we may write

$$\phi(z) - \phi(\infty) = \frac{a_{d-1}z^{d-1} + \cdots + a_0}{z^d + b_{d-1}z^{d-1} + \cdots + b_0}.$$

Therefore, we may choose $R > 0$ large enough that $\|\phi(z) - \phi(\infty)\|_\zeta < r$ whenever $\|z\|_\zeta > R$. Thus,

$$\phi(\mathbb{P}^1_{\mathrm{an}} \smallsetminus \overline{D}_{\mathrm{an}}(0, R)) \subseteq D_{\mathrm{an}}(\phi(\infty), r) = D_{\mathrm{an}}(c, r) = U,$$

completing the proof that $\phi^{-1}(U)$ is open in $\mathbb{P}^1_{\mathrm{an}}$.

Case 2. Otherwise, $U = \mathbb{P}^1_{\mathrm{an}} \smallsetminus \overline{D}_{\mathrm{an}}(c, r)$. Then

$$\phi^{-1}(U) \cap \mathbb{A}^1_{\mathrm{an}} = V \cup (\phi^{-1}(\infty) \cap \mathbb{C}_v),$$

where

$$V := \{\zeta \in \mathbb{A}^1_{\mathrm{an}} : \|f\|_\zeta > r\|g\|_\zeta\}.$$

However, for any $a \in \phi^{-1}(\infty) \cap \mathbb{C}_v$, we have $f(a) \neq 0$ and $g(a) = 0$. Thus, $\|f\|_a > 0 = r\|g\|_a$, and hence $\phi^{-1}(\infty) \cap \mathbb{C}_v \subseteq V$. That is, $\phi^{-1}(U) \cap \mathbb{A}^1_{\mathrm{an}} = V$, which is open by Lemma 16.1.

In addition, if $\phi(\infty) \in U \smallsetminus \{\infty\}$, then by considering $\phi(z) - \phi(\infty)$ as in Case 1, we may choose $R > 0$ large enough that $\|\phi(z) - \phi(\infty)\|_\zeta < r$ whenever $\|z\|_\zeta > R$, whence

$$\phi(\mathbb{P}^1_{\mathrm{an}} \smallsetminus \overline{D}_{\mathrm{an}}(0, R)) \subseteq D_{\mathrm{an}}(\phi(\infty), r) \subseteq \mathbb{P}^1_{\mathrm{an}} \smallsetminus \overline{D}_{\mathrm{an}}(c, r) = U.$$

Finally, if $\phi(\infty) = \infty$, then we may write

$$\phi(z) - c = \frac{z^d + a_{d-1}z^{d-1} + \cdots + a_0}{b_{d-1}z^{d-1} + \cdots + b_0},$$

and therefore we may choose $R > 0$ large enough that $\|\phi(z) - c\|_\zeta > r$ whenever $\|z\|_\zeta > R$. Thus,

$$\phi(\mathbb{P}^1_{\mathrm{an}} \smallsetminus \overline{D}_{\mathrm{an}}(0, R)) \subseteq \mathbb{P}^1_{\mathrm{an}} \smallsetminus \overline{D}_{\mathrm{an}}(c, r) = U,$$

completing the proof that $\phi^{-1}(U)$ is open. □

Proof of Proposition 7.6. By Theorem 3.15, $\phi(D(a, r))$ is indeed an open disk $D(b, s)$ of the same type (rational or irrational) as $D(a, r)$. Pick a sequence of points $\{x_n\}_{n \geq 1} \subseteq D(a, r)$ such that $\lim_{n \to \infty} |x_n - a| = r$. Then because $\phi(z) - b$ has a power series expansion on $D(a, r)$, we have

$$\begin{aligned}\lim_{n\to\infty} |\phi(x_n) - b| &= \|\phi - b\|_{\zeta(a,r)} = \sup\{|\phi(x) - b| : x \in D(a, r)\} \\ &= \sup\{|y - b| : y \in D(b, s)\} = s,\end{aligned}$$

where the first equality is by Proposition 3.19, and the second is by Proposition 3.17. In particular, the sequence

$$\{y_n\}_{n\geq 1} \quad \text{given by} \quad y_n := \phi(x_n)$$

satisfies $\lim_{n\to\infty} |y_n - b| = s$.

Thus, for any polynomial $h \in \mathbb{C}_v[z]$, noting that $h \circ \phi$ has no poles in $D(a, r)$, we have

$$\|h\|_{\phi(\zeta)} = \|h \circ \phi\|_\zeta = \lim_{n\to\infty} \big|h\big(\phi(x_n)\big)\big| = \lim_{n\to\infty} \big|h(y_n)\big| = \|h\|_{\zeta(b,s)},$$

where the first equality is by Definition 7.2, and the second and fourth are by Proposition 3.19. Hence, $\phi(\zeta) = \zeta(b, s)$. □

Proof of Proposition 7.7. Because ϕ has only finitely many poles, there is some $N \geq 1$ such that ϕ has no poles in D_n for $n \geq N$. By Theorem 3.15, $\phi(D_n) \subseteq \mathbb{C}_v$ is a disk for all $n \geq N$, and $\phi : D_n \to \phi(D_n)$ is a d_n-to-1 map, for some integer $d_n \geq 1$. Clearly, $d_n \leq d_m$ for $n \geq m$, and hence this decreasing sequence of positive integers is eventually constant. By increasing N if necessary, we may assume that $d_n = d$ is constant for all $n \geq N$.

It is immediate that $\phi(D_N) \supseteq \phi(D_{N+1}) \supseteq \cdots$ is a decreasing sequence of disks with empty intersection. This sequence therefore defines a Type IV point $\xi \in \mathbb{P}^1_{\text{an}}$. It remains to show that $\phi(\zeta) = \xi$.

For each $n \geq N$, let $\zeta_n \in \mathbb{P}^1_{\text{an}}$ be the point of Type II or III associated with the disk D_n. Bearing in mind that $\phi(D(a,r)) = D(b,s)$ implies that $\phi(\overline{D}(a,r)) = \overline{D}(b,s)$ provided ϕ has no poles in $\overline{D}(a,r)$, Proposition 7.6 implies that $\phi(\zeta_n) \in \mathbb{P}^1_{\text{an}}$ is the point of Type II or III associated with the disk $\phi(D_n)$. Thus, by the continuity of ϕ from Theorem 7.4, we have

$$\phi(\zeta) = \lim_{n\to\infty} \phi(\zeta_n) = \xi.$$ □

Proof of Theorem 7.8.

Step 1. We first prove the statements about the forward image $\phi(W)$ in the case that W is closed and hence compact. Since $\phi : \mathbb{P}^1_{\text{an}} \to \mathbb{P}^1_{\text{an}}$ is continuous, it follows that $\phi(W)$ is also compact. Therefore, since $\mathbb{P}^1_{\text{an}}$ is Hausdorff, $\phi(W)$ is a closed subset of $\mathbb{P}^1_{\text{an}}$ containing $\phi(W_{\text{I}})$.

On the other hand, by Exercise 6.28(a), W is the closure of W_{I} in $\mathbb{P}^1_{\text{an}}$. Thus, given any closed set $C \subseteq \mathbb{P}^1_{\text{an}}$ containing $\phi(W_{\text{I}})$, its inverse image $\phi^{-1}(C)$ contains W. It follows that C contains $\phi(W)$, which is therefore the closure of $\phi(W)$. Again by Exercise 6.28(a), $\phi(W)$ is the closed Berkovich connected affinoid corresponding to $\phi(W_{\text{I}})$. In addition, one is clearly rational (respectively, irrational) if and only if the other is.

Step 2. Still assuming W is closed, the proof that $\phi^{-1}(W)$ is the closed Berkovich affinoid corresponding to $\phi^{-1}(W_{\mathrm{I}})$, and that statement (b) holds, is similar. We leave the details to the reader; see Exercise 16.1.

Step 3. The proof of the foregoing statements, including part (b), when W is not necessarily closed, follows from the fact that an arbitrary connected affinoid (either in $\mathbb{P}^1(\mathbb{C}_v)$ or in $\mathbb{P}^1_{\mathrm{an}}$) can be written as a union of an increasing sequence of closed affinoids. This, in turn, is because an arbitrary disk can be written as a union of an increasing sequence of closed disks.

Step 4. We now turn to part (a). Given $\zeta \in \partial(\phi(W))$, choose a sequence of Type I points $\{y_n\}_{n\geq 1} \subseteq \phi(W_{\mathrm{I}})$ approaching ζ, and for each n, choose $x_n \in \phi^{-1}(y_n) \cap W$. Replacing $\{x_n\}_{n\geq 1}$ by a convergent subsequence, let ξ be the limit point; then $\xi \in \overline{W}$, and $\phi(\xi) = \zeta$. For any open set U containing ξ, we need to show that $U \not\subseteq W$. Given any such U, however, we may shrink U if necessary to ensure that U is an open connected affinoid. By Step 3, $\phi(U)$ is also open and contains $\phi(\xi) = \zeta$. Because $\zeta \in \partial(\phi(W))$, there is some $y \in \phi(U) \smallsetminus \phi(W)$. Write $y = \phi(x)$ for some $x \in U$. Since $y \notin \phi(W)$, we have $x \notin W$, and hence $x \in U \smallsetminus W$, as desired. Thus, $\xi \in \partial W$, and hence $\zeta \in \phi(\partial W)$, proving statement (a).

Step 5. We now turn to part (c). To show that $\phi(\partial V_i) \subseteq \partial W$, given a point $\zeta \in \partial V_i$ and an open set U containing $\phi(\zeta)$, then $\phi^{-1}(U)$ is an open set containing ζ. There must be an open connected Berkovich affinoid U' containing ζ and contained in $\phi^{-1}(U)$. Since $\zeta \in \partial V_i$, there is some $x \in V_i \cap U'$, and hence $\phi(x) \in W \cap U$. On the other hand, if $U' \subseteq \phi^{-1}(W)$, then $U' \cap V_i$ is connected and contained in $\phi^{-1}(W)$. Since V_i is a connected component of $\phi^{-1}(W)$, we must have $\zeta \in U' \subseteq V_i$, contradicting the fact that $\zeta \in \partial V_i$. Thus, U intersects both W and its complement, and hence $\phi(\zeta) \in \partial W$, as desired. The reverse inclusion is immediate from part (a).

We leave the similar (and simpler) proof that $\phi(\operatorname{int} V_i) = \operatorname{int} W$ to the reader; see Exercise 16.1. Finally, part (d) is immediate from part (c). □

Proof of Corollary 7.10. If $\phi(U) \cap \phi(\partial U) = \varnothing$, let V be the connected component of $\phi^{-1}(\phi(U))$ containing U. If $V \neq U$, then because U and V are both connected and open, there must be some point $\zeta \in V \cap \partial U$. Thus, by Theorem 7.8(b),

$$\phi(\zeta) \in \phi(V) \cap \phi(\partial U) = \phi(U) \cap \phi(\partial U) = \varnothing,$$

a contradiction. Hence, $V = U$, as desired.

Conversely, if U is a connected component of $\phi^{-1}(\phi(U))$, then by Theorem 7.8(c), the two sets

$$\phi(U) = \phi(\operatorname{int} U) = \operatorname{int} \phi(U) \quad \text{and} \quad \phi(\partial U) = \partial(\phi(U))$$

are disjoint, as desired. □

Proof of Theorem 7.12. For any $\lambda < 1$ sufficiently close to 1, ϕ has no poles in the annulus

$$U := \{\lambda r < |z-a| < r\} = D(a,r) \smallsetminus \overline{D}(a,\lambda r).$$

By Proposition 3.32, we may write ϕ as a Laurent series

$$\phi(z) = \sum_{n\in\mathbb{Z}} c_n(z-a)^n$$

converging on U. Increasing $\lambda < 1$ if necessary, we may further assume that $\phi - c_0$ has no zeros in U. By the same proposition, the inner and outer Weierstrass degrees of $\phi - c_0$ on U both equal the same nonzero integer d. The statement characterizing the image $\phi(U)$ is now immediate from Theorem 3.33: by part (b) if $d > 0$, or by part (c) if $d < 0$. Moreover, the integer d does not change if we move λ even closer to 1. In addition, since $d \neq 0$, we have $r \in |\mathbb{C}_v^\times|$ if and only if $|c_d|r^d \in |\mathbb{C}_v^\times|$, yielding the final statement of the theorem.

It remains only to show that $\phi(\zeta) = \zeta(c_0, s)$. We consider the case that $d > 0$; the case that $d < 0$ is similar. Pick a sequence of points

$$\{x_n\}_{n\geq 1} \subseteq U = \{\lambda r < |z-a| < r\}$$

such that $\lim_{n\to\infty} |x_n - a| = r$, and observe that

$$\lim_{n\to\infty} |\phi(x_n) - c_0| = \lim_{n\to\infty} \left|c_d(x_n-a)^d\right| = |c_d|r^d = s.$$

Thus, for any $h \in \mathbb{C}_v[z]$,

$$\|h\|_{\phi(\zeta)} = \|h\circ\phi\|_\zeta = \lim_{n\to\infty} \big|h\big(\phi(x_n)\big)\big| = \|h\|_{\zeta(c_0,s)},$$

where the first equality is by Definition 7.2, and the second and third are by Proposition 3.19. Hence, $\phi(\zeta) = \zeta(c_0, s)$. ☐

16.2. Proofs on local degrees

Proof of Theorem 7.22. If equation (7.3) holds, then ϕ would map $[\zeta,\xi]$ bijectively onto its image; and by unique path-connectedness, that image would have to be $[\phi(\zeta),\phi(\xi)]$. Since ϕ is continuous and the interval $[\zeta,\xi]$ is compact, the mapping

$$\phi : [\zeta,\xi] \to [\phi(\zeta),\phi(\xi)]$$

would be a homeomorphism. Thus, it suffices to show equation (7.3), which we now do on a type-by-type basis. Note, by Proposition 7.17, that we may freely change coordinates via linear fractional transformations on either the domain or range.

If ζ is of Type I, then after (possibly different) changes of coordinates on both the domain and range, we may assume that $\zeta = \phi(\zeta) = 0$, and hence that the power series expansion of ϕ on some small enough disk $D(0,r)$ is of the form

$$\phi(z) = \sum_{n\geq m} c_n z^n \quad \text{with} \quad c_m \neq 0.$$

(Recall that $m = \deg_{\zeta,\boldsymbol{v}}(\phi) \geq 1$.) Shrinking r if necessary, we may assume that

$$|c_m| r^m \geq |c_n| r^n \quad \text{for all } n \geq m.$$

Let $\xi := \zeta(0,r)$. Given $\xi_1, \xi_2 \in [0,\xi]$, we may write $\xi_i = \zeta(0,s_i)$ for $s_i \in [0,r]$, yielding

$$\phi(\xi_i) = \zeta\big(0, |c_m| s_i^m\big).$$

Hence, assuming $s_1 \leq s_2$,

$$d_{\mathbb{H}}\big(\phi(\xi_1), \phi(\xi_2)\big) = \log\left(\frac{|c_m|s_2^m}{|c_m|s_1^m}\right) = m\log\left(\frac{s_2}{s_1}\right) = m \cdot d_{\mathbb{H}}(\xi_1,\xi_2), \tag{16.1}$$

where we understand both sides to be $+\infty$ if $s_1 = 0$.

If ζ if of Type II or III, then after changing coordinates on domain and range, we may write

$$\zeta = \zeta(0,r) \quad \text{and} \quad \phi(\zeta) = \zeta(0,s),$$

and we may assume that $\boldsymbol{v}$ is the direction at ζ containing 0, and that $\phi_{\#}(\boldsymbol{v})$ is the direction at $\phi(\zeta)$ containing 0. By Theorem 7.12, there is some $\lambda < 1$ sufficiently close to 1 such that ϕ has inner and outer Weierstrass degrees equal to m on the annulus

$$U := \{\lambda r < |z-a| < r\} = D(a,r) \smallsetminus \overline{D}(a,\lambda r).$$

Let $\xi := \zeta(a,\lambda r)$.

Given $\xi_1, \xi_2 \in [\zeta,\xi]$, we may write $\xi_i = \zeta(0,s_i)$ with $s_i \in [\lambda r, r]$ for $i = 1,2$. Writing $\phi(z) = \sum_{n\in\mathbb{Z}} c_n z^n$ as a Laurent series converging on U, by the final statement of Theorem 3.33, we have

$$\phi(\xi_1) = \zeta\big(0, |c_m| s_1^m\big) \quad \text{and} \quad \phi(\xi_2) = \zeta\big(0, |c_m| s_2^m\big).$$

Equation (16.1) then applies as before, and we are done.

Finally, if ζ is of Type IV, let $D_1 \supsetneq D_2 \supsetneq \cdots$ be a chain of disks representing ζ. By Proposition 7.7, there is an integer $N \geq 1$ such that

$$\phi : D_n \to \phi(D_n) \quad \text{is} \quad m\text{-to-1} \quad \text{for all } n \geq N.$$

Write $D_n = \overline{D}(a_n, r_n)$, and let $\xi := \zeta(a_N, r_N)$.

Given $\xi_1, \xi_2 \in [\zeta,\xi]$, the desired conclusion holds trivially if $\xi_1 = \xi_2$. Thus, we may assume the two points are distinct. Consider first the case that neither point coincides with ζ, and write

$$\xi_i = \zeta(b_i, s_i), \quad \text{where} \quad \operatorname{diam}(\zeta) < s_i \leq r_N \quad \text{for } i = 1,2.$$

Pick $\ell \geq N$ large enough that $s_i > r_\ell$ for $i = 1, 2$. Then ξ_1 and ξ_2 both lie between $\zeta(a_\ell, r_\ell)$ and $\xi = \zeta(a_\ell, r_N)$.

After changing coordinates on both the domain and range, we may assume that $a_\ell = \phi(a_\ell) = 0$, and thus we may expand ϕ as a power series on $\overline{D}(0, r_N)$ of the form

$$\phi(z) = \sum_{i \geq 1} c_i z^i \quad \text{with} \begin{cases} |c_i| r_\ell^i \leq |c_m| r_\ell^m & \text{for all } i \leq m, \quad \text{and} \\ |c_i| r_N^i \leq |c_m| r_N^m & \text{for all } i \geq m. \end{cases}$$

That is, ϕ has inner and outer Weierstrass degrees both equal to m on the annulus $\{r_\ell < |z| < r_N\}$. Equation (16.1) then applies as before, and we are done. The same conclusion immediately extends to the case that one of ξ_1 or ξ_2 is the point ζ, by the continuity of the diam function on the interval $[\zeta, \xi]$, according to Lemma 15.7. □

Proof of Lemma 7.27. Because the quantities in equation (7.5) are positive integers, we may shrink U if necessary to guarantee that

$$\sup_y \left(\sum_{x \in U \cap \phi^{-1}(y)} \deg_x(\phi) \right) = \deg_\zeta(\phi), \tag{16.2}$$

where the supremum is over all $y \in \phi(U) \cap \mathbb{P}^1(\mathbb{C}_v)$. Moreover, equation (16.2) remains true if we replace U by any smaller open set containing ζ, since U attains the infimum of equation (7.5).

By Theorem 6.17, we may shrink U further and assume that U is an open connected Berkovich affinoid. Its boundary $S := \partial U$ is a finite set of points of Types II and III, and U is the connected component of $\mathbb{P}^1_{\text{an}} \smallsetminus S$ containing ζ. If $\phi(\zeta) = \phi(\xi)$ for some point $\xi \in S$, then we may yet again shrink U slightly by moving ξ a small distance toward ζ along the interval $[\zeta, \xi] \subseteq \mathbb{P}^1_{\text{an}}$. Thus, we may assume that $\phi(\zeta) \notin \phi(S)$.

Let $T := \phi^{-1}(\phi(S))$. Then T is a finite set, because we claim that any point $\xi \in \mathbb{P}^1_{\text{an}}$ has at most $\deg(\phi)$ distinct preimages under ϕ. To see this, suppose that ξ had at least $n > \deg(\phi)$ distinct preimages $\eta_1, \ldots, \eta_n$. Because $\mathbb{P}^1_{\text{an}}$ is Hausdorff and the open affinoids form a basis for the topology, we may choose pairwise disjoint open connected Berkovich affinoids $U_1, \ldots, U_n$ with $\eta_i \in U_i$. Let

$$V := \bigcap_i \phi(U_i),$$

which is an open connected Berkovich affinoid containing ξ, by Theorem 7.8. Moveover, each Type I point in V has at least one preimage in each U_i, contradicting the fact that each can only have $\deg(\phi) < n$ preimages. Thus, our claim is proven, and T is indeed finite.

Note that $\zeta \notin T$ and that $S \subseteq T$. In addition, T consists solely of points of Types II and III. Let W be the connected component of $\mathbb{P}^1_{\mathrm{an}} \smallsetminus T$ containing ζ. Then W is an open connected Berkovich affinoid (by Exercise 6.19(c)) and $\zeta \in W \subseteq U$. We must show that W also satisfies properties (b) and (c).

Given $\xi \in \partial W \subseteq T$, the image $\phi(\xi) \in \phi(T)$ must lie in the closure of $\phi(W)$, by continuity. If $\phi(\xi)$ lies in the interior of $\phi(W)$, then since ξ is a point of Type II or III, there are points $x, y \in W$ such that $\phi(x)$ and $\phi(y)$ lie in distinct directions at $\phi(\xi)$. Since W and $\mathbb{P}^1_{\mathrm{an}}$ are each uniquely path-connected by Theorem 6.32, the interval $[x, y]$ lies entirely in W. Thus,

$$\phi(\xi) \in [\phi(x), \phi(y)] \subseteq \phi([x, y]),$$

where the inclusion is by continuity and unique path-connectedness. In particular, $[x, y]$ contains a point of

$$\phi^{-1}(\phi(\xi)) \subseteq \phi^{-1}(\phi(T)) = T.$$

But $[x, y] \subseteq W$, which is disjoint from T by definition. It follows from this contradiction that $\phi(\xi)$ must lie in the boundary of $\phi(W)$, proving property (b).

We claim that every Type I point in $\phi(W)$ has the same number of preimages in W, counted with multiplicity; property (c) then follows by equation (16.2). By Proposition 3.29(b) and Theorem 7.8(b), it suffices to show that W is a connected component of $\phi^{-1}(\phi(W))$. Clearly,

$$W \subseteq \phi^{-1}(\phi(W)),$$

and by definition, W is connected. If $W' \supsetneq W$ is a strictly larger connected set, then W' must contain at least one point of $\partial W \subseteq T$. Hence, $\phi(W') \cap S \neq \varnothing$, and therefore W' is not contained in $\phi^{-1}(\phi(W))$. Thus, no such W' exists, and W is indeed a connected component of $\phi^{-1}(\phi(W))$. □

Proof of Theorem 7.29.

(a) Let $U \subseteq \mathbb{P}^1_{\mathrm{an}}$ be an open set that contains ζ and attains the infimum of equation (7.5) defining $\deg_\zeta(\phi)$. For any $\xi \in U$, the same U is one of the open sets appearing in the corresponding infimum defining $\deg_\xi(\phi)$, and therefore $\deg_\xi(\phi) \leq \deg_\zeta(\phi)$. Thus, ζ has an open neighborhood on which $\deg_\xi(\phi)$ is never larger than $\deg_\zeta(\phi)$, as desired.

(b) Let $U_1, U_2 \subseteq \mathbb{P}^1_{\mathrm{an}}$ be a open sets such that

- $\zeta \in U_1$, and U_1 attains the infimum defining $\deg_\zeta(\psi \circ \phi)$, and
- $\phi(\zeta) \in U_2$, and U_2 attains the infimum defining $\deg_{\phi(\zeta)}(\psi)$.

Let $U := U_1 \cap \phi^{-1}(U_2)$, which is an open set containing ζ. Choose $W \subseteq U$ to be an open connected Berkovich affinoid having the property of Lemma 7.27 for the map ϕ at ζ.

Then for any $y \in \psi(\phi(W)) \cap \mathbb{P}^1(\mathbb{C}_v)$, we have

$$\sum_{x \in W \cap (\psi\circ\phi)^{-1}(y)} \deg_x(\psi \circ \phi) = \sum_{t \in \phi(W)\cap\psi^{-1}(y)} \ \sum_{x \in W \cap \phi^{-1}(t)} \deg_x(\phi) \cdot \deg_t(\psi)$$
$$= \deg_\zeta(\phi) \cdot \sum_{t \in \phi(W)\cap\psi^{-1}(y)} \deg_t(\psi),$$

where the last equality is because W has the property of Lemma 7.27 for ϕ. Thus,

$$\sup_y \Big(\sum_x \deg_x(\psi \circ \phi) \Big) = \deg_\zeta(\phi) \sup_y \Big(\sum_t \deg_t(\psi) \Big), \tag{16.3}$$

where the sum on the left is over all $x \in W \cap (\psi \circ \phi)^{-1}(y)$, the sum on the right is over all $t \in \phi(W) \cap \psi^{-1}(y)$, and the suprema on both sides are over all $y \in \psi(\phi(W)) \cap \mathbb{P}^1(\mathbb{C}_v)$. However, because U_1 attains the infimum defining $\deg_\zeta(\psi \circ \phi)$ and $\zeta \in W \subseteq U_1$, the supremum on the left-hand side of equation (16.3) is $\deg_\zeta(\psi \circ \phi)$. Similarly, because

$$\phi(\zeta) \in \phi(W) \subseteq U_2,$$

along with the facts that $\phi(W)$ is open (by Corollary 7.9) and that U_2 attains the infimum defining $\deg_{\phi(\zeta)}(\psi)$, the supremum on the right-hand side of equation (16.3) is $\deg_{\phi(\zeta)}(\psi)$. Thus,

$$\deg_\zeta(\psi \circ \phi) = \deg_\zeta(\phi) \cdot \deg_{\phi(\zeta)}(\psi).$$

(c) Since $\mathbb{P}^1_{\text{an}}$ is Hausdorff, for each $\xi \in \phi^{-1}(\zeta)$, we may choose an open set $U(\xi)$ containing ξ such that any two of the sets $U(\xi)$ are disjoint. For each such ξ, there is an open connected Berkovich affinoid $W(\xi) \subseteq U(\xi)$ satisfying the conclusion of Lemma 7.27 at ξ.

Let $V := \bigcap_\xi \phi(W(\xi))$, which is also an open connected Berkovich affinoid, by Theorem 7.8. We claim that each connected component of $\phi^{-1}(V)$ is contained in some $W(\xi)$. To see this, write $\phi^{-1}(V)$ as a disjoint union of finitely many open connected Berkovich affinoids $X_1, \ldots, X_m$, according to Theorem 7.8. Each X_i maps onto V, and hence each X_i contains some $\xi \in \phi^{-1}(\zeta)$. If X_i were not contained in $W(\xi)$, then being connected, X_i must contain some point of $\partial W(\xi)$. By property (b) of Lemma 7.27 for $W(\xi)$, $V = \phi(X_i)$ contains a point of $\partial\phi(W(\xi))$. However, V is contained in $\phi(W(\xi))$, and hence the open set $\phi(W(\xi))$ intersects its own boundary. This contradiction proves the claim.

Since each $W(\xi)$ contains exactly one point of $\phi^{-1}(\zeta)$, and since each connected component of $\phi^{-1}(V)$ contains at least one such point, the claim implies that each connected component of $\phi^{-1}(V)$ contains *exactly* one point of $\phi^{-1}(\zeta)$. (Here, we are counting distinct points, not with multiplicity.)

Thus, we may denote the components of $\phi^{-1}(V)$ by $W'(\xi)$, as ξ ranges through $\phi^{-1}(\zeta)$, with $W'(\xi) \cap \phi^{-1}(\zeta) = \{\xi\}$.

For fixed $\xi \in \phi^{-1}(\zeta)$, each Type I point of V has the same number of preimages in $W'(\xi)$, counting multiplicity, by Proposition 3.29. Because

$$W'(\xi) \subseteq W(\xi) \subseteq U(\xi),$$

this common number must be $\deg_\xi(\phi)$. Thus, picking any Type I point $y \in V \cap \mathbb{P}^1(\mathbb{C}_v)$, we have

$$\sum_{\phi(\xi)=\zeta} \deg_\xi(\phi) = \sum_{\phi(\xi)=\zeta} \sum_{\substack{x\in W'(\xi)\\ \phi(x)=y}} \deg_x(\phi) = \sum_{\phi(x)=y} \deg_x(\phi) = \deg(\phi). \qquad \square$$

Proof of Theorem 7.30. If ζ is of Type I, then there is only one direction $\boldsymbol{v}$ at ζ. We have $\phi_\#(\boldsymbol{v}) = \boldsymbol{w}$, and $\deg_\zeta(\phi) = \deg_{\zeta,\boldsymbol{v}}(\phi)$ both coincide with the algebraic multiplicity of ϕ at ζ.

If ζ is of Type IV, then let $\boldsymbol{v}$ be the unique direction at ζ, and let $D_1 \supseteq D_2 \supseteq \cdots$ be a chain of disks representing ζ. Discarding the first few terms and re-indexing if need be, we may assume that ϕ is a $\deg_{\zeta,\boldsymbol{v}}(\phi)$-to-1 map on each D_n. Any open set $U \subseteq \mathbb{P}^1_{\text{an}}$ containing ζ must contain some D_n, and hence Definition 7.26 gives

$$\deg_\zeta(\phi) = \deg_{\zeta,\boldsymbol{v}}(\phi).$$

If ζ is of Type II or III, then after a change of coordinates on the range, we may assume that $\phi(\zeta) = \zeta(0, r)$, and that $\boldsymbol{w}$ is the direction at $\phi(\zeta)$ containing 0. Because $\mathbb{P}^1_{\text{an}}$ is Hausdorff, for each of the finitely many preimages $\zeta' \in \phi^{-1}(\phi(\zeta))$ other than ζ itself, we may choose open sets separating ζ from ζ'. Let U be the intersection of the resulting finitely many open sets containing ζ; then

$$\overline{U} \cap \phi^{-1}(\phi(\zeta)) = \{\zeta\}.$$

Let $W \subseteq U$ be an open connected Berkovich affinoid containing ζ and satisfying the conclusions of Lemma 7.27.

Let $Y = \{s < |\zeta| < r\}$, where we have chosen the radius $s < r$ sufficiently large that $Y \subseteq \phi(W)$. (This can be done because $\zeta \in \phi(W)$. Note that $\phi(W)$ is open by Corollary 7.9.) Let S be the set of directions $\boldsymbol{v}$ at ζ for which $\phi_\#(\boldsymbol{v}) = \boldsymbol{w}$. Shrinking Y by increasing s if necessary, we can choose, for each $\boldsymbol{v} \in S$, a sufficiently small open annulus $V_{\boldsymbol{v}} \subseteq W$ according to Theorem 7.12 or Remark 7.14, so that

$$\phi : V_{\boldsymbol{v}} \to Y$$

is $\deg_{\zeta,\boldsymbol{v}}(\phi)$-to-1 on Type I points. (Technically, we can only guarantee the mapping for finitely many $\boldsymbol{v} \in S$ at a time, but we will see in a moment that S is finite.) See Figure 16.1, which represents each annulus as a line segment connecting its two boundary points.

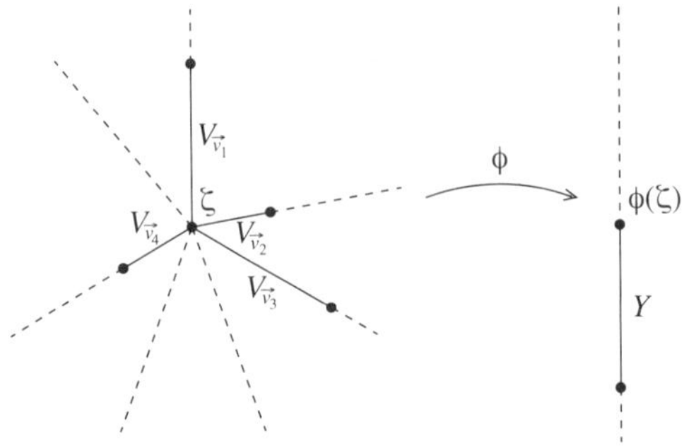

Figure 16.1. Proving Theorem 7.30: each annulus $V_{\boldsymbol{v}}$ maps onto the annulus Y.

Counting the number of preimages in W of Type I points in Y,

$$\deg_\zeta(\phi) \geq \sum_{\boldsymbol{v}\in S} \deg_{\zeta,\boldsymbol{v}}(\phi), \tag{16.4}$$

since W contains exactly $\deg_\zeta(\phi)$ preimages of any Type I point in Y, counting multiplicity. (In addition, inequality (16.4) shows that S must be finite, as promised above; in fact, $\#S \leq \deg_\zeta(\phi) \leq \deg(\phi)$.) It remains to show that inequality (16.4) is in fact an equality.

Given a connected component X of $\phi^{-1}(Y)$ such that $X \cap W \neq \varnothing$, we claim that $X = V_{\boldsymbol{v}}$ for some $\boldsymbol{v} \in S$. To prove the claim, first note that X is a connected affinoid mapping onto Y by Theorem 7.8(b). Moreover, we have $\phi(\partial W) \subseteq \partial(\phi(W))$ by property (b) of Lemma 7.27, and

$$\partial\big(\phi(W)\big) \quad \text{is disjoint from} \quad \phi(X) = Y \subseteq \phi(W).$$

Thus, $X \cap \partial W = \varnothing$, and since X and W are connected open sets, we must have $X \subseteq W$.

However, $\phi(\overline{X})$ must be a compact set containing $\phi(X) = Y$, and hence containing $\overline{Y}$. In particular, $\overline{X}$ contains a preimage of $\phi(\zeta)$, but on the other hand, we have

$$\overline{X} \subseteq \overline{W} \subseteq \overline{U}.$$

By our contruction of U, the preimage of $\phi(\zeta)$ in $\overline{X}$ must be ζ. But $\zeta \notin X$ (since $X \subseteq W$), and hence $\zeta \in \partial X$.

That is, ζ is one of the finitely many boundary points of the open connected affinoid X. Therefore, X must contain a thin annulus emanating from ζ, in the sense of Theorem 7.12 or Remark 7.14, corresponding to a direction $\boldsymbol{v}$ that maps to an annulus contained in Y. That is, X intersects $V_{\boldsymbol{v}}$. However, Corollary 7.10 tell us that $V_{\boldsymbol{v}}$, like X, is a full connected component of $\phi^{-1}(Y)$, since the two boundary points of $V_{\boldsymbol{v}}$ map to the two boundary points of Y. Therefore, $X = V_{\boldsymbol{v}}$, proving the claim.

Thus, the sum on the right-hand side of inequality (16.4) counts all preimages in W of any given Type I point in Y. Since the left-hand side counts the same thing, we are done. $\square$

16.3. Proving Rivera-Letelier's reduction theorem

Proof of Lemma 7.35. Write $\phi = f/g$, where $f, g \in \mathcal{O}[z]$, and at least one coefficient of f or g has absolute value 1. In particular,

$$\|f\|_{\zeta(0,1)} \leq 1 \quad \text{and} \quad \|g\|_{\zeta(0,1)} \leq 1$$

with equality for at least one of f or g. Observe that

$$\|f - ag\|_{\zeta(0,1)} = 1 \quad \text{for any } a \in \overline{D}(0,1).$$

Indeed, otherwise $\|f - ag\|_{\zeta(0,1)} < 1$, and hence $\overline{\phi} = \overline{a}$ would be constant, violating the hypothesis. Similarly, we have $\|g\|_{\zeta(0,1)} = 1$, or else $\overline{g} = 0$ and hence $\overline{\phi} = \infty$ would be constant. It follows that

$$\|f - ag\|_{\zeta(0,1)} = |a| \quad \text{for } |a| > 1.$$

Thus, by Definition 7.2, for any $a \in \mathbb{C}_v$, we have

$$\|z - a\|_{\phi(\zeta(0,1))} = \frac{\|f(z) - ag(z)\|_{\zeta(0,1)}}{\|g(z)\|_{\zeta(0,1)}} = \begin{cases} 1 & \text{if } |a| \leq 1, \\ |a| & \text{if } |a| > 1. \end{cases}$$

That is, $\|z - a\|_{\phi(\zeta(0,1))} = \|z - a\|_{\zeta(0,1)}$ for all $a \in \mathbb{C}_v$. By Lemma 15.2(c), we have $\phi(\zeta(0,1)) = \zeta(0,1)$, proving part (a).

If $\eta \in \mathrm{PGL}(2, \mathcal{O})$ is a linear fractional transformation of explicit good reduction, then η maps the direction $\vec{v}(a)$ at $\zeta(0,1)$ containing a bijectively onto the direction $\vec{v}(\eta(a))$; see Exercise 16.2. In addition, we leave it to the reader to check that

$$\overline{\eta \circ \phi} = \overline{\eta} \circ \overline{\phi} \quad \text{and} \quad \overline{\phi \circ \eta} = \overline{\phi} \circ \overline{\eta}$$

for any such η; see Exercise 16.3. Thus, after replacing ϕ by $\theta \circ \phi \circ \eta$ for appropriately chosen $\eta, \theta \in \mathrm{PGL}(2, \mathcal{O})$, we may assume for the remainder of the proof that $\overline{a} = \overline{\phi}(\overline{a}) = \overline{0}$.

Factor $f = f_1 f_2$, where $f_1, f_2 \in \mathcal{O}[z]$, and f_1 is the unique monic polynomial whose roots are precisely the roots of f in $D(0,1)$, repeated with the same multiplicities. Note that f_2 is not necessarily monic, but all of its roots lie outside $D(0,1)$. Define $g_1, g_2 \in \mathcal{O}[z]$ similarly with respect to g. Clearly,

$$\|f_1\|_{\zeta(0,1)} = \|g_1\|_{\zeta(0,1)} = 1,$$

and hence also

$$\|f_2\|_{\zeta(0,1)} = \|g_2\|_{\zeta(0,1)} = 1,$$

since $\|f_1\|\|f_2\| = \|f\| = 1$, and similarly for g. By inspection of the Newton polygons of these four polynomials, all nonlead coefficients of f_1 and g_1 must have absolute value strictly less than 1; that is, $\overline{f}_1$ and $\overline{g}_1$ are both of the form z^j. Similarly, the constant terms of f_2 and g_2 must have absolute value 1; that is, the constant terms of the reductions $\overline{f}_2$ and $\overline{g}_2$ are nonzero.

Let $\phi_2 := f_2/g_2$ and $m := \deg(f_1) - \deg(g_1) \in \mathbb{Z}$. By the previous paragraph, we have

$$\overline{\phi}(z) = z^m \overline{\phi}_2(z).$$

Moreover, even after any cancellation that occurs when we write $\overline{\phi}_2 = \overline{f}_2/\overline{g}_2$, the constant terms of both numerator and denominator of $\overline{\phi}_2$ are nonzero. Therefore, because we assumed $\overline{\phi}(\overline{0}) = \overline{0}$, it follows that $m \geq 1$.

Let U be the annulus

$$U := \{\lambda < |z| < 1\} = D(0,1) \smallsetminus D(0,\lambda),$$

where $0 < \lambda < 1$ is chosen sufficiently large that U contains no roots of f or g. For any $x \in U$, we have

$$|\phi(x)| = \frac{|f_1(x)|}{|g_1(x)|} \cdot \frac{|f_2(x)|}{|g_2(x)|} = |x|^m \cdot 1 = |x|^m. \tag{16.5}$$

Thus, $\phi(U) \subseteq D(0,1)$, and hence $\phi_{\#}(\vec{v}(0)) = \vec{v}(0)$, proving the forward implication of part (b); the reverse implication is then immediate. In addition, by Theorem 3.33, equation (16.5) shows that

$$\deg_{\zeta(0,1),\vec{v}(0)}(\phi) = m = \deg_{\overline{0}}\left(\overline{\phi}\right),$$

proving part (c). □

Proof of Theorem 7.34. We follow Rivera-Letelier's argument in [**114**], with slight modifications. Write $\phi = f/g$ as in the statement of the theorem.

If $\overline{f}/\overline{g}$ is constant, then after a $\mathrm{PGL}(2,\mathcal{O})$-change of coordinates, we can assume that $\overline{f}/\overline{g} = 0$. We must therefore have $\|g\|_{\zeta(0,1)} = 1$ but $\|f\|_{\zeta(0,1)} < 1$. Hence,

$$\|z\|_{\phi(\zeta(0,1))} = \frac{\|f\|_{\zeta(0,1)}}{\|g\|_{\zeta(0,1)}} < 1,$$

and in particular, $\phi(\zeta(0,1)) \neq \zeta(0,1)$.

For the remainder of the proof, we assume that $\overline{\phi} = \overline{f}/\overline{g}$ is nonconstant. By Lemma 7.35(a), we have $\phi(\zeta(0,1)) = \zeta(0,1)$. In addition, for any direction $\boldsymbol{w}$ at $\zeta(0,1)$, let $b \in \mathbb{P}^1(\mathbb{C}_v)$ be a Type I point such that $\boldsymbol{w}$ is the direction containing b. Then by Lemma 7.35(b)–(c), summing over all directions $\boldsymbol{v}$ at $\zeta(0,1)$ for which $\phi_{\#}(\boldsymbol{v}) = \boldsymbol{w}$, we have

$$\sum_{\phi_{\#}(\boldsymbol{v})=\boldsymbol{w}} \deg_{\zeta(0,1),\boldsymbol{v}}(\phi) = \sum_{\overline{\phi}(\overline{a})=\overline{b}} \deg_{\overline{a}}\left(\overline{\phi}\right) = \deg\left(\overline{\phi}\right),$$

where the second sum is over all $\overline{a} \in \mathbb{P}^1(k)$ for which $\overline{\phi}(\overline{a}) = \overline{b}$. Equation (7.7) is then immediate from Theorem 7.30.

As in the statement of Theorem 7.34, let T be the set of all directions $\boldsymbol{v}$ at $\zeta(0,1)$ that contain at least one zero and at least one pole of ϕ. Since ϕ has only finitely many poles, T is certainly finite, proving part (a). In addition, ϕ has explicit good reduction if and only if $T = \varnothing$, proving the final statement of the theorem. It remains to show that ϕ satisfies properties (b) and (c).

For each $\boldsymbol{v} \in T$, we can assume by a $\mathrm{PGL}(2, \mathcal{O})$-change of coordinates on the domain that $\boldsymbol{v} = \vec{v}(0)$, and hence that f and g both have zeros in $D(0,1)$. Since $\|f\|_{\zeta(0,1)} = \|g\|_{\zeta(0,1)} = 1$, the constant terms of both f and g must have absolute value strictly less than 1. We claim that $\phi(\boldsymbol{v}) = \mathbb{P}^1_{\mathrm{an}}$.

For $y \in \overline{D}(0,1)$, the fact that $\overline{\phi}$ is nonconstant implies that

$$\|f - yg\|_{\zeta(0,1)} = 1.$$

However, the constant term of $f - yg$ has absolute value less than 1, and therefore the Newton polygon of $f - yg$ shows that there is a point $x \in D(0,1)$ such that $\phi(x) = y$.

On the other hand, for $y \in \mathbb{C}_v \smallsetminus \overline{D}(0,1)$, we have

$$\|f - yg\|_{\zeta(0,1)} = |y|,$$

but the constant term of $f - yg$ has absolute value less than $|y|$. Again, the Newton polygon of $f - yg$ shows that there is a point $x \in D(0,1)$ such that $\phi(x) = y$.

The previous two paragraphs show that $\phi(\boldsymbol{v} \cup \{\zeta(0,1)\})$ is a compact subset of $\mathbb{P}^1_{\mathrm{an}}$ containing $\mathbb{C}_v$. Therefore, $\phi(\boldsymbol{v} \cup \{\zeta(0,1)\}) = \mathbb{P}^1_{\mathrm{an}}$. It follows that $\phi(\boldsymbol{v})$ is a connected subset of $\mathbb{P}^1_{\mathrm{an}}$ containing $\mathbb{P}^1_{\mathrm{an}} \smallsetminus \{\zeta(0,1)\}$. Thus, $\phi(\boldsymbol{v}) = \mathbb{P}^1_{\mathrm{an}}$, proving the claim and hence property (b) in the statement of the theorem.

For each direction $\boldsymbol{v}$ at $\zeta(0,1)$ that is *not* in T, we can again assume after a $\mathrm{PGL}(2, \mathcal{O})$-changes of coordinates on the domain that $\boldsymbol{v} = \vec{v}(0)$. Write

$$f = b_0 + b_1 z + \cdots + b_d z^d \quad \text{and} \quad g = c_0 + c_1 z + \cdots + c_d z^d.$$

We consider three cases.

Case 1. If $|\phi(0)| < 1$, then $|b_0| < |c_0|$. However, because

$$\|f\|_{\zeta(0,1)} = \|g\|_{\zeta(0,1)} = 1,$$

and by the definition of T and the Newton polygons of f and g, we cannot have both $|b_0| < 1$ and $|c_0| < 1$. Therefore, $|b_0| < |c_0| = 1$. For any $\zeta \in D_{\mathrm{an}}(0,1)$, we must have $\|g\|_\zeta = 1$, and hence

$$\|z\|_{\phi(\zeta)} = \frac{\|f\|_\zeta}{\|g\|_\zeta} \le \max\left\{|b_0|, \|z\|_\zeta\right\} < 1.$$

That is, $\phi(\zeta) \in D_{\mathrm{an}}(0,1)$. We have shown that

$$\phi\big(D_{\mathrm{an}}(0,1)\big) \subseteq D_{\mathrm{an}}(0,1).$$

This image must be an open Berkovich disk, by Proposition 3.25 and Theorem 7.8. On the other hand, for any $y \in D(0,1)$, it is evident from the Newton polygon of $f - yg$ that the equation $\phi(z) = y$ has a solution in $D(0,1)$. Thus, the Berkovich disk $\phi(D_{\mathrm{an}}(0,1))$ contains $D(0,1)$, and hence $\phi(D_{\mathrm{an}}(0,1)) = D_{\mathrm{an}}(0,1)$. That is,

$$\phi_{\#}\big(\vec{v}(0)\big) = \vec{v}(0) = \vec{v}\big(\phi(0)\big) = D_{\mathrm{an}}(0,1) = \phi\big(D_{\mathrm{an}}(0,1)\big) = \phi\big(\vec{v}(0)\big),$$

proving the first statement of property (c). For the second statement, let $m \geq 1$ be the smallest integer such that $|b_m| = 1$. Then because $|c_0| = 0$, and again bearing the Newton polygons of f and g in mind, we have

$$\deg_{\overline{0}}\big(\overline{\phi}\big) = m = \mathrm{wdeg}_{D(0,1)}(f) = \mathrm{wdeg}_{D(0,1)}(\phi).$$

Case 2. If $|\phi(0)| > 1$, replace $\phi(z)$ by $1/\phi(z)$, which does not change the fact that $\vec{v}(0) \notin T$. We are then done by Case 1.

Case 3. Finally, suppose $|\phi(0)| = 1$. Then $|b_0| = |c_0|$. Since $b_0 \in \mathcal{O}$, we have $|b_0| \leq 1$; but as in Case 1, we cannot have $|b_0| = |c_0| < 1$. Hence, $|b_0| = |c_0| = 1$. In particular, ϕ has no poles in $D(0,1)$. Thus, we may replace ϕ by $\phi - b_0/c_0$, which does not change the fact that $\vec{v}(0) \notin T$, and we are done by Case 1. □

Exercises for Chapter 16

16.1. Complete the proof of Theorem 7.8. More precisely, with notation as in that proof:

(a) do Step 2—if W is a closed Berkovich connected affinoid, prove that $\phi^{-1}(W)$ is the closed Berkovich affinoid corresponding to $\phi^{-1}(W_{\mathrm{I}})$, and that statement (b) holds for W.

(b) finish Step 5—prove that $\phi(\mathrm{int}\, V_i) = \mathrm{int}\, W$.

16.2. Prove the following claim made in the proof of Lemma 7.35(b). Let $\eta \in \mathrm{PGL}(2,\mathcal{O})$ be a linear fractional transformation of explicit good reduction, and let $a \in \mathbb{P}^1(\mathbb{C}_v)$. Then η maps the direction $\vec{v}(a)$ at $\zeta(0,1)$ containing a bijectively onto the direction $\vec{v}(\eta(a))$. Obviously, you cannot use Theorem 7.34 or Lemma 7.35(b) itself.

16.3. Prove the following claim made in the proof of Lemma 7.35(b). Let $\phi \in \mathbb{C}_v(z)$ have nonconstant reduction, and let $\eta \in \mathrm{PGL}(2,\mathcal{O})$. Then

$$\overline{\eta \circ \phi} = \overline{\eta} \circ \overline{\phi} \quad \text{and} \quad \overline{\phi \circ \eta} = \overline{\phi} \circ \overline{\eta}.$$

Obviously, you cannot use Theorem 7.34 or Lemma 7.35(b) itself.

Appendices

Fatou Components without Berkovich Space

It is possible to define useful notions of Fatou components in $\mathbb{P}^1(\mathbb{C}_v)$ without the use of the Berkovich line $\mathbb{P}^1_{\text{an}}$. Indeed, several results in this book were first proven before Rivera-Letelier introduced Berkovich's theory to the study of non-archimedean dynamics. The more elementary notions of components discussed in this appendix require less background preparation than is required for the study of Berkovich Fatou components and hence may provide a more accessible introduction to the subject.

A.1. Analytic components and D-components

We begin with an echo of Definition 8.1, modified for $\mathbb{P}^1(\mathbb{C}_v)$ instead of $\mathbb{P}^1_{\text{an}}$.

Definition A.1. Let $\phi \in \mathbb{C}_v(z)$ be a rational function, and let $U \subseteq \mathbb{P}^1(\mathbb{C}_v)$ be an open set. We say that U is *dynamically stable* under ϕ if $\bigcup_{n\geq 0} \phi^n(U)$ omits infinitely many points of $\mathbb{P}^1(\mathbb{C}_v)$.

Using Theorem 5.19, one can show that any dynamically stable open set U is contained in the (Type I) Fatou set $\mathcal{F}_{\phi,\text{I}}$.

Even though $\mathbb{P}^1(\mathbb{C}_v)$ is totally disconnected, we saw in Chapter 3 that disks and connected affinoids behave like connected sets, in that the image of such an object under a rational function is another such object. This observation motivates the following definition.

Definition A.2. Let $\phi \in \mathbb{C}_v(z)$ be a rational function with Fatou set $\mathcal{F}_{\phi,\mathrm{I}}$, and let $\mathcal{B}$ be the set of all dynamically stable

(i) connected affinoids in $\mathbb{P}^1(\mathbb{C}_v)$, or
(ii) $\mathbb{P}^1(\mathbb{C}_v)$-disks, or
(iii) open connected affinoids in $\mathbb{P}^1(\mathbb{C}_v)$, or
(iv) open $\mathbb{P}^1(\mathbb{C}_v)$-disks.

Let $x \in \mathcal{F}_{\phi,\mathrm{I}}$. The *$\mathcal{B}$-component* of $\mathcal{F}_{\phi,\mathrm{I}}$ containing x is the set of all $y \in \mathcal{F}_{\phi,\mathrm{I}}$ for which there are sets $U_1, \ldots, U_n \in \mathcal{B}$ such that $x \in U_1$, $y \in U_n$, and $U_i \cap U_{i+1} \neq \varnothing$ for all $i = 1, \ldots, n-1$.

If $\mathcal{B}$ is defined by (i)–(iv) above, we call the $\mathcal{B}$-components *analytic components*, *D-components*, *open analytic components*, or *open D-components*, respectively.

The relation "y belongs to the $\mathcal{B}$-component containing x" is clearly an equivalence relation. Less obviously, if one stated Definition A.2 for $\mathbb{C}$ instead of $\mathbb{C}_v$, with $\mathcal{B}$ the set of open disks in $\mathbb{P}^1(\mathbb{C})$, the resulting complex $\mathcal{B}$-components are just the usual connected components of the Fatou set. The same is true if $\mathcal{B}$ is the set of all disks, all open connected sets, or all connected sets in $\mathbb{P}^1(\mathbb{C})$; see Exercise A.1.

Theorem A.3. *Let $\phi \in \mathbb{C}_v(z)$ be a rational function of degree $d \geq 2$ with Fatou set $\mathcal{F}_{\phi,\mathrm{I}}$, and let $\mathcal{B}$ be one of the four sets in Definition* A.2. *If ϕ does not have potential good reduction, then for any $x \in \mathcal{F}_{\phi,\mathrm{I}}$, the $\mathcal{B}$-component of $\mathcal{F}_{\phi,\mathrm{I}}$ containing x is the union of all (dynamically stable) $U \in \mathcal{B}$ for which $x \in U \subseteq \mathcal{F}_{\phi,\mathrm{I}}$. Moreover,*

(a) *every open D-component of $\mathcal{F}_{\phi,\mathrm{I}}$ is an open $\mathbb{P}^1(\mathbb{C}_v)$-disk, and*
(b) *every D-component of $\mathcal{F}_{\phi,\mathrm{I}}$ is a $\mathbb{P}^1(\mathbb{C}_v)$-disk.*

The full proof of Theorem A.3 requires some results on Berkovich dynamics; see Exercise A.2. However, the proof is much simpler in the still very general case that $\mathcal{J}_{\phi,\mathrm{I}} \neq \varnothing$.

Proof of Theorem A.3 when $\mathcal{J}_{\phi,\mathrm{I}} \neq \varnothing$. For the first statement, it suffices to show that for each choice of $\mathcal{B}$, and for any $U_1, U_2 \in \mathcal{B}$ for which $U_1 \cap U_2 \neq \varnothing$, we have $U_1 \cup U_2 \in \mathcal{B}$. We know from Propositions 3.26 and 3.28 that the union of two intersecting (open) connected affinoids is again an (open) connected affinoid, and that the union of two intersecting (open) $\mathbb{P}^1(\mathbb{C}_v)$-disks is again either an (open) disk or else all of $\mathbb{P}^1(\mathbb{C}_v)$. Thus, we only need to show that $U_1 \cup U_2$ is still dynamically stable.

However, all iterates of $U_1 \cup U_2 \subseteq \mathcal{F}_{\phi,\mathrm{I}}$ avoid the Julia set $\mathcal{J}_{\phi,\mathrm{I}}$. Because we assume $\mathcal{J}_{\phi,\mathrm{I}}$ is nonempty, and hence uncountable by Corollary 5.25, it follows that $U_1 \cup U_2$ is dynamically stable, as desired.

We now prove statement (b); statement (a) (for open D-components) is similar. Consider the D-component containing some point $x \in \mathcal{F}_{\phi,\mathrm{I}}$. Because $\mathcal{J}_{\phi,\mathrm{I}}$ has more than one point, we may assume, after changing coordinates, that $x = 0$ and that $1, \infty \in \mathcal{J}_{\phi,\mathrm{I}}$. Thus, every dynamically stable disk containing $x = 0$ is contained in $\mathcal{F}_{\phi,\mathrm{I}}$ and hence is contained in $D(0,1)$. Therefore, every such disk is of the form either $\overline{D}(0,r)$ for $0 < r < 1$ or $D(0,r)$ for $0 < r \leq 1$. Taking the supremum R of all such r, it follows that the Fatou D-component containing x is either $\overline{D}(0,R)$ or $D(0,R)$. □

For each of the four types (i)–(iv), the map ϕ acts on the set of $\mathcal{B}$-components, as follows. By Theorem 3.15 and Proposition 3.29, the image $\phi(U)$ of $U \in \mathcal{B}$ is again an element of $\mathcal{B}$. Thus, for any $\mathcal{B}$-component W, the image $\phi(W)$ is a subset of some (probably different) $\mathcal{B}$-component. We can therefore define an action of ϕ on the $\mathcal{B}$-components by

$$\phi[W] := \text{ the } \mathcal{B}\text{-component containing } \phi(W).$$

Remark A.4. If W is an (open) analytic component, then in fact we have $\phi[W] = \phi(W)$, because by Proposition 3.29, the inverse image $\phi^{-1}(V)$ of each of the connected affinoids V contained in $\phi[W]$ is itself a finite union of connected affinoids, and those affinoids are clearly dynamically stable because V is.

However, if W is a D-component or an open D-component, then it is possible that $\phi(W) \subsetneq \phi[W]$; see Example A.6 below. Put another way, there may be (open) D-components V such that $\phi^{-1}(V)$ consists of more than one (and in fact, infinitely many) D-components. In that case, however, there are at most $\deg(\phi) - 1$ D-components V whose inverse images are not also D-components, by Lemma 10.14.

A.2. Examples of non-archimedean Fatou components

Example A.5. Let $\phi(z) \in \mathbb{C}_v(z)$ be a map of explicit good reduction. The reader can check that any residue class is dynamically stable, but any strictly larger open connected affinoid is not dynamically stable. Thus, the open analytic components and the open D-components are each precisely the residue classes, and ϕ maps components to each other exactly as prescribed by the way the reduced map $\overline{\phi}$ maps points of $\mathbb{P}^1(\overline{k})$ to each other.

We now turn to the (a priori larger) D-components and analytic components. If $\overline{\phi}$ is not conjugate over $\overline{k}$ to a polynomial, to z^{-d}, or to a totally inseparable map, then any connected affinoid (even a closed one) strictly larger than a residue class is also not dynamically stable. After all, such a connected affinoid U contains all but finitely many residue classes; and by the above assumptions about $\overline{\phi}$, the exceptional set of $\overline{\phi}$ is empty, by

Theorem 1.19. Therefore, the analytic components and D-components, like their open counterparts, are also precisely the residue classes.

For the remaining cases (polynomials, etc.), see Exercise A.3.

Example A.6. Let $\phi(z) = z^2 + \lambda z \in \mathbb{C}_v[z]$, as in Example 5.29, and assume that $|\lambda| > 1$. (The case $|\lambda| \leq 1$ is subsumed by Example A.5 above.) Recall that the complement V_0 of the disk $U_0 := \overline{D}(0, |\lambda|)$ is attracted to ∞, and that the Julia set is the Cantor set $\mathcal{J}_{\phi,\mathrm{I}} = \bigcap_{n \geq 0} \phi^{-n}(U_0)$. Although the Fatou set $\mathcal{F}_{\phi,\mathrm{I}}$ is not itself a connected affinoid, it is a union of open connected affinoids $\phi^{-n}(V_0)$. Thus, $\mathcal{F}_{\phi,\mathrm{I}}$ is a single (open) analytic component.

On the other hand, V_0 is itself the (open) D-component containing ∞, as any strictly larger $\mathbb{P}^1(\mathbb{C}_v)$-disk must intersect $\mathcal{J}_{\phi,\mathrm{I}}$. (Note that $\phi^{-1}(V_0)$ is not a finite union of disks; see Remark A.4.) Finally, for any other Fatou point $x \in \mathcal{F}_{\phi,\mathrm{I}} \smallsetminus V_0$, the (open) D-component containing x is the largest open disk $D(x, r)$ centered at x and contained in $\mathcal{F}_{\phi,\mathrm{I}}$; see Definition 9.4.

Example A.7. Assume $\mathbb{C}_v$ does not have residue characteristic 2, and let

$$\phi(z) = az^3 + z^2 + bz + c$$

as in Example 5.31, where $0 < |a| < 1$ and $|b|, |c| \leq 1$. As in Example A.6, the D-component and open D-component at ∞ are both the open $\mathbb{P}^1(\mathbb{C}_v)$-disk $V_0 := \mathbb{P}^1(\mathbb{C}_v) \smallsetminus U_0$, where $U_0 := \overline{D}(0, |a|^{-1})$, and the analytic component and open analytic component are the Cantor-type domain $W := \bigcup_{n \geq 0} \phi^{-n}(V_n)$. All the D-components again map onto D-components, with the exception of those that ϕ maps into V_0. Moreover, each D-component in W lands in V_0 after finitely many iterations.

This time, however, there are also Fatou components that are not attracted to ∞. In particular, the closed disk $\overline{D}(0, 1)$ maps into itself, but the iterates of any larger disk grow without bound. Thus, $\overline{D}(0, 1)$ is a fixed D-component and a fixed analytic component. On the other hand, it consists of infinitely many *open* D-components and *open* analytic components, which are all the open disks $D(x, 1)$ in $\overline{D}(0, 1)$. These open disks map to one another as prescribed by the reduced map $\overline{\phi}(z) = z^2 + \overline{b}z + \overline{c}$. If the residue field $\overline{k}$ is algebraic over a finite field, then all such open components are preperiodic; but otherwise, many of them wander.

Finally, the components in $\overline{D}(0, 1)$ have preimages outside $\overline{D}(0, 1)$. Indeed, because we assumed the residue characteristic is not 2, it can be shown that all Fatou points eventually map into either V_0 or into $\overline{D}(0, 1)$.

Example A.7 shows that the four different types of components are truly different but are closely related. The next example shows that they can sometimes all coincide.

Example A.8. Assume $\mathbb{C}_v$ does not have residue characteristic 2, and let

$$\phi(z) = z^2 + \frac{az^2}{z+1} = \frac{z^3 + (a+1)z^2}{z+1},$$

where $0 < |a| < 1$, as in Example 5.32. Recall that the Julia set intersects the residue class containing -1, as well as all of its preimages under the squaring map. All other residue classes are contained in the Fatou set.

Let $W \subseteq \mathcal{F}_{\phi,\mathrm{I}}$ be a Fatou residue class. Because of the infinitely many non-Fatou residue classes, any connected affinoid properly containing W cannot be dynamically stable. Thus, the residue class W is a component of all four types, and the dynamics of these components is described by the squaring map acting on $\mathbb{P}^1(\overline{k})$.

Moreover, as noted in Example 5.32, all Fatou points in the bad residue classes (that is, in $D(-1,1)$ and its preimages) eventually land in one of the good residue classes. As a result, all of the components (of any of the four types) for such Fatou points are again open disks.

A.3. Open analytic components

We now present the following close relationship between open analytic components and connected components of the Berkovich Fatou set.

Theorem A.9. *Let $\phi \in \mathbb{C}_v(z)$ be a rational function of degree $d \geq 2$. Then for every connected component U of $\mathcal{F}_{\phi,\mathrm{an}}$, the set $U \cap \mathbb{P}^1(\mathbb{C}_v)$ of Type I points in U is an open analytic component of $\mathcal{F}_{\phi,\mathrm{I}}$. Conversely, for every open analytic component V of $\mathcal{F}_{\phi,\mathrm{I}}$, there is a unique connected component U of $\mathcal{F}_{\phi,\mathrm{an}}$ such that $V = U \cap \mathbb{P}^1(\mathbb{C}_v)$.*

To prove Theorem A.9, we need the following technical lemma.

Lemma A.10. *Let $\phi \in \mathbb{C}_v(z)$ be a rational function of degree $d \geq 2$ with Berkovich Fatou set $\mathcal{F}_{\phi,\mathrm{an}}$, and let $W_{\mathrm{an}} \subseteq \mathcal{F}_{\phi,\mathrm{an}}$ be a connected Berkovich affinoid. Then both W_{an} and the associated classical affinoid $W_{\mathrm{an}} \cap \mathbb{P}^1(\mathbb{C}_v)$ are dynamically stable under ϕ.*

Proof. By Theorem 12.5, the Berkovich Julia set $\mathcal{J}_{\phi,\mathrm{an}}$ contains a repelling fixed point of Type I or II. In the former case, the classical Julia set $\mathcal{J}_{\phi,\mathrm{I}}$ is uncountable, by Corollary 5.25. Since

$$\phi^n(W) \cap \mathcal{J}_{\phi,\mathrm{I}} = \varnothing \quad \text{for all} \quad n \geq 0$$

(by Proposition 8.2(b) and Theorem 8.3), it follows not only that the iterates of W_{an} omit an uncountable subset of $\mathbb{P}^1_{\mathrm{an}}$, but also that the iterates of $W_{\mathrm{an}} \cap \mathbb{P}^1(\mathbb{C}_v)$ omit an uncountable subset of $\mathbb{P}^1(\mathbb{C}_v)$, as desired.

Thus, we may assume that there is a periodic Type II point, which, after a change of coordinates, we may further assume is $\zeta(0,1)$, fixed by some iterate ϕ^m of ϕ. By Theorem 7.34, $\overline{\phi^m} \in \overline{k}(z)$ is nonconstant, and because $\zeta(0,1)$ is Julia, no positive iterate of $\overline{\phi^m}$ is the identity function. Let T be the finite set of bad directions for ϕ^m from Theorem 7.34.

We leave it to the reader to prove that there is a residue class $\overline{a}$ whose forward orbit under ψ avoids T; see Exercise A.4. After a change of coordinates, we may assume that the residue class $\psi(\overline{a})$ is the residue class $\overline{0}$ of $\zeta(0,1)$. It is immediate that $D_{\mathrm{an}}(0,1)$ is contained in $\mathcal{F}_{\phi,\mathrm{an}}$. Since $\zeta(0,1) \notin \mathcal{F}_{\phi,\mathrm{an}}$, the disk $D_{\mathrm{an}}(0,1)$ is in fact a component of $\mathcal{F}_{\phi,\mathrm{an}}$.

If no forward iterate of W_{an} is contained in $D_{\mathrm{an}}(0,1)$, then no such iterate can even intersect $D_{\mathrm{an}}(0,1)$. Since $D_{\mathrm{an}}(0,1)$ contains infinitely many Type I points, W_{an} and $W_{\mathrm{an}} \cap \mathbb{P}^1(\mathbb{C}_v)$ are both dynamically stable.

We may therefore assume that $\phi^n(W_{\mathrm{an}}) \subseteq D_{\mathrm{an}}(0,1)$ for some $n \geq 0$. Let U_{an} be the Berkovich Fatou component containing W_{an}; by Lemma 9.2,

$$\phi^n(U_{\mathrm{an}}) = D_{\mathrm{an}}(0,1).$$

If U_{an} is periodic, then the forward iterates of U_{an} consists of finitely many disjoint rational open Berkovich disks, which necessarily omit infinitely many points of $\mathbb{P}^1(\mathbb{C}_v)$. If U_{an} is not periodic, then $\phi^{-1}(U_{\mathrm{an}})$ includes a component V_{an} not in the forward orbit of U_{an}. Either way, then, the iterates of W_{an} omit an infinite subset of $\mathbb{P}^1(\mathbb{C}_v)$, and hence W_{an} and $W_{\mathrm{an}} \cap \mathbb{P}^1(\mathbb{C}_v)$ are both dynamically stable. □

Proof of Theorem A.9. Let U_{an} be a component of $\mathcal{F}_{\phi,\mathrm{an}}$, and let $x \in U_{\mathrm{an}}$ be a Type I point. (Such x exists, because U_{an} is open, and the Type I points are dense in $\mathbb{P}^1_{\mathrm{an}}$.) By Theorem 8.3, x lies in $\mathcal{F}_{\phi,\mathrm{I}}$. Let V be the open analytic component containing x. We must show that $V = U_{\mathrm{an}} \cap \mathbb{P}^1(\mathbb{C}_v)$.

Step 1. To show that $V \subseteq U_{\mathrm{an}} \cap \mathbb{P}^1(\mathbb{C}_v)$, given a dynamically stable (classical) open connected affinoid $W \subseteq \mathcal{F}_{\phi,\mathrm{I}}$ containing x, we need to show that $W \subseteq U_{\mathrm{an}} \cap \mathbb{P}^1(\mathbb{C}_v)$. Define W_{an} to be the open connected Berkovich affinoid associated to W, as in Remark 6.23. Since W is dynamically stable, its iterates $\phi^n(W)$ omit infinitely many points of $\mathbb{P}^1(\mathbb{C}_v)$. The iterates $\phi^n(W_{\mathrm{an}})$ also omit the same infinitely many points of $\mathbb{P}^1(\mathbb{C}_v)$ and hence of $\mathbb{P}^1_{\mathrm{an}}$. Therefore, W_{an} is dynamically stable. Since $x \in W_{\mathrm{an}} \subseteq \mathcal{F}_{\phi,\mathrm{an}}$ and W_{an} is connected, it follows that $W_{\mathrm{an}} \subseteq U_{\mathrm{an}}$. Thus,

$$W = W_{\mathrm{an}} \cap \mathbb{P}^1(\mathbb{C}_v) \subseteq U_{\mathrm{an}} \cap \mathbb{P}^1(\mathbb{C}_v).$$

Step 2. Next, we must show that $V \supseteq U_{\mathrm{an}} \cap \mathbb{P}^1(\mathbb{C}_v)$. That is, by Lemma 9.2, given any open connected Berkovich affinoid $W_{\mathrm{an}} \subseteq \mathcal{F}_{\phi,\mathrm{an}}$ containing x, we must show that the associated classical open connected affinoid $W := W_{\mathrm{an}} \cap \mathbb{P}^1(\mathbb{C}_v)$ is contained in V. Fortunately, this is immediate from

Lemma A.10, because W is a dynamically stable open connected affinoid in $\mathbb{P}^1(\mathbb{C}_v)$, and therefore $W \subseteq V$ by the definition of open analytic components.

Step 3. For the converse, let V be an open analytic component of $\mathcal{F}_{\phi,\mathrm{I}}$, and let $x \in V$. Then $x \in \mathcal{F}_{\phi,\mathrm{I}} \subseteq \mathcal{F}_{\phi,\mathrm{an}}$, and hence we can define U_{an} to be the connected component of $\mathcal{F}_{\phi,\mathrm{an}}$ containing x. By Steps 1 and 2 of the proof, $U_{\mathrm{an}} \cap \mathbb{P}^1(\mathbb{C}_v)$ is an open analytic component of $\mathcal{F}_{\phi,\mathrm{I}}$ containing x, and therefore $U_{\mathrm{an}} \cap \mathbb{P}^1(\mathbb{C}_v) = V$. Finally, the uniqueness statement is immediate from the fact that two intersecting connected components of an arbitrary topological space must coincide. □

Exercises for Appendix A

A.1. Let X be a locally path-connected topological space (such as an open subset of the Riemann sphere). Let $\mathcal{B}$ be a set of open connected subsets of X such that

$$\bigcup_{U \in \mathcal{B}} U = X.$$

For any point $x \in X$, define the $\mathcal{B}$-component Y of X containing x to be the set of all $y \in U$ for which there are sets $U_1, \ldots, U_n \in \mathcal{B}$ such that

$$x \in U_1, \quad y \in U_n, \quad \text{and} \quad U_i \cap U_{i+1} \subseteq X \quad \text{for all } i = 1, \ldots, n-1.$$

Prove that Y is precisely the connected component of X containing x.

A.2. In this exercise you will complete the proof of Theorem A.3. Let ϕ, $\mathcal{B}$, and x be as in that theorem.

(a) Follow the partial proof in Section A.1 to show that the theorem would follow from the following lemma:
For all $U_1, U_2 \in \mathcal{B}$ with nonempty intersection, the union $U_1 \cup U_2$ is still dynamically stable.

(b) Prove the lemma of part (a) if ϕ has potential good reduction. Theorem 1.19 will be useful.

(c) Let $U_{1,\mathrm{an}}$ and $U_{2,\mathrm{an}}$ be the Berkovich versions of U_1 and U_2 (including boundary points corresponding to rational closed disks). Show that the interiors $\operatorname{int} U_{1,\mathrm{an}}$ and $\operatorname{int} U_{2,\mathrm{an}}$ are contained in $\mathcal{F}_{\phi,\mathrm{an}}$.

(d) Assume for the remainder of this exercise that ϕ does not have potential good reduction. Use Theorem 8.15(e) and Corollary 12.6 to show that $\mathcal{J}_{\phi,\mathrm{an}}$ is uncountable.

(e) Show that the iterates of not only $(\operatorname{int} U_{1,\mathrm{an}}) \cup (\operatorname{int} U_{2,\mathrm{an}})$ but, in fact, $U_{1,\mathrm{an}} \cup U_{2,\mathrm{an}}$ omit uncountably many points of $\mathbb{P}^1_{\mathrm{an}}$.

(f) Conclude that $U_1 \cup U_2 \subseteq \mathbb{P}^1(\mathbb{C}_v)$ is dynamically stable, according to Definition A.1.

A.3. Finish the analysis of D-components and analytic components in Example A.5 for $\phi \in \mathbb{C}_v(z)$ of explicit good reduction and degree $d \geq 2$, as follows.

(a) If $\overline{\phi}$ is conjugate to a polynomial but not to z^d, show that there are two analytic components, both of which are also D-components; and that (after the conjugation to z^d) they are $\overline{D}(0,1)$ and its complement.

(b) If $\overline{\phi} = z^d$, show that there is only one analytic component, which is also a D-component, namely $\mathbb{P}^1(\mathbb{C}_v)$.

(c) If $\overline{\phi} = z^{-d}$ is not totally inseparable, show that the D-components are exactly the residue classes of $\mathbb{P}^1(\mathbb{C}_v)$, and that there are three analytic components:

$$D(0,1), \quad \mathbb{P}^1(\mathbb{C}_v) \smallsetminus \overline{D}(0,1), \quad \text{and} \quad \overline{D}(0,1) \smallsetminus D(0,1).$$

(d) If $\overline{\phi}$ is totally inseparable, show that it is conjugate to z^d.

A.4. Let k be an algebraically closed field, let $T \subseteq \mathbb{P}^1(k)$ be a finite set of points, and let $\psi \in k(z)$ be a nonconstant rational function. Suppose that no positive iterate of ψ is the identity function. Show that there is some $a \in \mathbb{P}^1(k)$ whose forward orbit does not intersect T.

Other Constructions of Berkovich Spaces

In Chapter 6, we constructed $\mathbb{P}^1_{\text{an}}$ by first defining $\mathbb{A}^1_{\text{an}}$ to be the set of multiplicative seminorms on $\mathbb{C}_v[z]$. In this appendix, we present some alternative approaches to the construction of $\mathbb{P}^1_{\text{an}}$.

B.1. Berkovich disks via power series

In a more complete presentation of Berkovich's theory, an (affine) Berkovich space is defined to be the set of bounded seminorms on a Banach ring. Here, a *Banach ring* is a commutative ring $\mathcal{A}$ with unity that is complete with respect to a submultiplicative norm. (A norm $\|\cdot\|$ is *submultiplicative* if we replace the axioms $\|1\| = 1$ and $\|fg\| = \|f\| \cdot \|g\|$ by $\|1\| \leq 1$ and $\|fg\| \leq \|f\| \cdot \|g\|$, respectively, in Definition 6.1.) The associated Berkovich space $\mathcal{M}(\mathcal{A})$ is the set of multiplicative seminorms on $\mathcal{A}$ that are bounded with respect to $\|\cdot\|$. We equip $\mathcal{M}(\mathcal{A})$ with the Gel'fand topology, which is the weakest topology such that, for every $f \in \mathcal{A}$, the real-valued function $\zeta \mapsto \|f\|_\zeta$ is continuous on $\mathcal{M}(\mathcal{A})$. More generally, a Berkovich space is formed by gluing affine Berkovich spaces, just as a scheme is formed by gluing affine schemes. For further details, see Berkovich's original monograph [**26**].

In this approach, a rational closed Berkovich disk $\overline{D}_{\text{an}}(a,r)$ is defined intrinsically rather than as a subset of $\mathbb{A}^1_{\text{an}}$. Specifically, consider the Banach ring $\mathcal{A}_D = \overline{\mathcal{A}}(a,r)$ of convergent power series on the rational closed disk $\overline{D}(a,r) \subseteq \mathbb{C}_v$, which we equip with the sup-norm $\zeta(a,r) = \|\cdot\|_{\zeta(a,r)}$. Recall that $\zeta(a,r)$ is indeed a multiplicative seminorm; see Definition 3.6, Proposition 3.7, and Proposition 3.17.

Definition B.1. Let $a \in \mathbb{C}_v$, and let $r \in |\mathbb{C}_v^\times|$. The *(intrinsic) rational closed Berkovich disk* $\overline{D}_{\mathrm{an}}^{\mathrm{in}}(a,r)$ is the set of all multiplicative seminorms on $\overline{\mathcal{A}}(a,r)$ that are bounded with respect to $\zeta(a,r)$. We equip $\overline{D}_{\mathrm{an}}^{\mathrm{in}}(a,r)$ with the *Gel'fand topology*, which is the weakest topology such that for every $f \in \overline{\mathcal{A}}(a,r)$ the real-valued function $\zeta \mapsto \|f\|_\zeta$ is continuous.

All points of $\overline{D}_{\mathrm{an}}^{\mathrm{in}}(a,r)$ are non-archimedean and extend $|\cdot|$, as the following simple result shows.

Lemma B.2. *Let $D \subseteq \mathbb{C}_v$ be a rational closed disk, with corresponding analytic Berkovich disk $D_{\mathrm{an}}^{\mathrm{in}}$. Then for every $\zeta \in D_{\mathrm{an}}^{\mathrm{in}}$,*

(a) *$\|\cdot\|_\zeta$ is non-archimedean, and*

(b) *$\|c\|_\zeta = |c|$ for every $c \in \mathbb{C}_v$.*

Proof.

(a) Writing $D = \overline{D}(a,r)$, the hypotheses specify that the seminorm $\|\cdot\|_\zeta$ is bounded by $\|\cdot\|_{\zeta(a,r)}$, which is non-archimedean. Thus, by Lemma 15.1(c), $\|\cdot\|_\zeta$ is also non-archimedean.

(b) If $c = 0$, then $\|c\|_\zeta = 0 = |c|$ by Definition 6.1(a). Otherwise, we have $c \in \mathbb{C}_v^\times$, and hence c is a unit in the ring $\mathcal{A}_D$ of convergent power series on D. Writing $D = \overline{D}(a,r)$ then, we have

$$\|c\|_\zeta = \|c\|_{\zeta(a,r)} = |c|$$

by Lemma 15.1(b) and the definition of $\zeta(a,r)$. □

For any rational closed disk $D = \overline{D}(a,r) \subseteq \mathbb{C}_v$, let $D_{\mathrm{an}}^{\mathrm{in}} := \overline{D}_{\mathrm{an}}^{\mathrm{in}}(a,r)$. The inclusion $\mathbb{C}_v[z] \hookrightarrow \mathcal{A}_D$ induces a map $F_D : D_{\mathrm{an}}^{\mathrm{in}} \to \mathbb{A}_{\mathrm{an}}^1$, since for any $\zeta \in D_{\mathrm{an}}^{\mathrm{in}}$, the seminorm $\|\cdot\|_\zeta\big|_{\mathbb{C}_v[z]}$ does indeed extend $|\cdot|$ by Lemma B.2(b). In fact, we have the following result.

Theorem B.3. *Given $a \in \mathbb{C}_v$ and $r \in |\mathbb{C}_v^\times|$, let D be the disk $D := \overline{D}(a,r)$. Define $F_D : D_{\mathrm{an}}^{\mathrm{in}} \to \mathbb{A}_{\mathrm{an}}^1$ to be the map given by*

$$\|\cdot\|_{F_D(\zeta)} := \|\cdot\|_\zeta\big|_{\mathbb{C}_v[z]}.$$

Then F_D is a topological embedding, with image $\overline{D}_{\mathrm{an}}(a,r) \subseteq \mathbb{A}_{\mathrm{an}}^1$.

Proof. Given $\zeta \in \overline{D}_{\mathrm{an}}^{\mathrm{in}}(a,r)$ and $f \in \mathbb{C}_v[z]$, we have

$$\|f\|_{F_D(\zeta)} = \|f\|_\zeta \le \|f\|_{\zeta(a,r)}.$$

Thus, $\|\cdot\|_{F_D(\zeta)} \in \overline{D}_{\mathrm{an}}(a,r)$, according to Definition 6.4. That is, the image of $F_D(\zeta)$ lies in $\overline{D}_{\mathrm{an}}(a,r)$.

It is immediate from the definition of the Gel'fand topology on both $\overline{D}_{\mathrm{an}}^{\mathrm{in}}(a,r)$ and $\overline{D}_{\mathrm{an}}(a,r)$ that F_D is continuous. Thus, it suffices to construct a continuous inverse $G_D : \overline{D}_{\mathrm{an}}(a,r) \to \overline{D}_{\mathrm{an}}^{\mathrm{in}}(a,r)$. We define G_D as follows.

Given $\zeta \in \overline{D}_{\mathrm{an}}(a,r)$, we must construct a bounded multiplicative seminorm $G_D(\zeta)$ on $\overline{\mathcal{A}}(a,r)$. For any $f = \sum_{n\geq 0} c_n(z-a)^n \in \overline{\mathcal{A}}(a,r)$ and any integer $N \geq 0$, let $f_N \in \mathbb{C}_v[z]$ be the degree-N truncation

$$f_N := \sum_{n=0}^{N} c_n(z-a)^n.$$

We then define

$$\|f\|_{G_D(\zeta)} := \lim_{N\to\infty} \|f_N\|_\zeta. \tag{B.1}$$

By Theorem 3.2, we have $\lim_{n\to\infty} |c_n| r^n = 0$, and therefore

$$\big\|f_N - f_M\big\|_{\zeta(a,r)} \to 0 \quad \text{as} \quad \min\{M,N\} \to \infty.$$

Hence, the real sequence $\{\|f_N\|_\zeta\}_{N\geq 0}$ is Cauchy, because

$$\Big|\|f_N\|_\zeta - \|f_M\|_\zeta\Big| \leq \big\|f_N - f_M\big\|_\zeta \leq \big\|f_N - f_M\big\|_{\zeta(a,r)}.$$

Thus, the limit in equation (B.1) converges. Moreover,

$$\|f\|_{G_D(\zeta)} = \lim_{N\to\infty} \|f_N\|_\zeta \leq \lim_{N\to\infty} \|f_N\|_{\zeta(a,r)} = \|f\|_{\zeta(a,r)}.$$

In addition, clearly $\|f\|_{G_D(\zeta)} \geq 0$, with $\|0\|_{G_D(\zeta)} = 0$ and $\|1\|_{G_D(\zeta)} = 1$. It remains to show that $G_D(\zeta)$ respects addition and multiplication. Given $f, g \in \overline{\mathcal{A}}(a,r)$,

$$\begin{aligned}\big\|f+g\big\|_{G_D(\zeta)} &= \lim_{N\to\infty} \big\|f_N + g_N\big\|_\zeta \leq \lim_{N\to\infty} \max\big\{\|f_N\|_\zeta, \|g_N\|_\zeta\big\} \\ &= \max\big\{\|f\|_{G_D(\zeta)}, \|g\|_{G_D(\zeta)}\big\},\end{aligned}$$

verifying the triangle inequality. Furthermore, note that

$$\big\|(fg)_N - f_N g_N\big\|_\zeta \leq \big\|(fg)_N - f_N g_N\big\|_{\zeta(a,r)} \to 0 \quad \text{as} \quad N \to \infty,$$

since the coefficients of f and g each satisfy $\lim_{n\to\infty} |c_n| r^n = 0$. Thus, $\lim_{N\to\infty} \|(fg)_N\|_\zeta - \|f_N g_N\|_\zeta = 0$, and therefore

$$\begin{aligned}\big\|fg\big\|_{G_D(\zeta)} &= \lim_{N\to\infty} \big\|(fg)_N\big\|_\zeta = \lim_{N\to\infty} \big\|f_N g_N\big\|_\zeta \\ &= \lim_{N\to\infty} \|f_N\|_\zeta \|g_N\|_\zeta = \|f\|_{G_D(\zeta)} \|g\|_{G_D(\zeta)}.\end{aligned}$$

Hence, $G_D(\zeta)$ is indeed an element of $\overline{D}_{\mathrm{an}}^{\mathrm{in}}(a,r)$.

To see that G_D is continuous, given an open subset $W \subseteq \mathbb{R}$ and a convergent power series $f \in \overline{\mathcal{A}}(a,r)$, we must show that

$$U_{f,W} := \big\{\zeta \in \overline{D}_{\mathrm{an}}(a,r) : \|f\|_{G_D(\zeta)} \in W\big\}$$

is open in $\overline{D}_{\mathrm{an}}(a,r)$. If $f=0$, this is clear; $U_{0,W}$ is empty if $0 \notin W$, or all of $\overline{D}_{\mathrm{an}}(a,r)$ if $0 \in W$. If $f \neq 0$, write $f = gh$ according to Theorem 14.2, where $g \in \mathbb{C}_v[z]$ is a polynomial and h is a unit in $\overline{\mathcal{A}}(a,r)$. By Lemma 15.1(b), we have $\|h\|_{G_D(\zeta)} = \|h\|_{\zeta(a,r)} \neq 0$. Hence, by multiplicativity, it follows that

$$U_{f,W} = \left\{\zeta \in \overline{D}_{\mathrm{an}}(a,r) : \|g\|_\zeta \in \|h\|^{-1}_{\zeta(a,r)} W\right\},$$

which is open in the Gel'fand topology on $\overline{D}_{\mathrm{an}}(a,r)$. Thus, G_D is indeed continuous.

It remains to show that $G_D : \overline{D}_{\mathrm{an}}(a,r) \to \overline{D}^{\mathrm{in}}_{\mathrm{an}}(a,r)$ is in fact the inverse of $F_D : \overline{D}^{\mathrm{in}}_{\mathrm{an}}(a,r) \to \overline{D}_{\mathrm{an}}(a,r)$. Given any $\zeta \in \overline{D}_{\mathrm{an}}(a,r)$ and $g \in \mathbb{C}_v[z]$, note that $g_N = g$ for all $N \geq \deg(g)$, and hence we have $\|g\|_{G_D(\zeta)} = \|g\|_\zeta$, according to the definition of G_D in equation (B.1). Thus,

$$\|g\|_{F_D \circ G_D(\zeta)} = \|g\|_{G_D(\zeta)} = \|g\|_\zeta,$$

and therefore $F_D \circ G_D$ is the identity on $\overline{D}_{\mathrm{an}}(a,r)$.

Similarly, given any $\zeta \in \overline{D}^{\mathrm{in}}_{\mathrm{an}}(a,r)$ and $f \in \overline{\mathcal{A}}(a,r)$,

$$\|f\|_{G_D \circ F_D(\zeta)} = \lim_{N\to\infty} \|f_N\|_{F_D(\zeta)} = \lim_{N\to\infty} \|f_N\|_\zeta = \|f\|_\zeta.$$

Thus, $G_D \circ F_D$ is the identity on $\overline{D}^{\mathrm{in}}_{\mathrm{an}}(a,r)$. ☐

B.2. Seminorms in homogeneous coordinates

In [**8**, Section 2.2], Baker and Rumely give a construction of $\mathbb{P}^1_{\mathrm{an}}$ that mirrors the Proj construction of $\mathbb{P}^1$ in algebraic geometry. We present their approach in this section.

Definition B.4. Let $\mathbb{A}^{2*}_{\mathrm{an}}$ be the set of all multiplicative seminorms on $\mathbb{C}_v[X,Y]$ that

(a) extend the absolute value $|\cdot|$ on $\mathbb{C}_v$, and

(b) are not identically zero on the maximal ideal $\langle X, Y\rangle$.

Equip $\mathbb{A}^{2*}_{\mathrm{an}}$ with the *homogeneous Gel'fand topology*, which is the weakest topology such that for every homogeneous $F \in \mathbb{C}_v[X,Y]$, the real-valued function $\zeta \mapsto \|F\|_\zeta$ is continuous.

Define $\sim$ to be the equivalence relation on $\mathbb{A}^{2*}_{\mathrm{an}}$ given by $\zeta \sim \xi$ if and only if there is a real constant $C > 0$ such that for all homogeneous polynomials $F \in \mathbb{C}_v[X,Y]$, $\|F\|_\zeta = C^{\deg(F)}\|F\|_\xi$. The *Berkovich projective line* over $\mathbb{C}_v$, denoted $\mathbb{P}^1_{\mathrm{an}}$, is the set of equivalence classes of $\mathbb{A}^{2*}_{\mathrm{an}}$ with respect to $\sim$, equipped with the quotient topology.

We leave to the reader the simple verification that $\sim$ is indeed an equivalence relation. To explain Definition B.4, we note that the set $\mathbb{A}^2_{\mathrm{an}}$ of multiplicative seminorms on $\mathbb{C}_v[X,Y]$ extending $|\cdot|$ is the Berkovich affine plane, and $\mathbb{A}^{2*}_{\text{an}}$ is the plane with the point at the origin removed, although we are using a slightly weaker topology. The quotient by $\sim$ is therefore analogous to the construction of the classical projective line $\mathbb{P}^1$ as a quotient of $\mathbb{A}^2 \smallsetminus \{(0,0)\}$ by the action of scalar multiplication.

The following lemma gives a more useful characterization of the topology on $\mathbb{P}^1_{\text{an}}$.

Lemma B.5. *The topology on $\mathbb{P}^1_{\text{an}}$ is the weakest topology such that for any real number $c \in \mathbb{R}$ and any two homogeneous polynomials $F, G \in \mathbb{C}_v[X,Y]$ of the same degree $d \geq 0$, the set*

$$\{\zeta \in \mathbb{P}^1_{\text{an}} : \|F\|_\zeta > c\|G\|_\zeta\}$$

is open.

Proof. This is left to the reader; see Exercise B.2. □

The reader may also check that the inequality $\|F\|_\zeta > c\|G\|_\zeta$ appearing in Lemma B.5 is independent of the choice of a seminorm $\zeta \in \mathbb{A}^{2*}_{\text{an}}$ representing the given equivalence class under $\sim$.

We can embed the Berkovich affine line $\mathbb{A}^1_{\text{an}}$ into $\mathbb{P}^1_{\text{an}}$, in much the same way that one embeds the classical affine line $\mathbb{A}^1(\mathbb{C}_v)$ into $\mathbb{P}^1(\mathbb{C}_v)$. More precisely, $\mathbb{A}^1(\mathbb{C}_v)$ is homeomorphic to $\mathbb{P}^1_{\text{an}} \smallsetminus \{\infty\}$, where $\infty \in \mathbb{P}^1_{\text{an}}$ is the equivalence class of seminorms represented by

$$\|F\|_\infty := |F(1,0)| \quad \text{for all} \quad F \in \mathbb{C}_v[X,Y]. \tag{B.2}$$

We leave it to the reader to show the following alternative characterization of the same point: ∞ is the unique element $\zeta \in \mathbb{P}^1_{\text{an}}$ for which $\|Y\|_\zeta = 0$; see Exercise B.3.

Theorem B.6. *Define a function $\psi : \mathbb{P}^1_{\text{an}} \smallsetminus \{\infty\} \to \mathbb{A}^1_{\text{an}}$ by*

$$\|f\|_{\psi(\zeta)} := \frac{\|Y^{\deg(f)} f(X/Y)\|_\zeta}{\|Y\|_\zeta^{\deg(f)}} \quad \textit{for all} \quad f \in \mathbb{C}_v[z] \smallsetminus \{0\} \tag{B.3}$$

and $\|0\|_{\psi(\zeta)} := 0$. Define a function $\theta : \mathbb{A}^1_{\text{an}} \to \mathbb{P}^1_{\text{an}} \smallsetminus \{\infty\}$ by setting $\theta(\zeta)$ to be the equivalence class of the seminorm $\|\cdot\|_{\Theta(\zeta)}$ on $\mathbb{C}_v[X,Y]$ given by

$$\|F\|_{\Theta(\zeta)} := \|F(z,1)\|_\zeta.$$

Then θ and ψ are homeomorphisms and inverse to one another.

Proof. First, we prove that ψ is well-defined. The denominator in equation (B.3) is nonzero because of our assumption that $\zeta \neq \infty$, and hence that $\|Y\|_\zeta \neq 0$, by the alternative characterization of the point $\infty \in \mathbb{P}^1_{\text{an}}$ above.

Given $\zeta, \xi \in \mathbb{A}^{2*}_{\text{an}}$ with $\zeta \sim \xi$, let $C > 0$ be the associated constant from Definition B.4. Then for any $f \in \mathbb{C}_v[z]$,

$$\|f\|_{\psi(\zeta)} = \frac{\|Y^{\deg(f)} f(X/Y)\|_\zeta}{\|Y\|_\zeta^{\deg(f)}} = \frac{C^{\deg(f)} \|Y^{\deg(f)} f(X/Y)\|_\xi}{C^{\deg(f)} \|Y\|_\xi^{\deg(f)}}$$
$$= \frac{\|Y^{\deg(f)} f(X/Y)\|_\xi}{\|Y\|_\xi^{\deg(f)}} = \|f\|_{\psi(\xi)}.$$

Moreover, $\psi(\zeta)$ is indeed a point of $\mathbb{A}^1_{\text{an}}$. After all, $\|c\|_{\psi(\zeta)} = |c|$ for all $c \in \mathbb{C}_v$, and the reader may easily check that $\psi(\zeta)$ respects multiplication and addition.

Second, we show that ψ is continuous. That is, given any $f \in \mathbb{C}_v[z]$ and any $r \in \mathbb{R}$, we must show that

$$\{\zeta \in \mathbb{P}^1_{\text{an}} \smallsetminus \{\infty\} : \|f\|_{\psi(\zeta)} < r\}$$
$$= \left\{\zeta \in \mathbb{P}^1_{\text{an}} \smallsetminus \{\infty\} : \left\|Y^{\deg(f)} f(X/Y)\right\|_\zeta < r\left\|Y^{\deg(f)}\right\|_\zeta\right\}$$

and

$$\{\zeta \in \mathbb{P}^1_{\text{an}} \smallsetminus \{\infty\} : \|f\|_{\psi(\zeta)} > r\}$$
$$= \left\{\zeta \in \mathbb{P}^1_{\text{an}} \smallsetminus \{\infty\} : \left\|Y^{\deg(f)} f(X/Y)\right\|_\zeta > r\left\|Y^{\deg(f)}\right\|_\zeta\right\}$$

are open in $\mathbb{P}^1_{\text{an}} \smallsetminus \{\infty\}$. However, by Lemma B.5, both are indeed open.

Third, given $\zeta \in \mathbb{A}^1_{\text{an}}$, we must prove that $\Theta(\zeta) \in \mathbb{A}^{2*}_{\text{an}}$. Clearly,

$$\|c\|_{\Theta(\zeta)} = \|c\|_\zeta = |c|$$

for any $c \in \mathbb{C}_v$; and $\|Y\|_{\Theta(\zeta)} = 1 \neq 0$. Again, the reader may easily check that $\|\cdot\|_{\Theta(\zeta)}$ respects addition and multiplication. Thus, $\Theta(\zeta) \in \mathbb{A}^{2*}_{\text{an}}$.

Fourth, we note that $\|Y\|_{\Theta(\zeta)} = |1| = 1$, whereas $\|Y\|_\infty = |0| = 0$, according to equation (B.2). Taking equivalence classes with respect to $\sim$, we have $\theta(\zeta) \neq \infty$. Thus, θ is indeed a function from $\mathbb{A}^1_{\text{an}}$ into $\mathbb{P}^1_{\text{an}} \smallsetminus \{\infty\}$.

Fifth, we must show that θ is continuous. By Lemma B.5, it suffices to show that for any $c \in \mathbb{R}$, any integer $d \geq 0$, and any homogeneous polynomials $F, G \in \mathbb{C}_v[X, Y]$ of degree d, the set

$$\{\zeta \in \mathbb{A}^1_{\text{an}} : \|F\|_{\Theta(\zeta)} > c\|G\|_{\Theta(\zeta)}\} = \{\zeta \in \mathbb{A}^1_{\text{an}} : \|F(z,1)\|_\zeta > c\|G(z,1)\|_\zeta\}$$

is open in $\mathbb{A}^1_{\text{an}}$. Fortunately, for any polynomial $f(z) \in \mathbb{C}_v[z]$, including $F(z,1)$ or $G(z,1)$, the function $\zeta \mapsto \|f\|_\zeta$ maps $\mathbb{A}^1_{\text{an}}$ continuously into $\mathbb{R}$. Thus, the desired set is indeed open.

Finally, we show that θ and ψ are inverses. For any $\zeta \in \mathbb{P}^1_{\text{an}} \smallsetminus \{\infty\}$, set $C := \|Y\|_\zeta > 0$. Then for any homogeneous $F \in \mathbb{C}_v[X, Y]$, we have

$$\|F\|_{\theta(\psi(\zeta))} = \|F(z,1)\|_{\psi(\zeta)} = \frac{\|F\|_\zeta}{\|Y^{\deg(F)}\|_\zeta} = (C^{-1})^d \|F\|_\zeta,$$

and hence $\theta(\psi(\zeta)) \sim \zeta$. Similarly, for any $\xi \in \mathbb{A}^1_{\text{an}}$ and any $f \in \mathbb{C}_v[z]$,

$$\|f\|_{\psi(\theta(\zeta))} = \frac{\|Y^{\deg(f)} f(X/Y)\|_{\theta(\zeta)}}{\|Y^{\deg(f)}\|_{\theta(\zeta)}} = \frac{\|f\|_\zeta}{1} = \|f\|_\zeta. \qquad \square$$

Remark B.7. By composing the embedding $i : \mathbb{C}_v \hookrightarrow \mathbb{A}^1_{\text{an}}$ of Theorem 6.21 with $\theta : \mathbb{A}^1_{\text{an}} \hookrightarrow \mathbb{P}^1_{\text{an}} \smallsetminus \{\infty\}$ of Theorem B.6, we see that any point $x \in \mathbb{C}_v$, when viewed as a point in $\mathbb{P}^1_{\text{an}}$, is the equivalence class of the seminorm on $\mathbb{C}_v[X, Y]$ given by

$$\|F\|_{(x,1)} := |F(x,1)|.$$

Combining this fact with the definition of $\infty \in \mathbb{P}^1_{\text{an}}$ in equation (B.2), it follows that any point $P = [x, y] \in \mathbb{P}^1(\mathbb{C}_v)$, when viewed as a point in $\mathbb{P}^1_{\text{an}}$, is the equivalence class of the seminorm on $\mathbb{C}_v[X, Y]$ given by

$$\|F\|_{(x,y)} := |F(x,y)|.$$

B.3. Berkovich morphisms via homogeneous coordinates

Definition B.8. Let $\phi = f/g \in \mathbb{C}_v(z)$ be a rational function of degree $d \geq 0$, where $f, g \in \mathbb{C}_v[z]$ are relatively prime polynomials. Set

$$F(X,Y) := Y^d f\left(\frac{X}{Y}\right) \quad \text{and} \quad G(X,Y) := Y^d g\left(\frac{X}{Y}\right),$$

so that $F, G \in \mathbb{C}_v[X, Y]$ are homogeneous of degree d, and define

$$\Phi(X,Y) := \big(F(X,Y), G(X,Y)\big) \in \mathbb{C}_v[X,Y] \times \mathbb{C}_v[X,Y].$$

(a) Extend Φ to a function $\Phi : \mathbb{A}^{2*}_{\text{an}} \to \mathbb{A}^{2*}_{\text{an}}$ as follows: given $\zeta \in \mathbb{A}^{2*}_{\text{an}}$, define $\Phi(\zeta)$ by

$$\|H\|_{\Phi(\zeta)} := \|H \circ \Phi\|_\zeta = \big\|H\big(F(X,Y), G(X,Y)\big)\big\|_\zeta \tag{B.4}$$

for all $H \in \mathbb{C}_v[X, Y]$.

(b) Extend ϕ to a function $\phi : \mathbb{P}^1_{\text{an}} \to \mathbb{P}^1_{\text{an}}$ by setting $\phi(\zeta)$ to be the equivalence class of the seminorm $\|\cdot\|_{\Phi(\zeta)}$.

Theorem B.9. *Let $\phi \in \mathbb{C}_v(z)$ be a rational function. Then the functions $\Phi : \mathbb{A}^{2*}_{\text{an}} \to \mathbb{A}^{2*}_{\text{an}}$ and $\phi : \mathbb{P}^1_{\text{an}} \to \mathbb{P}^1_{\text{an}}$ of Definition* B.8 *are well-defined. Moreover, $\phi : \mathbb{P}^1_{\text{an}} \to \mathbb{P}^1_{\text{an}}$ is a continuous extension of $\phi : \mathbb{P}^1(\mathbb{C}_v) \to \mathbb{P}^1(\mathbb{C}_v)$.*

Proof. Write $\phi = f/g$, and define d, F, G, and Φ as in Definition B.8. Because f and g are relatively prime, there exist polynomials $a, b \in \mathbb{C}_v[z]$ such that $af + bg = 1$. Let $m \geq 0$ be the larger of $\deg(a)$ and $\deg(b)$, and define $A := Y^m a(X/Y)$ and $B := Y^m b(X/Y)$. Thus, evaluating $af + bg = 1$ at $z = X/Y$ and multiplying by Y^{m+d}, we have

$$AF + BG = Y^{d+m}. \tag{B.5}$$

On the other hand, the polynomials

$$\tilde{f}(z) := z^d f(1/z) \quad \text{and} \quad \tilde{g}(z) := z^d g(1/z)$$

are also relatively prime, since $\tilde{f}(z)/\tilde{g}(z) = \phi(1/z)$ also has degree d. Moreover, we have $Y^d \tilde{f}(X/Y) = F[Y, X]$ and $Y^d \tilde{g}(X/Y) = G[Y, X]$. Hence, there are homogeneous polynomials $\tilde{A}, \tilde{B} \in \mathbb{C}_v[X, Y]$ of the same degree $n \geq 0$ such that

$$\tilde{A}F + \tilde{B}G = X^{d+n}. \tag{B.6}$$

Given $\zeta \in \mathbb{A}^{2*}_{\text{an}}$, note first that for any $c \in \mathbb{C}_v$, we have

$$\|c\|_{\Phi(\zeta)} = \|c \circ \Phi\|_\zeta = \|c\|_\zeta = |c|.$$

In addition, for any $H_1, H_2 \in \mathbb{C}_v[X, Y]$,

$$\begin{aligned} \big\|H_1 + H_2\big\|_{\Phi(\zeta)} &= \big\|H_1 \circ \Phi + H_2 \circ \Phi\big\|_\zeta \leq \max\left\{\|H_1 \circ \Phi\|_\zeta, \|H_2 \circ \Phi\|_\zeta\right\} \\ &= \max\left\{\|H_1\|_{\Phi(\zeta)}, \|H_2\|_{\Phi(\zeta)}\right\}. \end{aligned}$$

Similarly, $\Phi(\zeta)$ is multiplicative. Finally, if $\|\cdot\|_{\Phi(\zeta)}$ were identically zero on the ideal $\langle X, Y \rangle$, then we would have

$$\|F\|_\zeta = \|X\|_{\Phi(\zeta)} = 0 \quad \text{and} \quad \|G\|_\zeta = \|Y\|_{\Phi(\zeta)} = 0.$$

Therefore, by equation (B.5), it would follow that

$$\begin{aligned} \|Y\|_\zeta^{d+m} &= \big\|Y^{d+m}\big\|_\zeta = \|AF + BG\|_\zeta \\ &\leq \max\left\{\|A\|_\zeta \|F\|_\zeta, \|B\|_\zeta \|G\|_\zeta\right\} = 0, \end{aligned}$$

and hence $\|Y\|_\zeta = 0$. Similarly, by equation (B.6), we would also obtain $\|X\|_\zeta = 0$. Then $\|\cdot\|_\zeta$ would be trivial on $\langle X, Y \rangle$, a contradiction. Thus, $\Phi(\zeta)$ is in fact a point of $\mathbb{A}^{2*}_{\text{an}}$.

To show that $\phi : \mathbb{P}^1_{\text{an}} \to \mathbb{P}^1_{\text{an}}$ is also well-defined, suppose $\zeta \sim \xi \in \mathbb{A}^{2*}_{\text{an}}$, where $\sim$ is the relation of Definition B.4, and let $C > 0$ be the associated real constant. For any homogeneous $H \in \mathbb{C}_v[X, Y]$, the polynomial $H \circ \Phi$ is also homogeneous, of degree $d \cdot \deg(H)$, and therefore

$$\|H\|_{\Phi(\zeta)} = \|H \circ \Phi\|_\zeta = C^{d \cdot \deg(H)} \|H \circ \Phi\|_\xi = (C^d)^{\deg(H)} \|H\|_{\Phi(\xi)}.$$

Thus, $\Phi(\zeta) \sim \Phi(\xi)$, with associated constant $C^d > 0$.

Next, we show that ϕ is continuous. By Lemma B.5, it suffices to show that for any $c \in \mathbb{R}$, $d' \geq 0$, and homogeneous $F, G \in \mathbb{C}_v[X,Y]$ of degree d', the set

$$\{\zeta \in \mathbb{P}^1_{\mathrm{an}} : \|F\|_{\Phi(\zeta)} > c\|G\|_{\Phi(\zeta)}\}$$

is open in $\mathbb{P}^1_{\mathrm{an}}$. But this same set is exactly

$$\{\zeta \in \mathbb{P}^1_{\mathrm{an}} : \|F \circ \Phi\|_\zeta > c\|G \circ \Phi\|_\zeta\},$$

which is open in $\mathbb{P}^1_{\mathrm{an}}$ by Lemma B.5, because $F\circ\Phi$ and $G\circ\Phi$ are homogeneous of degree $d \cdot d'$.

Last, we must show that the function $\phi : \mathbb{P}^1_{\mathrm{an}} \to \mathbb{P}^1_{\mathrm{an}}$ extends the function $\phi : \mathbb{P}^1(\mathbb{C}_v) \to \mathbb{P}^1(\mathbb{C}_v)$. To distinguish these two maps, let us refer to the second as ψ, for the moment. Given a point $P = [x,y] \in \mathbb{P}^1(\mathbb{C}_v)$, we have $\psi(P) = [F(x,y), G(x,y)]$. These two points, viewed as elements of $\mathbb{P}^1_{\mathrm{an}}$ via Remark B.7, are the equivalence classes of the seminorms on $\mathbb{C}_v[X,Y]$ given by

$$\|H\|_P := \big|H(x,y)\big| \quad \text{and} \quad \|H\|_{\psi(P)} := \big|H(F(x,y), G(x,y))\big|,$$

respectively, for $H \in \mathbb{C}_v[X,Y]$. Hence,

$$\|H\|_{\psi(P)} = \big|H \circ \Phi(x,y)\big| = \|H \circ \Phi\|_{(x,y)} = \|H\|_{\Phi(x,y)}.$$

That is, $\psi(P) = \phi(P)$, as desired. □

The map $\Phi : \mathbb{A}^{2*}_{\mathrm{an}} \to \mathbb{A}^{2*}_{\mathrm{an}}$ of Definition B.8 is also continuous, but this fact is peripheral to our needs, and we say no more about it.

B.4. Rivera-Letelier's construction of $\mathbb{P}^1_{\mathrm{an}}$

In [**114**, **115**, **117**], Rivera-Letelier constructed $\mathbb{P}^1_{\mathrm{an}}$ by a completely different method and used different terminology. He soon realized that his space was essentially the same as Berkovich's projective line, and therefore his later work uses the seminorms we have described in Chapter 6. Nevertheless, Rivera-Letelier's purely geometric construction conveys some important insights, and the fact that he built it by different means without knowledge of Berkovich's work underscores how natural a setting the space $\mathbb{P}^1_{\mathrm{an}}$ is for studying non-archimedean dynamics. In this section, we give a rough sketch of his construction, including a dictionary between his terminology and the standard Berkovich space terminology, for the convenience of readers wishing to study those papers.

Rivera-Letelier begins by considering a strictly increasing sequence

$$D_1 \subsetneq D_2 \subsetneq \cdots$$

of closed $\mathbb{P}^1(\mathbb{C}_v)$-disks, and he sets $B := \bigcup_{n\geq 1} D_n$, which is either the complement of a single point, a rational open $\mathbb{P}^1(\mathbb{C}_v)$-disk, an irrational $\mathbb{P}^1(\mathbb{C}_v)$-disk, or all of $\mathbb{P}^1(\mathbb{C}_v)$. Ultimately, those possibilities correspond, of course, to points of Type I, II, III, or IV, respectively. (Actually, he disallows the first case, because his goal is to construct the hyperbolic space $\mathbb{H} := \mathbb{P}^1_{\mathrm{an}} \setminus \mathbb{P}^1(\mathbb{C}_v)$, which we discuss in Section 6.8. Moreover, in [**114**] he also disallows the last case, because in that paper he does not make use of Type IV points.)

He then considers the decreasing sequence $\{B \setminus D_n\}_{n \geq 1}$, which he calls an *evanescent chain* ("chaine évanescente"), because the intersection

$$\bigcap_{n \geq 1} (B \setminus D_n)$$

is empty. The sets $C_n := B \setminus D_n$ are either disks with a single point removed (later to correspond to Type I) or annuli ("couronnes", for Types II and III) or $\mathbb{P}^1(\mathbb{C}_v)$-disks (Type IV). He then declares two evanescent chains $\{C_n\}$, $\{C'_n\}$ to be equivalent if for every $N \geq 1$ there is some $M \geq N$ such that $C_M \subseteq C'_N$ and $C'_M \subseteq C_N$. Next, he defines an *end* ("bout") to be an equivalence class of evanescent chains. (Rivera-Letelier's bouts correspond to the *directions* we introduced in Definition 6.39.) He calls an end *singular* if its corresponding evanescent chains are chains of disks (ultimately, Type IV points); *irrational* if its evanescent chains are annuli all of the form $B \setminus D_n$ where B is an irrational disk (ultimately, Type III points); or *rational* if its evanescent chains are annuli all of the form $B \setminus D_n$ where B is a rational open disk (ultimately, Type II points). Again, he does not discuss the remaining case of Type I points in these three papers.

Given an end $\mathcal{P}$, Rivera-Letelier declares $B_{\mathcal{P}}$ to be the set B used to define one of the chains associated with $\mathcal{P}$. Thus, B is of one of the following four forms, in order corresponding to type: either $\mathbb{P}^1(\mathbb{C}_v) \setminus \{b\}$ for some point $b \in \mathbb{P}^1(\mathbb{C}_v)$, or a rational open $\mathbb{P}^1(\mathbb{C}_v)$-disk, or an irrational $\mathbb{P}^1(\mathbb{C}_v)$-disk, or all of $\mathbb{P}^1(\mathbb{C}_v)$. He notes that $B_{\mathcal{P}}$ is independent of the choice of chain representing $\mathcal{P}$. With that preparation, he can then define the actual points of $\mathbb{P}^1_{\mathrm{an}}$ as certain sets of ends. (In [**114**] he calls such a set of ends a *projective system* ("système projectif"), but in later papers he simply calls such a set a point.) His definition of such points, which he denotes $\mathcal{S}$, proceeds as follows.

The Type IV, or singular, points are singleton sets $\{\mathcal{P}\}$, where $\mathcal{P}$ is a singular end. For the Type II and III, or rational and irrational, points, he declares two open $\mathbb{P}^1(\mathbb{C}_v)$-disks D_1 and D_2 to be *associated* if $D_1 \cap D_2 = \varnothing$ and they are maximal with respect to that property. (That is, if $E_1 \supseteq D_1$ and $E_2 \supseteq D_2$ are open $\mathbb{P}^1(\mathbb{C}_v)$-disks for which $E_1 \cap E_2 = \varnothing$, then $E_1 = D_1$ and $E_2 = D_2$.) The Type III points are then equivalence classes of irrational ends, where two ends are said to be equivalent if the corresponding sets $B_{\mathcal{P}}$ are associated. Similarly, the Type II points are equivalence classses of rational ends. Finally, the Type I points, had Rivera-Letelier actually defined them in this construction, would again be singleton sets of ends $\{\mathcal{P}\}$

for which $B_{\mathcal{P}}$ is of the form $\mathbb{P}^1(\mathbb{C}_v) \setminus \{b\}$; in our notation, the corresponding Type I point is b itself.

Exercises for Appendix B

B.1. Define a function $\|\cdot\|_{(0,0)} : \mathbb{C}_v[X,Y] \to \mathbb{R}$ by $\|f\|_{(0,0)} := |f(0,0)|$.

(a) Show that $\|\cdot\|_{(0,0)}$ is a multiplicative seminorm on $\mathbb{C}_v[X,Y]$ extending $|\cdot|$ on $\mathbb{C}_v$ but is identically zero on the ideal $\langle X, Y\rangle$.

(b) Show that the only multiplicative seminorm on $\mathbb{C}_v[X,Y]$ extending $|\cdot|$ and being identically zero on $\langle X, Y\rangle$ is $\|\cdot\|_{(0,0)}$.

B.2. Prove Lemma B.5, including the fact that the truth of the condition $\|F\|_\zeta > c\|G\|_\zeta$ is independent of the choice of ζ in its equivalence class under the relation $\sim$ of Definition B.4.

B.3. Show that there is a unique element $\zeta \in \mathbb{P}^1_{\mathrm{an}}$ for which $\|Y\|_\zeta = 0$, and that this point ζ is precisely the point $\infty \in \mathbb{P}^1_{\mathrm{an}}$ defined in equation (B.2).

B.4. Define a map $i : \mathbb{P}^1(\mathbb{C}_v) \to \mathbb{P}^1_{\mathrm{an}}$ by setting $i([x,y])$ to be the equivalence class of the seminorm $\|\cdot\|_{(x,y)}$ on $\mathbb{C}_v[X,Y]$ given by

$$\|F\|_{(x,y)} := |F(x,y)|.$$

(a) Prove that i is an embedding of topological spaces.

(b) Show that $i([x,y])$ is the unique element $\zeta \in \mathbb{P}^1_{\mathrm{an}}$ for which

$$\|yX - xY\|_\zeta = 0.$$

Bibliography

[1] S. Albeverio, A. Khrennikov, and B. Tirozzi, *p-adic dynamical systems and neural networks*, Math. Models Methods Appl. Sci. **9** (1999), no. 9, 1417–1437, DOI 10.1142/S0218202599000634. MR1725825

[2] Y. Amice, *Les nombres p-adiques* (French), Presses Universitaires de France, Paris, 1975. Préface de Ch. Pisot; Collection SUP: Le Mathématicien, No. 14. MR0447195

[3] D. K. Arrowsmith and F. Vivaldi, *Geometry of p-adic Siegel discs*, Phys. D **71** (1994), no. 1-2, 222–236, DOI 10.1016/0167-2789(94)90191-0. MR1264116

[4] D. Bajpai, R. L. Benedetto, R. Chen, E. Kim, O. Marschall, D. Onul, and Y. Xiao, *Non-archimedean connected Julia sets with branching*, Ergodic Theory Dynam. Systems **37** (2017), no. 1, 59–78, DOI 10.1017/etds.2015.42. MR3590494

[5] M. Baker, personal communication, 2017.

[6] M. Baker and L. DeMarco, *Preperiodic points and unlikely intersections*, Duke Math. J. **159** (2011), no. 1, 1–29, DOI 10.1215/00127094-1384773. MR2817647

[7] M. H. Baker and R. Rumely, *Equidistribution of small points, rational dynamics, and potential theory* (English, with English and French summaries), Ann. Inst. Fourier (Grenoble) **56** (2006), no. 3, 625–688. MR2244226

[8] M. Baker and R. Rumely, *Potential theory and dynamics on the Berkovich projective line*, Mathematical Surveys and Monographs, vol. 159, American Mathematical Society, Providence, RI, 2010. MR2599526

[9] A. F. Beardon, *Iteration of rational functions*, Graduate Texts in Mathematics, vol. 132, Springer-Verlag, New York, 1991. Complex analytic dynamical systems. MR1128089

[10] S. Ben-Menahem, *p-adic iterations*, preprint, TAUP 1627–88, Tel-Aviv University (1988).

[11] J. J. Benedetto and W. Czaja, *Integration and modern analysis*, Birkhäuser Advanced Texts: Basler Lehrbücher. [Birkhäuser Advanced Texts: Basel Textbooks], Birkhäuser Boston, Inc., Boston, MA, 2009. MR2559803

[12] R. L. Benedetto, *Fatou components in p-adic dynamics*, ProQuest LLC, Ann Arbor, MI, 1998. Thesis (Ph.D.)–Brown University. MR2697379

[13] R. L. Benedetto, *p-adic dynamics and Sullivan's no wandering domains theorem*, Compositio Math. **122** (2000), no. 3, 281–298, DOI 10.1023/A:1002067315057. MR1781331

[14] R. L. Benedetto, *Hyperbolic maps in p-adic dynamics*, Ergodic Theory Dynam. Systems **21** (2001), no. 1, 1–11, DOI 10.1017/S0143385701001043. MR1826658

[15] R. L. Benedetto, *Reduction, dynamics, and Julia sets of rational functions*, J. Number Theory **86** (2001), no. 2, 175–195, DOI 10.1006/jnth.2000.2577. MR1813109

[16] R. L. Benedetto, *Components and periodic points in non-Archimedean dynamics*, Proc. London Math. Soc. (3) **84** (2002), no. 1, 231–256, DOI 10.1112/plms/84.1.231. MR1863402

[17] R. L. Benedetto, *Examples of wandering domains in p-adic polynomial dynamics* (English, with English and French summaries), C. R. Math. Acad. Sci. Paris **335** (2002), no. 7, 615–620. MR1941304

[18] R. L. Benedetto, *Wandering domains and nontrivial reduction in non-Archimedean dynamics*, Illinois J. Math. **49** (2005), no. 1, 167–193. MR2157374

[19] R. L. Benedetto, *Wandering domains in non-Archimedean polynomial dynamics*, Bull. London Math. Soc. **38** (2006), no. 6, 937–950, DOI 10.1112/S0024609306019126. MR2285248

[20] R. L. Benedetto, *An Ahlfors islands theorem for non-Archimedean meromorphic functions*, Trans. Amer. Math. Soc. **360** (2008), no. 8, 4099–4124, DOI 10.1090/S0002-9947-08-04546-7. MR2395165

[21] R. L. Benedetto, *A criterion for potentially good reduction in non-archimedean dynamics*, Acta Arith. **165** (2014), no. 3, 251–256, DOI 10.4064/aa165-3-4. MR3263950

[22] R. L. Benedetto, *Attaining potentially good reduction in arithmetic dynamics*, Int. Math. Res. Not. IMRN **22** (2015), 11828–11846. MR3456704

[23] R. Benedetto, J.-Y. Briend, and H. Perdry, *Dynamique des polynômes quadratiques sur les corps locaux* (French, with English and French summaries), J. Théor. Nombres Bordeaux **19** (2007), no. 2, 325–336. MR2394889

[24] R. L. Benedetto, D. Ghioca, P. Kurlberg, and T. J. Tucker, *A case of the dynamical Mordell-Lang conjecture*, Math. Ann. **352** (2012), no. 1, 1–26, DOI 10.1007/s00208-010-0621-4. With an appendix by Umberto Zannier. MR2885573

[25] R. Benedetto, P. Ingram, R. Jones, and A. Levy, *Attracting cycles in p-adic dynamics and height bounds for postcritically finite maps*, Duke Math. J. **163** (2014), no. 13, 2325–2356, DOI 10.1215/00127094-2804674. MR3265554

[26] V. G. Berkovich, *Spectral theory and analytic geometry over non-Archimedean fields*, Mathematical Surveys and Monographs, vol. 33, American Mathematical Society, Providence, RI, 1990. MR1070709

[27] J.-P. Bézivin, *Sur les points périodiques des applications rationnelles en dynamique ultramétrique* (French), Acta Arith. **100** (2001), no. 1, 63–74, DOI 10.4064/aa100-1-5. MR1864626

[28] J.-P. Bézivin, *Sur la compacité des ensembles de Julia des polynômes p-adiques* (French, with English summary), Math. Z. **246** (2004), no. 1-2, 273–289, DOI 10.1007/s00209-003-0599-7. MR2031456

[29] J.-P. Bézivin, *Dynamique des Fractions Rationnelles p-adiques*, Lecture notes, 2003. Available at `http://jp.bezivin.pagesperso-orange.fr/dealatex.pdf`

[30] S. Bosch, U. Güntzer, and R. Remmert, *Non-Archimedean analysis: A systematic approach to rigid analytic geometry*, Grundlehren der Mathematischen Wissenschaften [Fundamental Principles of Mathematical Sciences], vol. 261, Springer-Verlag, Berlin, 1984. MR746961

[31] P. Billingsley, *Convergence of probability measures*, 2nd ed., Wiley Series in Probability and Statistics: Probability and Statistics, John Wiley & Sons, Inc., New York, 1999. A Wiley-Interscience Publication. MR1700749

[32] L. E. Böttcher, *The principal laws of convergence of iterates and their applications to analysis* (Russian), Izv. Kazan. Fiz.-Mat. Obshch. **14** (1904), 155-234.

[33] N. Bourbaki, *Éléments de mathématique. Fasc. XIII. Livre VI: Intégration. Chapitres 1, 2, 3 et 4: Inégalités de convexité, Espaces de Riesz, Mesures sur les espaces localement compacts, Prolongement d'une mesure, Espaces L^p* (French), Deuxième édition revue et augmentée. Actualités Scientifiques et Industrielles, No. 1175, Hermann, Paris, 1965. MR0219684

[34] H. Brolin, *Invariant sets under iteration of rational functions*, Ark. Mat. **6** (1965), 103–144 (1965), DOI 10.1007/BF02591353. MR0194595

[35] J. Bryk and C. E. Silva, *Measurable dynamics of simple p-adic polynomials*, Amer. Math. Monthly **112** (2005), no. 3, 212–232, DOI 10.2307/30037439. MR2125384

[36] A. D. Bryuno, Analytical form of differential equations, *Trans. Moscow Math. Soc.* **25** (1971), 131–288.

[37] G. S. Call and J. H. Silverman, *Canonical heights on varieties with morphisms*, Compositio Math. **89** (1993), no. 2, 163–205. MR1255693

[38] S. Cantat and A. Chambert-Loir, *Dynamique p-adique (d'après les exposés de Jean-Christophe Yoccoz)* (French, with English and French summaries), Quelques aspects des systèmes dynamiques polynomiaux, Panor. Synthèses, vol. 30, Soc. Math. France, Paris, 2010, pp. 295–341. MR2932436

[39] L. Carleson and T. W. Gamelin, *Complex dynamics*, Universitext: Tracts in Mathematics, Springer-Verlag, New York, 1993. MR1230383

[40] A. Chambert-Loir, *Mesures et équidistribution sur les espaces de Berkovich* (French), J. Reine Angew. Math. **595** (2006), 215–235, DOI 10.1515/CRELLE.2006.049. MR2244803

[41] R. F. Coleman, *p-adic Banach spaces and families of modular forms*, Invent. Math. **127** (1997), no. 3, 417–479, DOI 10.1007/s002220050127. MR1431135

[42] J. B. Conway, *Functions of one complex variable*, 2nd ed., Graduate Texts in Mathematics, vol. 11, Springer-Verlag, New York-Berlin, 1978. MR503901

[43] L. De Marco and X. Faber, *Degenerations of complex dynamical systems*, Forum Math. Sigma **2** (2014), e6, 36, DOI 10.1017/fms.2014.8. MR3264250

[44] L. DeMarco and X. Faber, *Degenerations of complex dynamical systems II: analytic and algebraic stability*, Math. Ann. **365** (2016), no. 3-4, 1669–1699, DOI 10.1007/s00208-015-1331-8. With an appendix by Jan Kiwi. MR3521102

[45] A. Douady, *Disques de Siegel et anneaux de Herman* (French), Astérisque **152-153** (1987), 4, 151–172 (1988). Séminaire Bourbaki, Vol. 1986/87. MR936853

[46] J. R. Doyle, K. Jacobs, and R. Rumely, *Configuration of the crucial set for a quadratic rational map*, Res. Number Theory **2** (2016), Art. 11, 16, DOI 10.1007/s40993-016-0041-y. MR3518909

[47] V. A. Dremov, *On a p-adic Julia set* (Russian), Uspekhi Mat. Nauk **58** (2003), no. 6(354), 151–152, DOI 10.1070/RM2003v058n06ABEH000682; English transl., Russian Math. Surveys **58** (2003), no. 6, 1194–1195. MR2054097

[48] V. Dremov, G. Shabat, and P. Vytnova, *On the chaotic properties of quadratic maps over non-Archimedean fields*, *p*-adic mathematical physics, AIP Conf. Proc., vol. 826, Amer. Inst. Phys., Melville, NY, 2006, pp. 43–54, DOI 10.1063/1.2193109. MR2258672

[49] J. Écalle, *Théorie itérative: introduction à la théorie des invariants holomorphes* (French), J. Math. Pures Appl. (9) **54** (1975), 183–258. MR0499882

[50] M. Einsiedler and T. Ward, *Ergodic theory with a view towards number theory*, Graduate Texts in Mathematics, vol. 259, Springer-Verlag London, Ltd., London, 2011. MR2723325

[51] A. Escassut, *The ultrametric spectral theory*, Period. Math. Hungar. **11** (1980), no. 1, 7–60, DOI 10.1007/BF02019983. MR571136

[52] X. Faber, *Topology and geometry of the Berkovich ramification locus for rational functions, I*, Manuscripta Math. **142** (2013), no. 3-4, 439–474, DOI 10.1007/s00229-013-0611-4. MR3117171

[53] X. Faber, *Topology and geometry of the Berkovich ramification locus for rational functions, II*, Math. Ann. **356** (2013), no. 3, 819–844, DOI 10.1007/s00208-012-0872-3. MR3063898

[54] P. Fatou, *Sur les équations fonctionnelles* (French), Bull. Soc. Math. France **48** (1920), 208–314. MR1504797

[55] C. Favre and T. Gauthier, *Distribution of postcritically finite polynomials*, Israel J. Math. **209** (2015), no. 1, 235–292, DOI 10.1007/s11856-015-1218-0. MR3430241

[56] C. Favre and T. Gauthier, *Classification of special curves in the space of cubic polynomials*, Int. Math. Res. Not. IMRN **2** (2018), 362–411, DOI 10.1093/imrn/rnw245. MR3801434

[57] C. Favre, J. Kiwi, and E. Trucco, *A non-Archimedean Montel's theorem*, Compositio Math. **148** (2012), no. 3, 966–990, DOI 10.1112/S0010437X11007470. MR2925406

[58] C. Favre and M. Jonsson, *The valuative tree*, Lecture Notes in Mathematics, vol. 1853, Springer-Verlag, Berlin, 2004. MR2097722

[59] C. Favre and M. Jonsson, *Eigenvaluations* (English, with English and French summaries), Ann. Sci. École Norm. Sup. (4) **40** (2007), no. 2, 309–349, DOI 10.1016/j.ansens.2007.01.002. MR2339287

[60] C. Favre and J. Rivera-Letelier, *Théorème d'équidistribution de Brolin en dynamique p-adique* (French, with English and French summaries), C. R. Math. Acad. Sci. Paris **339** (2004), no. 4, 271–276, DOI 10.1016/j.crma.2004.06.023. MR2092012

[61] C. Favre and J. Rivera-Letelier, *Équidistribution quantitative des points de petite hauteur sur la droite projective* (French, with English and French summaries), Math. Ann. **335** (2006), no. 2, 311–361, DOI 10.1007/s00208-006-0751-x. MR2221116

[62] Charles Favre and Juan Rivera-Letelier, *Corrigendum à l'article "Équidistribution quantitative des points de petite hauteur sur la droite projective"*, Math. Ann. **339** (2007), 799–801.

[63] C. Favre and J. Rivera-Letelier, *Théorie ergodique des fractions rationnelles sur un corps ultramétrique* (French, with English and French summaries), Proc. Lond. Math. Soc. (3) **100** (2010), no. 1, 116–154, DOI 10.1112/plms/pdp022. MR2578470

[64] G. B. Folland, *Real analysis: Modern techniques and their applications; A Wiley-Interscience Publication*, 2nd ed., Pure and Applied Mathematics (New York), John Wiley & Sons, Inc., New York, 1999. MR1681462

[65] A. Freire, A. Lopes, and R. Mañé, *An invariant measure for rational maps*, Bol. Soc. Brasil. Mat. **14** (1983), no. 1, 45–62, DOI 10.1007/BF02584744. MR736568

[66] J. Fresnel and M. van der Put, *Rigid analytic geometry and its applications*, Progress in Mathematics, vol. 218, Birkhäuser Boston, Inc., Boston, MA, 2004. MR2014891

[67] T. Gauthier, Y. Okuyama, and G. Vigny, *Approximation of non-archimedean Lyapunov exponents and applications over global fields*, `arXiv:1803.06859` (2018).

[68] D. Ghioca, H. Krieger, K. D. Nguyen, and H. Ye, *The dynamical André-Oort conjecture: unicritical polynomials*, Duke Math. J. **166** (2017), no. 1, 1–25, DOI 10.1215/00127094-3673996. MR3592687

[69] D. Goss, *A short introduction to rigid analytic spaces*, The arithmetic of function fields (Columbus, OH, 1991), Ohio State Univ. Math. Res. Inst. Publ., vol. 2, de Gruyter, Berlin, 1992, pp. 131–141. MR1196516

[70] F. Q. Gouvêa, *p-adic numbers: An introduction*, Universitext, Springer-Verlag, Berlin, 1993. MR1251959

[71] M. Gundlach, A. Khrennikov, and K.-O. Lindahl, *On ergodic behavior of p-adic dynamical systems*, Infin. Dimens. Anal. Quantum Probab. Relat. Top. **4** (2001), no. 4, 569–577, DOI 10.1142/S0219025701000632. MR1876486

[72] R. Hartshorne, *Algebraic geometry*, Springer-Verlag, New York-Heidelberg, 1977. Graduate Texts in Mathematics, No. 52. MR0463157

[73] M.-R. Herman, *Exemples de fractions rationnelles ayant une orbite dense sur la sphère de Riemann* (French, with English summary), Bull. Soc. Math. France **112** (1984), no. 1, 93–142. MR771920

[74] M. Herman and J.-C. Yoccoz, *Generalizations of some theorems of small divisors to non-Archimedean fields*, Geometric dynamics (Rio de Janeiro, 1981), Lecture Notes in Math., vol. 1007, Springer, Berlin, 1983, pp. 408–447, DOI 10.1007/BFb0061427. MR730280

[75] M. Hindry and J. H. Silverman, *Diophantine geometry: An introduction*, Graduate Texts in Mathematics, vol. 201, Springer-Verlag, New York, 2000. MR1745599

[76] L.-C. Hsia, *A weak Néron model with applications to p-adic dynamical systems*, Compositio Math. **100** (1996), no. 3, 277–304. MR1387667

[77] L.-C. Hsia, *Closure of periodic points over a non-Archimedean field*, J. London Math. Soc. (2) **62** (2000), no. 3, 685–700, DOI 10.1112/S0024610700001447. MR1794277

[78] P. Ingram, *The critical height is a moduli height*, Duke Math. J. **167** (2018), no. 7, 1311–1346, DOI 10.1215/00127094-2017-0053. MR3799700

[79] K. Jacobs, *Lower bounds for non-Archimedean Lyapunov exponents*, `arXiv:1510.02440` (2015).

[80] N. Jacobson, *Basic algebra. II*, 2nd ed., W. H. Freeman and Company, New York, 1989. MR1009787

[81] M. Jonsson, *Dynamics of Berkovich spaces in low dimensions*, Berkovich spaces and applications, Lecture Notes in Math., vol. 2119, Springer, Cham, 2015, pp. 205–366, DOI 10.1007/978-3-319-11029-5_6. MR3330767

[82] G. Julia, *Mémoire sur l'itération des fonctions rationelles*, J. Math. Pures Appl. (7) **4** (1918), 47–245.

[83] J. Kiwi, *Puiseux series polynomial dynamics and iteration of complex cubic polynomials* (English, with English and French summaries), Ann. Inst. Fourier (Grenoble) **56** (2006), no. 5, 1337–1404. MR2273859

[84] J. Kiwi, *Puiseux series dynamics of quadratic rational maps*, Israel J. Math. **201** (2014), no. 2, 631–700, DOI 10.1007/s11856-014-1047-6. MR3265299

[85] J. Kiwi, *Rescaling limits of complex rational maps*, Duke Math. J. **164** (2015), no. 7, 1437–1470, DOI 10.1215/00127094-2916431. MR3347319

[86] N. Koblitz, *p-adic numbers, p-adic analysis, and zeta-functions*, 2nd ed., Graduate Texts in Mathematics, vol. 58, Springer-Verlag, New York, 1984. MR754003

[87] G. Kœnigs, *Recherches sur les intégrales de certaines équations fonctionnelles* (French), Ann. Sci. École Norm. Sup. (3) **1** (1884), 3–41. MR1508749

[88] S. Lang, *Fundamentals of Diophantine geometry*, Springer-Verlag, New York, 1983. MR715605

[89] S. Lang, *Algebra*, 3rd edition, Addison-Wesley Publishing Co., Inc., Reading, Mass., 1965. MR0197234

[90] J. Lee, *J-stability of expanding maps in non-Archimedean dynamics*, Ergodic Theory Dynam. Systems (to appear). `https://doi.org/10.1017/etds.2017.53`

[91] H.-C. Li, *p-adic periodic points and Sen's theorem*, J. Number Theory **56** (1996), no. 2, 309–318, DOI 10.1006/jnth.1996.0020. MR1373554

[92] H.-C. Li, *Counting periodic points of p-adic power series*, Compositio Math. **100** (1996), no. 3, 351–364. MR1387670

[93] H.-C. Li, *p-adic dynamical systems and formal groups*, Compositio Math. **104** (1996), no. 1, 41–54. MR1420709

[94] D. A. Lind and T. Ward, *Automorphisms of solenoids and p-adic entropy*, Ergodic Theory Dynam. Systems **8** (1988), no. 3, 411–419, DOI 10.1017/S0143385700004545. MR961739

[95] K.-O. Lindahl, *On Siegel's linearization theorem for fields of prime characteristic*, Nonlinearity **17** (2004), no. 3, 745–763, DOI 10.1088/0951-7715/17/3/001. MR2057125

[96] K.-O. Lindahl, *Divergence and convergence of conjugacies in non-Archimedean dynamics*, Advances in p-adic and non-Archimedean analysis, Contemp. Math., vol. 508, Amer. Math. Soc., Providence, RI, 2010, pp. 89–109, DOI 10.1090/conm/508/09993. MR2597687

[97] K.-O. Lindahl, *The size of quadratic p-adic linearization disks*, Adv. Math. **248** (2013), 872–894, DOI 10.1016/j.aim.2013.07.020. MR3107530

[98] K.-O. Lindahl and J. Rivera-Letelier, *Generic parabolic points are isolated in positive characteristic*, Nonlinearity **29** (2016), no. 5, 1596–1621, DOI 10.1088/0951-7715/29/5/1596. MR3481344

[99] J. Lubin, *Non-Archimedean dynamical systems*, Compositio Math. **94** (1994), no. 3, 321–346. MR1310863

[100] M. Ju. Ljubich, *Entropy properties of rational endomorphisms of the Riemann sphere*, Ergodic Theory Dynam. Systems **3** (1983), no. 3, 351–385, DOI 10.1017/S0143385700002030. MR741393

[101] R. Mañé, P. Sad, and D. Sullivan, *On the dynamics of rational maps*, Ann. Sci. École Norm. Sup. (4) **16** (1983), no. 2, 193–217. MR732343

[102] M. Manes and Y. Yasufuku, *Explicit descriptions of quadratic maps on* $\mathbb{P}^1$ *defined over a field* K, Acta Arith. **148** (2011), no. 3, 257–267, DOI 10.4064/aa148-3-3. MR2794930

[103] C. T. McMullen, *The Mandelbrot set is universal*, The Mandelbrot set, theme and variations, London Math. Soc. Lecture Note Ser., vol. 274, Cambridge Univ. Press, Cambridge, 2000, pp. 1–17. MR1765082

[104] J. Milnor, *Dynamics in one complex variable: Introductory lectures*, 2nd edition, Friedr. Vieweg & Sohn, Braunschweig, 1999. MR1721240

[105] J. Milnor, *Geometry and dynamics of quadratic rational maps*, Experiment. Math. **2** (1993), no. 1, 37–83. With an appendix by the author and Lei Tan. MR1246482

[106] P. Morton and J. H. Silverman, *Rational periodic points of rational functions*, Internat. Math. Res. Notices **2** (1994), 97–110, DOI 10.1155/S1073792894000127. MR1264933

[107] P. Morton and J. H. Silverman, *Periodic points, multiplicities, and dynamical units*, J. Reine Angew. Math. **461** (1995), 81–122, DOI 10.1515/crll.1995.461.81. MR1324210

[108] Y. Okuyama, *Repelling periodic points and logarithmic equidistribution in non-archimedean dynamics*, Acta Arith. **152** (2012), no. 3, 267–277, DOI 10.4064/aa152-3-3. MR2885787

[109] Y. Okuyama, *Quantitative approximations of the Lyapunov exponent of a rational function over valued fields*, Math. Z. **280** (2015), no. 3-4, 691–706, DOI 10.1007/s00209-015-1443-6. MR3369346

[110] R. Oselies and H. Zieschang, *Ergodische Eigenschaften der Automorphismen p-adischer Zahlen* (German), Arch. Math. (Basel) **26** (1975), 144–153, DOI 10.1007/BF01229718. MR0369301

[111] T. Pezda, *Polynomial cycles in certain local domains*, Acta Arith. **66** (1994), no. 1, 11–22, DOI 10.4064/aa-66-1-11-22. MR1262650

[112] M. Rees, *Ergodic rational maps with dense critical point forward orbit*, Ergodic Theory Dynam. Systems **4** (1984), no. 2, 311–322, DOI 10.1017/S0143385700002467. MR766108

[113] M. Rees, *Positive measure sets of ergodic rational maps*, Ann. Sci. École Norm. Sup. (4) **19** (1986), no. 3, 383–407. MR870689

[114] J. Rivera-Letelier, *Dynamique des fonctions rationnelles sur des corps locaux* (French, with English and French summaries), Astérisque **287** (2003), xv, 147–230. Geometric methods in dynamics. II. MR2040006

[115] J. Rivera-Letelier, *Espace hyperbolique p-adique et dynamique des fonctions rationnelles* (French, with English summary), Compositio Math. **138** (2003), no. 2, 199–231, DOI 10.1023/A:1026136530383. MR2018827

[116] J. Rivera-Letelier, *Sur la structure des ensembles de Fatou p-adiques*, `arxiv:math.DS/0412180` (2004).

[117] J. Rivera-Letelier, *Points périodiques des fonctions rationnelles dans l'espace hyperbolique p-adique* (French, with English summary), Comment. Math. Helv. **80** (2005), no. 3, 593–629, DOI 10.4171/CMH/27. MR2165204

[118] J. Rivera-Letelier, *Wild recurrent critical points*, J. London Math. Soc. (2) **72** (2005), no. 2, 305–326, DOI 10.1112/S0024610705006885. MR2156656

[119] J. E. Rivera-Letelier, *There are rational and irrational ultrametric wandering domains*, preprint (2011).

[120] J. E. Rivera-Letelier, personal communication, 2012.

[121] P. Robba, *Fonctions analytiques sur les corps valués ultramétriques complets* (French, with English summary), Prolongement analytique et algèbres de Banach ultramétriques, Société Mathématique de France, Paris, 1973, pp. 109–218, 219–220. Astérisque, No. 10. MR0357841

[122] A. M. Robert, *A course in p-adic analysis*, Graduate Texts in Mathematics, vol. 198, Springer-Verlag, New York, 2000. MR1760253

[123] H. L. Royden and P. Fitzpatrick, *Real Analysis*, 4th ed., Prentice Hall, New York, 2010.

[124] W. Rudin, *Real and complex analysis*, 3rd ed., McGraw-Hill Book Co., New York, 1987. MR924157

[125] R. Rumely, *The minimal resultant locus*, Acta Arith. **169** (2015), no. 3, 251–290, DOI 10.4064/aa169-3-3. MR3361223

[126] W. H. Schikhof, *Ultrametric calculus: An introduction to p-adic analysis*, Cambridge Studies in Advanced Mathematics, vol. 4, Cambridge University Press, Cambridge, 1984. MR791759

[127] J.-P. Serre, *Algebraic groups and class fields*, Graduate Texts in Mathematics, vol. 117, Springer-Verlag, New York, 1988. Translated from the French. MR918564

[128] A. N. Shiryayev, *Probability*, Graduate Texts in Mathematics, vol. 95, Springer-Verlag, New York, 1984. Translated from the Russian by R. P. Boas. MR737192

[129] M. Shishikura, *On the quasiconformal surgery of rational functions*, Ann. Sci. École Norm. Sup. (4) **20** (1987), no. 1, 1–29. MR892140

[130] J. H. Silverman, *The arithmetic of elliptic curves*, Graduate Texts in Mathematics, vol. 106, Springer-Verlag, New York, 1986. MR817210

[131] J. H. Silverman, *The arithmetic of dynamical systems*, Graduate Texts in Mathematics, vol. 241, Springer, New York, 2007. MR2316407

[132] T. Silverman, A non-archimedean λ-lemma, `arXiv:1712.01372` (2017).

[133] N. P. Smart and C. F. Woodcock, *p-adic chaos and random number generation*, Experiment. Math. **7** (1998), no. 4, 333–342. MR1678087

[134] D. Sullivan, *Quasiconformal homeomorphisms and dynamics. I. Solution of the Fatou-Julia problem on wandering domains*, Ann. of Math. (2) **122** (1985), no. 3, 401–418, DOI 10.2307/1971308. MR819553

[135] J. Tate, *Rigid analytic spaces*, Invent. Math. **12** (1971), 257–289, DOI 10.1007/BF01403307. MR0306196

[136] E. Thiran, D. Verstegen, and J. Weyers, *p-adic dynamics*, J. Statist. Phys. **54** (1989), no. 3-4, 893–913, DOI 10.1007/BF01019780. MR988564

[137] E. Trucco, *Wandering Fatou components and algebraic Julia sets* (English, with English and French summaries), Bull. Soc. Math. France **142** (2014), no. 3, 411–464, DOI 10.24033/bsmf.2670. MR3295719

[138] A. Thuillier, *Théorie du potential sur les courbes en géométrie analytique non archimédienne.* Applications à la théorie d'Arakelov, University of Rennes Ph.D. thesis, 2005.

[139] D. Verstegen, *p-adic dynamical systems*, Number theory and physics (Les Houches, 1989), Springer Proc. Phys., vol. 47, Springer, Berlin, 1990, pp. 235–242, DOI 10.1007/978-3-642-75405-0_25. MR1058468

[140] P. Walters, *An introduction to ergodic theory*, Graduate Texts in Mathematics, vol. 79, Springer-Verlag, New York-Berlin, 1982. MR648108

[141] J.-C. Yoccoz, *Linéarisation des germes de difféomorphismes holomorphes de* $(\mathbf{C}, 0)$ (French, with English summary), C. R. Acad. Sci. Paris Sér. I Math. **306** (1988), no. 1, 55–58. MR929279

[142] M. Zieve, *Non-Archimedean dynamical systems and their applications*, Berkeley Ph.D. thesis, 1996.

Index

Selected Published Titles in This Series

198 **Robert L. Benedetto,** Dynamics in One Non-Archimedean Variable, 2019
197 **Walter Craig,** A Course on Partial Differential Equations, 2018
196 **Martin Stynes and David Stynes,** Convection-Diffusion Problems, 2018
195 **Matthias Beck and Raman Sanyal,** Combinatorial Reciprocity Theorems, 2018
194 **Seth Sullivant,** Algebraic Statistics, 2018
193 **Martin Lorenz,** A Tour of Representation Theory, 2018
192 **Tai-Peng Tsai,** Lectures on Navier-Stokes Equations, 2018
191 **Theo Bühler and Dietmar A. Salamon,** Functional Analysis, 2018
190 **Xiang-dong Hou,** Lectures on Finite Fields, 2018
189 **I. Martin Isaacs,** Characters of Solvable Groups, 2018
188 **Steven Dale Cutkosky,** Introduction to Algebraic Geometry, 2018
187 **John Douglas Moore,** Introduction to Global Analysis, 2017
186 **Bjorn Poonen,** Rational Points on Varieties, 2017
185 **Douglas J. LaFountain and William W. Menasco,** Braid Foliations in Low-Dimensional Topology, 2017
184 **Harm Derksen and Jerzy Weyman,** An Introduction to Quiver Representations, 2017
183 **Timothy J. Ford,** Separable Algebras, 2017
182 **Guido Schneider and Hannes Uecker,** Nonlinear PDEs, 2017
181 **Giovanni Leoni,** A First Course in Sobolev Spaces, Second Edition, 2017
180 **Joseph J. Rotman,** Advanced Modern Algebra: Third Edition, Part 2, 2017
179 **Henri Cohen and Fredrik Strömberg,** Modular Forms, 2017
178 **Jeanne N. Clelland,** From Frenet to Cartan: The Method of Moving Frames, 2017
177 **Jacques Sauloy,** Differential Galois Theory through Riemann-Hilbert Correspondence, 2016
176 **Adam Clay and Dale Rolfsen,** Ordered Groups and Topology, 2016
175 **Thomas A. Ivey and Joseph M. Landsberg,** Cartan for Beginners: Differential Geometry via Moving Frames and Exterior Differential Systems, Second Edition, 2016
174 **Alexander Kirillov Jr.,** Quiver Representations and Quiver Varieties, 2016
173 **Lan Wen,** Differentiable Dynamical Systems, 2016
172 **Jinho Baik, Percy Deift, and Toufic Suidan,** Combinatorics and Random Matrix Theory, 2016
171 **Qing Han,** Nonlinear Elliptic Equations of the Second Order, 2016
170 **Donald Yau,** Colored Operads, 2016
169 **András Vasy,** Partial Differential Equations, 2015
168 **Michael Aizenman and Simone Warzel,** Random Operators, 2015
167 **John C. Neu,** Singular Perturbation in the Physical Sciences, 2015
166 **Alberto Torchinsky,** Problems in Real and Functional Analysis, 2015
165 **Joseph J. Rotman,** Advanced Modern Algebra: Third Edition, Part 1, 2015
164 **Terence Tao,** Expansion in Finite Simple Groups of Lie Type, 2015
163 **Gérald Tenenbaum,** Introduction to Analytic and Probabilistic Number Theory, Third Edition, 2015
162 **Firas Rassoul-Agha and Timo Seppäläinen,** A Course on Large Deviations with an Introduction to Gibbs Measures, 2015
161 **Diane Maclagan and Bernd Sturmfels,** Introduction to Tropical Geometry, 2015
160 **Marius Overholt,** A Course in Analytic Number Theory, 2014